Protein Targeting Protocols

METHODS IN MOLECULAR BIOLOGY™

John M. Walker, SERIES EDITOR

METHODS IN MOLECULAR BIOLOGY™

Protein Targeting Protocols

Second Edition

Edited by

Mark van der Giezen

School of Biological and Chemical Sciences, Queen Mary,
University of London, London, UK

HUMANA PRESS ✳ TOTOWA, NEW JERSEY

Preface

Cells are separated from their environments by a physical barrier. This barrier is either a plasma membrane alone, or a plasma membrane surrounded by a cell wall. From the outside, a cell might seem a serene and static entity. However, nothing could be further from the truth - the interior of most cells is as busy as Times Square on New Year's Eve. There is a continuous turnover of proteins which need to be replenished, signals are transduced, and the cell is in a constant state of maintenance and repair. At first sight this may not seem to be a big deal; ribosomes continuously produce new proteins which could easily find their way to their appropriate destinations in the cell, considering its tiny dimensions. Nonetheless, placing a protein in the right place at the right time is not a trivial task. Cells have to make sure that mitochondrial proteins don't end up in secretory vesicles, nuclear proteins in vacuoles, and cytosolic proteins in any organelle at all. Protein targeting is not trivial for bacteria, either. Despite their 'simpler' nature as prokaryotes, bacteria, too, have subcellular compartments; apart from the cytosol they can have membrane systems, a periplasmic space, and might need to excrete certain proteins and not others.

So, how do cells make sure that all proteins end up in the place where they are supposed to go? As we shall discover in the chapters of this book, targeting mechanisms exist to ensure the correct localization of all proteins in all cells. Experts from all over the world have provided their latest protocols to enable us to isolate different organelles, and to localize particular proteins using a variety of methods such as light, confocal or electron microscopy. We can become experts ourselves.

In contrast to the previous edition of *Protein Targeting Protocols*, the emphasis in this edition is on protein targeting to cellular compartments in both prokaryotic and eukaryotic systems. Reviews on the import machinery of mitochondria and plastids have been written by world-leading authorities, and provide us with the state-of-the-art in these fast-moving fields. Bacterial protein targeting protocols using the Sec-system, type-V secretion apparatus and the Tat-pathway are described, as well as a periplasmic targeting protocol. In addition, a detailed protocol is described to follow the movement of protein complexes in bacterial membranes using fluorescent recovery after photo-bleaching. For the eukaryotic systems, virtually every cellular compartment is treated in a total of twenty-one chapters. Care has been taken to include targeting protocols from different systems, including animal, plant, fungal and protist models. This diversity gives insight into the intricate challenges presented by different systems. Finally, a few chapters describe more generally applicable techniques. Although they might use specific experimental models, these techniques will aid scientists working with many different organisms. These chapters also include bioinformatic methods that guide us through the ever increasing number of in silico tools available to the modern scientist.

I am grateful for the time and effort all authors have put into the production of this book. Without them, this excellent collection of organelle purification protocols and targeting experiments would still be hidden in primary litera-ture. The authors allow us, the readers, to take a look into their laboratories. In addition, I am indebted to my colleague, Prof. John F. Allen, whose input and stimulation at various points has been extremely valuable. Finally, I would like to dedicate this book to my son Daan who doesn't want to be a fire-fighter, as his friends do, but seems to be heading for a career in science at a frighteningly young age. However, to continue the excellent scientific work described in this book, we need eager young people to go on with our work. The future is theirs!

Mark van der Giezen

Contents

Contributors

NIHAL ALTAN-BONNET • *Rutgers University, Department of Biological Sciences, Newark, New Jersey.*

SHASHI BHUSHAN • *Department of Biochemistry and Biophysics, Arrhenius Laboratories for Natural Science, Stockholm University, Stockholm, Sweden.*

JEFFREY P. BOCOCK • *Department of Biochemistry and Biophysics, University of North Carolina, Chapel Hill, NC.*

BETTINA BÖLTER • *Department Biologie I, Botanik 3, Ludwig-Maximilians-Universität München, Menzingerstr. München, Germany.*

FEDERICA BRANDIZZI • *Department of Biology, University of Saskatchewan, Saskatoon, Canada and Department of Energy, Plant Research Laboratory, Michigan State University, East Lansing, MI*

DARREN S. CARNEY • *Biochemistry and Molecular Biology Department, Mayo Clinic College of Medicine, Rochester, MN.*

BALBIR K. CHAAL • *Botany Department, University of British Columbia, Vancouver, Canada. School of Biological Sciences, Nanyang Technological University, Singapore.*

SIÂN S. E. COX • *School of Biological Sciences, Royal Holloway, University of London, Egham, UK.*

BRIAN A. DAVIES • *Biochemistry and Molecular Biology Department, Mayo Clinic College of Medicine, Rochester, MN.*

JEANINE DE KEYZER • *Department of Molecular Microbiology, Groningen Biomolecular Sciences and Biotechnology Institute and the Materials Science Center Plus, University of Groningen, The Netherlands. Faculty of Life Sciences, Michael Smith Building, University of Manchester, Manchester, UK.*

GRAŻYNA DOMAŃSKA • *Institut für Physiologische Chemie, Ruhr-Universität Bochum, Bochum, Germany.*

ARNOLD J. M. DRIESSEN • *Department of Molecular Microbiology, Groningen Biomolecular Sciences and Biotechnology Institute and the Materials Science Center Plus, University of Groningen, The Netherlands.*

ANN H. ERICKSON • *Department of Biochemistry and Biophysics, University of North Carolina, Chapel Hill, NC.*

JADE K. FORWOOD • *Institute for Molecular Bioscience and School of Molecular and Microbial Sciences, The University of Queensland, Brisbane, Australia.*

BERNARDO J. FOTH • *School of Biological Sciences, Nanyang Technological University, Singapore.*

KIPROS GABRIEL • *Institut für Biochemie und Molekularbiologie, Universität Freiburg, Freiburg, Germany. Department of Genetics, University of Melbourne, Australia.*

ELZBIETA GLASER • *Department of Biochemistry and Biophysics, Arrhenius Laboratories for Natural Science, Stockholm University, Stockholm, Sweden.*

BEVERLEY R. GREEN • *Botany Department, University of British Columbia, Vancouver, Canada.*

SALLY L. HANTON • *Department of Biology, University of Saskatchewan, Saskatoon, Canada.*

CHRIS HAWES • *School of Life Sciences, Oxford Brookes University, Oxford, UK.*

KORET HIRSCHBERG • *Department of Pathology, Sackler School of Medicine, Tel Aviv University, Tel Aviv, Israel.*

EVA HOLZAPFEL • *Institut fiir Biochemie und Molekularbiologie, Zentrum fiir Biochemie und Molekulare Zellforschung, Universität Freiburg, Freiburg, Germany.*

BRUCE F. HORAZDOVSKY • *Biochemistry and Molecular Biology Department, Mayo Clinic College of Medicine, Rochester, MN.*

PAUL HORTON • *AIST Computational Biology Research Center, Tokyo, Japan.*

FRIEDERIKE HÖRMANN • *Department Biologie I, Botanik 3, Ludwig-Maximilians-Universität München, Menzingerstr., München, Germany.*

DAVID A. JANS • *Department of Biochemistry and Molecular Biology, Nuclear Signalling Laboratory, Monash University, Clayton, Australia and ARC Centre of Excellence for Biotechnology and Development, Canberra, Australia.*

GURPREET KAUR • *Department of Biochemistry and Molecular Biology, Nuclear Signalling Laboratory, Monash University, Clayton, Australia.*

DANIEL J. KLIONSKY • *Departments of Molecular, Cellular and Developmental Biology and of Biological Chemistry, Life Sciences Institute, University of Michigan, Ann Arbor, MI.*

PETER G. KROTH • *Department of Biology, University of Konstanz, Konstanz, Germany.*

SMITA KURUP • *CPI Division, Rothamsted Research, Hertfordshire, UK.*

LOREN A. MATHESON • *Department of Biology, University of Saskatchewan, Saskatoon, Canada.*

FILIPE J. MERGULHÃO • *LEPAE, Faculty of Engineering of the University of Porto, Chemical Engineering Department, Porto, Portugal.*

GABRIEL A. MONTEIRO • *Centre for Biological and Chemical Engineering, Instituto Superior Técnico, Lisbon, Portugal.*

MICHAEL MOSER • *Institut für Biochemie und Molekularbiologie, Zentrum für Biochemie und Molekulare Zellforschung, Universität Freiburg, Freiburg, Germany.*

CHRISTIAN MOTZ • *Institut für Physiologische Chemie, Ruhr-Universität Bochum, Bochum, Germany.*

MATTHIAS MÜLLER • *Institut für Biochemie und Molekularbiologie, Zentrum für Biochemie und Molekulare Zellforschung, Universität Freiburg, Freiburg, Germany.*

CONRAD W. MULLINEAUX • *School of Biological and Chemical Sciences, Queen Mary, University of London, London, UK.*

KENTA NAKAI • *Laboratory of Functional Analysis in silico, Human Genome Center, The Institute of Medical Science, The University of Tokyo, Tokyo, Japan.*

TETSUAKI OSAFUNE • *Department of Life Science, Nippon Sport Science University, Yokohama, Japan.*

TAKEAKI OZAWA • *Department of Molecular Structure, Institute for Molecular Science, Okazaki, Aichi, and Precursory Research for Embryonic Science and Technology, Japan Science and Technology Agency, Saitama, Japan.*

SASCHA PANAHANDEH • *Institut für Biochemie und Molekularbiologie, Zentrum für Biochemie und Molekulare Zellforschung, Universität Freiburg, Freiburg, Germany.*

PANAGIOTIS PAPATHEODOROU • *Institut für Physiologische Chemie, Ruhr-Universität Bochum, Bochum, Germany.*

PAVEL F. PAVLOV • *Department of Biochemistry and Biophysics, Arrhenius Laboratories for Natural Science, Stockholm University, Stockholm, Sweden.*

NIKOLAUS PFANNER ● *Institut für Biochemie und Molekularbiologie, Universität Freiburg, Freiburg, Germany.*

OLGA RANDELJ ● *Institut für Physiologische Chemie, Ruhr-Universität Bochum, Bochum, Germany.*

JOACHIM RASSOW ● *Institut für Physiologische Chemie, Ruhr-Universität Bochum, Bochum, Germany.*

CHARLOTTA RUDHE ● *Department of Biochemistry and Biophysics, Arrhenius Laboratories for Natural Science, Stockholm University, Stockholm, Sweden.*

JOHN RUNIONS ● *School of Life Sciences, Oxford Brookes University, Oxford, UK.*

JEANNE SHEPSHELOVICH ● *Department of Pathology, Sackler School of Medicine, Tel Aviv University, Tel Aviv, Israel.*

STEVEN D. SCHWARTZBACH ● *Department of Biology, 201 Life Sciences, University of Memphis, Memphis, TN.*

SILVIA SLÁVIKOVÁ ● *Department of Plant Sciences, The Weizmann Institute of Science, Rehovot, Israel.*

JÜRGEN SOLL ● *Department Biologie I, Botanik 3, Ludwig-Maximilians-Universität München, Menzingerstr. München, Germany.*

CHRISTOS STATHOPOULOS ● *Department of Biological Sciences, California State Polytechnic University, Pomona, CA.*

JORGE TOVAR ● *School of Biological Sciences, Royal Holloway, University of London, Egham, UK.*

YOSHIO UMEZAWA ● *Department of Chemistry, School of Science, The University of Tokyo, Tokyo, Japan.*

ROSTISLAV VACULA ● *Institute of Cell Biology, Faculty of Natural Sciences, Comenius University, Bratislava, Slovakia.*

MARK VAN DER GIEZEN ● *School of Biological and Chemical Sciences, Queen Mary, University of London, London, UK.*

IDA VAN DER KLEI ● *Eukaryotic Microbiology, Groningen Biomolecular Sciences and Biotechnology Institute, University of Groningen, Haren, The Netherlands.*

MARTIN VAN DER LAAN ● *Department of Molecular Microbiology, Groningen Biomolecular Sciences and Biotechnology Institute and the Materials Science Center Plus, University of Groningen, The Netherlands. Institut für Biochemie und Molekularbiologie, Universität Freiburg, Hermann-Herder-Straße 7, Freiburg, Germany.*

MARTEN VEENHUIS • *Eukaryotic Microbiology, Groningen Biomolecular Sciences and Biotechnology Institute, University of Groningen, Haren, The Netherlands.*

THERESA H. WARD • *Department of Infectious & Tropical Diseases, London School of Hygiene & Tropical Medicine, London, UK.*

YIHFEN T. YEN • *Department of Biological Sciences, California State Polytechnic University Pomona, CA and Department of Biology and Biochemistry, University of Houston, TX.*

I

BACTERIAL SYSTEMS

1

Localization and Mobility of Bacterial Proteins by Confocal Microscopy and Fluorescence Recovery After Photobleaching

Conrad W. Mullineaux

Summary

This chapter describes the use of laser-scanning confocal fluorescence microscopy for determining the localization of fluorescently tagged proteins within bacterial cells, discussing the problems caused by the limited resolution of an optical microscope. It also explains a relatively simple method for using fluorescence recovery after photobleaching (FRAP) to observe and quantify the diffusion of fluorescently tagged proteins in bacterial cells. The techniques are illustrated with reference to measurements on green fluorescent protein (GFP)-tagged proteins in *Escherichia coli.*

Key Words: Bacterium; confocal microscopy; cytoplasm; diffusion; *Escherichia coli*; fluorescence recovery after photobleaching; green fluorescent protein; periplasm; plasma membrane.

1. Introduction

The use of green fluorescent protein (GFP) tagging and fluorescence microscopy to observe the localization of specific proteins in eukaryotic cells is becoming routine. There have also been numerous studies using fluorescence recovery after photobleaching (FRAP) to probe the mobility and distribution of GFP-tagged proteins in eukaryotic cells. Both types of measurement can be performed under rather similar conditions in a standard laser-scanning confocal microscope. The two approaches give complementary information. High-resolution fluorescence imaging shows the distribution of the protein, but does not reveal whether the protein is stationary or whether what is observed is in fact the steady-state dynamic distribution of a mobile protein. This important

From: *Methods in Molecular Biology, Vol. 390: Protein Targeting Protocols: Second Edition*
Edited by: M. van der Giezen © Humana Press Inc., Totowa, NJ

information can be provided by a FRAP measurement, in which the confocal laser spot is used to bleach fluorescence in a small region of the sample. Subsequent imaging of the sample reveals whether the tagged protein is mobile. If it is, the bleached spot will spread and fill in as the tagged protein diffuses *(1–3)*. The same methods can be applied to bacterial cells. The only additional problem posed by bacteria is their generally smaller cell size. The resolution of a standard fluorescence microscope depends on the excitation and emission wavelengths and the numerical aperture of the objective lens. With a typical oil-immersion lens, the resolution when imaging GFP fluorescence is about 230 nm in the XY plane. The resolution in the Z-direction is harder to predict, because it depends on the details of the optics and the diameter of the confocal pinhole. The Z-resolution should therefore be determined experimentally: one method is described here. The Z-resolution is invariably lower than the XY-resolution. In our microscope, the best Z-resolution achievable when imaging GFP is about 750 nm.

Limited optical resolution obviously puts some constraints on the precision with which proteins can be localized. These constraints tend to appear more severe in bacterial cells, where the overall cell dimensions are so much smaller than in eukaryotes. **Fig. 1** shows images of *E. coli* cells in which GFP is localized in different cell compartments. In each case, the localization was confirmed by cell fractionation and immunoblotting *(4)*. The cells shown in **Fig. 1** were greatly elongated by growth in the presence of the cell division inhibitor cephalexin *(4–6)*. We find this very helpful for FRAP measurements (*see* below) but it generally does not help with protein localization because the cell diameter is unaffected. In **Fig. 1A**, GFP was fused to TatA, an *E. coli* plasma membrane protein *(4,5)*. Because of the limited XY-resolution, the plasma membrane is visualized as a layer about 500 nm thick (**Fig. 1A**). GFP fluorescence in the cytoplasm appears significantly higher than in the background (**Fig. 1A**). In fact, cell fractionation showed that there was no significant cytoplasmic GFP *(4)*. The fluorescence that appears to originate from the cytoplasm comes from the plasma membrane regions above and below the confocal plane. Its appearance in the cytoplasmic region of the image is mainly a result of the low Z-resolution. This problem means that fluorescence imaging alone is not sufficient to distinguish the case where 100% of GFP is at the membrane from the case where GFP is in the cytoplasm as well. In **Fig. 1C**, GFP is localized in the periplasm. This case can clearly be distinguished from **Fig. 1B**, where the GFP is in the cytoplasm. However, periplasmic localization (**Fig. 1C**) cannot be distinguished from plasma membrane localization (**Fig. 1A**) at optical resolution. The images shown in **Fig. 1** are essentially raw images

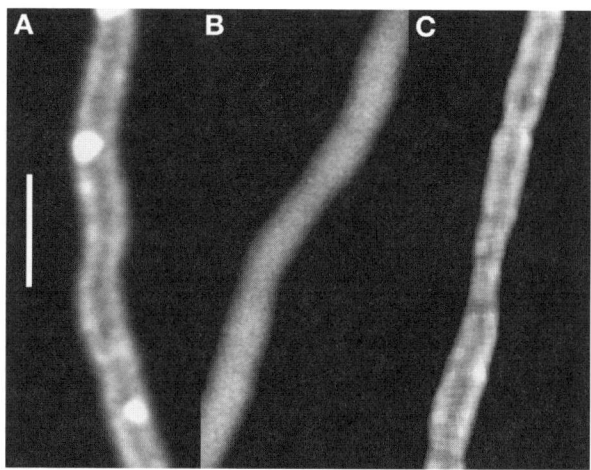

Fig. 1. Imaging green fluorescent protein (GFP) localized in different cell compartments in *E. coli* cells. In this case, cells have been greatly elongated by growth in the presence of cephalexin. (**A**) GFP fused to TatA, a plasma membrane protein. (**B**) GFP in the cytoplasm. (**C**) GFP exported to the periplasm by the TAT pathway: GFP was fused to the TorA signal peptide. Bar = 5 µm. (From **ref. 4** by permission of the American Society for Microbiology.)

as obtained directly from the confocal microscope. Deconvolution routines of various kinds can be used to "sharpen" images. This can improve the appearance of the images, but generally does not increase the information content.

GFP tagging can reveal other forms of protein localization in bacterial cells, including clusters of protein at the poles of the cell and rings of protein at the developing cell division site *(7,8)*. Where the protein has low copy number in the cell and hence the number of GFP fluorophores is very small, a standard confocal microscope may not be sensitive enough to visualize the protein. Sensitivity is greatly improved in a total internal reflection of fluorescence (TIRF) microscope *(9)*.

If fluorescently tagged proteins can be visualized in a laser-scanning confocal microscope, it is generally possible to use FRAP to probe their mobility. However, the small size of bacterial cells can pose some particular problems for FRAP measurements. In bacteria it is generally easier to bleach a line across the cell, rather than a spot. Diffusion can then be followed in one dimension (e.g., along the long axis of a rod-shaped bacterial cell) *(4–6,10,11)*. Ideally the bleached line should be narrow compared to the length of the cell. The theoretical lower limit for the bleach width depends on optical resolution, so it should be possible to bleach a line of less than 500 nm in diameter. In practice, however, the bleached line tends to be wider than this. This is usually because diffusion has already broadened the line during the bleach and in the time interval between bleaching and recording the first postbleach image. **Fig. 2** shows an extreme example of this. In this case, GFP was diffusing freely in the cytoplasm of an *E. coli* cell. Diffusion was rapid: the diffusion coefficient (D) was eventually estimated to be approx $9\,\mu m^2/s$ *(4)*. Bleaching required about 0.5 s, and there was a further delay of 1–2 s between bleaching and recording the first postbleach image. During this time, diffusion broadened the diameter ($1/e^2$) of the bleach to about $18\,\mu m$ **(Fig. 2)**. In the experiment shown in **Fig. 2** this did not preclude an accurate estimate of the diffusion coefficient, because the cell had been greatly elongated by growth in the presence of cephalexin *(4)*. However, in a normal-sized *E. coli* cell (length approx $3\,\mu m$) this would have been impossible because fluorescence would have re-equilibrated completely before imaging could begin. The problem is much less severe in cases where diffusion is slower (with membrane proteins, for example). However, small cell size can still make data analysis more difficult *(12)*.

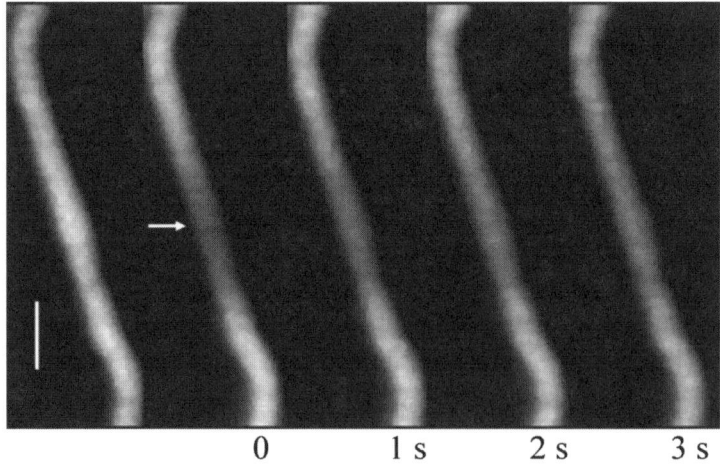

$$0 \qquad 1\,s \qquad 2\,s \qquad 3\,s$$

Fig. 2. Fluorescence recovery after photobleaching image sequence showing diffusion of green fluorescent protein (GFP) in the *E. coli* cytoplasm. A cell elongated by growth in the presence of cephalexin was used. The bleach was carried out in a narrow line across the center of the cell, indicated by the arrow. Rapid diffusion of GFP in the cytoplasm had already greatly widened the bleached area before the first postbleach image was recorded. Bar $= 5\,\mu m$. (From **ref. 4** by permission of the American Society for Microbiology.)

2. Materials

2.1. GFP Genes and Expression Vectors

1. GFP genes. The use of red-shifted, fluorescence-enhanced, versions of GFP *(13)* is recommended. We use GFP-mut3* *(13)*. This has a red-shifted absorption maximum at about 490 nm, which can conveniently be excited by the 488-nm line of an Argon laser. Cell autofluorescence generally decreases at longer excitation wavelengths. Therefore, there is less background autofluorescence with 488 nm excitation than with the short-wavelength excitation required for native GFP.
2. Expression vectors. We have successfully used both IPTG- and arabinose-inducible expression vectors in *E. coli (4,6)*.

2.2. Immobilizing Bacterial Cells

1. A suitable growth medium for the bacteria being studied.
2. Difco Bacto-agar (BD Biosciences, Franklin Lakes, NJ).
3. Glass cover slips of 0.1 mm thickness (VWR International, West Chester, PA, USA).
4. Cell division inhibitor. For *E. coli* use cephalexin (Sigma-Aldrich, St. Louis, Mo): prepare a 10 mg/mL stock solution in water and filter-sterilize.
5. Temperature-controlled microscope sample holder, with well to accommodate a small block of agar, with a cover slip on top. Ours is laboratory built. VWR supply disposable slides with wells, but these do not have temperature control.
6. Silicone vacuum grease (VWR International, West Chester, PA, USA).

2.3. Imaging and FRAP

We use a Nikon PCM2000 laser scanning confocal microscope (Nikon Instech Co., Kanagawa, Japan) equipped with a 100 mW Argon laser (Spectra-Physics, Mountain View, CA). For imaging GFP we use the 488-nm line from the Argon laser, selected by an interference bandpass filter. Fluorescence is selected by a 505-nm dichroic mirror and a bandpass filter transmitting at about 500–527 nm. FRAP measurements require suitable acquisition software for switching between XY and X-scanning modes and recording a series of images at defined time intervals. We use the EZ2000 software supplied with the PCM2000 microscope. It must be possible to rapidly change the light output from the laser (e.g., with neutral density filters).

2.4. Data Analysis

1. Suitable software for extracting quantitative data from images, for example, Image-ProPlus 5.1 (MediaCybernetics, Silver Spring, MD).
2. Suitable software for data analysis and curve fitting. SigmaPlot 9.0 (Systat, Point Richmond, CA) is suitable for most of the applications described here.

2.5. Estimating the Z-Resolution of the Confocal Microscope

1. Egg yolk phosphatidyl choline (Sigma-Aldrich) made into a 10 mM solution in 2,2,2-trifluoroethanol (TFE) (Sigma-Aldrich).
2. BODIPY® FL C_{12} (4,4-difluoro-5,7-dimethyl-4-bora-3a,4a-diaza-*s*-indacene-3-dodecanoic acid) supplied by the Molecular Probes division of Invitrogen (http://probes.invitrogen.com/). Prepare a 10 mM stock solution in dimethylsulfoxide and store at −20°C.
3. Buffer solution: 30 mM 2-(N-Morpholino) ethanesulfonic acid (MES) (pH 6.5), 5 mM MgCl$_2$, 40 mM NaCl.

3. Methods

3.1. Expressing GFP-Tagged Proteins

Where the gene is expressed from an expression vector rather than from the native promoter on the chromosome, the optimum level of expression must be determined by trial and error. This is most easily done by varying the time for which the cells are exposed to the inducer *(4,6,14)* (*see* **Notes 1** and **2**).

3.2. Immobilizing Bacterial Cells

1. Grow a cell culture in liquid medium. For example, for *E. coli* grow a culture overnight at 37°C in Luria-Bertani medium. Induce expression of the tagged protein if necessary (*see* **Subheading 3.1.**).
2. FRAP measurements are much easier to perform on bacterial cells that have been elongated by growth in the presence of cell division inhibitors. In *E. coli* this can be achieved by growth for 3 h in the presence of 30 μg/ml cephalexin prior to the measurement *(4,6)*.
3. Take the cell suspension and check the cell concentration by measuring apparent absorbance at 750 nm in a spectrophotometer. The optimal cell concentration must be determined by trial and error. For *E. coli* an overnight culture diluted 10-fold is about right.
4. Take a Petri dish with 1.5% Bacto-agar made up in growth medium. Put drops of the diluted cell culture onto the agar surface and allow to dry down. This can be achieved in about 15 min by placing the open Petri dish in a laminar flow hood.
5. When all the liquid has been adsorbed, use a scalpel to cut out a small block of agar with cells adsorbed to the surface. Place in the well of the sample holder.
6. Smear a thin layer of vacuum grease around the rim of the well, and use this to stick down a glass cover slip so that the cover slip is gently pressing onto the agar surface.
7. Place on the microscope stage.

3.3. Imaging and FRAP

1. Focus the microscope on the sample. This is usually most easily done in simple transmission or epifluorescence mode. We use a 60× oil-immersion objective lens (numerical aperture 1.4).

2. Switch to laser-scanning confocal mode and image GFP fluorescence (*see* **Note 3**). Trial and error must be used to establish conditions under which the fluorescence can be imaged without significant bleaching (*see* **Note 4**). Record a series of successive images, and see if the fluorescence intensity decreases from image to image. If it does, it will be necessary to decrease the laser intensity. We routinely use a 100 mW Argon laser with intensity decreased by a factor of 32 with neutral density filters.

3. For high-resolution imaging, use a small confocal pinhole to increase Z-resolution. Signal-to-noise is best improved by averaging a series of images. The method described in **Subheading 3.2.** should ensure that the cell does not move during imaging.

4. For FRAP, select a suitable cell of interest (*see* **Note 5**). In the case of elongated bacterial cells, select a long cell aligned roughly in the Y-direction (**Fig. 2**)

5. For FRAP measurements, we usually open the confocal pinhole to reduce the Z-resolution. The confocal spot should extend through the full depth of the cell *(11)* (*see* **Note 6**).

6. Zoom in by reducing the scanning area so that the cell chosen fills most of the field of view. Record a prebleach image of the cell. Check that the cell can still be imaged without bleaching, since scanning the laser over a smaller area increases the exposure of the cell.

7. Bleach fluorescence in a line across the cell. Where possible the bleach should always be done away from the ends of the cell (*see* **Note 7**). Switch the confocal microscope from XY-scanning mode to X-scanning mode so that the line of the X-scan passes through the center of the sample. Start the X-scan, if possible increasing the laser power at the same time. We generally increase the laser power by a factor of 32 by briefly removing neutral density filters. After about a second, the fluorescence signal should be significantly decreased (*see* **Note 8**). It is not necessary or desirable to bleach away all the fluorescence in the center of the bleached zone. Generally something between 50 and 90% bleaching is ideal.

8. Quickly replace the neutral density filters, switch back to XY scanning mode and record a series of images at set time intervals. The choice of time interval depends on the rate and the size of the cell. Generally it is sufficient to record a series of about 10 images at intervals of 1–10 s. One or two longer time points are useful to check whether recovery is complete (recovery will be incomplete if there is an immobile population of the fluorophore). In a successful FRAP measurement, the initial bleach will appear as a clearly defined dark zone in the fluorescence image (*see* **Note 9**). With time, the dark zone will spread and fill in as the bleached fluorophore diffuses away and unbleached fluorophore diffuses into the bleached area (*see* **Note 10**).

3.4. FRAP Data Analysis

The following procedure relates to the estimation of the diffusion coefficient from an elongated bacterial cell. In this case, a line is bleached across the cell (**Fig. 2**) and diffusion is observed in one dimension along the long axis of the cell *(4,5,11)* (*see* **Note 11**). The same procedure may be applied whether the GFP is localized in the cytoplasm, the periplasm, or the plasma membrane of an elongated cell.

1. Import the images into Image ProPlus or similar software.
2. Use the software to extract a one-dimensional fluorescence profile along the long axis of the cell, summing the fluorescence values across the width of the cell. This is done by setting a sampling line width greater than the width of the cell, then drawing the line down the long axis of the cell and extracting the data *(15)*.
3. Repeat for each time point in the sequence, including the prebleach image.
4. Import the fluorescence profiles into SigmaPlot or similar. Subtract each postbleach profile from the prebleach profile to generate a series of difference profiles, which should be approximately Gaussian in shape (**Fig. 3**).
5. Use SigmaPlot to fit the profiles to Gaussian curves, noting the parameters C (the peak height of the Gaussian curve) and R (the $1/e^2$ radius of the Gaussian curve). If the changes observed are a result of diffusion, then C should decrease with time and R should increase with time, but the area under the Gaussian curve should remain constant (**Fig. 3**).
6. In this one-dimensional case, C is related to the diffusion coefficient D as follows:

$$C = C_0 R_0 (R_0^2 + 8Dt)^{-0.5}$$

where t is time and C_0 and R_0 are the values for C and R extracted from the first postbleach image *(11)*. D can be obtained by plotting $1/C^2$ vs time. The plot should give a straight line (**Fig. 3**). The gradient can be obtained by linear regression and is equal to $8D/C_0^2 R_0^2$ (*see* **Note 12**). C_0 and R_0 are known, so D can be calculated (*see* **Note 13**). If t is in s and R_0 is in μm, then D is in units of $\mu m^2/s$.

3.5. Estimating Z-Resolution of the Confocal Microscope

This requires the generation of a layer of fluorophore that is extremely thin compared to the Z-resolution. The method described uses an artificial lipid bilayer stained with a lipophilic green fluorophore.

1. Take $100\,\mu$L phosphatidyl choline solution ($10\,mM$ in TFE). Mix with $1\,\mu$L BODIPY FL C_{12} solution ($10\,mM$ in dimethylsulfoxide).
2. Spot 20μL onto a glass microscope slide. Evaporate the TFE and dry the lipid film under a stream of nitrogen for 1 h.

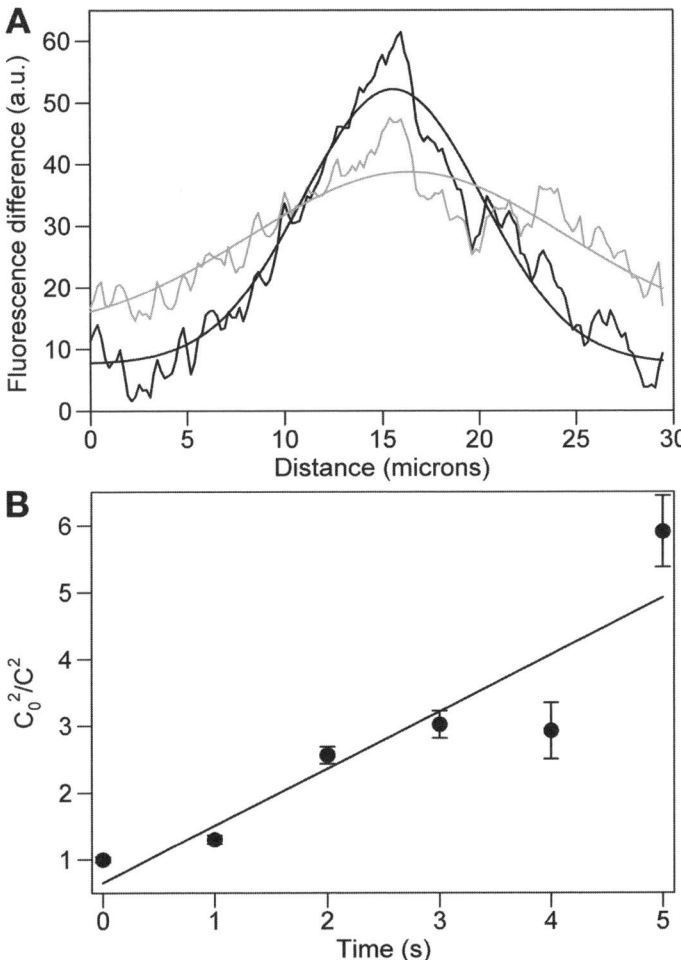

Fig. 3. Estimation of the diffusion coefficient from the image sequence shown in **Fig. 2**. (**A**) One-dimensional fluorescence profiles extracted as described in **Subheading 3.4.** The black line shows the fluorescence profile from the first postbleach image and the grey line shows the profile 4 s later. Raw data and fitted Gaussian curves are shown. (**B**) Plot of C_0^2/C^2 vs time, where C is the bleach depth (C_0 in the first postbleach image). The gradient of the line is $8D/R_0^2$, where R_0 is the half-width ($1/e^2$) of the bleach in the first postbleach image. (From **ref. *4*** by permission of the American Society for Microbiology.)

3. Warm the dried lipid film to 40°C for 1 min. Add 30 μL of buffer solution. Incubate for 2 min at 40°C.
4. Seal with a glass cover slip.
5. Image the fluorescent lipid bilayer in the confocal microscope, with the same settings used for cell imaging. Find a region with a single bilayer (multiple stacks of bilayer show higher fluorescence).
6. Record images at a series of Z-positions (100-nm Z-intervals are appropriate).
7. Use image analysis software to sum the total fluorescence in each image.
8. Plot fluorescence against Z-position: this normally gives an approximately Gaussian curve from which Z-resolution may be estimated.

Notes

1. The level of GFP-tagged protein tends to vary considerably from cell to cell. It may be possible to find a suitable individual cell even when the overall level of expression in the culture is too low or too high.
2. Increased expression of the GFP-tagged protein gives better signal-to-noise ratio and a higher ratio of GFP fluorescence to background autofluorescence. However, excessive expression may cause artefacts. Excessive expression of a membrane protein may result in artificial aggregates and inclusion bodies *(6)*. If the protein is normally exported to the periplasm, overexpression may saturate the export machinery leading to accumulation in the cytoplasm as well *(14)*.
3. All cells show some background autofluorescence in the green region. To check that this is insignificant compared to GFP fluorescence, record images under identical conditions with uninduced cells or a reference strain in which the GFP-tagged protein is not expressed.
4. Photobleaching is increased by high laser power, low scanning speeds, and a small scanning area (field of view).
5. The longer the cell, the easier the measurement will be. Data analysis is easier and more robust if you have a tightly localized bleach in the middle of a long cell—it becomes difficult if the bleach removes a large proportion of the total fluorescence. A longer cell makes it much easier to follow rapid diffusion. With a small cell and rapid diffusion, the bleach might re-equilibrate over the whole sample before you have time to record any images.
6. For FRAP measurements we recommend the use of the highest possible scanning speeds, because this also increases the time-resolution of the measurement. Generally high Z-resolution is not required for FRAP measurements, so it is best to use a larger confocal pinhole, which gives a higher signal-to-noise ratio. This in turn allows images to be obtained at lower laser intensity, which minimizes the problem of photobleaching during imaging.

7. It is not always possible to the bleach in the center of the cell. With a small cell it may be necessary to bleach at one end to prevent the whole sample from being bleached away. This is feasible, but it makes the data analysis more difficult *(13)*.

8. We lift the filters by hand. This can be done very quickly with practice. Shutters would allow shorter exposure times, but in our experience very short exposure times do not give sufficient bleaching in any case. With manual control you can watch the fluorescence signal decreasing and drop the filters when the bleach is deep enough. It is not necessary to time the bleach accurately as long as the postbleach image sequence is timed accurately. The first postbleach image provides the first time point for the data analysis.

9. The minimum width of the bleach could theoretically be about half the wavelength of the laser light. In practice, the width of the bleach is generally increased by light scattering by the sample and diffusion occurring during the bleach and before the first image is recorded. A common problem with small samples and rapid diffusion is that the bleach spreads over the whole sample before the first image can be recorded. Then the first postbleach image shows decreased fluorescence overall, but a similar fluorescence distribution to the prebleach image. This problem can be tackled in various ways:

    a. Do the bleach more quickly. This may require more laser power.
    b. Record the images more quickly. This can often be achieved by reducing the pixel resolution of the images.
    c. Try to find a longer cell.

10. It is best to use a simpie control to check that fluorescence recovery is because of diffusion, rather than reversible fluorescence quenching of the fluorophore. Try bleaching out an entire small cell and check that fluorescence does not recover.

11. Analytical solutions depend on simplifying assumptions, and the easiest assumption to make is that cell is long compared to the bleach width. If it isn't, then you have to factor in the exact length of the cell and the width, depth, and position of the bleach. Under those conditions, analytical solutions often are not worthwhile. We recommend as a more practical approach measuring the fluorescence profile after the bleach and then using an iterative computer routine to predict how it will evolve with time, assuming simple random diffusion. Comparison of the simulation to the experimental data allows estimation of the diffusion coefficient.

12. Take care to weight the linear regression accurately (weight each point according to the reciprocal of the estimated variance). In a plot of $1/C^2$ vs time, the variance increases sharply as C decreases.

13. Individual measurements of D can be very accurate if done with care. But real biological samples often show variation: every cell is different. Always do enough replicates to check that the result is representative.

Acknowledgments

The *E. coli* transformants shown in **Figs. 1–3** were produced by Colin Robinson (University of Warwick). The method used for estimating Z-resolution was developed in collaboration with Helmut Kirchhoff (University of Münster) during a stay in CWM's lab funded by the Deutsche Forschungsgemeinschaft. Work in CWM's laboratory is supported by the Biotechnology and Biological Sciences Research Council and the Wellcome Trust.

References

1. Reits, E. A. J. and Neefjes, J .J. (2001) From fixed to FRAP: measuring protein mobility and activity in living cells. *Nature Cell Biol.* **3**, 145–147.
2. Blonk, J. C. G., van Aalst, D. H., and Birmingham, J. J. (1993) Fluorescence photobleaching recovery in the confocal scanning light microscope. *J. Microsc.* **169**, 363–374.
3. Kubitscheck, U., Wedekind, P., and Peters, R. (1994) Lateral diffusion measurements at high spatial resolution by scanning microphotolysis in a confocal microscope. *Biophys. J.* **67**, 948–956.
4. Mullineaux, C. W., Nenninger, A., Ray, N., and Robinson, C. (2006) Diffusion of green fluorescent protein in three cell environments in *Escherichia coli. J. Bacteriol.* **188**, 3442–3448.
5. Elowitz, M. B., Surette, M. G., Wolf, P.-E., Stock, J. B., and Leibler, S. (1999) Protein mobility in the cytoplasm of *Escherichia coli. J. Bacteriol.* **181,** 197–203.
6. Ray, N., Nenninger, A., Mullineaux, C. W., and Robinson, C. (2005) Location and mobility of twin-arginine translocase subunits in the *Escherichia coli* plasma membrane. *J. Biol. Chem.* **280,** 17961–17968.
7. Ma, X., Ehrhardt, D. W., and Margolin, W. (1996) Colocalization of cell division proteins FtsZ and FtsA to cytoskeletal structures in living *Escherichia coli coli* cells by using green fluorescent protein. *Proc. Natl. Acad. Sci. USA* **93,** 12998–123003.
8. Sourjik, V. and Berg, H. C. (2000) Localization of components of the chemotaxis machinery of *Escherichia coli* using fluorescent protein fusions. *Mol. Microbiol.* **37,** 740–751.
9. Mashanov, G. I., Tacon, D., Knight, A. E., Peckham, M., and Molloy, J. E. (2003) Visualizing single molecules inside living cells using total internal reflection fluorescence microscopy. *Methods* **29**, 142–152.
10. Brass, J. M., Higgins, C. F., Foley, M., Rugman, P. A, and Birmingham, J. (1986) Lateral diffusion of proteins in the periplasm of *Escherichia coli. J. Bacteriol.* **165,** 787–795.
11. Mullineaux, C. W., Tobin, M. J. and Jones, G. R. (1997) Mobility of photosynthetic complexes in thylakoid membranes. *Nature* **390**, 421–424.

12. Cowan, A. E., Koppel, D. E., Setlow, B., and Setlow, P. (2003) A soluble protein is immobile in dormant spores of *Bacillus subtilis* but is mobile in germinated spores: implications for spore dormancy. *Proc. Nat. Acad. Sci. USA* **100,** 4209–4214.

13. Cormack, B. P., Valdivia, R. H. and Falkow, S. (1996) FACS-optimized mutants of the green fluorescent protein (GFP). *Gene* **173,** 33–38.

14. Barrett, C. M. L., Ray, N., Thomas, J. D., Robinson, C., and Bolhuis, A. (2003) Quantitative export of a reporter protein, GFP, by the twin-arginine translocation pathway in *Escherichia coli. Biochem. Biophys. Res. Comm.* **304,** 279–284.

15. Mullineaux, C. W. (2004) FRAP analysis of photosynthetic membranes. *J. Exp. Bot.* **55,** 1207–1211.

2

Membrane Protein Insertion and Secretion in Bacteria

Jeanine de Keyzer, Martin van der Laan, and Arnold J. M. Driessen

Summary

Export of secretory proteins across and insertion of membrane proteins into the cytoplasmic membrane of *Escherichia coli* and other bacteria is mediated by the enzyme complex translocase. The last decade has seen a major advance in the understanding of the mechanism of these processes. A large part of this progress can be attributed to the development of general and powerful methods to study the translocase activity in vitro. Here we describe a transcription–translation method used to analyze the insertion of membrane proteins into *E. coli* inner membrane vesicles and a rapid and quantitative fluorescent method to analyze the translocation of secretory proteins.

Key Words: *E. coli* translocase; protein translocation; membrane protein insertion; transcription; translation; fluorescent imaging; SecA; SecYEG.

1. Introduction

In *Escherichia coli*, the majority of secretory proteins and membrane proteins are translocated across or inserted into the inner membrane with the aid of a multimeric enzyme complex called translocase (for reviews see **refs. *1*** and *2*). The core of this enzyme complex is formed by the SecY, SecE, and SecG proteins *(3)*, which form a highly conserved protein-conducting channel in the inner membrane, and the ATPase SecA *(4)*, a motor protein that is peripherally bound to the SecYEG complex *(5)*. Secretory and membrane proteins are targeted to the membrane via two distinct pathways that converge at the SecYEG complex *(6)*. Protein secretion occurs mostly via a posttranslational pathway wherein proteins are translocated as fully synthesized preproteins with an N-terminal signal sequence that is needed for targeting. Protein translocation is driven by cycles of ATP binding and hydrolysis at SecA *(7)*. Insertion into

From: *Methods in Molecular Biology, Vol. 390: Protein Targeting Protocols: Second Edition*
Edited by: M. van der Giezen © Humana Press Inc., Totowa, NJ

the inner membrane occurs while the membrane proteins are still translated at the ribosome (co-translational). Ribosome nascent chains are targeted to the SecYEG complex by the signal recognition particle (SRP) and its receptor FtsY. Membrane insertion is driven by chain elongation at the ribosome, while large, hydrophilic, periplasmic loops require the assistance of SecA.

In the last decade, efficient methods have been developed to study protein translocation and membrane protein insertion in vitro. The SecYEG complex can be overexpressed to high levels in *E. coli (8)*, and when membranes isolated from these SecYEG overexpressing strains are supplemented with purified SecA, they support translocation of purified preproteins and the insertion of newly synthesized membrane proteins. This chapter describes in vitro methods to assay the translocation and membrane insertion of two model proteins, proOmpA and FtsQ, that are commonly used in studies on the mechanism of the *E. coli* translocase.

2. Materials

2.1. Preparation of S135 Lysate

1. Luria-Bertani (LB) broth (Sigma-Aldrich, St. Louis, MO) supplemented with 0.2% glucose.
2. *E. coli* MC4100 *(9)*.
3. 100 mM Phenylmethylsulfonylfluoride (PMSF) (Sigma-Aldrich) in 96% ethanol. 1 M dithiothreitol (DTT) (Roche Diagnostics, Mannheim, Germany). Store at $-20°C$.
4. One Shot cell disrupter (Constant Systems Ltd, Daventry, UK).
5. Lysis buffer: 50 mM HEPES KOH pH 7.5, 50 mM KCl, 15 mM MgCl$_2$, 1 mM DTT in double distilled water. Prepare without DTT, autoclave, and store at 4°C. Add DTT prior to use.
6. Dialysis buffer: 10 mM HEPES KOH pH 7.5, 10 mM MgCl$_2$, 22 mM NH$_4$ acetate, 1 mM DTT in double distilled water. Prepare without DTT, autoclave, and store at 4°C. Add DTT prior to use.
7. Dialysis tubing (MWCO 12–14 kDa) (Medicell International Ltd, London, UK).

2.2. Isolation of Inner Membrane Vesicles

1. LB broth (*see* **Subheading 2.1., item 1**) supplemented with 0.1 mg/mL ampicillin (Roche Diagnostics, Mannheim, Germany).
2. *E. coli* SF100 *(10)* transformed with plasmid pET610 (overexpression of the SecYEG complex with a histidine tag at the amino-terminus of the SecY subunit, Ap^R) *(11)*.
3. 1 M isopropyl-β-D-thiogalactopyranoside (IPTG) (Roche Diagnostics). Filter-sterilize and store at $-20°C$.

4. 50 mM HEPES KOH pH 8.0 supplemented with 20% (w/v) sucrose (HEPES-sucrose) or 20% (w/v) glycerol (HEPES-glycerol).
5. One Shot cell disrupter (Constant Systems Ltd).
6. Two 36–54% sucrose step gradients prepared in TLA100.4 tubes (Beckman Coulter, Fullerton, CA). Each gradient consist of layers of 54% (0.55 mL), 51% (1 mL), 45% (0.45 mL) and 36% (0.45 mL) (w/v) sucrose in 50 mM HEPES-KOH pH 8.0.

2.3. Purification of SecA

1. *E. coli* DH5α transformed with plasmid pMKL18 (unpublished, gift of R. Freudl, SecA gene cloned in pUC19 vector, expression of SecA, Ap^R) *(12)*.
2. Tris-sucrose: 50 mM Tris-HCl, pH 7.6, and 20% sucrose.
3. One Shot cell disrupter (Constant Systems Ltd).
4. SecA buffer: 50 mM Tris-HCl, pH 7.6, 10% (v/v) glycerol, and 1 mM DTT.
5. SecA buffer supplemented with 150 mM NaCl.
6. SecA buffer supplemented with 1 M NaCl.
7. FPLC system (ÄKTAexplorer™ or equivalent [GE Healthcare Biosciences AB, Uppsala, Sweden]).
8. Amicon Centriprep® YM-30 centrifugal filter, MCWO 30 kDa (Millipore, Amsterdam, Netherlands).
9. Superdex 200 XK26/60 column (GE Healthcare Biosciences AB).
10. Sodium dodecyl sulfate (SDS)–polyacrylamide gel electrophoresis (PAGE) loading buffer and materials for running a 12% SDS-PAGE gel (*see* **Subheading 2.8., items 1–9**).

2.4. ProOmpA Purification

1. LB broth (*see* **Subheading 2.1., item 1**) supplemented with 0.1 mg/mL ampicillin.
2. *E. coli* MM52 *(13)* transformed with pET503 (overexpression of proOmpAC290S, Ap^R) *(14)*.
3. MSE Soniprep 150 (Sanyo Biomedical Europe, Loughborough, UK).
4. proOmpA-buffer: 8 M urea, 50 mM Tris-HCl pH 7.0.

2.5. Fluorescent Labeling of proOmpA

1. 100 mM Tri(2-carboxyethyl)phosphine (TCEP) (Invitrogen, Carlsbad, CA) in 100 mM Tris-HCl pH 7.0. Store at −20°C.
2. 40 mM Fluorescein-5-maleimide (Invitrogen) in dimethylformamide (DMF). Wrap containers in aluminum foil to protect the fluorescein probe from light. Aliquots can be stored at −20°C for several months.
3. 1 M DTT (*see* **Subheading 2.1., item 3**).
4. MicroBio-Spin 6 chromatography column (BioRad, Hercules, CA).
5. 10 mg/mL Bovine serum albumin (BSA) (Sigma-Aldrich).

2.6. In Vitro Transcription–Translation and Insertion of FtsQ

1. All solutions for the in vitro transcription–translation are prepared in sterilized double distilled water. Spatulas, pipet tips, and vials are sterilized to prevent RNase activity.
2. 0.3–1 mg/mL pBSKftsQ *(15)* plasmid DNA in sterilized double distilled water.
3. *E. coli* SF100 × pET610 inner membrane vesicles (*see* **Subheading 3.2.**).
4. 20X Transcription–translation buffer: $0.8\,M$ HEPES-KOH pH 8.0, $0.56\,M$ KCl, $40\,mM$ DTT. Prepare without the DTT, autoclave, add the DTT, and store at $-20°C$.
5. Enzyme mix: a component of the RiboMAX™ large-scale RNA production system T-7 (Promega, Madison, WI).
6. RNAguard™ RNase inhibitor (Amersham Biosciences, Piscatway, NY).
7. 10 mg/mL *E. coli* tRNA (Roche Diagnostics). Store in aliquots at $-20°C$.
8. 15 m*M* Folinic acid (Sigma-Aldrich). Store in aliquots at $-20°C$.
9. 320 m*M* Phosphocreatine (Sigma-Aldrich) in 200 m*M* HEPES-KOH pH 7.6. Store in aliquots at $-20°C$.
10. 1.6 mg/mL Creatine kinase (Roche Diagnostics) in 200 m*M* HEPES-KOH pH 7.6. Store in aliquots at $-20°C$.
11. 100 m*M* Stock solutions of ATP, GTP, CTP, and UTP (Sigma-Aldrich). Adjust the pH to 7.0 and store in aliquots at $-20°C$.
12. 19-Amino-acid mix: all amino acids except for methionine, separately dissolved as 100 m*M* stock solutions, mixed and stored in aliquots at $-20°C$.
13. Redivue Pro-mix l-[^{35}S] in vitro cell labeling mix (Amersham Biosciences).
14. 2 mg/mL Protease K (Merck, Darmstad, DE). Store at $-20°C$.
15. 10% Triton X100 in double distilled water.
16. 20% Trichloroacetic acid (TCA) in double distilled water. Store at 4°C.
17. Acetone cooled to $-20°C$.
18. 2X SDS-PAGE loading buffer and materials for running a 12% SDS-PAGE gel (*see* **Subheading 2.8., items 1–9**).
19. Rainbow™ [^{14}C] methylated protein molecular weight marker (Amersham Biosciences).

2.7. In Vitro Translocation of Fluorescently Labeled proOmpA

1. 10X Translocation buffer: 500 m*M* HEPES-KOH, pH 7.5, 300 m*M* KCl, 5 mg/mL BSA (Sigma-Aldrich), 100 m*M* DTT (Roche Diagnostics), and 20 m*M* Mg(OAc)$_2$.
2. *E. coli* SF100 × pET610 inner membrane vesicles (*see* **Subheading 3.2.**).
3. Purified *E. coli* SecA (*see* **Subheading 3.3.**).
4. Stock solutions of creatine phosphate, creatine kinase, protease K, and TCA (*see* **Subheading 2.6., item 16**). Acetone cooled to $-20°C$.
5. 2X SDS-PAGE loading buffer and materials for running a 12% SDS-PAGE gel (*see* **Subheading 2.8., items 1–9**).
6. Roche Lumi-imager F1 (Roche Diagnostics).

2.8. SDS-PAGE

1. 5X SDS-PAGE sample buffer: 250 mM Tris-HCl pH 6.8, 10% SDS, 10% DTT, 50% glycerol, 0.05% bromophenol blue. Store at −20°C.
2. 4X SDS-PAGE running gel buffer: 1.5 M Tris-HCl pH 8.7, 0.4% SDS. Store at room temperature.
3. 4X SDS-PAGE stacking gel buffer: 0.5 M Tris-HCl pH 6.8, 0.4% SDS. Store at room temperature.
4. 10% Ammonium persulfate (APS) in double distilled water. Store at −20°C.
5. 40% Acrylamide/bis solution (37.5:1 with 2.6% C) and N,N,N,N'-tetramethyl-ethylenediamine (TEMED) (Bio-Rad, Hercules, CA).
6. Water-saturated isobutanol. Mix equal volumes of water and isobutanol, allow the layers to separate, and use the top layer. Store at room temperature.
7. 1X Electrophoresis buffer: 25 mM Tris, 192 mM glycine, 0.1% (w/v/) SDS. This buffer can also be prepared as a 10X concentrated stock solution. Store at room temperature.
8. CBB-R: 25% ethanol, 55% water, 20% acetic acid, 0.5% (w/v) CuSO$_4$, 0.5% (w/v) Coomassie Brilliant Blue R250 (Sigma-Aldrich). Filter before use and store at room temperature.
9. 5X SDS-PAGE loading buffer: 250 mM Tris-HCl pH 6.8, 10% (w/v) SDS, 10% (w/v) DTT, 50% glycerol, 0.05% (w/v) bromophenol blue.
10. Filter paper and cellophane.
11. Gel dryer (BioRad, Hercules, CA).
12. MultiSensitive Phosphor Screen (Perkin Elmer, Boston, MA).
13. Film exposure cassette.
14. Cyclone™ Storage Phosphor System (Perkin Elmer).
15. Optiquant™ software (Perkin Elmer).

3. Methods

FtsQ is a monotopic membrane protein with a short cytoplasmic domain and a globular periplasmic domain (*see* **Fig. 1A**). It can be synthesized radioactively in a cell-free expression system in the presence of *E. coli* inside-out inner membrane vesicles (IMVs). Upon correct insertion into the membrane, the periplasmic domain crosses the membrane and becomes inaccessible to proteases added to the outside of the IMVs. The exposed cytoplasmic tail can be cleaved by the protease resulting in a truncation of 24 amino acids (Δ1 − 24FtsQ). Correctly inserted FtsQ migrates on SDS-PAGE as a band with a slightly smaller molecular weight than wild-type FtsQ (*see* **Fig. 1B**).

The assay for the in vitro translocation of the precursor of outer membrane protein A (proOmpA) into IMVs also relies on protease protection. proOmpA is overexpressed in *E. coli*, purified, and labeled with a fluorescent dye for visualization. proOmpA is dissolved in a high-molar urea solution to keep

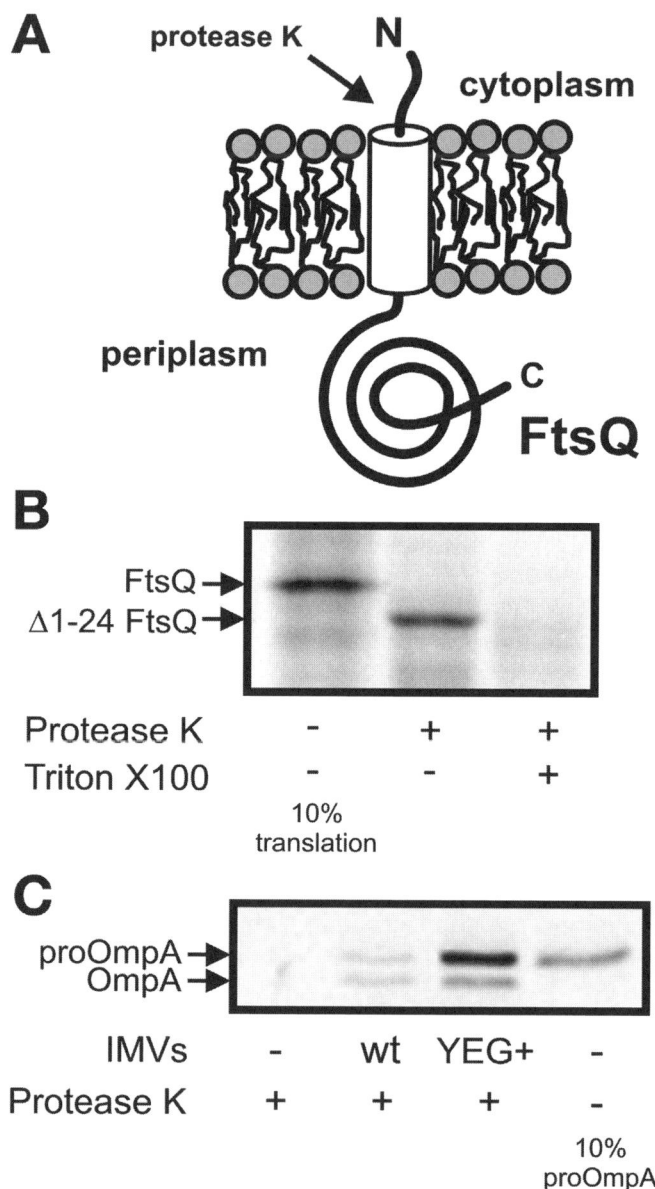

Fig. 1. In vitro membrane insertion of FtsQ and translocation of proOmpA. (A) Schematic representation of the model protein FtsQ. (B) In vitro co-translational membrane insertion of FtsQ as described in **Subheading 3.6.** Ten percent translation represents 10% control of the in vitro synthesized FtsQ. (C) In vitro translocation of

it in an unfolded state. Incubation of unfolded fluorescent proOmpA in the presence of IMVs, SecA, and ATP results in the translocation of the labeled protein into the lumen of the IMVs. After protease treatment of the reaction, the protected (translocated) proOmpA can be visualized by fluorescent imaging of an SDS-PAGE gel (*see* **Figure 1C**).

3.1. Preparation of S135 Lysate

1. An overnight culture of *E. coli* MC4100 (*9*) is diluted to an OD_{660} of ~ 0.05 into 4 L LB supplemented with 0.2% glucose and grown at 37°C.
2. At an OD_{660} of 0.25, the cells are harvested by centrifugation (15 min 5000*g*), washed with 250 mL cold lysis buffer, and re-collected by centrifugation.
3. The cell pellet is resuspended in 10 mL lysis buffer, supplemented with 0.5 m*M* PMSF, and lysed by French Press treatment (two passes at 8000 psi). After the first lysis step, the DTT and PMSF concentrations are increased to 2 and 1 m*M*, respectively.
4. Unbroken cells are removed by centrifugation (5000*g*, 15 min), after which the supernatant is centrifuged twice at 30,000g for 30 min.
5. Endogenous mRNA is inactivated by incubation of the lysate for 90 min at 37°C (*see* **Note 1**).
6. Membranes are removed by two 30-min centrifugation steps at 135,000g.
7. The supernatant is transferred to a dialysis tube and dialyzed for 20 h against 2 L dialysis buffer. The buffer is replaced after the first 16 h.
8. The cell lysate is subsequently aliquoted, snap-frozen in liquid nitrogen, and stored at −80°C. Each aliquot should not be refrozen more than once.

3.2. Isolation of Inner Membrane Vesicles Containing Overexpressed SecYEG

1. One liter LB containing 0.1 mg/mL ampicillin is inoculated with 25 mL overnight culture of *E. coli* SF100 (*10*) transformed with plasmid pET610 (*11*) and grown at 37°C.
2. At an OD_{660} of approx 0.6, the cells are induced with 0.5 m*M* IPTG and grown for 2 h longer.

◀━━

Fig. 1. fluorescently labeled proOmpA as described in **Subheading 3.7.** Translocation was done in the absence of membranes (-) or in the presence of membranes containing wild-type (wt) or overexpression (YEG⁺) levels of SecYEG. Part of the translocated proOmpA is processed by the membrane-embedded leader peptidase, which cleaves off the signal sequence, resulting in the mature OmpA protein. Ten percent proOmpA represents 10% of the proOmpA added to the translocation reaction.

3. Cells are collected by centrifugation (5000*g*, 10 min, 4°C), resuspended in 10 mL HEPES-sucrose, and rapidly frozen as nuggets in liquid nitrogen. Frozen cells can be stored for several months at −20°C.

4. The cells are slowly thawed in water and supplemented with 0.5 m*M* PMSF. Subsequent steps are performed at 4°C to prevent proteolysis.

5. The cells are lysed by French Press treatment (two passes at 8000 psi). After the first lysis step, the PMSF concentration is increased to 2 m*M* (*see* **Note 2**).

6. Unbroken cells are removed by centrifugation (5000*g*, 15 min.) and the membranes are collected by ultracentrifugation (125,000g, 60 min.).

7. The membrane pellet is resuspended in 2 mL 50 m*M* HEPES-KOH, pH 8.0, and divided over two 36–54% sucrose step gradients (*see* **Note 3**). The gradients are spun for 30 min at 4°C at 250,000g.

8. The brownish inner membrane fraction is collected from the 45% sucrose layer, diluted 30-fold with 50 m*M* HEPES-KOH, pH 8.0, and re-collected by centrifugation (125,000g, 60 min).

9. The pellet is resuspended in 0.5 mL HEPES-glycerol, aliquoted, snap-frozen in liquid nitrogen, and stored at −80°C. One liter of culture yields approx 10–15 mg of membrane protein.

3.3. Purification of SecA

1. Two liters of LB supplemented with 0.1 mg/mL ampicillin are inoculated with 20 mL of an overnight culture of *E. coli* DH5α transformed with plasmid pMKL18 and grown at 37°C.

2. At an OD_{660} of approx 0.6, overexpresssion of SecA is induced by the addition of 0.5 m*M* IPTG. To facilitate further growth, 0.4% arabinose can be added.

3. After a further 2 h of growth, the cells are collected by centrifugation and resuspended in Tris-sucrose. The suspension is supplemented with 1 m*M* DTT, 0.5 m*M* PMSF, and 1 mg/mL DNase and RNase each.

4. Cells are lysed by two passes through a cell disrupter at 8000 psi. Unbroken cells and membranes are removed by two centrifugation steps at 30,000g (15 min) and at 125,000g (60 min), respectively.

5. The pumps of the ÄKTAexplorer™ (or equivalent) are washed with SecA buffer or SecA buffer with 1 *M* NaCl.

6. Two 5-mL HiTrap Q HP columns are connected in series, attached to the ÄKTA explorer™ and equilibrated with SecA buffer containing 150 m*M* NaCl at a flow rate of 1 mL/min.

7. The supernatant of **step 4** is diluted to 50 mL with SecA buffer supplemented with 150 m*M* NaCl and loaded onto the HiTrap Q HP columns at a flow rate of 1 mL/min.

8. The columns are washed with 8 column volumes of SecA buffer with 180 m*M* NaCl at a flow rate of 1 mL/min, whereafter the SecA protein is

eluted at the same flow rate with a 14-column volume of a linear gradient of 180–400 mM NaCl in SecA buffer.

9. Of each fraction, an aliquot of 10 μL is mixed with 5 μL 5X SDS-sample buffer and analyzed on a 10% SDS-PAGE gel (*see* **Subheading 3.8.**). SecA migrates as a band of approx 100 kDa.

10. The SecA containing fractions are pooled and concentrated to 5 mL using Amicon ultrafiltration.

11. After ultrafiltration a Superdex 200 XK26/60 column is equilibrated with SecA buffer at a flow rate of 1 mL/min.

12. The concentrated SecA fraction is loaded onto the gel filtration column at a flow rate of 0.5 mL/min, after which SecA is eluted from the column with the same buffer at a flow rate of 1 mL/min.

13. Aliquots of the collected fractions are analyzed on 10% SDS-PAGE gel as described in **step 9**. SecA-containing fractions are pooled, aliquoted, and snap-frozen in liquid nitrogen. The purified SecA is stored at −80°C. One liter of culture yields approx 6 mg of purified protein.

3.4. Purification of proOmpA

1. 140 mL LB broth supplemented with 0.5% glucose and 0.1 mg/mL ampicillin is inoculated with 10 mL of an overnight culture of *E. coli* strain MM52 transformed with pET503 (overexpression of proOmpAC290S, Ap^R) (*see* **Notes 4–6**) and grown at 30°C to an OD$_{660}$ of approx 1.0.

2. The remaining concentration of glucose is reduced by diluting the culture with 750 mL LB supplemented with 0.1 mg/mL ampicillin prewarmed at 37°C, and growth is continued at 37°C.

3. High-level expression of the *ompA* gene is induced by the addition of 1 mM IPTG, and growth is continued for 2 h longer.

4. Cells are harvested by centrifugation (10 min at 5000g), washed with 100 mL 50 mM Tris-HCl, pH 7.0, and re-collected by centrifugation.

5. The cells are resuspended in 5 mL 50 mM Tris-HCl, pH 7.0, and lysed by sonication for 10 min with pulses of 10 s on a 50% duty cycle.

6. Inclusion bodies and cell debris are collected by low-speed centrifugation (1800g, 10 min).

7. The pellet is dissolved in proOmpA buffer, and nonsoluble material is removed by centrifugation (200,000g, 60 min) (*see* **Note 7**).

8. The urea-dissolved proOmpA is frozen in liquid nitrogen and stored at −20°C. One liter of culture yields approx 40 mg of purified protein.

3.5. Fluorescent Labeling of Purified proOmpA

1. Urea-dissolved proOmpA (1–3.5 mg/mL) is reduced with a 10-fold molar excess of TCEP and incubated for 30 min at room temperature.

2. A 20-fold molar excess of fluorescein-maleimide (*see* **Note 8**) is added, and the mixture is incubated for 30 min at room temperature. Containers are wrapped in aluminum foil to protect the fluorescein probe from light. Exposure to light is kept to a minimum in subsequent steps.
3. Labeling is terminated by the addition of a 10-fold molar excess of DTT followed by 15-min incubation at room temperature.
4. The excess fluorescein is removed using a Micro-Biospin 6 chromatography column (*see* **Note 9**). Before use, the column is drained and washed once with 500 μL of 10 mg/mL BSA and three times with 500 μL proOmpA buffer. After each wash step the column is spun for 2 min at 1000g in an Eppendorf centrifuge.
5. A volume of 75 μL labeled proOmpA is applied to the Biospin column, and the column is subsequently spun for 2 min at 1000g in an Eppendorf centrifuge.
6. The proOmpA containing flowthrough is collected and snap-frozen in liquid nitrogen. The vial is wrapped in aluminum foil and stored at $-20°C$.

3.6. In Vitro Synthesis and Membrane Insertion of FtsQ

1. A 25-μL In vitro transcription translation reaction mixture as described in **Table 1** is prepared without the addition of ^{35}S ProMix. The reaction mixture is placed at 37°C and translation is started by the addition of the ^{35}S ProMix.
2. After 20 min of incubation, two aliquots of 10 μL of the reaction are transferred to two vials containing 2.5 μL 2 mg/mL protease K. To one of the vials 1.5 μL 10% Triton X-100 is added. The vials are subsequently incubated on ice for 30 min to digest the noninserted FtsQ. Of the remaining reaction mix, 1 μL is transferred to a vial containing 15 μL 2X SDS-PAGE sample buffer. This sample serves as a synthesis control.
3. Protease-protected proteins are precipitated by the addition of 15 μL 20% TCA followed by 30-min incubation on ice.
4. Precipitated proteins are collected by 10-min centrifugation at maximum speed in an Eppendorf table centrifuge. The pellets are washed with 0.5 mL ice cold acetone and recollected by 5-min centrifugation. The acetone is removed and the pellets are dried by placing the open vials for 2 min at 37°C.
5. The pellets are resuspended in 15 μL 2X SDS-PAGE sample buffer, incubated for 2 min at 90°C, and analyzed on a 12% SDS-PAGE (*see* **Subheading 3.8.**). The [^{14}C]-marker can be used as a reference. An example of an FtsQ insertion reaction is shown in **Fig. 1B**.

3.7. In Vitro Translocation of Fluorescently Labeled proOmpA

1. In all steps, prolonged exposure of the samples to light is avoided to protect the fluorescein probe.
2. Fluorescein-labeled proOmpA(C290S) (5 μg/mL) is added to a 50-μL reaction mixture containing translocation buffer, 10 mM phosphocreatine, 50 μg/mL

Table 1
In Vitro Transcription Translation Mix

Component	Final concentration
Transcription-translation buffer	×1
MgCl$_2$	0–20 mM (*see* **Note 10**)
Creatine kinase	40 μg/mL
Creatine phosphate	8 mM
pBSKftsQ plasmid DNA	20 μg/mL
19 Amino acids	0.1 mM of each amino acid
Folinic acid	0.3 mM
E. coli tRNA	100 μg/mL
ATP	3 mM
GTP	2.25 mM
CTP	2 mM
UTP	2 mM
Enzyme mix	×50 diluted
RNA guard	540 units/mL
Inner membrane vesicles	80 μg/mL
S135 lysate	×3 diluted
^{35}S Pro-mix	×17 diluted

creatine kinase, 50 μg/mL IMVs, and 20 μg/mL SecA. The mixture is incubated at 37°C, and translocation is started by the addition of 2 mM ATP.

3. After 5–30 min of translocation, 45 μL of the reaction mixture is transferred to a vial containing 5 μL 1 mg/mL proteinase K and incubated for 15 min on ice to digest the nontranslocated proOmpA.

4. Protease-protected proOmpA is precipitated by the addition of 150 μL 10% (w/v) TCA followed by 30-min incubation on ice.

5. Precipitated proteins are pelleted for 10 min at maximal speed in an Eppendorf centrifuge. After removal of the supernatant, pellets are washed with 0.5 mL ice-cold acetone and recollected by centrifugation. The acetone is removed and the pellets are dried by placing the open vials for 5 min at 37°C.

6. The dried pellets are dissolved in 10 μL twofold concentrated SDS-PAGE sample buffer, incubated for 2 min at 90°C, and analyzed on a 12% SDS-PAGE gel. The fluorescence of the protease-protected proOmpA is visualized *in gel* using the highly sensitive Roche Lumi-imager F1 (*see* **Subheading 3.8.**). An example of a translocation reaction imaged by fluorescence is shown in **Fig. 1C**.

3.8. SDS-PAGE

1. These instructions are written for the BioRad Miniprotein II system (BioRad, Hercules, CA) using 0.8-mm spacers but can easily be modified for other gel systems.
2. Ten or 12% running gels are prepared according to **Table 2**. The gel is poured, leaving enough space for the stacking gel and covered with 0.5 mL water saturated isobutanol. The gel is allowed to polymerize at room temperature for 20–30 min.
3. The isobutanol is removed and the top of the gel is rinsed with distilled water. The remaining water is removed with filter paper.
4. The stacking gel is prepared by mixing 1.77 mL double distilled water, 0.75 mL SDS-PAGE stacking gel buffer, 0.48 mL 40% acrylamide/bis solution, 30 μL 10% APS, and 3 μL TEMED. After the gel is poured, the comb is inserted and polymerization is continued at room temperature.
5. When the gel is polymerized, the comb is removed and the wells are rinsed with electrophoresis buffer. After assembly of the gel system, electrophoresis buffer is added to the inner and outer compartments of the gel system and the samples loaded into the wells.
6. The gel system is covered with the lid, connected to a power supply and run at a constant 180 V until the blue dye front reaches the bottom of the gel.
7. The following instructions **(steps 8–10)** apply for samples prepared in the in vitro membrane insertion assay; for samples prepared in the in vitro protein translocation assay, proceed to **step 11**; for the analysis of the SecA purification, proceed to **step 12**.
8. For samples prepared in the in vitro membrane insertion assay: After running, the gel is washed with distilled water (three times for 1 min) and transferred to a piece of filter paper. The gel is covered with cellophane and dried for 1 h at 70°C using a Bio-Rad gel dryer. The dried gel is then transferred, without removal of

Table 2
SDS-PAGE Running Gels

	10% SDS-PAGE gel	12% SDS-PAGE gel
Double distilled water	4.35 mL	3.87 mL
SDS-PAGE running gel buffer	2.25 mL	2.25 mL
40% Acrylamide/bis solution	2.40 mL	2.88 mL
10% APS	45 μL	45 μL
TEMED	4 μL	4 μL

SDS-PAGE, sodium dodecyl sulfate–polyacrylamide gel electrophoresis; APS, ammonium persulfate.

the cellophane, to a film exposure cassette containing a MultiSensitive Phosphor Screen and exposed for 16 h.

9. The Phosphor Screen is analyzed in a Cyclone™ Storage Phosphor System, using the Optiquant™ software.

10. For the fluorescent proOmpA translocation assay, the gel is analyzed without fixation in a highly sensitive Roche Lumi-imager F1 with the emission filters set at 520 nm (*see* **Note 11**).

11. For the analysis of the SecA purification, the gel is incubated in CCB-R for 15 min at room temperature. The gel is rinsed with distilled water and destained by heating the gel in distilled water, using a microwave oven, until the protein bands are visible.

4. Notes

1. The 37°C incubation step can reduce background synthesis from endogenous mRNA, but is not absolutely required.

2. When the inner membranes are not used for in vitro membrane insertion assays, 1 mg/mL DNase and RNase each can be added to facilitate French Press treatment.

3. The sucrose gradient is set up by gently layering the different sucrose solutions with a pipet, starting with the highest concentration.

4. proOmpA is expressed in the temperature-sensitive *E. coli* strain MM52 *(13)* that contains a conditional lethal mutation in the *secA* gene. At the nonpermissive temperature of 37°C, this strain exhibits a major secretion defect, which prevents the cleavage of the signal sequence and causes the accumulation of the precursor form of OmpA in the cytosol.

5. The overnight culture of *E. coli* MM52 transformed with pET503 is also grown at 30°C.

6. Wild-type OmpA contains two cysteine residues at positions 290 and 302. To obtain proOmpA labeled at a single position, a single cysteine mutant of proOmpA (proOmpA[C290S]) is used. The protocol, however, also works with the wild-type proOmpA molecule, in which case two cysteines are labeled.

7. The purification of proOmpA from inclusion bodies usually results in proOmpA that is sufficiently pure for the experiments described in this chapter. Further purification can be obtained by applying the proOmpA on a 1-mL HiTrap™ Q HP column (GE Healthcare Biosciences AB), equilibrated with proOmpA -buffer. Wash the column with proOmpA buffer and analyze the flowthrough and wash fractions on 12% SDS-PAGE. proOmpA elutes with the nonbound protein fraction, while the contaminating proteins remain bound to the column.

8. The fluorescein-5-maleimide can be replaced by other fluorescent probes such as Oregon Green 488 maleimide or Texas Red maleimide (Invitrogen) *(16)*.

9. Alternatively, the remaining nonreacted probe can be removed by TCA precipitation.

10. The synthesis yield of the in vitro transcription-translation reaction is highly dependent on the magnesium concentration. The optimal magnesium concentration usually lies between 0 and $20\,\text{m}M$ MgCl$_2$ but needs to be determined for each lysate preparation.

11. Alternatively, imaging can be done with a regular gel documentation system equipped with a CCD camera using an UV tray for excitation. Care should be taken that the excitation light is distributed evenly over the slab gel.

Acknowledgments

The authors would like to thank Francois du Plessis for careful reading of the manuscript. This work was supported by European Community Grant LSHG-CT-2004-504601 (E-Mep).

References

1. Osborne, A. R., Rapoport, T. A., and van den Berg, B. (2005) Protein translocation by the Sec61/SecY channel. *Annu. Rev. Cell Dev. Biol.* **21**, 529–550.

2. de Keyzer, J., van der Does, C., and Driessen, A. J. M. (2003) The bacterial translocase: a dynamic protein channel complex. *Cell Mol. Life Sci.* **60**, 2034–2052.

3. Brundage, L., Hendrick, J. P., Schiebel, E., Driessen, A. J., and Wickner, W. (1990) The purified *E. coli* integral membrane protein SecY/E is sufficient for reconstitution of SecA-dependent precursor protein translocation. *Cell* **62**, 649–657.

4. Economou, A. and Wickner, W. (1994) SecA promotes preprotein translocation by undergoing ATP-driven cycles of membrane insertion and deinsertion. *Cell* **78**, 835–843.

5. Hartl, F. U., Lecker, S., Schiebel, E., Hendrick, J. P., and Wickner, W. (1990) The binding cascade of SecB to SecA to SecY/E mediates preprotein targeting to the *E. coli* plasma membrane. *Cell* **63**, 269–279.

6. Valent, Q. A., Scotti, P. A., High, S., et al. (1998) The *Escherichia coli* SRP and SecB targeting pathways converge at the translocon. *EMBO J.* **17**, 2504–2512.

7. Schiebel, E., Driessen, A. J. M., Hartl, F. U., and Wickner, W. (1991) $\Delta\mu_{H+}$ and ATP function at different steps of the catalytic cycle of preprotein translocase. *Cell* **64**, 927–939.

8. van der Does, C., Manting, E. H., Kaufmann, A., Lutz, M., and Driessen, A. J. M. (1998) Interaction between SecA and SecYEG in micellar solution and formation of the membrane-inserted state. *Biochemistry* **37**, 201–210.

9. Casadaban, M. J. (1976) Transposition and fusion of the *lac* genes to selected promoters in *Escherichia coli* using bacteriophage lambda and Mu. *J. Mol. Biol.* **104**, 541–555.

10. Baneyx, F. and Georgiou, G. (1990) *In vivo* degradation of secreted fusion proteins by the *Escherichia coli* outer membrane protease OmpT. *J. Bacteriol.* **172**, 491–494.

11. Kaufmann, A., Manting, E. H., Veenendaal, A. K. J., Driessen, A. J. M., and van der Does, C. (1999) Cysteine-directed cross-linking demonstrates that helix 3 of SecE is close to helix 2 of SecY and helix 3 of a neighboring SecE. *Biochemistry* **38**, 9115–9125.

12. van der Wolk, J. P., Klose, M., de Wit, J. G., den, B. T., Freudl, R., and Driessen, A. J. M. (1995) Identification of the magnesium-binding domain of the high-affinity ATP-binding site of the *Bacillus subtilis* and *Escherichia coli* SecA protein. *J. Biol. Chem.* **270**, 18975–18982.

13. Oliver, D. B. and Beckwith, J. (1981) *E. coli* mutant pleiotropically defective in the export of secreted proteins. *Cell* **25**, 765–772.

14. Manting, E. H., van der Does, C., Remigy, H., Engel, A., and Driessen, A. J. M. (2000) SecYEG assembles into a tetramer to form the active protein translocation channel. *EMBO J.* **19**, 852–861.

15. van der Laan, M., Nouwen, N., and Driessen, A. J. M. (2004) SecYEG proteoliposomes catalyze the $\Delta\psi$-dependent membrane insertion of FtsQ. *J. Biol. Chem.* **279**, 1659–1664.

16. de Keyzer, J., van der Does, C., and Driessen, A. J. M. (2002) Kinetic analysis of the translocation of fluorescent precursor proteins into *Escherichia coli* membrane vesicles. *J. Biol. Chem.* **277**, 46059–46065.

3

Identification of Autotransporter Proteins Secreted by Type V Secretion Systems in Gram-Negative Bacteria

Yihfen T. Yen and Christos Stathopoulos

Summary

Autotransporters belong to a group of virulence factors secreted by Gram-negative bacteria using a simple mechanism termed type V or autotransporter secretion. These large proteins have diverse virulence functions, and many are found to play relevant roles in bacterial infections. An autotransporter polypeptide is equipped with two translocator domains (signal peptide and β-domain), which enable its own export across bacterial membranes. Because of significant sequence conservation in the translocator domains among various species, genes of putative autotransporters can be easily identified in bacterial genomic sequences. Thereafter, gene expression can be determined and protein localization elucidated. Such a method for identifying autotransporter virulence proteins may be an important first step in understanding bacterial pathogenicity or discovering new targets for antimicrobial and vaccine development.

Key Words: Gram-negative bacteria; protein secretion; protein localization; type V secretion; autotransporter; *in silico* analysis; RT-PCR.

1. Introduction

Separating the cytoplasm of a Gram-negative bacterium from its external environment is the cell envelope formed by an inner/cytoplasmic membrane (IM), an outer membrane (OM), and a periplasm sandwiched in between. For the purpose of survival, bacteria export numerous proteins across the two membranes. These proteins include enzymes, organelle components, and/or virulence factors. Most proteins rely on the Sec translocase to travel across the IM; on the other hand, translocation across the OM can be achieved by more diverse means, which involve transport through various OM pores or channels.

From: *Methods in Molecular Biology, Vol. 390: Protein Targeting Protocols: Second Edition*
Edited by: M. van der Giezen © Humana Press Inc., Totowa, NJ

One family of proteins secreted by Gram-negative bacteria is the autotransporters, many of which have been associated with bacterial pathogenicity *(1–3)*.

An autotransporter polypeptide consists of three domains: an N-terminal signal peptide, an internal passenger/effector domain, and a C-terminal β-domain. The interaction between the signal peptide and the Sec translocase allows an autotransporter to be transported across the IM. Thereafter, the signal peptide is cleaved, and the C-terminal β-domain forms a β-barrel pore structure in the OM. Export of the internal passenger/effector domain to the cell surface is mediated by this β-barrel structure. Once externalized, the passenger domain folds into a functional protein and either remains on the cell surface or is released into the extracellular milieu by cleavage *(4–6)*.

In silico identification of autotransporter proteins can be performed by BLAST searching a bacterial genome for possible protein candidates and screening each candidate for the presence of an N-terminal signal peptide, C-terminal consensus residues, and other motifs located in the internal passenger domain *(7,8)*. *In vivo* expression of autotransporter genes in their native species can be assessed by the technique of reverse transcription–polymerase chain reaction (RT-PCR). For protein secretion studies, putative autotransporter genes can be amplified from the bacterial genome and cloned into *E. coli* strain K12. Subsequently, OM isolation and protein precipitation of the culture medium can be performed to allow detection of autotransporter proteins that localize to the OM or the extracellular milieu *(8)*.

2. Materials

2.1. Bioinformatics: Screening Genomes for Autotransporter Genes

1. A computer with Internet access.

2.2. RT-PCR: Assessing Autotransporter Gene Expression

2.2.1. Isolation of Total RNA

1. Protoplasting buffer: 150 mM Tris-HCl, pH 8.0, 80 mM EDTA, pH 8.0, and 4.5 M sucrose. Store at room temperature.
2. Lysozyme (Sigma, St. Louis, MO): prior to use, dissolve 50 mg in 1 mL of 10 mM Tris-HCl buffer, pH 8.0.
3. Lysis buffer: 2 M Tris-HCl, pH 8.0, 2 M NaCl, 200 mM sodium citrate, and 3% sodium dodecyl sulfate (SDS). Store at room temperature.
4. 5 M NaCl. Store at room temperature.
5. 100% Ethanol. Store at −20°C.
6. 70% Ethanol. Store at room temperature.

7. Water treated with DEPC (diethyl pyrocarbonate; Sigma): dissolve 1 mL of DEPC in 1 L of water. Incubate at 37°C with agitation over night. Autoclave.
8. Phenol, pH 4.3 (Fisher Scientific, Hampton, MA).
9. Chloroform.
10. Isopropanol.
11. 5X TBE buffer (stock): 0.5 M Tris base, 0.45 M boric acid, and 100 mM EDTA, pH 8.0. Prepare the working concentration (1X) prior to use.

2.2.2. RNA Cleaning

1. DNase I (Promega, Madison, WI). A reaction stop solution and buffer are supplied with purchase of the enzyme.
2. RNeasy Cleanup kit (Qiagen, Valencia, CA).
3. dNTP mix (Promega).
4. Taq DNA polymerase (Promega). A 10X reaction buffer and a 25 mM MgCl$_2$ solution are supplied with purchase of the enzyme.
5. Forward and reverse primers targeting the 16S rRNA gene.
6. 50X TAE buffer (stock): 2 M Tris base, 5.71% (v/v) glacial acetic acid, and 5 mM EDTA, pH 8.0. The working concentration is 1X.

2.2.3. RT-PCR

1. Random hexameric primers (Promega).
2. ImPromII reverse transcriptase (Promega). A 5X reaction buffer without MgCl$_2$ and a 25 mM MgCl$_2$ solution are supplied with purchase of the enzyme.
3. dNTP mix (Promega).
4. RNase inhibitor (Promega).
5. Dithiothreitol (DTT; Invitrogen, Carlsbad, CA).
6. Taq DNA polymerase (Promega). A 10X reaction buffer and a 25 mM MgCl$_2$ solution are supplied with purchase of the enzyme.
7. Forward and reverse primer sets.
8. 50X TAE buffer (stock): 2 M Tris base, 5.71% (v/v) glacial acetic acid, and 5 mM EDTA, pH 8.0. The working concentration is 1X.

2.3. Cloning: Transforming E. coli K12 With Autotransporter Genes for Secretion Studies

2.3.1. Gene Amplification by PCR

1. DyNAzyme Ext DNA polyermase (Finnzymes, Espoo, Finland). A 10X reaction buffer and a 50 mM MgCl$_2$ solution are supplied with purchase of the enzyme.
2. dNTPs (Promega).
3. Forward and reverse primer sets.
4. TAE buffer: see **Subheading 2.2.3., item 8**.
5. PCR gel extraction kit (Qiagen).

2.3.2. Cloning and Transformation

1. TOPO TA cloning kit (Invitrogen).
2. Competent *E. coli* K12 cells.
3. X-Gal solution: dissolve 40 mg of X-Gal in 1 mL of DMF (*N, N*-dimethylformamide; Sigma). Store at 4°C, protected from light. To make X-Gal agar plates, first make medium plates supplemented with appropriate antibiotics, and then smear ∼ 40 μL of X-Gal on each plate prior to use. X-Gal should be prepared fresh each time.
4. 100 μg/mL ampicillin; 100 μg/mL kanamycin: for transformant selection and plasmid (TOPO cloning vector) maintenance.
5. Appropriate digestive enzyme(s): the optimal enzyme(s) should cleave both the gene insert and the vector.

2.4. Cell Fractionation: Localizing Autotransporter Proteins in E. coli K12

1. LB medium: 1% (w/v) tryptone, 0.5% (w/v) yeast, and 1% (w/v) NaCl.
2. 100 μg/mL Ampicillin; 100 μg/mL kanamycin.
3. Ultracentrifuge (Beckman L7-55; Beckman, Fullerton, CA).
4. Acetone: store at −20°C.
5. Resuspension buffer: 20 m*M* Tris-HCl, pH 7.4, 1 m*M* EDTA, and 0.1 m*M* PMSF (Sigma).
6. French press cell (Thermo Spectronic).
7. Sarkosyl (Sigma), 0.5% (w/v). Store at room temperature.
8. Tris buffer: 20 m*M* Tris-HCl, pH 7.4.

2.5. SDS-PAGE: Analyzing Autotransporter Proteins

1. SDS-PAGE electrophoresis system (Miniprotean 3; Bio-Rad, Hercules, CA).
2. 30% Acrylamide/bis solution (Bio-Rad).
3. 1.5 *M* Tris-HCl buffer, pH 8.8.
4. 0.5 *M* Tris-HCl buffer, pH 6.8.
5. 10% SDS (w/v).
6. 1% SDS (w/v).
7. Ammonium persulfate (Promega), 10% (w/v).
8. TEMED (*N, N, N′, N′*-tetramethylethylenediamine; Bio-Rad).
9. 4X Sample loading buffer (5 mL): 625 μL of 0.5 *M* Tris-HCl, pH 6.8, 1.25 mL glycerol, 1 mL of 10% SDS, bromophenol blue (Bio-Rad), 100 μL of 0.5% (w/v); and β-mercaptoethanol (Bio-Rad), 250 μL. Working concentrations are 1, 2, and 4X. Short-term storage at 4°C; long-term at −80°C.
10. 10X Gel running buffer, pH 8.3: 250 m*M* Tris-Base, 2 *M* glycine, and 1% SDS. Do not adjust pH. The working concentration is 1X.

2.6. Silver Staining: Visualizing Proteins on SDS–Polyacrylamide Gels

1. Fixing solution: 30% methanol and 10% acetic acid.
2. 10% Ethanol.
3. Sodium thiosulfate (Sigma), 2% (w/v). To make a sodium thiosulfate working solution, add 1 mL of 2% sodium thiosulfate in 100 mL of water prior to use.
4. Silver nitrate (Sigma), 5% (w/v). To make a silver nitrate working solution, add 2 mL of 5% silver nitrate in 100 mL of water prior to use.
5. Development solution: 113 mM sodium carbonate, 0.01% formaldehyde, and 0.004% (v/v) of 2% sodium thiosulfate (e.g., 20 µL of 2% sodium thiosulfate in 500 mL of the development solution).
6. 1% Acetic acid.

3. Methods

3.1. Bioinformatics: Screening Genomes for Autotransporter Genes

1. Retrieve protein sequences of desired autotransporters from a genome database, for example, GenBank of the National Center for Biotechnology Information (NCBI): http://www.ncbi.nlm.nih.gov/ (*see* **Note 1**).
2. Search microbial genomes by BLAST using the selected autotransporter sequences as queries (*see* **Note 2**).
3. Screen each hit by analyzing its C terminus (last 250 amino acid residues). The query and the subject should be similar in this region, rather than in the N-terminal or passenger domain (*see* **Note 3**).
4. Screen the selected autotransporter sequences for the presence of a signal peptide using a prediction program, for example, SignalP (http://cbs.dtu.dk/services/SignalP/) (*see* **Note 4**).
5. Examine the end of an autotransporter's C-terminal domain for the presence of consensus residues: [Y/V/I/V/W]-[X]-[F/W], where X is a polar amino acid.
6. Analyze the autotransporter sequences for unique characteristics or motifs *(8)*. This can be done manually or by searching in available databases, for example, http://motif.genome.jp/. Particularly, look for large sizes and scarcity of cysteine residues in the passenger domain. Some autotransporters may also harbor the GDSGSP serine protease motif, a BrkA junction, or RGD integrin-binding motifs. See **Table 1** for an example of the results.

3.2. RT-PCR: Assessing Autotransporter Gene Expression

3.2.1. Isolation of Total RNA (see **Note 5**)

1. Centrifuge a 25-mL culture at 2300g for 12 min at 4°C.
2. Resuspend the pellet in 25 mL protoplasting buffer.
3. Add to the suspension 1 mL of lysozyme. Incubate on ice for 15 min.
4. Centrifuge the protoplast suspension at 3800g for 10 min at 4°C.

Table 1
Data of *In Silico* Analyses Showing Two Putative Autotransporters Identified in *Yersinia pestis* KIM Strain

Protein	GenBank accession no.	Open reading frame	No. of amino acid residues	MW	Signal peptide residues; prediction score	Cysteine residues	Motifs
YapA	NP_668668	y1346	1458	149634.03	28 (1.000)	0	1RGD Autotransporter-like β-domain Conserved BrkA junction
YapG	NP_670888	y3591	994	106869.11	49 (1.000)	1	Autotransporter-like β-domain

5. Resuspend the pellet in 6 mL lysis buffer. Incubate at 37°C for 5 min. Place on ice for an additional 2 min.
6. Add to the suspension 3 mL of 5 M NaCl. Incubate on ice for 10 min. Transfer samples into microfuge tubes. Centrifuge at maximum speed for 10 min at 4°C.
7. Recover the supernatant and add 1 mL of ice-cold 100% ethanol into each tube. Centrifuge for at maximum speed 15 min at 4°C.
8. Discard the supernatant and rinse each pellet with 1 mL of 70% ethanol.
9. Air-dry the pellets (*see* **Note 6**). Dissolve each pellet in a small amount (10–20 μL) of DEPC-treated water. Combine samples. At this point, nucleic acids are extracted.
10. To extract RNA, the combined sample is mixed with 500 μL of acid phenol. Vortex the suspension for 15 s and incubate for 5 min at room temperature.
11. Add to the phenol suspension 300 μL of chloroform. Shake the mixture vigorously. Incubate the mixture for 10 min at room temperature while shaking it every minute to keep it mixed.
12. Centrifuge the mixture at maximum speed for 15 min at 4°C. Then transfer the top, clear, aqueous layer containing RNA to a new microfuge tube.
13. Add an equal volume of DEPC-treated water to the RNA-containing suspension. After mixing, add an equal volume of isopropanol to the mixture. For example, if 500 μL of suspension are drawn from the aqueous layer, add 500 μL of DEPC-treated water and then 1 mL of isopropanol to the entire mixture.
14. Incubate the mixture at room temperature for 10 min. Centrifuge the mixture at maximum speed for 15 min at 4°C.
15. Discard the supernatant. Add 1 mL of ice-cold 75% ethanol to the pellet. Centrifuge at maximum speed for 15 min at 4°C.
16. Remove ethanol and air-dry the RNA pellet ($\sim$ 5–10 min). Dissolve the pellet in a small amount ($\sim$ 30–50 μL) of DEPC-treated water (*see* **Note 6**).
17. Run 2 μL of the RNA sample on a 1% TBE agarose gel to check for RNA integrity (*see* **Note 7**).
18. Measure the concentration and assess the purity of extracted RNA by taking the optical density (OD) at 260 and 280 nm (*see* **Note 8**).

3.2.2. RNA Cleaning

1. Vortex RNA samples for 1 min to shear genomic DNA (gDNA).
2. To digest remaining DNA fragments, perform DNase treatment according to the protocol provided by Promega.
3. To remove salts, residual DNA, and DNase, pass the extracted RNA samples through Qiagen cleanup columns according to the manufacturer's instructions.
4. PCR assessment of DNA contamination: include 2.5 μg of RNA in a 25-μL, 35-cycle reaction containing primers that target the 16S rRNA gene. Resolve PCR

products on a TAE agarose gel. No bands should be visible on the gel if the sample
is indeed free of DNA contamination. *See* **Subheading 3.2.3.2.** for PCR conditions
and **Fig. 1**, lane 9, for a sample result.

3.2.3. RT-PCR

3.2.3.1. Step-1 RT-PCR: Reverse Transcription Reaction

1. In a thin-walled PCR tube placed on ice, set up a 5-μL reaction containing 1 μg of
   RNA, 0.5 μg of random primers, and DEPC-treated water. Incubate the reaction at
   70°C for 5 min and place it on ice immediately for at least 5 min.
2. In another thin-walled PCR tube placed on ice, set up a 15-μL reaction containing
   DEPC-treated water, 4 μL of 5X reaction buffer, 1.5–8.0 mM MgCl$_2$, 0.5 mM
   dNTP mix, 20 U RNase inhibitor, 2 μL DTT, and 1 μL reverse transcriptase (*see*
   Note 9).
3. Add the 15 μL reaction from **step 2** to 5 μL denatured RNA primers from **step 1**.
   Incubate at 25°C for 5 min to allow annealing, at 37–42°C for 1 h to allow extension,
   and 70°C for 15 min to inactivate reverse transcriptase. This step generates first
   strand cDNA, which can be stored at −20°C.
4. To be performed in parallel is a negative RT control reaction, which includes
   everything listed in **steps 1–3** except that total RNA is to be substituted with water.

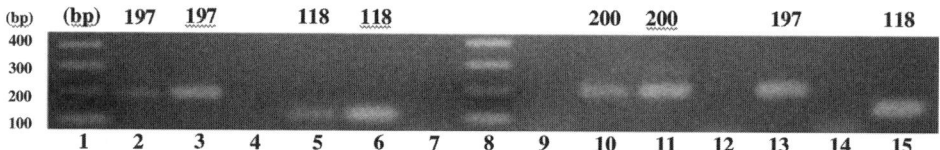

Fig. 1. Products of RT-PCR resolved on a 2% TAE agarose gel. Sizes of amplicons
are shown above the figure in base pairs (bp). Lanes: 1, molecular markers; 2, RT-PCR
amplicon *yapA*; 3, positive PCR control containing *yapA* primers and gDNA template;
4, negative PCR control using *yapA* primers and no template; 5, RT-PCR amplicon
yapG; 6, positive PCR control including *yapG* primers and gDNA template; 7, negative
PCR control containing *yapG* primers and no template; 8, molecular markers; 9, RNA
control containing 2.5 μg of total RNA as template and primers targeting the gene of
16S rRNA; 10, positive RT control containing cDNA as template and 16S rRNA-gene
primers; 11, positive PCR control containing gDNA template and 16S rRNA-gene
primers; 12, negative RT control using *yapA* primers and the negative RT reaction as
template; 13, positive PCR control using *yapA* primers and gDNA as a template; 14,
negative RT control containing *yapG* primers and the negative RT reaction as template;
15, positive PCR control using *yapG* primers and gDNA as a template. In each lane,
5 μL of the total (25 μL) reaction was loaded for analysis.

5. PCR assessment of RT controls: for a positive RT control, set up a 25-μL PCR reaction containing 1 μL of cDNA and 100 ng of each 16S rRNA-gene primer. For a negative control, set up a 25-μL PCR reaction containing the negative RT reaction from **step 4** and 100 ng of each 16S rRNA-gene primer. Resolve 5–10 μL of each reaction on a TAE agarose gel. A strong band from the positive RT control reaction indicates that reverse transcription has worked (*see* **Fig. 1**, lane 10). No amplification products should be generated from the negative RT control reaction (*see* **Fig. 1**, lane 12). *See* **Subheading 3.2.3.2.** for PCR conditions.

3.2.3.2. Step-2 RT-PCR: Amplifying Specific Gene Products

1. In thin-walled PCR tubes placed on ice, set up 25-μL reactions, each containing water, 2.5 μL of 10X buffer without MgCl$_2$, 2.0 mM MgCl$_2$, 0.5 mM dNTP mix, 100 ng of each gene-specific primer, 1–2 μL of cDNA, and 5 U of Taq polymerase.
2. Carry out the amplification step with an initial denaturation period at 94°C for 2 min, a denaturation period at 94°C for 1 min per cycle, an annealing period at an appropriate annealing temperature (*see* **Note 10**) for 30 s per cycle, a polymerization period at 72°C for 30 s per cycle, and finally an extended polymerization period at 72°C for 2 min. Perform 35 cycles. Analyze PCR products by resolving 5–10 μL on a TAE agarose gel. *See* **Fig. 1**, lanes 2 and 5, for examples of two RT-PCR amplicons.
3. Positive and negative PCR controls: both controls are to be run in parallel with amplifications of gene products, except that the positive control is to contain gDNA in place of cDNA and the 16S rRNA-gene primers in place of gene specific primers, and that the negative control is to contain water instead of cDNA. Analyze PCR products by resolving 5–10 μL on a TAE agarose gel. *See* **Fig 1**, lanes 3, 4, 6, 7, 11, and 13, for examples of the results.

3.3. Cloning: Transforming E. coli K12 With Autotransporter Genes for Secretion Studies

3.3.1. Gene Amplification by PCR

1. In thin-walled PCR tubes placed on ice, set up 50-μL reactions, each containing water, 5 μL of 10X buffer without MgCl$_2$, 2 mM of MgCl$_2$, 200 μM of each dNTP, 1 μM of each primer, 50 ng of gDNA template, and 0.5 U of DyNAzyme (*see* **Note 11**).
2. Carry out each reaction with an initial denaturation period at 94°C for 3 min, a denaturation period at 94°C for 1 min per cycle, an annealing period at an appropriate temperature for 1 min per cycle, a polymerization period at 72°C for 5 min per cycle (*see* **Note 12**), and finally an extended polymerization period at 72°C for 10 min. Perform 45 cycles per reaction.
3. Analyze PCR products by running 5–10 μL on a TAE agarose gel.
4. Clean PCR products using a gel extraction kit.

3.3.2. Cloning and Transformation

1. Perform cloning using the TA-overhang method by following the protocol provided by Invitrogen: incubate the cloning reaction containing 0.5–4 μL extracted PCR gene product, 1 μL of the TOPO vector (Invitrogen), 1 μL of salt solution (Invitrogen), and 5 μL of water for 30 min.
2. Transform competent *E. coli* K12 cells with serial dilutions of the cloning reaction. We used approx 1 μL of the cloning reaction for 200 μL of competent cells (OD$_{600}$ of 0.2).
3. Plate the bacteria on X-gal/antibiotic-containing plates. After overnight incubation, select white colonies and screen for the insert by performing plasmid extraction and restriction enzyme digestion.

3.4. Cell Fractionation: Localizing Autotransporter Proteins in E. coli K12

1. Dilute overnight culture 1:100 in LB medium supplemented with 100 μg/mL ampicillin and kanamycin antibiotics. Grow culture to OD$_{600}$ of 0.8–0.9.
2. Centrifuge cells at 11,250g for 10 min at 4°C to separate the supernatant from the cell pellet. Save both fractions.
3. In microfuge tubes placed on ice, acetone-precipitate proteins in the supernatant by mixing one part of the supernatant with four parts of chilled (−20°C) acetone. Incubate the mixture at −20°C for 30 min and centrifuge at maximum speed for 30 min. Decant the supernatant and air-dry the protein pellets.
4. To obtain OM proteins, resuspend the cell pellet obtained from **step 2** in resuspension buffer. Pass the resuspension through a French press column twice and lyse the cells at 600 psi. Centrifuge the cell lysate at 3214g for 10 min at 4°C and discard the pellet containing unbroken cells. Ultracentrifuge the cell lysate at 114,000g for 1 h at 4°C to pellet the cell envelopes. To remove IMs, resuspend the pellet in Tris buffer, incubate it on ice with 0.5% sarkosyl *(9)* for 5 min, and ultracentrifuge at 114,000g for 1 h at 4°C. Finally, resuspend the OM pellet in Tris buffer (20–200 μL) and store at −20°C.

3.5. SDS-PAGE: Analyzing Autotransporter Proteins (see Note 13)

1. To cast an 8%, 1.5-mm-thick mini-resolving gel, mix 2.3 mL of water, 1.3 mL of 30% acrylamide mix, 1.3 mL of 1.5 *M* Tris-HCl, pH 8.8, 50 μL of 10% SDS, 50 μL of 10% ammonium persulfate, and 3 μL TEMED. The percentage of a gel depends on the sizes of proteins to be resolved. Load the mixture in between the glass plates and leave room at the top for the stacking gel.
2. To insulate the resolving gel, immediately overlay 1% SDS (~ 1 mL) on the top. Allow 30 min to 1 h for the resolving gel to solidify.
3. Pour off 1% SDS and rinse the top of the resolving gel with water.
4. To cast a stacking gel (5%), mix 1.4 mL of water, 0.33 mL of 30% acrylamide mix, 0.25 mL of 1.0 *M* Tris-HCl, pH 6.8, 20 μL of 10% SDS, 20 μL of 10% ammonium

persulfate, and 2 µL of TEMED. Put in the comb to create wells and allow 30 min to 1 h for the stacking gel to set.

5. Mix protein samples from acetone precipitation with 1X loading buffer. Mix protein samples from OM isolation with 2X loading buffer. Denature the samples at 100°C on a hot plate for 5 min.

6. Assemble the gel apparatus. Remove the comb from the gel. Load markers and samples into wells.

7. Run samples at 15–20 mA until the dye front runs off the gel. Disassemble the apparatus and remove the gel. Cell fractionation results can be seen in **Fig. 2**.

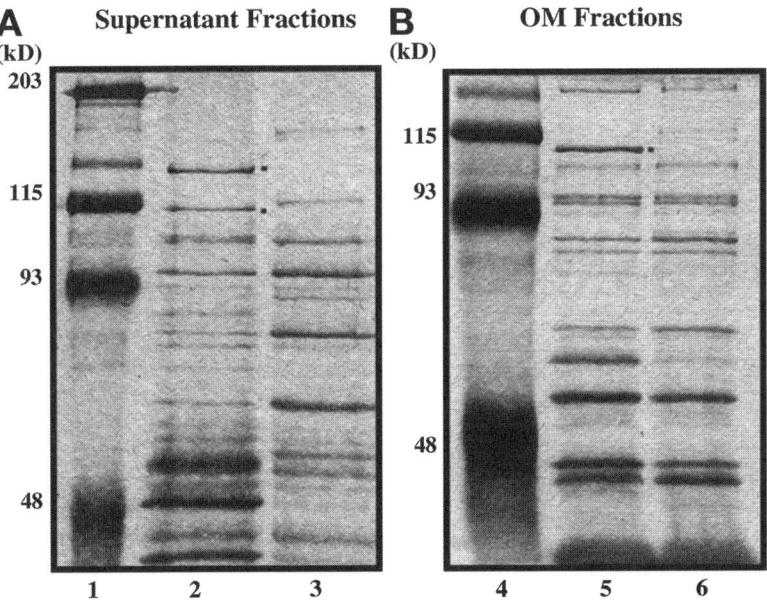

Fig. 2. SDS-PAGE analysis of autotransporter protein expression. (**A**) Supernatant fractions were resolved on an 8% SDS–polyacrylamide gel. Lanes: 1, markers (2 µL); 2, expression of the cloning vector containing the *yapA* insert (125 µL of supernatant); 3, expression of the empty cloning vector (125 µL of supernatant). (**B**) Outer membrane (OM) fractions were resolved on an 8% SDS–polyacrylamide gel. Lanes: 4, markers (2 µL); 5, expression of the cloning vector containing the *yapG* insert (0.4 µg of OM proteins); 6, expression of the empty cloning vector (0.4 µg of OM proteins). Comparing the expression profile of the vector carrying the *yap* insert to that of the empty cloning vector allowed for identification of unique bands (dotted), which represent putative autotransporter proteins. Subsequently N-terminal sequencing or Western analysis can be performed to verify protein identity.

3.6. Silver Staining: Visualizing Proteins on SDS–Polyacrylamide Gels

All steps are to be carried out at room temperature:

1. Immerse the gel in 100 mL of the fixing solution for at least 45 min.
2. Wash the gel with 100 mL of 10% ethanol for 5 min. Repeat the step.
3. Wash the gel with 100 mL of water for 5 min. Repeat the step.
4. Wash the gel with 100 mL of the sodium thiosulfate working solution for 1 min.
5. Incubate the gel in 100 mL of the silver nitrate working solution for 20 min.
6. Rinse the gel with approx 50 mL of the development solution for 10 s to remove unbound silver.
7. Develop with 100 mL of the development solution until protein bands are visible.
8. Stop development with 100 mL of 1% acetic acid, approx 5 min incubation.
9. Rinse the gel with water three times, 5 min each. Sample gels stained with silver nitrate are shown in **Fig. 2**.

Notes

1. In the BLAST search of *Yersinia pestis* KIM autotransporters, we used the protein sequences of seven previously characterized autotransporters as queries *(8)*: BrkA (*Bordetella pertussis*), Hap and Hia (*Haemophilus influenzae*), IcsA (*Shigella flexneri*), Tsh (avian pathogenic *E. coli*), UspA (*Moraxella catarrhalis*), VacA (*Helicobacter pylori*), and rOmpB (*Rickettsia typhi*).
2. We used the BLAST engine provided by the NCBI website. Select protein-to-protein sequence BLAST or "BLASTp."
3. If too many hits are generated, a limit can be set at a particular E(expected) value. The smaller the E value, the more significant a hit is, that is, the hit will be more similar to the query sequence, and it is also less likely that such similarity occurs by chance.
4. Some autotransporters may contain an extra N-terminal extension *(5,7)*. Such an extra peptide renders the signal peptide different from a typical one recognized by most computer prediction programs. To circumvent this problem, use truncated signal peptides, that is, omit the first 25–30 residues in prediction.
5. To prevent RNA degradation, glassware should be autoclaved and baked at 250°C for at least 3 h, plastics should be treated with RNase Away (Molecular BioProducts, San Diego, CA), and pipet tips should have filter barriers. Use RNase-free centrifuge tubes that are autoclaved by manufacturers. Gloves should be worn at all times. Working solutions should be autoclaved or filtered.
6. A dry nucleic-acid pellet becomes transparent. After resuspension with DEPC-treated water, transfer the sample to a new microfuge tube to prevent ethanol contamination.
7. On the TBE gel, two thick bands migrating between 1 and 3 Kb that correspond to 16S and 23S rRNA should be visible. If the two bands are not present but instead a smear is observed, RNA degradation has occurred and the sample should not be used.

8. Use a quartz cuvette. Dilute the RNA sample in 10 mM Tris-HCl, pH 7.5. The concentration of RNA (g/mL) is equal to: 40X A260 X dilution factor. A reading of A260 below 0.15 is insignificant; more RNA is needed for the reading. The ratio of A260/A280 determines the purity of RNA. If this reading is below 1.9, reprecipitate RNA with isopropanol and repeat **Subheading 3.2.1., steps 14–18**.
9. A high concentration of MgCl$_2$ is preferred if shorter cDNA products are desired.
10. The annealing temperature of a primer varies with its length and is usually set at 3–5°C lower than the melting temperature.
11. DyNAzyme DNA polymerase is a high-fidelity enzyme capable of amplifying large gene sizes (> 4.5 Kb), which is why it is a more suitable enzyme for autotransporter gene amplification than Taq DNA polymerase.
12. The extension time depends on the size of the amplicon. Allow 1 min per 1.5 Kb.
13. Wear gloves, because β-mercaptoethanol and unpolymerized acrylamide are known toxins.

Acknowledgments

We would like to thank two of our lab members: Maria Kostakioti for valuable suggestions and Aarthi Karkal for assisting in data collection. This work was funded by the Robert A. Welch Foundation (E-1548).

References

1. Stathopoulos, C., Hendrixson, D. R., Thanassi, D. G., Hultgren, S. J., St Geme, 3rd, J. W., and Curtiss, 3rd, R. (2000) Secretion of virulence determinants by the general sretory pathway in gram-negative pathogens: an evolving story. *Microbes Infect.* **2,** 1061–1072.
2. Kostakioti, M., Newman, C. L., Thanassi, D. G., and Stathopoulos, C. (2005) Mechanisms of protein export across the bacterial outer membrane. *J. Bacteriol.* **187,** 4306–4314.
3. Henderson, I. R. and Nataro, J. P. (2001) Virulence functions of autotransporter proteins. *Infect. Immun.* **69,** 1231–1243.
4. Newman, C. L. and Stathopoulos, C. (2004) Autotransporter and two-partner sretion: delivery of large-size virulence factors by gram-negative bacterial pathogens. *Crit. Rev. Microbiol.* **30,** 275–286.
5. Jacob-Dubuisson, F., Fernandez, R., and Coutte, L. (2004) Protein sretion through autotransporter and two-partner pathways. *Biochim. Biophys. Acta* **1694,** 235–257.
6. Thanassi, D. G., Stathopoulos, C., Karkal, A., and Li, H. (2005) Protein sretion in the absence of ATP: the autotransporter, two-partner sretion and chaperone/usher pathways of gram-negative bacteria. *Mol. Membr. Biol.* **22,** 63–72.
7. Henderson, I. R., Navarro-Garcia, F., and Nataro, J. P. (1998) The great escape: structure and function of the autotransporter proteins. *Trends Microbiol.* **6,** 370–378.

8. Yen, Y. T., Karkal, A., Bhattacharya, M., Fernandez, R., and Stathopoulos, C. (2007) Identification and characterization of autotransporter proteins of Yersinia pestis KIM. *Mol. Membr. Biol.* **24** (in press).
9. Filip, C., Fletcher, G., Wulff, J. L., and Earhart, C. F. (1973) Solubilization of the cytoplasmic membrane of Escherichia coli by the ionic detergent sodium-lauryl sarcosinate. *J. Bacteriol.* **115,** 717–722.

4

Periplasmic Targeting of Recombinant Proteins in *Escherichia coli*

Filipe J. Mergulhão and Gabriel A. Monteiro

Summary

Targeting recombinant protein production to the periplasmic space of *Escherichia coli* presents several advantages over cytoplasmic production in inclusion bodies and at the same time overcomes the low productivity problem often associated with culture medium secretion. This chapter presents a strategy for periplasmic production of recombinant proteins fused to synthetic Z domains derived from staphylococcal protein A. Expression, purification, and monitoring strategies are discussed using green fluorescent protein and human proinsulin as model proteins.

Key Words: *Escherichia coli*; recombinant proteins; periplasmic secretion; affinity purification; ZZ-tag; proinsulin; green fluorescent protein; ELISA.

1. Introduction

1.1. Escherichia coli as an Expression Host

Among the various hosts available for recombinant protein production, the Gram-negative bacterium *Escherichia coli* is one of the most versatile (*1*). This is a result of its ability to grow rapidly and at high cell density on inexpensive substrates, its well-characterized genetics, and the availability of a large number of cloning vectors and mutant host strains (*2*). This organism has the ability to accumulate many recombinant gene products up to 50% of the total cell protein (*2*) and to translocate them from the cytoplasm to the periplasm (*3*).

From: *Methods in Molecular Biology, Vol. 390: Protein Targeting Protocols: Second Edition*
Edited by: M. van der Giezen © Humana Press Inc., Totowa, NJ

1.2. Protein Secretion

Secretion of recombinant proteins to the culture medium or to the periplasm of *E. coli* has several advantages over intracellular production. These advantages include a simplified downstream, enhanced biological activity, higher product stability, enhanced product solubility and N–terminal authenticity of the expressed peptide (**Fig. 1**).

Because *E. coli* does not naturally secrete high amounts of proteins *(4)*, recovery of the gene product can be greatly simplified by a secretion strategy that minimizes contamination from host proteins. Additionally, if the product is secreted to the culture medium, cell disruption is not required for recovery and even in the case of periplasmic translocation, a simple osmotic shock or cell-wall permeabilization can be used to obtain the target product without the release of cytoplasmic protein contaminants *(1,5)*.

Biological activity is dependent on the folding state of the protein, and, particularly if disulfide bonds have to be formed, proper folding is unlikely to occur in the reducing environment of the cytoplasm. The *E. coli* periplasm contains a series of enzymes like disulfide-binding proteins (DsbA, DsbB, DsbC, and DsbD) and peptidyl-prolyl isomerases (SurA, RotA, FklB, FkpA) that promote the appropriate folding of thiol-containing proteins *(2,5)*.

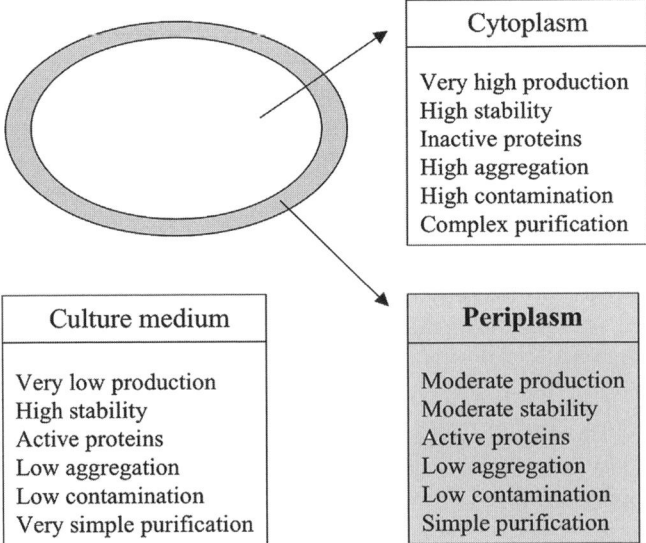

Fig. 1. Advantages and disadvantages of targeting recombinant protein production to the different cell compartments.

Protein aggregation can be caused by the limited amount of chaperones available when gene expression is performed at nonphysiological levels *(6)*. In this situation, the intra- or intermolecular association of hydrophobic surfaces exposed prior to folding can cause the precipitation of folding intermediates *(7)*. Periplasmic or extracellular secretion can increase the solubility of a gene product, and this may be a result of protein dilution, as the periplasm and the extracellular medium have lower protein contents than the cytoplasm *(8)*.

Also, secretion can be a way to guarantee the N-terminal authenticity of the expressed polypeptide because it often involves the cleavage of a signal sequence *(9)*, thus avoiding the unwanted presence of the initial methionine on a protein that does not normally contain it. This extra methionine residue can reduce the biological activity and stability of the molecule *(10)* or even elicit an immunogenic response in the case of therapeutic proteins. Protein secretion can also increase the stability of cloned gene products, probably because of the lower levels of *E. coli* proteases that can be found outside the cytoplasm *(1,11)*.

Protein secretion to the extracellular medium is advantageous regarding downstream strategy and product stability. However, production values tend to be much lower than those obtained by periplasmic secretion, and therefore these systems are not yet attractive to industrial application *(3)*. Periplasmic production values are usually lower than those obtained from cytoplasmic inclusion body production. However, by careful choice of strain, culturing optimization, and vector design, it is possible to obtain yields of at least 0.5 mg/L with most heterologous proteins *(12)*.

1.3. Fusion with ZZ Domains

The promoter and secretion signals of staphylococcal protein A (SpA) have been shown to be functional in *E. coli*, and therefore protein A fusions can be secreted to the periplasm of the bacteria and in some cases to the culture medium *(13)*.

The strong and specific interaction between the Fc part of IgG and SpA was the first described affinity fusion system for the purification of expressed gene products *(14)*. The dissociation constant for this interaction has been determined to be $2 \times 10^{-8}\,M$ *(13)*, which is within the affinity range of specific antibody–antigen interactions allowing protein purification through IgG affinity chromatography *(14)*.

The "Z" domain is an affinity tag engineered from the IgG-binding "B" domain of protein A. Its dimeric form "ZZ" was further demonstrated as the optimal fusion partner because of its strong IgG binding and efficient secretion. This fusion partner was also shown to increase the solubility of the target protein and to prevent its degradation *(15)*.

Two model proteins were used in this work: human proinsulin, with a molecular weight of 9.7 kDa, and green fluorescent protein (GFP) from *Aequorea victoria*, with 27.1 kDa. In the experiments described in this chapter, both proteins were fused to ZZ domains, and therefore the molecular weights of the resulting chimeras are 24.1 and 41.5 kDa, respectively.

2. Materials

2.1. Cloning

1. Plasmid pEZZ18 vector (Pharmacia Biotech/ GE Healthcare, Uppsala, Sweden).
2. Plasmid pEGFP-N1 vector (Clontech, Mountain View, CA).
3. Restriction enzymes (Promega, Madison, WI).

2.2. Cultivation

1. Strain *E. coli* JM109(DE3) (Promega).
2. LB medium (Sigma, St. Louis, MO).
3. Ampicillin (Sigma): 100 mg/mL stock solution (1000X), sterile filtered.
4. Kanamycin (Sigma): 30 mg/mL stock solution (1000X), sterile filtered.
5. IPTG (Sigma): 1 M stock solution (1000X) dissolved in milli-Q water, sterile filtered.

2.3. ZZ-Proinsulin Purification

1. Sucrose solution: 20% (w/v) sucrose, 1 mM EDTA, 0.3 M Tris-HCl, pH 8.0.
2. IgG Sepharose 6 Fast Flow (Pharmacia Biotech/ GE Healthcare).
3. FPLC column XK 16/12 (Pharmacia Biotech/ GE Healthcare).
4. Tris-Saline-Tween (TST) buffer: 1 mM EDTA, 0.2 M NaCl, 0.5 mL Tween® 20 (Bio-Rad, Hercules, CA), 25 mM Tris-HCl, pH 8.
5. Acetic acid solution 0.5 M, pH 2.8 (adjusted with ammonium acetate).
6. Ammonium acetate 5 mM, pH 5.0 (adjusted with acetic acid).
7. BCA assay kit (Pierce, Rockford, IL).

2.4. ELISA Procedure

1. Coating buffer: 15 mM Na$_2$CO$_3$, 35 mM NaHCO$_3$, pH 9.6.
2. Phosphate-buffered saline (PBS): 136 mM NaCl, 2.7 mM KCl, 10 mMNa$_2$PO$_4$, 1.8 mM KH$_2$PO$_4$, pH 7.4.
3. PBS-Tween (PBST): PBS-0.05% Tween® 20.
4. Maxisorp 96-well microplates (Nalge Nunc, Copenhagen, Denmark).
5. Primary antibodies: mouse anti-GFP monoclonal antibody (Santa Cruz, Santa Cruz, CA), mouse anti-human proinsulin monoclonal antibody (Advanced Immuno-chemical, Long Beach, CA).

6. Secondary antibody: anti-mouse Ig, horseradish peroxidase-conjugated from sheep (Amersham, Freiburg, Germany).
7. Substrate solution: 3.6 mM O-phenylenediamine (OPD), 0.036% (v/v) H_2O_2 in 5 mM Na_2HPO_4 and 2 mM citric acid, pH 5.5.
8. 3, 3′, 5, 5′-Tetramethylbenzidine (TMB) (Calbiochem, San Diego, CA).
9. Stop solution: 2 M H_2SO_4.

3. Methods

3.1. Cloning

The human proinsulin gene was obtained as described before *(16)* and cloned into the pEZZ18 vector as described *(9)*. The resulting vector, called pFM7 was then used to create a cloning fragment that was later inserted on a plasmid bearing a *lacUV5* promoter as described in *(17)* yielding plasmid pFM15 (*see* **Notes 1** and **2**).

The GFP gene was amplified by PCR using plasmid pEGFP-N1 as a template. The resulting fragment was then cloned into pFM7 replacing the proinsulin gene and yielding plasmid pFM20 *(18)*.

3.2. Cultivation

1. Glycerol stocks, 20% (v/v), of the *E. coli* strain harboring the relevant plasmids had been previously prepared and kept at −80°C.
2. Up to 100 µL of the glycerol stock (thawed in ice) is used to inoculate 25 mL of LB medium in a 100-mL shake-flask, supplemented with ampicillin in the case of plasmids pFM7 and pFM20 or kanamycin for pFM15.
3. The shake-flask is incubated overnight with agitation (220 rpm) at 37°C on an orbital shaker.
4. The optical density (O.D.) is read at 600 nm, and the culture is used as inoculum for the production flasks.
5. Calculations are made ensuring that the volume of the inoculum yields a starting O.D. below 0.1 (*see* **Note 3**).
6. Cultivation is performed on an orbital shaker (220 rpm) at 37°C (*see* **Note 4**).
7. When using plasmid pFM15 harboring the inducible *lacUV5* promoter, induction is performed by adding 1 mM IPTG in mid-exponential phase (*see* **Note 5**).
8. Cells are harvested 5–6 h after induction (*see* **Note 6**).

3.3. ZZ-Proinsulin Purification

3.3.1. Periplasmic Extraction

1. The contents of two shake flasks (each containing 125 mL of culture broth) are transferred to two centrifuge bottles (capacity 250 mL).

2. Cells are harvested by centrifugation at room temperature for 5 min at 3000*g* (*see* **Note 7**).

3. The supernatant is discarded, and each pellet is resuspended in 15 mL of sucrose solution, and the suspensions are transferred to the same centrifugation tube (capacity 50 mL).

4. After incubation for 15 min at room temperature, the suspension is centrifuged for 10 min at 4000*g* (*see* **Note 7**).

5. The pellet is then carefully resuspended in the small residual volume (*see* **Note 8**).

6. A volume of 30 mL of ice-cold water is added, and extensive mixing is performed by vortex (*see* **Note 9**).

7. The mixture is incubated for 15 min on ice and centrifuged for 10 min at 25,000*g*, 4°C (*see* **Note 10**).

8. The supernatant is carefully removed and equilibrated with 3 mL of 10X TST buffer (*see* **Notes 11** and **12**).

3.3.2. FPLC Purification

The flow rate used throughout this protocol is 1 mL/min (*see* **Note 13**).

1. A volume of 10 mL of IgG Sepharose® 6 Fast Flow is packed by gravity onto a FPLC column using TST as liquid phase (*see* **Note 14**).

2. Column is washed with 50 mL of TST prior to equilibration.

3. Equilibration is performed by washing the column with 30 mL of acetic acid followed by washing with 30 mL of TST.

4. **Step 3** is repeated.

5. Periplasmic extract is loaded maintaining a flow rate of 1 mL/min.

6. Column is washed with 100 mL of TST.

7. Column is washed with 100 mL ammonium acetate.

8. Purified protein is eluted in 10 mL of acetic acid and 1 mL fractions are collected.

9. Column is re-equilibrated with TST until the pH of the effluent is around 7.

10. Fractions are analyzed by SDS-PAGE (16% acrylamide, 1.7% bisacrylamide) stained with silver nitrate (*see* **Fig. 2** and **Note 15**).

11. Protein fractions are quantitated using the bicinchoninic acid method (BCA assay) according to the instructions by the manufacturer.

3.4. Monitoring of Periplasmic Secretion

3.4.1. Preparation of Periplasmic Extract

1. Samples of 1 mL of culture broth are periodically removed from the flasks.

2. Cells are harvested by centrifugation for 6 min at 2500*g* in 1.5-mL Eppendorf tubes (*see* **Note 7**).

3. The pellet is resuspended on a volume of 120 μL of sucrose solution and incubated for 15 min at room temperature.

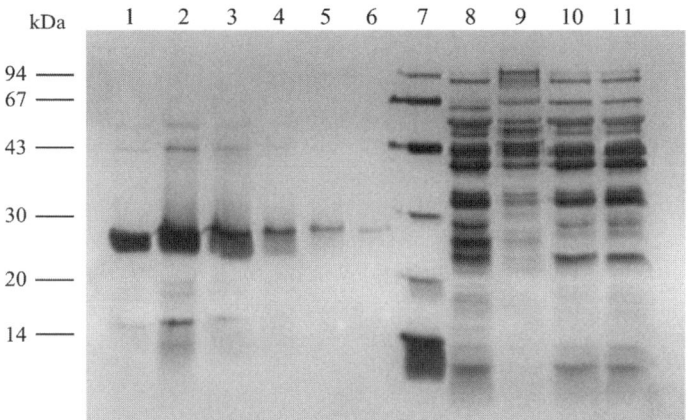

Fig. 2. SDS-PAGE showing ZZ-proinsulin purification. Lanes 1–6: Affinity purified ZZ-proinsulin fractions (0.9–0.05 µg); lane 7: molecular weight markers; lane 8: periplasmic extract (5 µg); lane 9: cell culture (≈ 2 µg); lanes 10 and 11: column flowthrough (≈ 5 µg). The localization and weight of each band of the molecular weight markers (lane 7) are depicted on the left.

4. Cells are harvested by centrifugation for 6 min at 2500*g* (*see* **Note 7**).
5. Cells are resuspended in the small residual volume (*see* **Note 8**).
6. Ice-cold water is added (120 µL), and after mixing, the suspension is incubated for 15 min on ice (*see* **Note 9**).
7. The suspension is centrifuged for 10 min at 11,000*g* and the supernatant is quickly removed (*see* **Note 10** and **11**).

3.4.2. ELISA for ZZ-Proinsulin

1. Samples for a calibration curve are prepared by adding up to 540 ng of affinity purified ZZ-proinsulin spiked with 20 µL of periplasmic extract of cells without plasmid (*see* **Note 17**).
2. Samples of 20 µL of periplasmic extract and standards for the calibration curve are dissolved in 100 µL of coating buffer (*see* **Note 17**).
3. Samples are applied onto the wells of Maxisorp microplates.
4. Plate is incubated at 37°C for 1 h.
5. Plate is incubated overnight at 4°C.
6. Samples are removed by aspiration under vacuum.
7. The wells of the plates are filled with 350 µL of PBST and the plates incubated for 2 h at room temperature.
8. The wells are washed with PBST (one quick wash followed by four 2-min washes).
9. The first antibody, a mouse anti-human proinsulin monoclonal antibody, is added (170 ng in 100 µL of PBST).

10. Plate is incubated at room temperature for 2 h.
11. The wells are washed with PBST (one quick wash followed by four 2-min washes).
12. Anti-mouse Ig, horseradish peroxidase-conjugated is added (100 μL in a 1:1000 dilution in PBST).
13. Plate is incubated at room temperature for 2 h.
14. The wells are washed with PBST (one quick wash followed by four 2-min washes).
15. A volume of 200 μL of substrate solution is added.
16. Plate is incubated at room temperature in the dark for 1 h (*see* **Note 18**).
17. Reaction is stopped by adding 50 μL of stop solution to each well, and absorbance at 490 nm is read with a multiplate reader (**Fig. 3**).

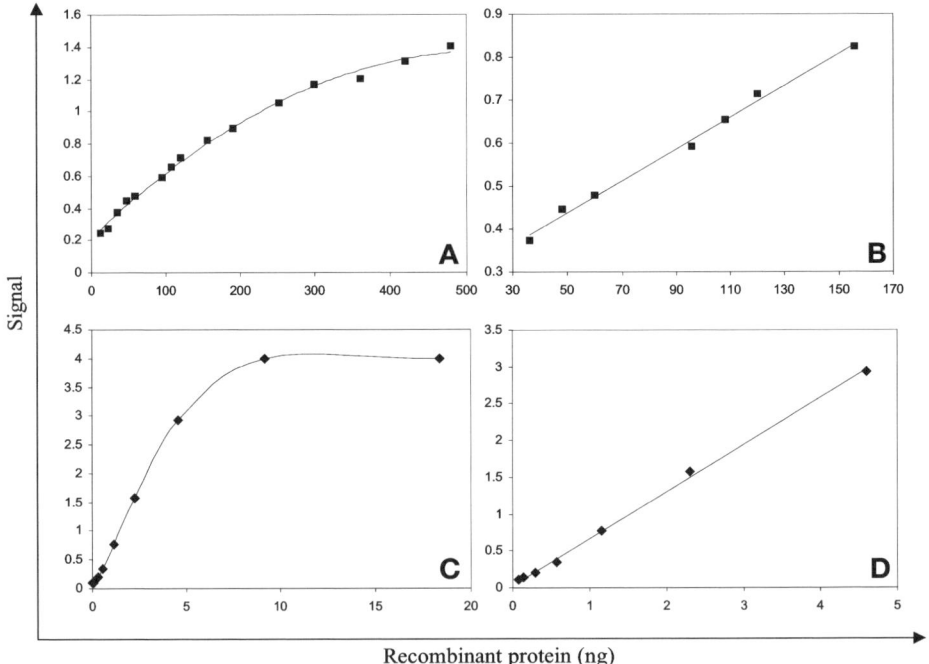

Fig. 3. Response curves for the developed ELISA tests: (**A, B**) ZZ-proinsulin assay, (**C, D**) ZZ-GFP assay. (**A**) Response curve for ZZ-proinsulin. (**B**) Linear range of the curve, line equation is $Y = 0.0037X + 0.254$ ($R^2 = 0.994$). (**C**) Response curve for ZZ-GFP. (**D**) Linear range of the curve, line equation is $Y = 0.6348X + 0.039$ ($R^2 = 0.998$).

3.4.3. ELISA for ZZ-GFP

1. Up to 10 ng of GFP is diluted in 100 μL of coating buffer to create a calibration curve (*see* **Note 16**).
2. Periplasmic extracts are prepared as described in **Subheading 3.4.1.**
3. Volumes of 0.1–0.05 μL of periplasmic extracts are dissolved in 100 μL of coating buffer (*see* **Note 17**).
4. Assay is conducted as described in **Subheading 3.4.2., steps 3–8**.
5. After washing, a mouse anti-GFP monoclonal antibody is added (100 μL in a 1:1000 dilution in PBST).
6. Plate is incubated at room temperature for 1 h.
7. The wells are washed with PBST (one quick wash followed by four 2-min washes).
8. Anti-mouse Ig, horseradish peroxidase-linked is added (100 μL in a 1:1000 dilution in PBST).
9. Plate is incubated at room temperature for 1 h.
10. The wells are washed with PBST (one quick wash followed by four 2-min washes).
11. A volume of 200 μL of TMB solution is added (*see* **Note 18**).
12. Plate is incubated at room temperature in the dark for 10 min.
13. Reaction is stopped by adding 50 μL of stop solution to each well, and absorbance at 490 nm is read with a multiplate reader (**Fig. 4**).

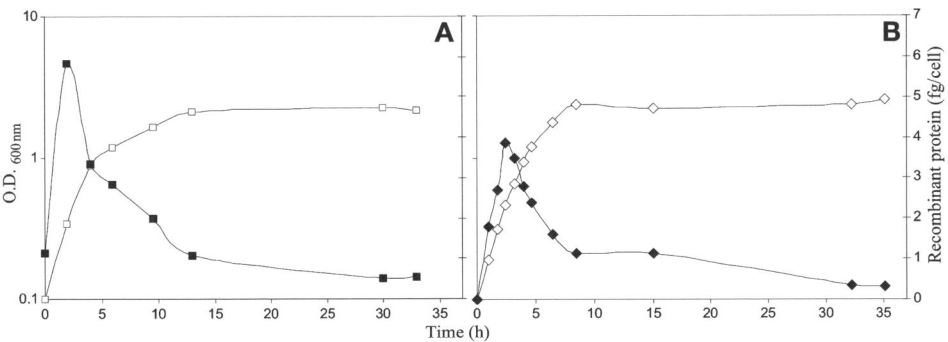

Fig. 4. Monitoring the periplasmic production of the fusion proteins: (**A**) ZZ-proinsulin results, (**B**) ZZ-GFP results. Open symbols represent growth curve (logarithmic scale on the left); closed symbols represent periplasmic concentrations (linear scale on the right). Values resulted from two independent cultivations; standard deviation for periplasmic values was lower than 13 and 27% for ZZ-proinsulin and ZZ-GFP, respectively.

Notes

1. When using plasmid pEZZ18 as an expression vector, inserts must conform to two basic specifications:

    a. The insert can be cloned into any unique site (although directional cloning using two restriction enzymes is recommended), but the gene sequence must be cloned maintaining the reading frame imposed by the leader peptide. When the insert originates from another expression vector and contains suitable restriction sites flanking the coding sequence, it might be necessary to introduce one or two bases to maintain the reading frame. These bases can be inserted by site-directed mutagenesis using any commercial kit (we have used the Quick-Change site-directed mutagenesis kit from Stratagene).

    b. Inserts must contain their own stop codon, otherwise transcription continues into the lacZ portion of the vector yielding a C-terminal fused peptide. The presence of this lacZ fragment is to enable blue/white screening of the clones, and this is particularly useful when cloning nondirectionally with a single restriction enzyme. If this is the case vector dephosphorylation can be performed by treating the digested vector with alkaline phosphatase prior to ligation.

2. Another vector was constructed with the aim of increasing the ZZ-proinsulin production value that could be reached with a pEZZ-derived vector. Vector pFM15 *(17)* was designed including the same leader sequence and fusion partner, but in this new vector, expression was controlled by the much stronger and inducible *lacUV5* promoter. In order not to saturate the protein export machinery, a medium copy number origin of replication (pMB1 derived) was chosen with predicted copy numbers of 15–60 copies per cell. The results obtained with this vector were similar to those obtained with pFM7 *(17)*.

3. When setting up the cultivation for the production of recombinant proteins, the initial O.D. of the cells should be kept below 0.1. A higher initial O.D. can be deleterious for production.

4. Although expression is controlled by the SpA promoter and is not inducible, it has been reported that expression can be increased by heat shock *(13,19)*. This can be accomplished by raising the culture temperature to 44°C for 2–6 h during the stationary phase, although we have not tested this procedure on this work because of potential degradation of the recombinant proteins.

5. When using the pFM15 plasmid (bearing the inducible *lacUV5* promoter), induction was performed in mid-exponential phase (O.D. $\approx$ 1.3). A standard growth curve should be obtained prior to production in order to determine the culture behavior in those particular conditions.

6. When using pFM7 or pFM20 vectors for production, one must bear in mind that periplasmic production values peak in mid-exponential phase in both cases and after the maximum value is reached periplasmic titers drop sharply **(Fig. 4)**. This

might be the result of a lower protein expression level upon entry into stationary phase, periplasmic degradation of the recombinant proteins, and/or leakage to the culture medium.

7. When preparing the periplasmic extracts, the two first centrifugation steps should be performed at relatively low centrifugal force. This prevents cells lysis and facilitates subsequent resuspension steps.

8. After cells are resuspended in sucrose solution and harvested again by centrifugation, the supernatant should be removed by inversion and gentle tapping of the tube against absorbent paper. Residual liquid should be left on the tube (e.g., in a 1.5-mL tube, a volume of approx 10 μL of liquid usually remains). The pellet should then be resuspended in this small residual volume prior to adding ice-cold water. Extensive vortex should be employed to disperse the cells efficiently. The performance of the osmotic shock is greatly affected by this step.

9. After cell resuspension on the small volume, ice-cold water is added and the suspension is thoroughly mixed by vortex. The water (we use milliQ water) should be very cold to enhance the shock. Thus, the procedure is a combination of a thermal and osmotic shocks.

10. After the osmotic shock, the periplasmic fraction is obtained by high-speed centrifugation at low temperature. This is particularly important in preparative experiments because some cell lysis may occur during the shock and therefore cell debris might be present on the suspension.

11. Following centrifugation, the supernatant containing the periplasmic extract should be carefully and quickly removed from the centrifuge tube. Bear in mind that cell debris contains cell membranes and lipopolysacharides, which may resuspend and contaminate the periplasmic extract if the sample is not carefully handled. At this point the extract should be clear and fluid. If the periplasmic extraction is not performed carefully, cell lysis may occur. This will release DNA into the supernatant, thus increasing the viscosity of the solution. The fact that the periplasmic extract is very fluid at this point (viscosity like water) is an indication that no significant lysis occurred.

12. The extract must be equilibrated in TST prior to loading onto the column. It is easier to prepare a 10-fold concentrated solution of TST that can be used in this step.

13. Although the resin that was used is capable of sustaining considerable pressure drops, we have maintained the flow at 1 mL/min using a column with 2 cm^2 of cross-sectional area. Depending on the protein, using higher flow rates might affect the resolution and yield of the chromatographic step.

14. The resin may be packed onto disposable columns and operated by gravity flow. We have chosen to use a FPLC system to monitor both conductivity and absorbance at 280 nm. Monitoring the absorbance of the eluate allows better control of the fraction collection procedure. Usually 10 fractions of 1 mL were collected.

15. Protein purity varies among different fractions. A second purification step by gel filtration can be used if necessary. Our goal was to obtain some pure protein to use as standard in the ELISAs. For that purpose we have chosen fraction 2 (lane 1 in **Fig. 2**), which has purity of greater than 99%. The performance of this step is very high, and a total recovery of 96% was obtained (**Table 1**).

16. ZZ-GFP purification was not performed by the described method (an IMAC column was used instead to purify GFP *(18)*. The purification conditions, namely the very low pH on the elution step, are probably to harsh for ZZ-GFP. The ZZ fusion strategy is not guaranteed to work for every single protein, although a considerable number of proteins has been successfully expressed and purified using this system *(14)*.

17. The conditions that were used on the two ELISAs are very different and allow distinct operational ranges. These conditions should be adapted to suit the particular applications of this protocol. On the assay developed to quantitate ZZ-proinsulin, a large volume of periplasmic extract (20 μL) was used. For this reason, when preparing the standards for the calibration curve, the affinity purified protein had to be spiked with the periplasmic extract of cells without the plasmid to mimic the protein content of the samples. At this stage it might be useful to prepare a 10X concentrated coating buffer solution to make up the final volume of 100 μL. In the method developed for ZZ-GFP, a highly sensitive ELISA was established.

Table 1
ZZ-Proinsulin Purification Data[a]

Protein load:

Total protein: 1.4 mg/mL
40.7 mg
ZZ-proinsulin: 20.7 mg/L
0.6 mg
1.6%

Protein recovery:

ZZ-proinsulin: 79.1 mg/L
0.6 mg
Recovery yield: 96%

[a]On the protein load, the amounts and concentrations of total protein and ZZ-proinsulin are indicated. The percentage of recombinant protein on the protein load is also shown. The recovered amount and the mean concentration of the eluted fractions are indicated along with a recovery yield. Results originated from two independent assays; mean variation was less than 19%.

This procedure uses between 0.1 and 0.05 μL of sample. Since production changes considerably during cultivation, we recommend using two volumes for each time point (e.g., 0.1 and 0.05 μL diluted in coating buffer) to ensure that at least one of them fits into the calibration curve. Because of the reduced sample volume it is not necessary to spike with the periplasmic extract of empty cells. The two methods that were presented are illustrative because they enable working ranges spanning two orders of magnitude (**Fig. 3**) and they can be adapted to any particular application.

18. The detection methods that were used with the horseradish peroxidase-conjugated antibody are also illustrative of the different available options. In the ZZ-proinsulin method, the system composed of OPD/H_2O_2 in citrate buffer was used. We have used freshly prepared citrate buffer and mixed OPD just prior to use. OPD tablets are available from several manufacturers and are easier to use than weighing the powder on every assay (OPD is extremely toxic). In our lab some researchers use the tablets and prepare a ready-to-use solution to which only H_2O_2 must be added prior to use. This solution is stable for at least a week at 4°C when kept in a glass bottle covered with aluminum foil to block the light. OPD solutions are particularly sensitive to light, so when preparing them light exposure should be restricted to a minimum. For the ELISA we have chosen a 1-h incubation for the endpoint assay because the reaction kinetics is slow. The incubation time and the concentration of the OPD solution can be optimized to obtain different linear ranges (**Fig. 3**). In the ZZ-GFP method we have used the TMB detection system. The system that we have used has faster kinetics than the OPD/H_2O_2 system and is presented as a one-ingredient ready-made solution. Other products are available from different manufacturers offering TMB solutions with faster or slower kinetics. The incubation time should then be optimized according to each particular application.

19. ELISA methods are highly prone to error resulting from inefficient liquid handling. Multichannel pipets are very useful, but operators must make sure that tips are firmly attached on all positions and no air bubbles are present in any of the tips on the suction stage. The reproducibility of washing steps is also difficult, but some automatic plate washers are available from different manufacturers, ensuring greater washing consistency between different wells. Furthermore, the reproducibility of the assay should be checked by loading reference samples on different locations on the same plate. These samples should be loaded on every plate to ensure plate-to-plate consistency. By loading the same sample in three different dilutions onto two different plates, we have obtained a standard error of less than 20% for the described assays.

20. The commercial pEZZ18 vector is used for the expression of ZZ-tagged proteins, but the number of Z domains to be used on each particular application is not restricted to two and may also be optimized (*20*). It has been reported (*13*) that the number of Z domains to be used depends on the steric interference with the target protein, the efficiency of secretion, and the desired strength of the IgG interaction.

The use of two Z domains was found to be optimal for the expression of several proteins, and analysis of different repeats of Z domains has shown that the binding strength of a single Z domain is lower than that attained with two Z domains *(21)* and that there is no increase in the binding capacity when using more than two Z domains.

Acknowledgments

Financial support from Fundação para a Ciência e Tecnologia, Ministério da Ciência Tecnologia e Ensino Superior, Portugal, is acknowledged.

References

1. Mergulhão, F. J. M., Monteiro, G. A., Cabral, J. M. S., and Taipa, M. A. (2004) Design of bacterial vector systems for the production of recombinant proteins in *Escherichia coli*. *J. Microb. Biotechnol.* **14**, 1–14.
2. Baneyx, F. and Mujacic, M. (2004) Recombinant protein folding and misfolding in *Escherichia coli*. *Nat. Biotechnol.* **22**, 1399–1408.
3. Mergulhão, F. J. M. and Monteiro, G. A. (2005) Recombinant protein secretion in *Escherichia coli*. *Biotechnol. Adv.* **23**, 177–202.
4. Sandkvist, M., and Bagdasarian, M. (1996) Secretion of recombinant proteins by Gram-negative bacteria. *Curr. Opin. Biotechnol.* **7**, 505–511.
5. Shokri, A., Sandén, A. M., and Larsson, G. (2003) Cell and process design for targeting of recombinant protein into the culture medium of *Escherichia coli*. *Appl. Microbiol. Biotechnol.* **60**, 654–664.
6. Hoffmann, F., van den Heuvel, J., Zidek, N., and Rinas, U. (2004) Minimizing inclusion body formation during recombinant protein production in *Escherichia coli* at bench and pilot plant scale. *Enzyme Microb. Tech.* **34**, 235–241.
7. Carrio, M. M., and Villaverde, A. (2002) Construction and deconstruction of bacterial inclusion bodies. *J. Biotechnol.* **96**, 3–12.
8. Makrides, S. C. (1996) Strategies for achieving high-level expression of genes in *Escherichia coli*. *Microbiol. Rev.* **60**, 512–538.
9. Mergulhão, F., Monteiro, G., Kelly, A., Taipa, M., and Cabral, J. (2000) Recombinant human proinsulin: A new approach in gene assembly and protein expression. *J. Microbiol. Biotechnol.* **10**, 690–693.
10. Liao, Y. D., Jeng, J. C., Wang, C. F., Wang, S. C., and Chang, S.T. (2004) Removal of N-terminal methionine from recombinant proteins by engineered *E. coli* methionine aminopeptidase. *Protein Sci.* **13**, 1802–1810.
11. Gottesman, S. (1996) Proteases and their targets in *Escherichia coli*. *Annu. Rev. Genet.* **30**, 465–506.
12. Georgiou, G. and Segatori, L. (2005) Preparative expression of secreted proteins in bacteria: status report and future prospects. *Curr. Opin. Biotechnol.* **16**, 538–545.

13. Moks, T., Abrahmsen, L., Holmgren, E., et al. (1987) Expression of human insulin-like growth factor I in bacteria: use of optimized gene fusion vectors to facilitate protein purification. *Biochemistry* **26**, 5239–5244.
14. Stahl, S. and Nygren, P.A. (1997) The use of gene fusions to protein A and protein G in immunology and biotechnology. *Pathol. Biol. (Paris)* **45**, 66–76.
15. Stahl, S., Nilsson, J., Hober, S., Uhlen, M., and Nygren, P. (1999) Affinity fusions, gene expression. In *Encyclopedia of Bioprocess Technology: Fermentation, Biocatalysis and Bioseparation* (Flickinger, M., and Drew, S., eds.), John Wiley & Sons, New York, pp. 49–63.
16. Mergulhão, F. J., Kelly, A. G., Monteiro, G. A., Taipa, M. A., and Cabral, J. M. (1999) Troubleshooting in gene splicing by overlap extension: a step-wise method. *Mol. Biotechnol.* **12**, 285–287.
17. Mergulhão, F. J. M., Monteiro, G. A., Larsson, G., et al. (2003) Evaluation of inducible promoters on the secretion of a ZZ-Proinsulin fusion protein. *Biotechnol. Appl. Biochem.* **38**, 87–93.
18. Mergulhão, F. J. M., and Monteiro, G. A. (2007) Analysis of factors affecting the periplasmic production of recombinant proteins in *Escherichia coli. J. Microbiol. Biotechnol.* **17**.
19. Abrahmsen, L., Moks, T., Nilsson, B., and Uhlen, M. (1986) Secretion of heterologous gene products to the culture medium of *Escherichia coli. Nucleic Acids Res.* **14**, 7487–7500.
20. Mergulhão, F. J., Taipa, M. A., Cabral, J. M., and Monteiro, G. A. (2004) Evaluation of bottlenecks in proinsulin secretion by *Escherichia coli. J. Biotechnol.* **109**, 31–43.
21. Ljungquist, C., Jansson, B., Moks, T., and Uhlen, M. (1989) Thiol-directed immobilization of recombinant IgG-binding receptors. *Eur. J. Biochem.* **186**, 557–561.

5

In Vitro Analysis of the Bacterial Twin-Arginine-Dependent Protein Export

Michael Moser, Sascha Panahandeh, Eva Holzapfel, and Matthias Müller

Summary

Prokaryotic organisms possess a specialized protein translocase in their cytoplasmic membranes that catalyzes the export of folded preproteins. Substrates for this pathway are distinguished by a twin-arginine consensus motif in their signal peptides (twin-arginine translocation [Tat] pathway). We have compiled detailed protocols for the preparation and operation of a cell-free system by which the bacterial Tat pathway can be fully reproduced in vitro. This system has proven useful and is being further exploited for the study of precursor–translocase interactions, assembly of the translocase, and the mechanism of transmembrane passage.

Key Words: Protein transport; twin-arginine translocation; in vitro transcription-translation system; inner membrane vesicles; inverted membrane vesicles; *Escherichia coli*.

1. Introduction

Bacteria export proteins from the cytoplasm to the cell envelope and the extracellular milieu by means of two major protein transport machineries in their cytoplasmic membrane: Sec and Tat translocases (for recent reviews, *see* **refs.** *1–3)*. The Tat (twin-arginine translocation) pathway is reserved to fully folded substrates that carry a characteristic twin-arginine (RR) motif in their signal sequences. It derives its energy from the transmembrane H^+ gradient. In *Escherichia coli*, three individual membrane proteins, TatA, TatB, and TatC, are required to allow RR-specific transport across the cytoplasmic membrane. Currently available data suggest that a pre-existing pore structure made of TatABC might not exist. TatC in concert with TatB were shown to be required for a specific targeting of substrates to the membrane, whereas TatA

From: *Methods in Molecular Biology, Vol. 390: Protein Targeting Protocols: Second Edition*
Edited by: M. van der Giezen © Humana Press Inc., Totowa, NJ

is believed to be a major constituent of the pore because of its propensity to form homo-oligomers of various sizes. A homologous transport machinery has been conserved in the thylakoid membranes of plant chloroplasts.

To lay the foundation for a biochemical analysis of the bacterial Tat pathway, a cell-free system has been developed that faithfully reproduces the synthesis of an RR precursor and its H^+-gradient-dependent transport into inverted inner membrane vesicles (INV) of *E. coli (4,5)*. This experimental system is largely identical to previous ones described for studying Sec-dependent translocation in bacteria *(6–8)*. It involves a cytosolic extract freed of endogenous mRNA and membranes, which is prepared by a method derived from the original Zubay protocol *(9)*. When supplemented with template DNA and RNA polymerase, the extract efficiently synthesizes radioactively labeled precursor proteins by coupled transcription-translation. In the presence of inside-out inner membrane vesicles, in vitro synthesized precursor proteins are found translocated into the lumen of the vesicles. Inside-out or inverted inner membrane vesicles are prepared by breaking *E. coli* cells under high pressure according to original protocols by Futai *(10)* and Schnaitman *(11)*. We use the original designation INV (<u>in</u>verted membrane <u>v</u>esicles), given to these vesicles *(6)*, which are now often synonymously called IMV (<u>in</u>ner <u>m</u>embrane <u>v</u>esicles). Translocation of preproteins into INV is monitored by *(1)* the size change of the precursor protein resulting from cleavage of the signal sequence by the vesicle-associated signal peptidase and *(2)* the precursor acquiring protease resistance exclusively in the presence of INV.

Here we describe the usage of this cell-free system to reproduce the Tat pathway in *E. coli*. We present detailed protocols for the preparation of the cytosolic extract and INV. We describe the design and demonstrate the results of an in vitro reaction, in which an RR precursor is synthesized and translocated into INV. Further developments of this technology allowing synthesis of ribosome-associated nascent chains (rather than full-size translation products) *(12)* and the incorporation of site-specific crosslinkers by a stop codon-suppressor approach have been published *(13)*.

2. Materials
2.1. Preparation of an S-135 Cell Extract from E. coli

1. Growth medium (S-30 medium): 9.0 g/L tryptone/peptone (pancreatic digest of casein, Carl Roth, Karlsruhe, Germany), 0.8 g/L yeast extract, 5.6 g/L NaCl, 1 mL/L 1*M* NaOH. Prepare 4–6 L in 1-L batches, each contained in a 5-L Erlenmeyer flask covered with aluminum foil and autoclave. Prepare an additional 100 mL of medium in a 0.5-L Erlenmeyer flask to be used as starter culture and autoclave (*see* **Note 1**).

2. 20% Glucose solution, autoclaved.
3. 1 *M* Triethanolamine acetate (TeaOAc) adjusted to pH 7.5 with acetic acid, filtered and stored at 4°C (*see* **Note 2**).
4. 1 *M* Magnesium acetate (Mg(OAc)$_2$), filtered and stored at 4°C.
5. 4 *M* Potassium acetate (KOAc) also adjusted to pH 7.5 with acetic acid, filtered and stored at 4°C.
6. 1 *M* Dithiothreitol (DTT) stored in 1-mL aliquots at −20°C.
7. S-30 buffer: 10 m*M* TeaOAc pH 7.5, 14 m*M* Mg(OAc)$_2$, 60 m*M* KOAc, 1 m*M* DTT, stored at 4°C.
8. Phenylmethylsulfonyl fluoride (PMSF; Roche): freshly prepare about 1 mL of a 0.1 *M* solution in ethanol before use (*see* **Note 3**).
9. A mix of 18 amino acids (without methionine and cysteine) in water, each at a concentration of 1 m*M*.
10. 1 m*M* Methionine.
11. 1 m*M* Cysteine.
12. 0.25 *M* ATP neutralized with 1 *M* KOH.
13. 0.2 *M* Phosphoenol pyruvate tri(cyclohexylammonium) salt.
14. 2 mg/mL Pyruvate kinase.
15. Supplemented S-30 for degradation of endogenous mRNA: per mL of S-30, add 60 μL 1 *M* TeaOAc pH 7.5, 0.6 μL 1 *M* DTT, 1.6 μL 1 *M* Mg(OAc)$_2$, 6 μL 1 m*M* 18 amino acid mix, 6 μL 1 m*M* methionine, 6 μL 1 m*M* cysteine, 2 μL 0.25 *M* ATP (neutralized), 27 μL 0.2 *M* phosphoenol pyruvate, and 2.4 μL 2 mg/mL pyruvate kinase.
16. Dialysis tubing with a width of 25 mm and a molecular weight cutoff of 14,000 Da (Visking; Carl Roth) (*see* **Note 4**). Two dialysis tubing clips.
17. For preparation of dialysis tubing: 2% NaHCO$_3$, 1 m*M* ethylenediamine tetraacetic acid (EDTA).

2.2. Preparation of Inverted Inner Membrane Vesicles

1. Growth medium (INV medium): 10 g/L each of yeast extract and tryptone/peptone (pancreatic digest of casein; Carl Roth), 28.9 g/L K$_2$HPO$_4$ anhydrous, 5.6 g/L KH$_2$PO$_4$ anhydrous, 10 g/L glucose. Prepare 4 × 5 L Erlenmeyer flasks, each containing 10 g yeast extract and 10 g tryptone/peptone dissolved in 753 mL H$_2$O, autoclave. In addition, prepare one 0.5-L Erlenmeyer flask containing 1 g yeast extract and 1 g tryptone/peptone dissolved in 75.3 mL H$_2$O, autoclave.
2. 1 *M* K$_2$HPO$_4$, autoclave.
3. 1 *M* KH$_2$PO$_4$, autoclave.
4. 25% Glucose, autoclave.
5. Starter culture medium (100 mL): to 75.3 mL yeast extract and tryptone/peptone (*see* **Subheading 2.2., item 1**) add 4.1 mL 1 *M* KH$_2$PO$_4$, 16.6 mL 1 *M* K$_2$HPO$_4$, and 4 mL 25% glucose.

6. Complete INV medium: to 753 mL yeast extract and tryptone/peptone (*see* **Subheading 2.2., item 1**) add 41 mL $1 M$ KH_2PO_4, 166 mL $1 M$ K_2HPO_4, and 40 mL 25% glucose.
7. $1 M$ TeaOAc adjusted to pH 7.5 with acetic acid, filtered and stored at 4°C.
8. $0.2 M$ EDTA-KOH, pH 7.0, filtered and stored at 4°C.
9. $2.5 M$ Sucrose ultrapure (MP Biomedicals, Solon, OH), heat slightly for better dissolution, store at room temperature.
10. $1 M$ DTT stored in 1-mL aliquots at -20°C.
11. $0.1 M$ PMSF freshly prepared in ethanol.
12. $1 M$ Isopropyl $-\beta$-D-thiogalactopyranoside (IPTG).
13. Buffer A: 50 mM TeaOAc, pH 7.5, 250 mM sucrose, 1 mM EDTA-KOH, pH 7.0, and 1 mM DTT. Prepare fresh.
14. Buffer B: $0.5 M$ TeaOAc, pH 7.5, 10 mM EDTA-KOH, pH 7 and 10 mM DTT. Prepare fresh.
15. Sucrose solutions for sucrose gradient centrifugation, freshly prepared. $0.77 M$ sucrose: 10 mL buffer B, 30.8 mL $2.5 M$ sucrose, H_2O to 99.5 mL and add 0.5 mL $0.1 M$ PMSF last; $1.44 M$ sucrose: 10 mL buffer B, 57.6 mL $2.5 M$ sucrose, H_2O to 99.5 mL and add 0.5 mL $0.1 M$ PMSF last; $2.02 M$ sucrose: 10 mL buffer B, 80.8 mL $2.5 M$ sucrose, H_2O to 99.5 mL and add 0.5 mL $0.1 M$ PMSF last.
16. INV buffer: 50 mM TeaOAc, pH 7.5, 250 mM sucrose, and 1 mM DTT. Cool on ice.

2.3. In Vitro Transcription-Translation Using an E. coli S-135 Cell Extract

2.3.1. In Vitro Transcription-Translation Reaction

1. Template DNA: plasmid DNA prepared by Qiagen plasmid maxi kit is suitable for in vitro synthesis (*see* **Note 5**). Prepare DNA in TE buffer (10 mM Tris-HCl, pH 8.0, 1 mM EDTA) at about 1 μg/μL and store at 4°C.
2. $1 M$ TeaOAc adjusted to pH 7.5 with acetic acid, filtered and stored at 4°C.
3. $4 M$ KOAc also adjusted to pH 7.5 with acetic acid, filtered and stored at 4°C.
4. $1 M$ Mg(OAc)$_2$, filtered and stored at 4°C.
5. $0.1 M$ Spermidine trihydrochloride (Sigma, St. Louis, MO), dissolved in water and stored in single-use aliquots at -20°C (*see* **Note 6**).
6. 40% (w/v) Polyethylene glycol 6000–8000 dissolved in water and stored in 1-mL aliquots at -20°C.
7. 1 mM (each) of 18 amino acids (without methionine and cysteine) dissolved in water and stored in 1-mL aliquots at -20°C.
8. $0.2 M$ DTT dissolved in water and stored in 10-μL aliquots at -20°C (*see* **Note 6**).
9. $0.2 M$ Phosphoenol pyruvate dissolved in water and stored in 50-μL aliquots at -20°C (*see* **Note 6**).

10. 0.5 *M* Creatine phosphate dissolved in water and stored in 10-µL aliquots at −20°C.
11. 10 mg/mL Creatine phosphokinase dissolved in water and stored in 10-µL aliquots at −20°C.
12. Neutralized nucleotide (NTP) stock (50 m*M* ATP and 10 m*M* each of GTP, CTP, UTP): prepare by mixing equal volumes of 250 m*M* ATP, 50 m*M* each of GTP, CTP, UTP, and 1 *M* KOH. Make all solutions in water and store the NTP stock in 10 µL aliquots at −20°C (*see* **Note 7**).
13. Pro-Mix, L-[^{35}S] in vitro cell-labeling mix, 530 MBq/mL (GE Healthcare). This mixture contains 70% [^{35}S]-methionine and 30% [^{35}S]-cysteine, stored in 50-µL aliquots at −80°C. (Radioactive material is hazardous. Avoid ingestion or contact with skin or clothing. Always wear gloves when handling. Monitor hands, equipment, and bench frequently.)
14. T7 RNA polymerase. Commercially available preparations (e.g., from Promega) are fine; large quantities are also reasonably easy to prepare from overproducing *E. coli* strains *(14)*.
15. Proteinase K: prepare a 20 mg/mL stock solution in water and store in single-use aliquots at −20°C. Dilute to 1 mg/mL before use.
16. 10% Trichloroacetic acid: prepare a 100% solution and use for further dilutions.

2.3.2. Sodium Dodecyl Sulfate–Polyacrylamide Gel Electrophoresis (SDS-PAGE)

1. Separating gel buffer: 2 *M* Tris-HCl, pH 8.8. Filter and store at 4°C.
2. Stacking gel buffer: 0.5 *M* Tris-HCl, pH 6.8. Filter and store at 4°C.
3. 25% (w/v) SDS. Store at room temperature.
4. 30% Acrylamide/0.8% bisacrylamide solution (Rotiphorese Gel 30; Carl Roth). Acrylamide is a neurotoxin; always wear gloves when handling acrylamide solutions and gels.
5. *N*, *N*, *N*, *N′*-Tetramethylethylene diamine (TEMED).
6. Ammonium peroxodisulfate. Prepare 10% solution in water and store at 4°C. It is stable for several days.
7. Running buffer (5x): dissolve 150 g Tris base and 720 g glycine in 5 L water including 100 mL 25% SDS and store at room temperature.
8. Solution 1: 2 mL 1 *M* Tris base, 1 mL 0.2 *M* EDTA, pH 8.0, 7 mL water.
9. Solution 2: 4 mL 25% SDS, 1 mL 1 *M* Tris base, 3.5 mL 100% glycerol, 3.5 mL 0.1% bromophenol blue.
10. Solution 3: 1 *M* DTT.
11. Prepare PAGE-loading buffer: mix five parts of solution 1, four parts of solution 2, and one part of solution 3. Always prepare fresh.
12. Prestained molecular weight marker: Precision Plus Protein Standards (Bio-Rad, Hercules, CA).
13. Fixing solution: 35% ethanol, 10% acetic acid.

3. Methods

3.1. Preparation of an S-135 Cell Extract from E. coli

1. Precool the French press cell by placing it at 4°C.
2. Supplement 100 mL of S-30 medium with 0.4 mL autoclaved 20% glucose solution and inoculate from plates or glycerol stocks (*see* **Note 8**). Grow cells overnight at 37°C with sufficient aeration in a rotary shaker (cover flask with aluminum foil). Use this culture to inoculate at a 1:100 ratio 4–6 L growth medium supplemented with 4 mL/L of 20% glucose and grow the cells in a rotary shaker to late log-phase (optical density at 600 nm = 1.0–1.2 U/mL).
3. Prepare 4 L of S-30 buffer and store at 4°C.
4. Chill the cell cultures quickly by placing the flasks in an ice water bath and harvest the cells at 4°C in a cooled SCL3000 rotor (Sorvall) for 10 min at 8650g (7000 rpm). All subsequent steps should be done at 4°C or on ice. Resuspend the cell pellets in S-30 buffer (*see* **Note 9**). Combine the cell suspensions in one or two tared centrifuge bottles and centrifuge again. Determine the wet weight of the cell pellet (approx 2 g/L medium).
5. Resuspend the cell pellet in 1 volume (1 mL/g wet cell mass) of S-30 buffer containing 0.5 mM PMSF (add PMSF from a fresh 0.1 M stock in ethanol).
6. For breakage of the cells pass the cell suspension two to three times through a French pressure 40k cell (Spectronic Unicam, Cambridge, UK) at 8000 psi. This corresponds to a gage pressure setting of 500 when using the 1-inch piston cell at the "high ratio" selection (*see* **Note 10**).
7. After cell breakage, centrifuge the suspension in a precooled SS34 rotor (Sorvall) for 30 min at 30, 000g (15,500 rpm) at 4°C. Remove supernatant (S-30) carefully (*see* **Note 11**).
8. To allow degradation of endogenous mRNA in the S-30, perform a readout of polysomal mRNA. To this end prepare supplemented S-30 according to **Subheading 2.1., item 15**. Incubate at 37°C for 1 h. Afterwards chill the S-30 on ice.
9. Dialyze the S-30 three times against 1 L of cold S-30 buffer for 1 h each at 4°C (*see* **Note 12**).
10. Prepare S-135 from the S-30 (*see* **Note 13**) by pipetting 1-mL aliquots of S-30 into tubes of a Beckman TLA 100.2 rotor and spin at 287, 600g (90,000 rpm) for 13 min at 4°C. Remove 750 μL (*see* **Note 14**) of each supernatant, combine (= S-135), and quick-freeze in aliquots of 50–100 μL in liquid nitrogen (*see* **Note 15**). Store the S-135 at −80°C.

3.2. Preparation of INV

1. Precool the French press cell by placing it at 4°C.
2. Inoculate 100 mL of starter culture medium from plates or glycerol stocks with an *E. coli* strain harboring the *tatABC* genes cloned under an inducible promoter (*see* **Note 16**). Grow cells over night at 37°C with sufficient aeration in a rotary shaker (cover flask with aluminium foil).

3. Inoculate 4 flasks containing 1 L complete INV medium with 20 mL of starter culture each. Grow cells at 37°C. Expression of the TatABC proteins is induced at an optical density of 0.5 by adding 1 mM isopropyl-β-D-thiogalactopyranoside (IPTG) and growth is continued for 3–4 h or until an optical density at 600 nm of 1.5–1.8 is reached (*see* **Note 16**).

4. Prepare 50 mL buffer A and cool on ice.

5. Chill the cell cultures quickly by placing the flasks in an ice water bath and harvest the cells at 4°C in a cooled SCL3000 rotor (Sorvall) for 10 min at 8650g (7000 rpm). All subsequent steps should be done at 4°C or on ice. Resuspend the cell pellets in buffer A (*see* **Note 9**). Combine the cell suspensions to one or two tared centrifuge bottles and centrifuge again. Determine the wet weight of the cell pellet (*see* **Note 17**).

6. Resuspend the cell pellet in 1 volume (1 mL/g wet cell mass) of buffer A containing 0.5 mM PMSF (add PMSF from a fresh 0.1 M stock in ethanol).

7. For breakage of the cells, pass the cell suspension two to three times through a French pressure 40k cell (Spectronic Unicam) at 8000 psi. This corresponds to a gage pressure setting of 500 when using the 1-in. piston cell at the "high ratio" selection (*see* **Note 10**).

8. To remove cell debris, the extract is centrifuged for 5 min in a precooled SS34 rotor (Sorvall) at 1954g (5000 rpm) at 4°C.

9. The supernatant is collected and centrifuged again for 2 h at approx 150,000g (40,000 rpm in a Beckman 50.2Ti Rotor) at 4°C to obtain a crude membrane pellet encompassing outer and inner membranes and ribosomes (*see* **Note 18**). The sticky pellets are carefully resuspended in buffer A by using a loosely fitting glass homogenizer (Fisher Scientific) to give a total volume of 8 mL. Crude membranes can be stored at −80°C after quick-freezing in liquid nitrogen.

10. Because of the different densities, the inner membrane vesicles can be separated from outer membranes and unbound ribosomes by sucrose gradient centrifugation. Prepare six sucrose gradients consisting of 12 mL 0.77 M, 12 mL 1.44 M, and 10 mL 2.02 M sucrose solution in polyallomer centrifuge tubes (38.5 mL, 25 × 89 mm, Herolab centrifuge labware). Start with the 0.77 M sucrose cushion and always underlay the denser solutions by using a smoothly running syringe equipped with a horizontally cut, wide-bore needle. Equilibrate the gradients at 4°C for about 1 h. Finally, load 2–2.5 mL of the crude membranes on top of each gradient and spin at 4°C for at least 16 h at approx 82,000g (25,000 rpm) in a swing-out rotor (Sorvall AH 629, eqivalent to Beckman type SW27/28).

11. After centrifugation the inner membrane fraction should be visible as a yellow layer at the interface between the 0.77 and 1.44 M sucrose steps. Recover the membranes with a syringe by introducing the needle from the top of the gradient or by carefully poking a hole into the tube wall at the height of the inner membrane layer. This is safely done (mind your fingers!) by use of a disposable hypodermic needle mounted on a syringe, which is slowly turned clockwise and counterclockwise between thumb and middle finger and thereby drilled across the tube wall. Push the vesicle suspension immediately into a tube placed on ice.

12. For subsequent collection, dilute the inner membrane vesicles with ice-cold 50 mM TeaOAc, pH 7.5 about fourfold (the fraction withdrawn from the gradients presumably stems to equal parts from both the 0.77 and 1.44 M sucrose layers, resulting in a calculated sucrose concentration of about 1.1 M).

13. Pellet the inner membranes by centrifugation for 2 h at approx 150, 000g (40,000 rpm) in a Beckman 50.2Ti Rotor at 4°C and carefully resuspend in INV buffer by using a loosely fitting glass homogenizer. The final desired volume of INV derived from 4 L of bacterial culture is about 1 mL. This will correspond to an absorption at 280 nm of about 30 U/mL (*see* **Note 19**).

14. Freeze the gradient-purified INV in small aliquots of about 15 μL in liquid nitrogen and store at −80°C (*see* **Note 20**).

3.3. In Vitro Transcription-Translation Using an E. coli S-135 Cell Extract

3.3.1. In Vitro Transcription-Translation Reaction: Synthesis and Transport of a Sec- (pOmpA) and a Tat- (pSufI) Specific Precursor Protein Into Inverted Inner Membrane Vesicles

1. Plan the experiment according to the table in **Fig. 1**: synthesize two precursor proteins (pOmpA and pSufI), each one in the absence of INV (samples 1 and 5), in the presence of INV containing high amounts of TatABC (samples 2 and 6), in the presence of wild-type INV (samples 3 and 7), and in the presence of Tat-deficient INV (samples 4 and 8). You will need 8 × 50 μL (i.e., 16 × 25 μL) reactions. In order to provide enough material, plan for three additional reactions, i.e., a total of 19 × 25 μL reactions. Next, calculate the reaction mixture according to **Table 1**. In order to prepare identically composed samples, combine all common ingredients in one tube (= reaction mixture) and distribute this between the reaction tubes that have received the individual components (according to **Fig. 1**).

2. Prepare 250 μL compensating buffer (CB) for the transcription-translation reaction on ice (*see* **Note 21**). Efficient in vitro transcription of template DNA and translation of transcripts into protein needs defined reaction conditions. The following final concentration of ions have proven optimal for the two DNAs used here: 40 mM TeaOAc pH 7.5, 140 mM KOAc, 11 mM Mg(OAc)$_2$ (*see* **Note 22**). The contributions of S-135 extract, membrane vesicles, and other components to the concentration of these ions are considered when calculating the CB. **Table 2** shows the calculation for a 25-μL single reaction containing 3 μL of S-135 extract, 1 μL of plasmid DNA, and 1 μL of purified INV. Calculate and prepare a new CB for each experiment.

3. Thaw all required components. This is best done by placing small aliquots simply on ice and larger ones in a water bath at room temperature. Set up a series of labeled 1.5-mL reaction tubes on ice and add buffer or INV as indicated in **Fig. 1**.

4. Prepare the reaction mixture on ice according to **Table 1**, strictly following the indicated order. Vortex before adding the first biological (creatine phosphokinase)

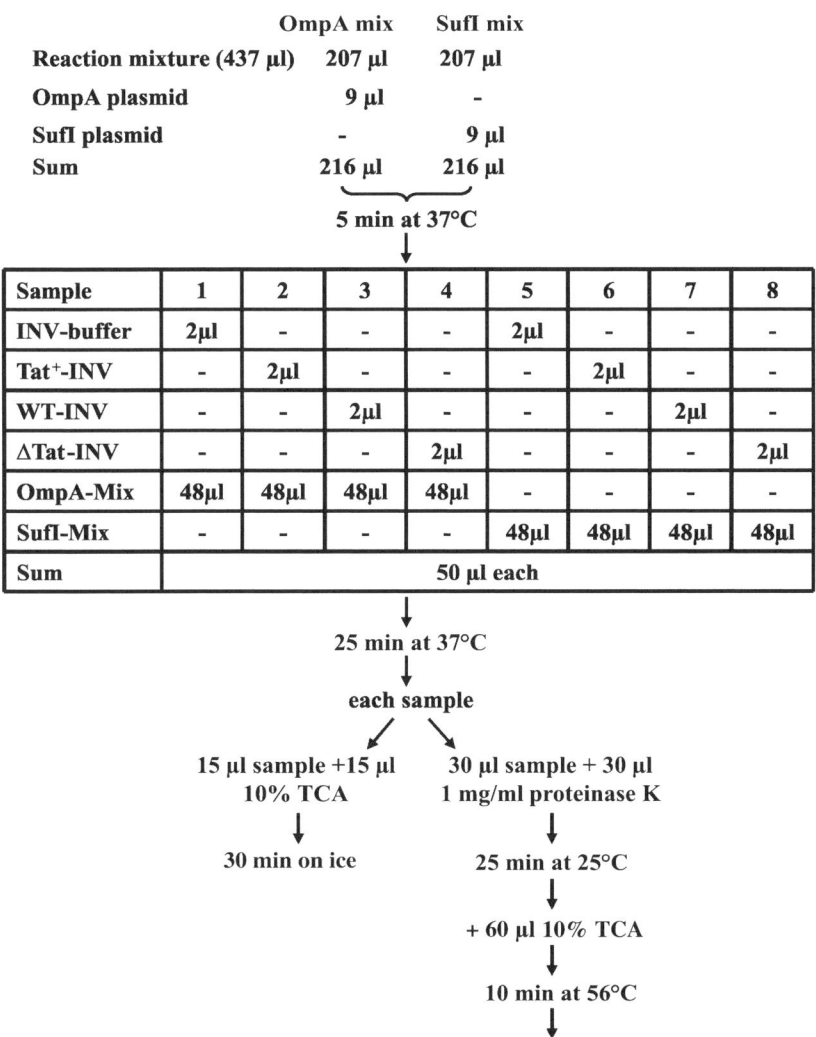

	OmpA mix	SufI mix
Reaction mixture (437 µl)	207 µl	207 µl
OmpA plasmid	9 µl	-
SufI plasmid	-	9 µl
Sum	216 µl	216 µl

5 min at 37°C

Sample	1	2	3	4	5	6	7	8
INV-buffer	2µl	-	-	-	2µl	-	-	-
Tat+-INV	-	2µl	-	-	-	2µl	-	-
WT-INV	-	-	2µl	-	-	-	2µl	-
ΔTat-INV	-	-	-	2µl	-	-	-	2µl
OmpA-Mix	48µl	48µl	48µl	48µl	-	-	-	-
SufI-Mix	-	-	-	-	48µl	48µl	48µl	48µl
Sum	50 µl each							

25 min at 37°C

each sample

15 µl sample +15 µl 10% TCA → 30 min on ice

30 µl sample + 30 µl 1 mg/ml proteinase K → 25 min at 25°C → + 60 µl 10% TCA → 10 min at 56°C → 30 min on ice

Fig. 1. In vitro synthesis of the two precursor proteins OmpA and SufI and analysis of their transport into inverted inner membrane vesicles (INV) of *E. coli*. The preparation of the reaction mixture is detailed in **Table 1**. OmpA plasmid is pDMB *(22)*, SufI plasmid is pKSMSuf-RR *(13)*. INV buffer is described in **Subheading 2.2., item 16.** Tat+-INV were prepared from strain BL21(DE3) pLysS p8737 *(5)*, WT-INV from strain MC4100 *(16)* and, ΔTat-INV from strain DADE *(19)*. TCA, trichloroacetic acid.

Table 1
Calculation of the Reaction Mixture

	Concentration of stock solution	Final concentration	μL/25 μL	μL × 19
Compensating buffer	5×	1×	5	95
H_2O		Up to 25 μL	7.9	150.1
Polyethylene glycol	40% (w/v)	3.2% (w/v)	2	38
18 Amino acids	1 mM each	0.04 mM each	1	19
DTT	200 mM	2 mM	0.25	4.75
NTP mixture			1.25	23.75
(ATP	50 mM	2.5 mM		
GTP, UTP, CTP)	10 mM each	0.5 mM each		
Phosphoenol pyruvate	200 mM	12 mM	1.5	28.5
Creatine phosphate	500 mM	8 mM	0.4	7.6
Creatine phospho- kinase	10 mg/mL	40 μg/mL	0.1	1.9
S-135			3	57
T7 RNA polymerase			0.1[a]	1.9
[^{35}S]-Met/Cys			0.5	9.5
Total			23 μL	437 μL
Added separately (*see* Fig. 1):				
DNA	1 μg/μL		1	
INV			1	

[a] Depends on activity; use 5–10 units of a commercial enzyme.

and after the last addition, each time briefly spinning to collect all liquid at the bottom of the tube again.

5. Divide the reaction mixture according to **Fig. 1**, and add the two DNAs. Vortex and spin briefly to collect all liquid at the bottom of the tube.
6. Start the reaction by incubating the two tubes at 37°C (*see* **Note 23**).
7. After 5 min of incubation (*see* **Note 24**), subdivide each of the reaction mixtures onto the eight reaction tubes as indicated in **Fig. 1**. Vortex briefly and spin to collect all liquid at the bottom of the tubes.

Table 2
Calculation of Compensating Buffer (CB)

	Tea/Tris (nmol)	K^+ (nmol)	Mg^{2+} (nmol)	Spermidine (nmol)	H_2O
$3\,\mu L$ S135 (10 mM Tea, 60 mM K^+, 14 mM Mg^{2+})[a]	30	180	42		
$1\,\mu L$ DNA (10 mM Tris)[a]	10				
$1\,\mu L$ INV (50 mM Tea)[a]	50				
Total (1)	90	180	42		
Desired final concentration: 40 mM Tea, 140 mM K^+, 11 mM Mg^{2+}, 0.8 mM spermidine					
→nmol desired in 25-μL reaction (2)	1000	3500	275	20	
Difference (1)-(2) →nmol in 25-μL reaction to be added via CB (3)	960	3360	264	20	
Required nmol (3) are added in 5 μL CB →required nmol/μL CB (4) (= mM concentration of CB)	192	672	52.8	4	
	μL	μL	μL	μL	μL
To prepare 1 mL of such CB from 1 M Tea, 4 M K^+, 1 M Mg^{2+}, 0.1 M spermidine stocks add	192	168	52.8	40	547.2
To prepare 250 μL of such CB add	48	42	13.2	10	136.8

Tea, triethanolamine acetate.
[a]Components that contribute relevantly to the ionic composition of the reaction mixture.

8. Incubate at 37°C for an additional 25 min (*see* **Note 25**).
9. Spin briefly and place tubes on ice to stop the translation reaction.
10. Split each sample. Digest one part with proteinase K to test for INV-protected translation products, the occurrence of which indicates translocation into the vesicle lumen. Precipitate the remaining part directly with 5% trichloroacetic acid. In the example given in **Fig. 1**, 30 μL is removed from each reaction tube and added to new labeled reaction tubes that contain 30 μL each of 1 mg/mL proteinase K.

Briefly vortex samples and incubate at 25°C for 20 min. Stop each protease digestion by the addition of 60 μL of 10% trichloroacetic acid. Mix and incubate for 10 min at 56°C to fully inactivate proteinase K, then place tubes on ice.

11. In the meantime, add 15 μL of the remaining in vitro reactions to 15 μL of 10% trichloroacetic acid, mix, and place on ice.
12. Precipitation with trichloroacetic acid should proceed for at least 30 min on ice but can as well be extended to an overnight incubation.
13. Pellet precipitated proteins by centrifugation for 10 min in a tabletop microcentrifuge at room temperature. Carefully remove supernatant by aspiration into the radioactive waste.
14. Add 20 μL PAGE-loading buffer to each sample and shake vigorously at room temperature to dissolve the pellet completely. The color of the loading buffer should remain dark blue. If it changes to yellow, add a few microliters of 1 *M* Tris base to neutralize residual trichloroacetic acid. Heat samples at 95°C for 5 min and analyze by SDS-PAGE and autoradiography.

3.3.2. SDS-PAGE and Autoradiography

1. Electrophoresis is carried out in large custom-made units. Dimensions of the gel are: 35 cm × 25 cm × 1 cm (W × L × T). These gels are made from about 80 and 20 mL of separating and stacking gel solutions, respectively.
2. To prepare 100 mL of a 12% separating gel, add 40 mL acrylamide/bisacrylamide solution, 20 mL separating gel buffer, 0.4 mL 25% SDS to a measuring cylinder and adjust volume to 100 mL with water. Add 0.04 mL TEMED and 0.6 mL ammonium peroxodisulfate to start polymerization, pour the solution into gel cassettes mounted in an upright position, and overlay with isobutanol.
3. After polymerization remove isobutanol, rinse with water, and prepare 30 mL of stacking-gel solution by adding 5 mL acrylamide/bisacrylamide solution, 3.6 mL stacking-gel buffer, 0.12 mL 25% SDS to a measuring cylinder and adjusting the volume to 30 mL with water. Add 0.012 mL TEMED and 0.2 mL ammonium peroxodisulfate to start polymerization pour the solution into the gel cassettes, and immediately insert a comb.
4. Prepare 2 L of running buffer by dilution from the 5x stock and add to upper and lower chambers of the electrophoresis apparatus.
5. Load the samples completely into the wells of the gel and include one lane for prestained molecular weight markers. Electrophoresis is usually carried out overnight at a constant current of 20 mA until the bromophenol blue dye has run to the bottom of the gel. Avoid its running off the gel in order to retain any radioactive substance of similarly low molecular mass on the gel.
6. Remove the stacking gel and incubate the separating gel in fixing solution for 20 min on a shaking platform. Discard the fixing solution and incubate the gel in water three times for 10 min each.

7. Transfer the gel onto a prewetted Whatman 3MM paper, cover with plastic wrap, and dry at 70°C for 2 h on a vacuum dryer (Biorad).
8. Expose the dried gel to a phosphorimaging screen overnight and analyze the autoradiogram on a PhosphorImager (e.g., Storm; GE Healthcare) using ImageQuant™ software.
9. Print an image of the autoradiogram at magnification 1 and transfer the positions of the prestained molecular weight markers to the printout. The autoradiogram of the experiment described here is shown in **Fig. 2**.

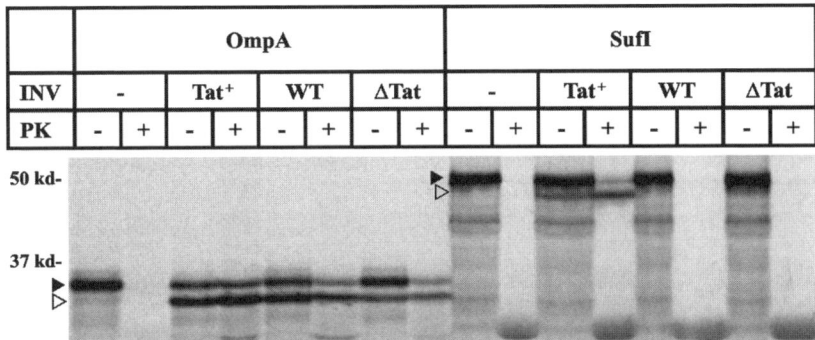

	OmpA				SufI			
INV	-	**Tat⁺**	**WT**	**ΔTat**	-	**Tat⁺**	**WT**	**ΔTat**
PK	- | +	- | +	- | +	- | +	- | +	- | +	- | +	- | +

Fig. 2. In vitro synthesis of the Sec-dependent OmpA and the Tat-specific SufI and analysis of their transport into inverted inner membrane vesicles (INV) of *E. coli*. Autoradiogram of the experiment described in the text. Precursor proteins were synthesized by a coupled in vitro transcription-translation system from *E. coli*. Where indicated, inverted inner membrane vesicles from wild-type strain (WT), a TatABC overproducing strain (Tat⁺), or a TatABC deletion strain (ΔTat) were present during synthesis. After synthesis, samples were split, one part being directly precipitated with trichloroacetic acid and the other part only after proteinase K (PK) digestion. Samples were separated by SDS-PAGE and visualized by phosphorimaging (closed arrowheads indicate precursor, open arrowheads mature form of OmpA and SufI, respectively). Irrespective of the type of INV, their presence leads to a proteolytic processing of the OmpA precursor by signal peptidase. For SufI this is observed only with INV prepared from a TatABC-overproducing strain. Protection of OmpA against proteinase K is observed in the presence of any INV reflecting translocation into the vesicle lumen. Likewise, protected SufI is seen only with Tat transport-proficient INV. The fact that translocation of SufI into INV is not seen with vesicles either containing low amounts of TatABC (WT) or none (ΔTat) indicates the specific requirement of SufI transport for the Tat machinery. Note that some precursor form is also always protected by transport-proficient INV, simply indicating that in vitro signal sequence cleavage and transmembrane transport are not strictly coupled.

Notes

1. Media and solutions are prepared using deionized water.
2. Stock solutions are usually freed of microorganisms and particles by filtration through 0.22 μm mixed cellulose ester filters (Millipore). Solutions 3–6, 9, 12, and 13 are also required for the in vitro transcription-translation reaction (*see* **Subheading 2.3.1.**).
3. Alternatively use Pefabloc SC (Roche), a water-soluble inhibitor of serine proteases at a final concentration of 0.5 mg/mL.
4. Use gloves to touch the dialysis tubing. Perform the following treatment before use: boil the dialysis tubing in 1 L 2% $NaHCO_3$, 1 mM EDTA for 10 min. Rinse the dialysis tubing with water and boil it again in 1 L water for 10 min. Store dialysis tubing in water at 4°C.
5. To obtain sufficient synthesis in vitro, the gene of interest preferably should be under the control of the T7 promoter. We have successfully used vectors such as pKSM717 (*15*) and pET derivatives.
6. If the efficiency of synthesis unexpectedly drops, it often can be overcome by preparing fresh stocks of spermidine, DTT, and phosphoenol pyruvate.
7. Polyethylene glycol, 18 amino acids, DTT, NTP stock, phosphoenol pyruvate, and creatine phosphate can be combined according to the ratios indicated in **Table 1** and stored in 100- to 200-μL aliquots at −20°C. This should be done only after proof has been obtained that these solutions allow efficient protein synthesis in vitro.
8. *E. coli* strains suitable for the preparation of S-135s are MC4100 (*16*), MRE600 (*17*), MZ9 (*6*), and SL119 (*18*). Because of a *recD* deletion, the extract of the latter strain lends itself to the expression of linearized DNA and truncated mRNA as generated for the synthesis of ribosome-associated nascent chains.
9. Fast resuspension is achieved by repeatedly forcing the cell suspension through the medium-bore opening of a ball-equipped glass pipet harboring a sufficiently large reservoir.
10. The best result is obtained by passing the cell suspension through the French pressure cell at a speed that allows a dropwise efflux. This requires more than one passage as the released DNA first causes high viscosity until it becomes fragmented by the applied shear forces.
11. Freeze the S-30 immediately in liquid nitrogen and store at −80°C or continue with the next step.
12. Use a volume ratio of S-30 to dialysis buffer of approx 1:100. One of the three steps can conveniently also be done overnight. After dialysis, the S-30 can be quick-frozen in liquid nitrogen and stored at −80°C.
13. High-speed centrifugation resulting in an S-135 extract is required to completely remove transport-competent inverted INV from the S-30 extract, which otherwise might display considerable background translocation activity. High-speed centrifugation also removes remaining polysomes from the extract.

14. Note that the time of spin and amount of supernatant withdrawn will have an influence on the performance of the S-135, the designation of which is an operational term rather than reflecting the actual *g*-force. Recovery of too much supernatant might still result in a contamination with endogenous membranes, whereas too little supernatant bears the risk of a shortage of monosomes. In the latter case try to reduce time of ultracentrifugation when preparing the S-135 or add separately isolated ribosomes. Low translation activity of an S-135 preparation can also result from residual cold methionine added for the readout of endogenous polysomes. This is effectively removed by repeated passages of the S-135 through an Amicon ultra centrifugal filter unit (Millipore, molecular weight cutoff of 10,000 Da), each time replacing the filtrate by fresh S-30 buffer.

15. Do not thaw and freeze the S-135 more than twice.

16. Translocation of twin-arginine-containing precursor proteins has thus far only been observed into INV derived from strains that overproduce the TatABC proteins, such as BL21(DE3) pLysS p8737TatABCD *(5)* or DADE (MC4100, Δ*tatABCD*Δ*tatE*) *(19)* transformed with plasmids pRep4 and pQE60-TatABC *(20)*. In the BL21 derivative, in which *tatABC* is under T7 promoter control, IPTG induces expression of T7 RNA polymerase, whereas in the other strain, IPTG directly enhances expression of *tatABC* from the *lac* promoter. High levels of the TatABC proteins in the vesicles prepared from these strains are verified via Western blotting.

17. Cells destined for the preparation of INV must not be frozen before breakage in the French press. If necessary, cell pellets can be kept on ice overnight.

18. If the protocol is to be directly continued beyond this step, prepare sucrose gradients during this 2-h centrifugation period.

19. For determining the absorbance of the vesicle suspension at 280 nm, prepare a 1:100 dilution in 2% SDS. With an absorbance of 30, usually 1–2 µL of INV are sufficient to observe transport in a single in vitro reaction. However, the optimal amount of INV needs to be titrated for each new batch of INV.

20. Do not freeze and thaw INV more than two or three times.

21. To avoid contamination with proteases and RNases, always wear gloves and preferentially use sterile disposable reaction tubes and pipet tips.

22. With every new preparation of S-135 it is necessary to re-adjust the reaction conditions. The variable with the strongest impact on expression efficiency is the concentration of Mg^{2+}, which even needs to be optimized for each particular DNA template. A titration is done by decreasing and increasing the Mg^{2+} concentration in 1 mM steps starting with 8 mM. For this titration, prepare the CB for the lowest Mg^{2+} concentration desired. Sometimes inclusion of 8 mM putrescin *(21)* into the reaction helps to improve expression. In this case the final Mg^{2+} concentration is usually lowered by about 3 mM and that of phosphoenol pyruvate by 6 mM. If in vitro expression remains unsatisfactory, try to vary the amount of S-135 in the range of 2–4 µL.

23. Incubation at 37°C is routinely used. In some cases (e.g., INV derived from *cs* mutants) it is necessary and possible to synthesize proteins also at lower temperatures.
24. INV very often exert an inhibitory effect on protein synthesis. This is largely reduced by adding the vesicles only 5 min after starting the synthesis reaction.
25. Although twin-arginine-dependent transport of folded substrates must be a posttranslational event, the best transport rates are observed when INV are present during the synthesis reaction. To analyze translocation in a posttranslational manner, stop synthesis after 30 min by the addition of puromycin (added from a neutralized 11 mM stock solution to a final concentration of 0.4 mM and incubated for 5 min at 37°C). Then add INV and an ATP regenerating system consisting of ATP, creatine phosphate, creatine phosphokinase, and DTT re-added to give the concentrations indicated in **Table 1**. Incubate at 37°C for an additional 20 min and proceed with proteinase K digestion.

Acknowledgments

This work was supported by grant LSHG-CT-2004-05257 of the European Union and grants from the Deutsche Forschungsgemeinschaft (Sonderforschungsbereich 388 and Graduiertenkolleg 434).

References

1. de Keyzer, J., van der Does, C., and Driessen, A. J. (2003) The bacterial translocase: a dynamic protein channel complex. *Cell. Mol. Life Sci.* **60**, 2034–2052.
2. Müller, M. (2005) Twin-arginine-specific protein export in *Escherichia coli*. *Res. Microbiol.* **156**, 131–136.
3. Müller, M. and Klösgen, R. B. (2005) The Tat pathway in bacteria and chloroplasts (review). *Mol. Membr. Biol.* **22**, 113–121.
4. Yahr, T. L. and Wickner, W. T. (2001) Functional reconstitution of bacterial Tat translocation in vitro. *EMBO J.* **20**, 2472–2479.
5. Alami, M., Trescher, D., Wu, L. F., and Müller, M. (2002) Separate analysis of twin-arginine translocation (Tat)-specific membrane binding and translocation in *Escherichia coli*. *J. Biol. Chem.* **277**, 20499–20503.
6. Müller, M. and Blobel, G. (1984) In vitro translocation of bacterial proteins across the plasma membrane of *Escherichia coli*. *Proc. Natl. Acad. Sci. USA* **81**, 7421–7425.
7. Rhoads, D. B., Tai, P. C., and Davis, B. D. (1984) Energy-requiring translocation of the OmpA protein and alkaline phosphatase of *Escherichia coli* into inner membrane vesicles. *J. Bacteriol.* **159**, 63–70.
8. Troschel, D. and Müller, M. (1990) Development of a cell-free system to study the membrane assembly of photosynthetic proteins of *Rhodobacter capsulatus*. *J. Cell Biol.* **111**, 87–94.

9. Zubay, G. (1973) In vitro synthesis of protein in microbial systems. *Annu. Rev. Genet.* **7**, 267–287.

10. Futai, M. (1974) Orientation of membrane vesicles from *Escherichia coli* prepared by different procedures. *J. Membr. Biol.* **15**, 15–28.

11. Schnaitman, C. A. (1970) Protein composition of the cell wall and cytoplasmic membrane of *Escherichia coli*. *J. Bacteriol.* **104**, 890–901.

12. Beck, K., Wu, L. F., Brunner, J., and Müller, M. (2000) Discrimination between SRP- and SecA/SecB-dependent substrates involves selective recognition of nascent chains by SRP and trigger factor. *EMBO J.* **19**, 134–143.

13. Alami, M., Lüke, I., Deitermann, S., et al. (2003) Differential interactions between a twin-arginine signal peptide and its translocase in *Escherichia coli*. *Mol. Cell* **12**, 937–946.

14. Davanloo, P., Rosenberg, A. H., Dunn, J. J., and Studier, F. W. (1984) Cloning and expression of the gene for bacteriophage T7 RNA polymerase. *Proc. Natl. Acad. Sci. USA* **81**, 2035–2039.

15. Maneewannakul, S., Maneewannakul, K., and Ippen-Ihler, K. (1994) The pKSM710 vector cassette provides tightly regulated lac and T7lac promoters and strategies for manipulating N-terminal protein sequences. *Plasmid* **31**, 300–307.

16. Casadaban, M. J. and Cohen, S. N. (1979) Lactose genes fused to exogenous promoters in one step using a Mu-lac bacteriophage: in vivo probe for transcriptional control sequences. *Proc. Natl. Acad. Sci. USA* **76**, 4530-4533.

17. Cammack, K. A. and Wade, H. E. (1965) The sedimentation behaviour of ribonuclease-active and -inactive ribosomes from bacteria. *Biochem. J.* **96**, 671–680.

18. Lesley, S. A., Brow, M. A., and Burgess, R. R. (1991) Use of in vitro protein synthesis from polymerase chain reaction-generated templates to study interaction of *Escherichia coli* transcription factors with core RNA polymerase and for epitope mapping of monoclonal antibodies. *J. Biol. Chem.* **266**, 2632–2638.

19. Wexler, M., Sargent, F., Jack, R. L., et al. (2000) TatD is a cytoplasmic protein with DNase activity. No requirement for TatD family proteins in sec-independent protein export. *J. Biol. Chem.* **275**, 16717–16722.

20. Halzapfel, E., Eisner, G., Alami, M., Barrett, C. M. L., Buchanan, G., Liike, I., Betton, J. M., Robinson, C., Palmer, T., Moser, M., and Miiller, M. The entire N-terminal half of TatC is involved in twin-arginine precursor binding. *Biochemistry* (in press).

21. Chen, L., Rhoads, D., and Tai, P. C. (1985) Alkaline phosphatase and OmpA protein can be translocated posttranslationally into membrane vesicles of *Escherichia coli*. *J. Bacteriol.* **161**, 973–980.

22. Behrmann, M., Koch, H. G., Hengelage, T., Wieseler, B., Hoffschulte, H. K., and Müller, M. (1998) Requirements for the translocation of elongation-arrested, ribosome-associated OmpA across the plasma membrane of *Escherichia coli*. *J. Biol. Chem.* **273**, 13898–13904.

II

EUKARYOTIC SYSTEMS

6

Nuclear Import Properties of the Sex-Determining Factor SRY

Jade K. Forwood, Gurpreet Kaur, and David A. Jans

Summary

The sex-determining factor SRY plays an important role in male sexual development, diverting primordial gonads from the ovarian pathway toward male differentiation to form testes. SRY is a DNA-binding protein and gains access to the nucleus through two independently acting nuclear localization signals (NLSs) that flank the high mobility group (HMG) DNA-binding domain. We have reconstituted the nuclear import of SRY using an in vitro nuclear transport assay, showing that nuclear import of SRY can occur in the absence of additional exogenous cytosolic factors, with a significant reduction in nuclear transport in the presence of antibodies to the nuclear transport protein importin (Imp) β1 but not Impα. We have also shown using in vitro binding assays that the C-terminal NLS of SRY binds directly to Impβ1. Finally, we have shown that SRY can target green fluorescent protein to the nucleus in a mammalian transfected cell line; importantly, mutations known to result in sex reversal that map to either NLS impair nuclear accumulation implying that SRY nuclear import is critical to its function.

Key Words: Nuclear import; importin; green fluorescent protein; SRY; nuclear localization sequence.

1. Introduction

1.1. Nuclear Protein Import

Bidirectional transport of macromolecules from the cytoplasm to the nucleus is an active process that occurs through nuclear pore complexes (NPCs) embedded within the nuclear envelope (NE). The transport process is mediated by a family of transporters or cytosolic receptor proteins known as importins (Imps) *(1)*, which work in concert with the guanine nucleotide-binding protein Ran and

From: *Methods in Molecular Biology, Vol. 390: Protein Targeting Protocols: Second Edition*
Edited by: M. van der Giezen © Humana Press Inc., Totowa, NJ

other regulatory proteins *(2)*. In conventional nuclear localization signal (NLS)-dependent nuclear protein import, Impα recognizes the NLS-containing protein and acts as an adapter to mediate binding of Impβ1 *(3)*. The latter then mediates docking of the importin-NLS-containing protein complex to NPC through binding of Impβ1 to hydrophobic repeat-containing nucleoporin components that make up the NPC, followed by translocation through the NPC via analogous docking events *(4)*. Upon entry into the nucleus, RanGTP dissociates the complex by binding to Impβ1, and the individual importins are recycled back to the cytoplasm *(5)*.

1.2. Sex-Determining Region-Y Protein

Sex determination is a chromosomally controlled process, with females having two X chromosomes and males having an X and a Y chromosome. The SRY (sex-determining region on the Y chromosome) protein plays an important role in sex determination and early gonadal differentiation of mammalian embryos. In humans, SRY is expressed at week 6 of embryonic development and acts as a genetic switch to divert primordial gonads from the ovarian pathway toward male differentiation to form testes *(6)*. SRY acts as genetic switch by binding specifically to an eight-base pair recognition sequence *(7)* via its 80-amino-acid high-mobility group (HMG) DNA-binding domain, located centrally within the protein. Binding of the HMG box to the minor groove of DNA *(8)* induces helix unwinding, minor groove expansion, and DNA bending *(9)*, which are believed to be the molecular switch that allows distantly bound proteins of the transcription machinery to attain close proximity, thereby permitting interaction in a way that can influence gene expression *(10)*. Underlining SRY's critical role in sex determination is the fact that 15% of all XY-females (Swyer syndrome individuals) possess mutations in SRY.

For SRY to interact with DNA/modulate gene transcription, it needs to be targeted to the nucleus. Targeting of SRY to the nucleus is mediated by two distinct NLSs that flank the HMG box, the N-terminal bipartite NLS (KRPMNAFIVWSRDQRRK[77]) and the monopartite C-terminal NLS (RPRRKAK[136]), both of which are able to act independently to target β-galactosidase to the nucleus *(11–14)*. Importantly, certain sex-reversing mutations in SRY map to these NLSs *(12,13)*.

Here we describe an in vitro nuclear transport assay and native gel mobility shift assay used to characterize the nuclear import properties of SRY in vitro. We demonstrate that antibodies to Impβ significantly reduce SRY nuclear accumulation and that SRY binds directly to Impβ rather than the conventional NLS receptor Impα. We also demonstrate that both full-length SRY protein and the HMG domain of SRY are capable of targeting green fluorescent protein

(GFP) to the nucleus in transfected mammalian cells and that clinical sex-reversing mutations in either of the SRY NLSs directly reduce nuclear accumulation. These results imply that nuclear import of SRY is critical to its role in sex determination, with the Impβ1-recognized C-NLS playing a critical role.

2. Materials

2.1. Preparation of SRY-GFP, Importin-α-GST, and Importin-β-GST

1. BL21DE3 competent cells (Invitrogen).
2. Luria-Bertani (LB) media: 10 g/L bacto-tryptone, 5 g/L bacto-yeast extract, 10 g/L NaCl; sterilized by autoclave.
3. LB agar (LB containing 1.5% agar): sterilized by autoclave, cool to 60°C, and add ampicillin to 100 µg/mL; add 30 mL each to a Petri dish and allow to set at RT for 1 h.
4. Isopropyl-β-D-thiogalactoside (IPTG): 1 M stock solutions.
5. His buffer A: 50 mM phosphate buffer, 10 mM Tris-HCl, pH 8.0, 300 mM NaCl, 20 mM imidazole.
6. His buffer B: 50 mM phosphate buffer, 10 mM Tris-HCl, pH 8.0, 300 mM NaCl, 500 mM imidazole.
7. GST buffer A: 50 mM Tris-HCl, pH 8.0, 100 mM NaCl.
8. GST buffer B: 50 mM Tris-HCl, pH 8.0, 100 mM NaCl, 10 mM glutathione.
9. Lysozyme: stock solutions of 50 mg/mL made up in water.
10. Ni-NTA agarose (Qiagen).
11. Glutathione sepharose matrix slurry (Amersham).
12. PD-10 column (Pharmacia).
13. Bio-Rad Protein Assay (Bio-Rad, Hercules, CA).

2.2. Cultivation of Rat Hepatoma Cells for In Vitro Nuclear Transport Assay

1. Cells of the rat hepatoma tissue culture (HTC) line, a derivative of Morris hepatoma 7288C cell line (Flow Laboratories, Bonn, Germany).
2. Dulbecco's modified Eagle's medium (DMEM; ICN Biomedicals, Costa Mesa, CA) supplemented with 10% heat-inactivated fetal calf serum (FCS, from CSL limited), 5 mL/L L-glutamine, 2.5 mL/L penicillin-streptamycin and 20 mM HEPES.
3. Humidified incubator with 5% CO_2 atmosphere at 37°C (Forma Scientific, Waltham, MA).
4. Cell culture flasks, 50 mL (Nunclon, Kamstrup, Denmark); 50-mL Falcon tubes.
5. 1× PBS: 137 mM NaCl, 6.25 mM Na_2HPO_4, 2.5 mM Na_2PO_4; sterilized in autoclave.
6. 0.25% Trypsin in ethylenediamine tetraacetic acid (EDTA) (Sigma, St. Louis, MO).
7. Autoclave sterilized 15 mm × 15 mm coverslips (Nunclon).
8. 12-Well cell culture plates (Nunclon).

2.3. In Vitro Nuclear Transport Assay

1. Untreated reticulocyte lysate (Promega cat. no. L415A) (*see* **Note 1**).
2. Transport substrate (1.0 μM GFP-tagged SRY fusion protein in IB+ buffer; *see* **Note 2**); store at −70°C.
3. Dextran 70 kDa (2 mg/mL in H_2O) labeled with Texas red (from Molecular Probes); store at −70°C.
4. Antibodies specific to importin-α and importin-β (Santa Cruz Biotechnology).
5. 4xIB buffer: 440 mM KCl, 20 mM $NaHCO_3$, 20 mM $MgCl_2$, 4 mM EGTA, 0.4 mM $CaCl_2$, 80 mM HEPES; pH to 7.4 with NaOH; store at 4°C.
6. 4xIB+: 1 mL 4 × IB, 8 μL DTT (0.5 M), 8 μL leupeptin (Sigma), 5 mg/mL, store at −20°C; make fresh each time.
7. ATP regenerator (make fresh each time): 10 μL 4 × IB+, 2 μL creatine phosphokinase (Sigma), 6.25 mg/mL; 1562 U/mL; store at −70°C, 6 μL creatine phosphate (Sigma), 500 mM; store at −70°C, 2 μL ATP (Sigma), 100 mM; store at −20°C.
8. BioRad-MRC500 CLSM microscope (equivalent or better).
9. DMEM without phenol red, supplemented 10% FCS and glutamine (ICN) in 1 M HEPES.

2.4. Native Gel Mobility Shift Assay

1. 10X TBE: 0.9 M Tris-HCl, 1.125 M boric acid, 20 mM EDTA.
2. 10X Gel shift binding buffer: 100 mM Tris-HCl, pH 8.0, 10 mM $MgCl_2$, 1% NP40, 100 mM DTT, 8 mM EDTA, 30% glycerol, 15% sucrose.
3. Acrylamide/bis-acrylamide (29:1) (Sigma).
4. 500 μL of 10% ammonium persulfate (APS).
5. 15 μL TEMED (Sigma).

2.5. Cultivation of African Green Monkey Kidney Cell Lines for Transfection Assays

1. Cells of the African green monkey kidney COS-7 cell line.
2. DMEM (ICN Biomedicals) supplemented with 10% heat-inactivated fetal bovine serum (FBS; CSL Ltd), 5 mL/L L-glutamine, 2.5 mL/L penicillin-streptamycin, 20 mM HEPES.
3. Humidified incubator with 5% CO_2 atmosphere at 37°C.
4. Cell culture flasks, 25 mL (Nunclon).
5. 1X PBS: 137 mM NaCl, 6.25 mM Na_2HPO_4, 2.5 mM Na_2PO_4; sterilized in autoclave.
6. 0.25% Trypsin in EDTA (Sigma).
7. Autoclave sterilized cover slips (15 mm × 15 mm) (Nunclon).
8. 12-Well culture plates (Nunclon).

2.6. Transfection of COS-7 Cells and Live Cell Imaging

1. DMEM (ICN Biomedicals) supplemented with 10% heat-inactivated FCS (CSL Ltd), 20 m*M* HEPES, 5 mL/L L-glutamine, and 2.5 mL/L penicillin-streptamycin (DMEM complete).
2. DMEM with 10% heat-inactivated FBS and 20 m*M* HEPES (DMEM serum-free).
3. Humidified incubator (Heraeus) with 5% CO_2 atmosphere at 37°C.
4. Lipofectamine 2000 (Invitrogen).
5. GFP-SRY fusion protein (wild-type and mutant derivatives) encoding plasmids for mammalian cell expression (*see* **Note 3**).
6. Phenol-free DMEM (ICN); prewarmed to 37°C.
7. Heating block assembled on BioRad-MRC500 CLSM microscope (equivalent or better).
8. 60-mm Culture dish (Nunclon).

3. Methods

3.1. Preparation of SRY-GFP, Importin-α-GST, and Importin-β-GST

1. Introduce expression clones encoding His_6-SRY-GFP, GST-tagged Impα and Impβ into *E. coli* BL21DE3 competent cells (*see* **Note 4**):

   a. Add 1 μL of the expression plasmids to 50 μL of BL21DE3 competent cells (Invitrogen), mix gently, and incubate on ice for 30 min.
   b. Heat shock by transferring cells to 42°C for 45 s, chill cells on ice for 2 min.
   c. Add 500 μL of LB to the culture and grow at 37°C for 2 h.
   d. Plate cells onto LB agar containing 100 μg/mL ampicillin and incubate overnight at 37°C.

2. Inoculate 5 mL of LB with a single colony and grow at 37°C for 12 h.
3. Transfer the culture to 2 L of LB containing 100 μg/μL ampicillin and grow at 37°C to an OD_{600} of 0.6.
4. Induce protein expression with addition of 1 m*M* IPTG for 8 h at 37°C.
5. Harvest the bacteria through centrifugation at 4000 *g* for 30 min and gently resuspend cell pellet in GST buffer A (*see* **Subheading 2.1.**).
6. Add 1 mg/mL of lysozyme to lyse cells and remove cellular debris by centrifugation at 23,000 *g* for 30 min.

3.1.1. Preparation of His-SRY-GFP

1. Add 3 mL of Ni-NTA agarose (Qiagen) (prewashed in His Buffer A; *see* **Subheading 2.1.**) to the protein extract in a column and incubate for 2 h at 4°C.
2. Remove nonspecifically bound proteins from the column by washing the resin three times with 50 mL His buffer A and elute His-SRY-GFP from the column using 3.5 mL His buffer B (*see* **Subheading 2.1.**).

3. Exchange buffer to 1X PBS using a PD-10 column (Pharmacia) and store proteins in aliquots at −70°C.
4. Determine the concentration of proteins using the Bio-Rad Protein Assay according to manufacturer's specifications:

   a. Set up solutions in Eppendorf tubes containing 2–14 µg of protein standard bovine serum albumin (BSA) made up to 0.8 mL in distilled water (dH$_2$O).
   b. Also set up the proteins (Impα- and Impβ-GST) up to 0.8 mL in dH$_2$O.
   c. Add 0.2 mL of the protein-binding reagent (Bio-Rad) to each Eppendorf tube and incubate for 10 min at room temperature.
   d. Measure the absorbance at 595 nm and subtract the values from a reagent blank prepared from 0.8 mL dH$_2$O and 0.2 mL of the above reagent.
   e. Calculate the protein concentration by extrapolating from the BSA standard curve.

3.1.2. Preparation of Impα- and Impβ-GST Fusion Proteins

1. Add 2 mL of glutathione sepharose matrix slurry (Amersham) (prewashed in GST buffer A; *see* **Subheading 2.1.**) to the protein extract in a column and incubate for 2 h at 4°C.
2. Remove nonspecifically bound proteins from the column by washing the matrix three times with 50 mL of GST buffer A and elute the protein from the column using 2.5 mL of GST buffer B (*see* **Subheading 2.1.**).
3. Exchange buffer to 1X PBS using PD-10 columns (Pharmacia) and store protein aliquots at −70°C.
4. Determine the concentration of proteins using the Bio-Rad Bradford according to manufacturer's specifications, as described in **Subheading 3.1.1., step 10**.

3.2. Maintenance and Preparation of Rat HTC Cells for In Vitro Nuclear Transport Assay

Cells of the rat HTC line, a derivative of Morris hepatoma 7288C cell line (Flow Laboratories), are routinely used in the in vitro nuclear transport assay; other cell lines have also been used with success, with the main difference being the confluency at which mechanical perforation is optimal. These cells are cultured in DMEM supplemented with 10% FCS and passaged at 95% confluency (*see* **Note 5**).

1. At 95% confluency, gently wash cells by replacing the DMEM medium with 2 mL of sterile PBS.
2. Trypsinize cells by replacing the PBS with 400 µL 0.25% trypsin in EDTA.
3. Incubate cells for 5 min at 37°C, giving one gentle tap to lift the adhered cells from the plate.

4. Harvest the cells through the addition of 5 mL DMEM supplemented with 10% heat-inactivated FCS in a 50-mL Falcon tube.
5. Centrifuge cells at 100 g for 5 min and carefully aspirate the supernatant without disturbing the cell pellet.
6. Gently resuspend the cell pellet in 5 mL of fresh DMEM supplemented with 10% heat-inactivated FCS (*see* **Note 6**).
7. Prepare a 12-well cell culture plate by placing sterile cover slips and adding 1 mL of DMEM in each well.
8. Add 120 μL of cells to each well and incubate for 2 d in the 37°C humidified incubator containing 5% CO_2; cells should reach 50–70% confluency in 48 h (*see* **Note 7**).
9. Maintain the cell line by making 1:10, 1:15, and 1:20 splits every 3–4 d.

3.3. In Vitro Nuclear Transport Assay

The nuclear transport assay itself is performed over a 40-min time period, with each scan ideally containing five or more nuclei, such that an average nuclear accumulation (the *Fn/c*) for each time point can be obtained. To avoid photobleaching and obtain a representative determination of the rate of nuclear accumulation, it is best to analyze different fields of view for each scan.

1. Just prior to use, replace the media of the cells with DMEM lacking phenol red (*see* **Subheading 2.3.**).
2. Prepare 3 mL of 1x IB+ in a Petri dish and incubate at 37°C.
3. Remove cover slip from the 12-well cell culture plate and immerse in 1x IB+.
4. Remove the cover slip and gently blot the excess liquid from the slide on tissue paper.
5. Add 3 μL of 1x IB+ to the cover slip and perforate by gently placing tissue on the cover slip, apply for 5 s, and quickly remove the tissue with forceps (*see* **Note 8**; *see* **Note 9** regarding optimization of microscopic settings).
6. Pipet onto a microscope slide: 1.5 μL cytosolic extract (untreated reticulocyte lysate; Promega) or 45 mg/mL BSA, 0.5 μL 1x IB+ solution, 1.0 μL ATP regenerating system, 1.0 μL of GFP-tagged SRY (1 μM), and 1.0 μL of 70 kDa Texas red-labeled dextran (Sigma) (*see* **Note 10**).
7. Mix briefly on the slide and place the cover slip with perforated cells face down onto the drop (start timer; represents time, $t = 0$), and seal with nail polish.
8. Take CLSM scans as often as possible over the first 5–6 min, and then every 3–5 min up to 40 min (*see* **Note 11**).

Nuclear accumulation of SRY is observed in the presence and absence of exogenously added cytosol, with intact nuclei indicated by exclusion of the control Texas red-labeled 70-kDa dextran (*see* **Fig. 1A**). Blocking of the activity of the nucleoporins that make up the NPC by wheat germ agglutinin

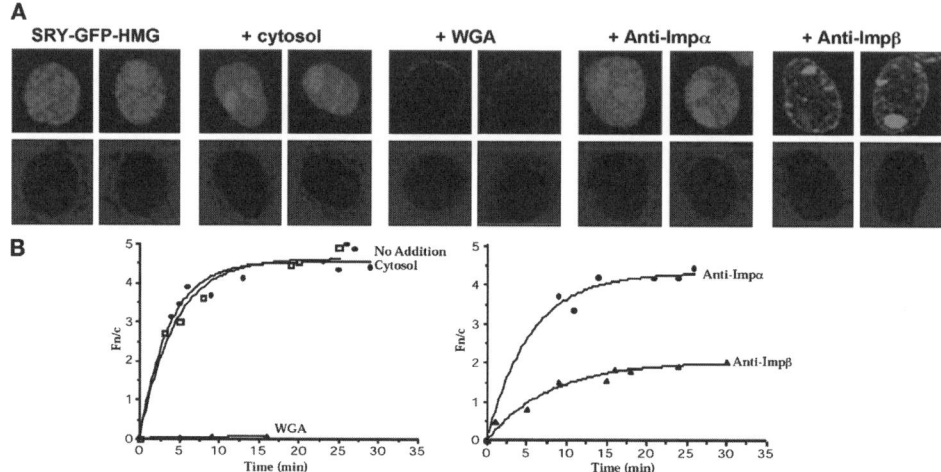

Fig. 1. Nuclear import in mechanically perforated hepatoma tissue culture (HTC) cells. **(A)** Confocal laser scanning microscope (CLSM) images of nuclear accumulation of GFP-SRY-HMG after 20 min in the absence or presence of cytosol, WGA, or anti-Impα or anti-Impβ1 antibodies as indicated. **(B)** Nuclear import kinetics from a single typical experiment performed as in **(A)** for GFP-SRY-HMG in the absence of cytosol with the additions as indicated, where each data point represents at least three separate measurements for each of Fn (nuclear fluorescence), Fc (cytoplasmic fluorescence), and background fluorescence. Data were fitted for the function $Fn/c = Fn/c_{max}(1 - e^{-kt})$, where Fn/c is the nuclear-to-cytoplasmic ratio, Fn/c_{max} represents the maximal level of nuclear accumulation, k is the rate constant, and t is time in min.

also abolishes nuclear transport, whereas the addition of antibodies specific to Impβ1, but not Impα, inhibits nuclear accumulation. The latter indicates that nuclear import of SRY is mediated by Impβ1 independent of Impα.

3.3.1. Image Analysis and Curve Fitting

1. Using the NIH ImageJ 1.62 public domain software, open the images obtained from the CLSM for each time point.
2. Measure the fluorescence in the nucleus, the cytoplasm, and the background for at least five nuclei in each time point.
3. Calculate the nuclear accumulation compared to the cytoplasm using the equation $Fn/c = (Fn - Fb)/(Fc - Fb)$, where Fn/c is the nucleocytoplasmic ratio; Fn, Fc, and Fb represent the fluorescence in the nucleus, cytoplasm, and background as a result of autofluorescence, respectively.

4. Fit the experimental data for the nuclear import kinetics over time using the equation $Fn/c(t) = Fn/c_{max}(1 - e^{-kt})$—using appropriate software (*see* **Note 12**)—where $Fn/c(t)$ represents the nuclear/cytoplasmic ratio at time t, Fn/c_{max} the maximal level of nuclear accumulation, and k the import rate constant.

Quantitative analysis of SRY nuclear accumulation is presented in **Fig. 1B** and summarized in **Table 1**. Maximal levels of SRY in the nucleus were approximately fourfold higher than in the cytoplasm within the first 20 min of the experiments, with the time required to reach half-maximal nuclear levels ($t_{1/2}$) approx 2.5 min.

3.4. Native Gel Mobility Shift Assay

1. Prepare a 5% TBE gel by first combining 18.8 mL H_2O, 4.31 mL acrylamide/bis (29:1), and 1.25 mL 10x TBE. Set the gel by adding 500 µL 10% ammonium persulfate and 15 µL TEMED, allow the gel to set for at least 1 h.
2. Prepare 0.5 × TBE running buffer and chill to 4°C.
3. Prerun the gel for 5 h at 100 V to remove traces of ammonium persulfate (used to polymerize the acrylamide) and to ensure an even distribution of electrophoresis buffer throughout the gel.

Table 1
Kinetics of Nuclear Import of GFP-SRY

Addition	Nuclear import parameter[a]		n
	Fn/c_{max}	$t_{1/2}$(min)	
No addition	4.4 ±0.2[b]	2.3±0.3	2
+ Cytosol	4.5 ±0.1	2.5±0.2	2
WGA	0.06±0.01[c]	ND[d]	1
Anti-Impα	4.3 ±0.2	3.9±0.7	2
Anti-Impβ	2.0 ±0.1[b]	5.5± 0.5	2

[a] Results represent the mean ± SEM (n indicated) from data (*see* **Fig. 1**) fitted to the function $Fn/c = Fn/c_{max}(1 - e^{-kt})$, where Fn/c_{max} is the maximal level of nuclear accumulation, k is the rate constant, and t is time in min.

[b] Significant differences were observed between Fn/c_{max} values for GFP-SRY(HMG) in the absence and presence of anti-Impβ antibody ($p = 0.0065$).

[c] SE was determined from the curve fit.

[d] ND, not able to be determined because of low nuclear accumulation.

4. For each binding condition, combine 1 μ*M* of GFP-tagged SRY, increasing amounts of GST-tagged mouse Impα or -β protein to give final concentrations of 25–3000 n*M*, 2.5 μL 10x gel shift binding buffer, and make up to a final volume of 25 μL with water.
5. Incubate the mixture for 20 min on ice.
6. Add 15 μL of 40% sucrose, carefully load the gel, and carry out electrophoresis at 100 V for 5 h at 4°C.
7. Analyze the gel by fluorescence imaging using a Fuji FLA-3000 gel imager using the excitation laser source 473 nm, Gain 1000, and the 520 LP emission filter.

The results of native gel electrophoresis for GFP-SRY in the presence of Impα and Impβ are shown in **Fig. 2**. While Impα at its maximal concentration was only able to induce a change in mobility of a small percentage of SRY, Impβ1 induced a change in mobility at low concentrations, indicating direct interaction of SRY with Impβ1.

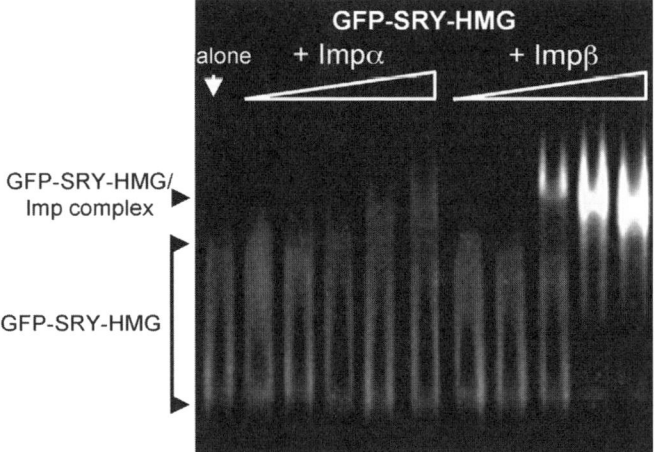

Fig. 2. Gel mobility shift assay for Impβ and Impα binding to GFP-SRY-HMG using native polyacrylamide gel electrophoresis (PAGE). One μM GFP-SRY-HMG-wt protein was incubated in the absence or presence of increasing concentrations of Impβ and Impα, as indicated (25–3000 n*M*), for 20 min on ice, and then electrophoresed on a 5% native polyacrylamide gel for 5 h at 100 V; fluoroimaging was then performed using a Fuji FLA-3000 gel imager, as described in **Subheading 3.4.**

3.5. Maintenance and Preparation of COS-7 Cells for Transfection Assays

Cos-7 cells are cultured in DMEM supplemented with 10% FBS, 20 m*M* HEPES, 5 mL/L L-glutamine, and 2.5 mL/L penicillin-streptomycin (DMEM complete) and passaged at 95% confluency.

1. At 95% confluency, wash cells gently by replacing existing DMEM complete medium with 2 mL of sterile PBS.
2. Trypsinize the cells by replacing the PBS with 400 μL of trypsin and incubating at 37°C in the humidified incubator for 4–5 min.
3. Resuspended the cells in 4.5 mL of DMEM complete.
4. Prepare a 12-well cell culture plate by placing sterile coverslips and 2 mL of DMEM in each well.
5. Add 100 μL of the cells to each well and incubate in the 37°C humidified incubator containing 5% CO_2, to be used for transfection the next day and imaging 48 h following seeding.
6. To maintain the cell line, make 1:10, 1:15, and 1:20 splits every 3–4 d.

3.6. Transfection of Cos-7 Cells

1. In Eppendorf tubes, set up 2 μg GFP-SRY fusion plasmids (including the wild-type as well as mutant derivatives) and make up to 50 μL using serum-free DMEM.
2. Add 50 μL of a 1:25 dilution of lipofectamine 2000 in serum-free DMEM to the DNA and incubate at RT for 20 min.
3. During this incubation period, starve cells (in the 12-well cell culture plates) by replacing the DMEM medium with 500 μL serum-free DMEM for 3–4 min.
4. Remove the serum-free DMEM from the cells and add 100 μL of the DNA/lipofectamine complex mixture dropwise onto each cover slip and incubated for a further 20 min at room temperature.
5. Gently add 1 mL each of serum-free DMEM and DMEM complete to each of the cover slips and incubate at 37°C, 5% CO_2 for imaging 24 h later.

3.6.1. Live Cell Imaging of Transfected Cos-7 Cells

1. Prior to imaging, prewarm phenol-free DMEM in a 60-mm dish.
2. To image cover slips, remove from the wells and gently place in the phenol-free DMEM.
3. Place the dish on the heating block.
4. Locate the cells using a ×40 water immersion objective to find green cells using a UV filters (×10 objective may also be used).
5. Place cells in the center of the microscopic field, turn off UV, and expose cells to the beam of the photomultiplier.
6. Focus and record images using the Comos imaging program.
7. Images are saved (>20 cells/construct) and later analyzed using Image J.

Confocal laser scanning microscope (CLSM) imaging of cells transfected **(Fig. 3A)** to express GFP-SRY-FL or HMG domain with and without the NLS mutated derivatives indicate that, compared to GFP itself, wild-type SRY localizes exclusively to the nucleus, with NLS-mutated constructs showing increased cytoplasmic localization.

3.6.2. Image Analysis

1. Each CLSM image is opened using the NIH ImageJ 1.62 public domain software.
2. All cells are analyzed for fluorescence in the nucleus (Fn), the cytoplasm (Fc), and the background (Fb), which is autofluorescence from a nontransfected cell.

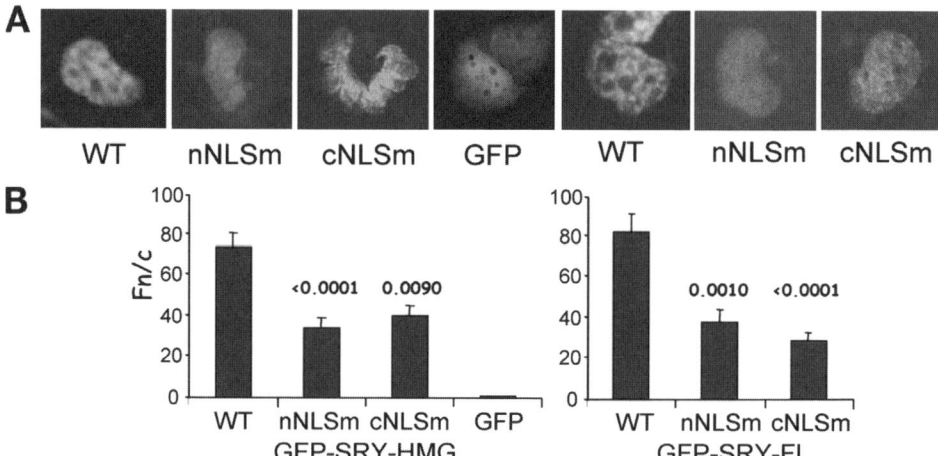

Fig. 3. Nuclear accumulation of GFP-SRY-HMG (left) or GFP-SRY-full-length (FL) fusion proteins with or without sex-reversing nuclear localization signal (NLS) mutations in transfected COS-7 cells. **(A)** Cells were transfected to express the indicated GFP-SRY-fusion expression constructs, or GFP alone as a control, and imaged live 24 h later using a BioRAD MRC500 CLSM with a ×40 water immersion objective. **(B)** The nuclear-to-cytoplasmic ratio (Fn/c), representing the ratio between the nuclear (Fn) and cytoplasmic (Fc) fluorescence subsequent to subtraction of background fluorescence, was determined by image analysis, with results shown for the mean +/− SEM, from a single typical experiment ($n > 21$). Values above the histograms denote the p-values for the test of significance for the Fn/c values for the mutant derivatives compared to that of wild type (WT).

3. The nucleocytoplasmic ratio (Fn/c) for each cell is calculated using the equation $Fn/c = (Fn - Fb)/(Fc - Fb)$.
4. For each construct, the Fn/c values are pooled and statistical significance of the differences between WT and NLS mutated constructs determined using the InStat software (or equivalent).

Wild-type SRY fusion constructs accumulate essentially in the nucleus to levels greater than 70-fold those in the cytoplasm as determined by quantitative analysis (**Fig. 3B**), with GFP alone showing equal levels in the nucleus and cytoplasm ($Fn/c \sim 1.5$). No significant difference in the level of nuclear accumulation is observed between the SRY-HMG domain or full-length SRY, implying that the HMG domain with its two NLSs accounts for SRY's nuclear targeting abilities. The sex-reversing N- and C-NLS mutated derivatives showed significantly ($p < 0.009$) reduced levels of nuclear accumulation, indicating that both NLSs contribute to SRY nuclear accumulation. These results also imply the critical function of nuclear localization of SRY, where threshold levels of SRY in the nucleus are essential for sex determination.

Notes

1. Reticulocyte lysate should be made up to 150 mg/mL by diluting with 1x IB, and then aliquoted and stored in liquid nitrogen.
2. It is also possible to use proteins labeled covalently with fluorescent dyes rather than only GFP-tagged fusion proteins.
3. In these transfection experiments, Gateway™ recombination technology was employed in order to generate SRY-HMG and full-length (FL) GFP-fusion plasmids for mammalian expression. Both wild-type and mutant derivatives were studied, including the sex-reversing N-terminal NLS mutation R62G and the C-terminal NLS mutation R133W. The sex-reversed patients carrying these mutations have 46,XY gonadal dysgenesis, where the patient carrying the R62G mutation presented with streak gonads and gonadoblastoma, and of two patients carrying the R133W mutations, both presented with streak gonads, but only one with bilateral gonadoblastoma *(12)*.
4. Expression clones encoding His-SRY-GFP, mImpα-GST, and mImpβ-GST are available on request from Prof. David Jans.
5. Generally, splitting confluent cells that are greater than 95% confluent gives poor quality cells for use in the in vitro assay and should be avoided.
6. If cells are 80% confluent, resuspend in 4 mL.
7. Add 110 μL of cells if you plan to use cells in the afternoon. Note: Growth of cells for more or less than 2 d greatly affects the assay and is not recommended.
8. It may be necessary to optimize the perforation procedure to maximize the number of perforated cells yielding intact, dextran-excluding nuclei.

9. Cells can also be perforated using chemical permeabilization rather than mechanical perforation. Permeabilizing method: prepare permeabilizing buffer (PB): 15 mM Tris-HCl pH 7.5, 60 mM KCl, 15 mM NaCl, 5 mM MgCl$_2$, 0.5 mM EGTA, 300 mM sucrose, and filter-sterilize. Add to 10 mL permeabilizing buffer in glass tube 0.025% lysolecithin (2.5 mg). Dilute 1 in 5 (0.005% lysolecethin) and place 4–5 mL in Petri dish at 4 °C (on ice). Place cover slip in lysolecithin/PB solution for 5 min. Blot dry and replace in IB briefly. Blot dry and place face down on slide.

10. In inhibition experiments, cytosolic extract was preincubated at room temperature for 15 min with antibody (40 mg/mL) specific to Impα or Impβ (Santa Cruz Biotechnology).

11. Image optimization. Set up a sample containing a 70 kDa FITC-labeled dextran (Sigma) alone (PMT2 = green channel) and adjust gain in the green panel (PMT2). Then set up a second sample containing a 70 kDa Texas red-labelled dextran alone (PMT1 = red channel) and monitor bleed-through from red into the green channel. Set maximum gain on PMT2 that shows essentially no signal from the red dextran.

12. We use the Kaleidagraph™ 2.13 software (Macintosh), but any comparable software that allows equation input and variable output can be used.

Acknowledgments

This work was supported by the Australian National Health and Medical Research Council (Fellowship and Project grants ID#333013/384105 and #143710, respectively, to DAJ).

References

1. Lusk, C. P., Makhnevych, T., and Wozniak, R. W. (2004) New ways to skin a kap: mechanisms for controlling nuclear transport. *Biochem. Cell. Biol.* **82,** 618–625.

2. Poon, I. K. and Jans, D. A. (2005) Regulation of nuclear transport: central role in development and transformation? *Traffic* **6,** 173–186.

3. Izaurralde, E. and Adam, S. (1998) Transport of macromolecules between the nucleus and the cytoplasm. *RNA* **4,** 351–364.

4. Jans, D. A. (1995) The regulation of protein transport to the nucleus by phosphorylation. *Biochem. J.* **311,** 705–716.

5. Matsuura, Y. and Stewart, M. (2004) Structural basis for the assembly of a nuclear export complex. *Nature* **432,** 872–877.

6. Goodfellow, P. N. and Lovell-Badge, R. (1993) SRY and sex determination in mammals. *Annu. Rev. Genet.* **27,** 71–92.

7. Haqq, C. M., King, C. Y., Ukiyama, E., et al. (1994) Molecular basis of mammalian sexual determination: activation of Mullerian inhibiting substance gene expression by SRY. *Science* **266,** 1494–1500.

8. Laudet, V., Stehelin, D., and Clevers, H. (1993) Ancestry and diversity of the HMG box superfamily. *Nucleic Acids Res.* **21,** 2493–2501.

9. Werner, M. H., Huth, J. R., Gronenborn, A. M., and Clore, G. M. (1995) Molecular basis of human 46X,Y sex reversal revealed from the three-dimensional solution structure of the human SRY-DNA complex. *Cell* **81,** 705–714.

10. Ross, A. J. and Capel, B. (2005) Signaling at the crossroads of gonad development. *Trends Endocrinol. Metab.* **16,** 19–25.

11. Forwood, J. K., Harley, V., and Jans, D. A. (2001) The C-terminal nuclear localization signal of the sex-determining region Y (SRY) high mobility group domain mediates nuclear import through importin beta 1. *J. Biol. Chem.* **276,** 46575–46582.

12. Harley, V. R., Layfield, S., Mitchell, C. L., et al. (2003) Defective importin beta recognition and nuclear import of the sex-determining factor SRY are associated with XY sex-reversing mutations. *Proc. Natl. Acad. Sci. USA* **100,** 7045–7050.

13. Sim, H., Rimmer, K., Kelly, S., Ludbrook, L. M., Clayton, A. H., and Harley, V. R. (2005) Defective calmodulin-mediated nuclear transport of the sex-determining region of the Y chromosome (SRY) in XY sex reversal. *Mol. Endocrinol.* **19,** 1884–1892.

14. Sudbeck, P. and Scherer, G. (1997) Two independent nuclear localization signals are present in the DNA-binding high-mobility group domains of SRY and SOX9. *J. Biol. Chem.* **272,** 27848–27852.

7

The Mitochondrial Machinery for Import of Precursor Proteins

Kipros Gabriel and Nikolaus Pfanner

Summary

Mitochondria contain a small genome that codes for approx 1% of the total number of proteins that reside in the mitochondria. Hence, most mitochondrial proteins are encoded for by the nuclear genome. After transcription in the nucleus these proteins are synthesized by cytosolic ribosomes. Like proteins destined for other organellar compartments, mitochondrially destined proteins possess targeting signals within their primary or secondary structure that direct them to the organelle with the assistance of cytosolic factors. Very specialized and discriminatory protein translocase complexes in the mitochondrial membranes, intermembrane space, and matrix are then engaged for the translocation, sorting, integration, processing, and folding of the newly imported proteins. The principles of protein targeting into mitochondria have been and are still being unraveled, mostly by studies with the yeast *Saccharomyces cerevisiae* and the fungus *Neurospora crassa*. In this chapter the major principles of mitochondrial protein targeting as currently understood will be discussed as a foundation for the experimental methods discussed later in this volume.

Key Words: Mitochondrial protein import; translocase of the outer mitochondrial membrane; translocase of the inner mitochondrial membrane; sorting and assembly machinery; presequence translocase-associated motor; mitochondrial intermembrane space import and assembly; preprotein; signal sequence.

1. Introduction

Mitochondria have long been labeled as "the powerhouses of the cell" because of their involvement in the production of ATP, but recent advances in cell and molecular biology have revealed that mitochondrial function is much

From: *Methods in Molecular Biology, Vol. 390: Protein Targeting Protocols: Second Edition*
Edited by: M. van der Giezen © Humana Press Inc., Totowa, NJ

more diverse than this initial label suggests. In addition to the Krebs cycle and oxidative phosphorylation, mitochondria are crucial for heme biosynthesis and the metabolism of amino acids, lipids, and iron. It has also been established that mitochondria play integral roles in apoptosis *(1)* and that mitochondrial defects can lead to many neurodegenerative diseases in higher eukaryotes *(2)*. It is therefore not surprising that mitochondrial biogenesis is the focus of many laboratories worldwide and that scientists not usually conducting research in this field may also want to study the biogenesis of a mitochondrial protein that is of interest to them.

There are four destinations to which mitochondrial proteins can be targeted: the outer membrane, the intermembrane space, the inner membrane, and the matrix. Elaborate protein translocating and folding machineries exists in all four of the mitochondrial subcompartments to either integrate or translocate an incoming protein into or across the compartment (**Fig. 1**). The first membranous barrier for an incoming preprotein destined for any of the mitochondrial compartments is the outer mitochondrial membrane, where the first point of contact is the translocase of the outer mitochondrial membrane, the TOM complex *(3–5)*. From there a preprotein must be sorted to the appropriate complex in order to be targeted and integrated into the correct submitochondrial location *(6)*. Polytopic proteins of the outer mitochondrial membrane are passed on to the sorting and assembly machinery, the SAM complex *(7–9)*. Classical preproteins with N-terminal cleavable signal sequences destined for the matrix are passed on to the presequence translocase of the inner mitochondrial membrane, the TIM23 complex *(3,9–15)*. The TIM23 complex translocates proteins into the matrix with the assistance of the presequence translocase-associated motor, the PAM complex *(9,16,17)*. Polytopic inner membrane proteins such as the carrier proteins are transported, with the help of some intermembrane space chaperone-like small TIM proteins, to the carrier protein translocase of the inner mitochondrial membrane, the TIM22 complex *(18)*. The soluble proteins of the intermembrane space use a newly identified pathway for their import and assembly, the mitochondrial intermembrane space import and assembly pathway, or MIA pathway *(19)* (**Fig. 1**). All of these protein machineries have been described in a variety of organisms, and their components are conserved throughout the eukaryotic kingdom. **Table 1** shows the name of each component required for mitochondrial protein import and maturation in *S. cerevisiae* followed by the aliases or alternate names of the homolog proteins in other organisms that have arisen in the literature (for further details, *see* the Saccharomyces Genome Database, http://yeastgenome.org).

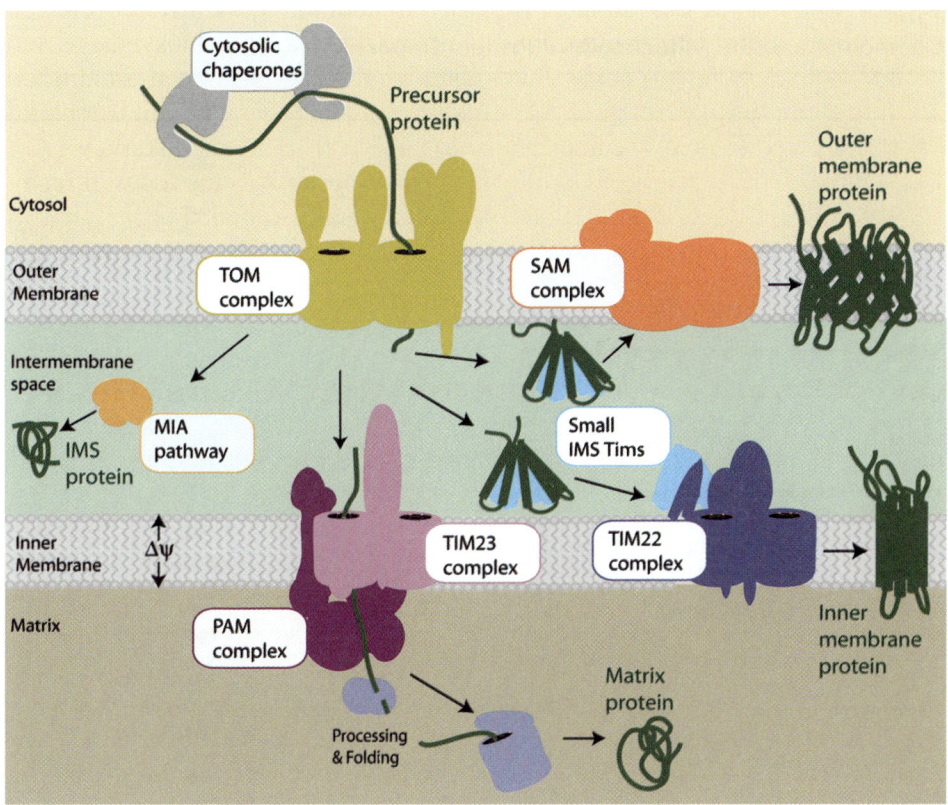

Fig. 1. The mitochondrial protein import pathways. Most mitochondrial proteins are synthesized in the cytosol and targeted to the mitochondria with the help of chaperones and other factors. All preproteins interact with the translocase of the outer mitochondrial membrane (TOM complex) before being sorted to the correct proceeding translocase complex. Polytopic outer membrane proteins are transferred to the sorting and assembly machinery (SAM complex). Intermembrane space (IMS)-destined proteins with cysteine repeat motifs are sorted to the mitochondrial IMS import and assembly (MIA) pathway and folded and assembled in the IMS. Proteins destined for the mitochondrial matrix are passed on to the presequence translocase of the inner membrane (TIM23 complex), where they are translocated into the matrix with the help of the presequence translocase-associated motor (PAM). Polytopic inner membrane proteins such as the carrier proteins are transported with the assistance of the soluble intermembrane space TIMs to the TIM22 complex, which integrates and folds these proteins in the membrane. Protein translocation into or across the inner membrane requires a membrane potential ($\Delta\psi$).

Table 1
Components of the Mitochondrial Protein Import Machinery

Name	Aliases
TOM complex	
Tom5	Mom8a
Tom6	Isp6, Mom8b
Tom7	Mom7, Yok22
Tom20	Mas20, Mom19, Pom13, Rir16
Tom22	Mas17, Mas22, Mom22
Tom40	Isp42, Mom38
Tom70	Mas70, Mom72, Omp1
Tom72	Tom71
Sorting and assembly machinery of the outer membrane	
Sam35	Fmp20, Tob38, Tom38
Sam37	Mas37, Pet3027, Tom37
Sam50	Omp85, Tob55
Mdm10	Fun37
Mim1	Tom13
Tim proteins of the intermembrane space	
Tim8	—
Tim9	—
Tim10	Mrs11, Ddp1
Tim13	—
Mitochondrial IMS assembly machinery	
Mia40	Fmp15, Tim40
Erv1	—
Hot13	—
TIM23 complex	
Tim17	Mim17, Mpi2, Sms1
Tim21	—
Tim23	Mas6, Mim23, Mpi3
Tim50	—
PAM complex	
Ssc1	mtHsp70, Ens1
Tim44	Isp45, Mim44, Mpi1
Pam16	Mia1, Tim16
Pam17	Fmp18
Pam18	Tim14
Mdj2	—
Mge1	Yge1
TIM22 complex	
Tim12	Mrs5
Tim18	—
Tim22	—
Tim54	—
Processing enzymes	
Mppα	Mas2, Mif2
Mppβ	Mas1, Mif1, Pep
Oct1	Mip1
Imp1	—
Imp2	—
Som1	—
Pcp1	Mdm37, Rbd1
Yta10	Afg3
Yta12	Rca1

2. The Beginning: Synthesis and Targeting in the Cytosol

The mitochondrial genome codes for only a handful of proteins; the remaining approx 1000 mitochondrial proteins are encoded for by the nuclear genome and synthesized in the cytosol with targeting motifs within their primary or secondary structure. There are eight currently identified classes of mitochondrial preproteins (**Fig. 2**), not all of which have clearly defined consensus

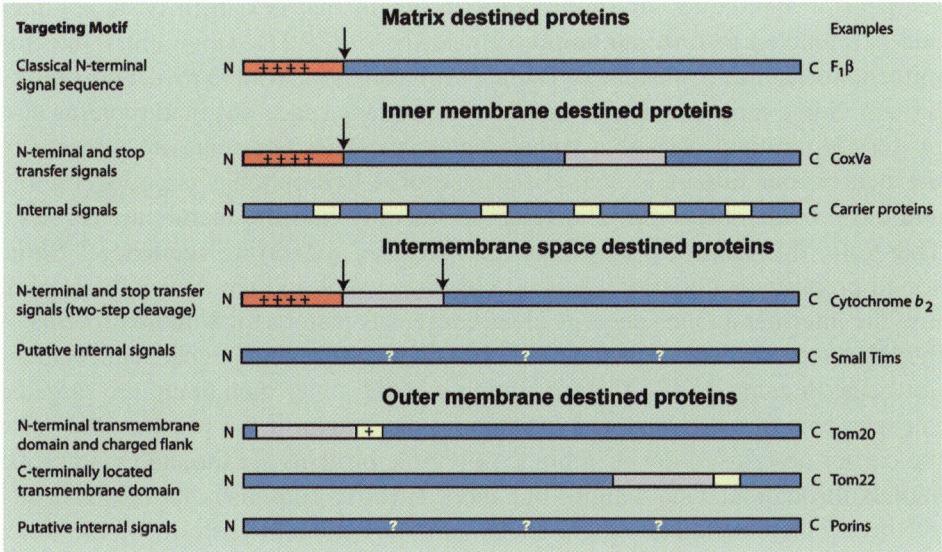

Fig. 2. Various classes of mitochondrial preproteins. Proteins destined for the mitochondrial matrix possess a cleavable N-terminal signal sequence. The signal sequence is predicted to form a positively charged amphipathic helix. Inner membrane proteins can be targeted via an N-terminal cleavable signal followed by a hydrophobic stop-transfer sorting signal in the case of single trans-membrane domain proteins or by more cryptic hydrophobic stretches throughout the length of the proteins, as is the case for carrier proteins. Intermembrane space proteins can be targeted via an N-terminal cleavable signal followed by a hydrophobic stop transfer signal, but a second cleavage site results in release into the intermembrane space. Other cysteine-rich intermembrane space proteins can be targeted via internal motifs. Three classes of outer membrane proteins exist. The single transmembrane domain proteins, both N- and C-terminally anchored, are targeted via their transmembrane domain and some charged residues flanking these domains. The polytopic outer membrane proteins, such as porins, possess internal cryptic signals throughout their length that are poorly characterized. Arrows denote sites of proteolytic cleavage.

motifs. Therefore, it is often difficult for a researcher to determine if a newly identified open reading frame encodes a protein that is destined for a mitochondrial compartment using bioinformatics-based approaches alone. An in vitro mitochondrial import reaction or subcellular fractionation experiment may be required to determine if a protein is indeed mitochondrial and, more specifically, in which mitochondrial compartment it resides. Both of these procedures will be described in the following experimental sections.

The classically described mitochondrial targeting sequence is an N-terminal segment of approx 15–70 amino acids in length that is rich in basic residues and is predicted to form an amphipathic α-helix *(20,21)*. Upon entry into the mitochondrial matrix the charged signal sequence is cleaved to reveal the mature protein. Some inner membrane and intermembrane space-destined proteins also possess N-terminal cleavable signal sequences. These proteins are targeted to the matrix, but import is stalled because of a hydrophobic targeting region, and the proteins are sorted and released laterally into the membrane *(22–24)*. Therefore, this targeting motif has been termed a "sorting sequence." Some sorted preproteins undergo a second cleavage event and are therefore released into the intermembrane space. A clear consensus pattern for soluble proteins of the intermembrane space that do not contain an N-terminal signal sequence has not been elucidated *(25)*. Carrier proteins of the inner membrane are targeted via segments that are scattered throughout the length of the protein *(26)*. Like the carrier proteins, outer membrane polytopic proteins are targeted via cryptic signals throughout their length, but more specific consensus motifs have not yet been elucidated. The single transmembrane domain proteins of the outer membrane are targeted via their transmembrane domain and some charged residues flanking the transmembrane domain *(27)*.

With the assistance of nucleotide-dependent cytosolic chaperones such as heat shock proteins of the 70 kDa class (Hsp70s) and targeting factors, such as presequence binding factor (PBF) and mitochondrial import stimulating factor (MSF), the mitochondrially targeted proteins are transported to the mitochondrial surface in an import competent state *(28)*. It is still unknown whether mRNA signals can direct the ribosomes to the mitochondrial surface before translation in the cytosol even begins *(29)*, but most proteins can be efficiently imported post translationally in vitro.

3. First Port: The TOM Complex

At the mitochondrial surface, the first point of contact for all classes of mitochondrial proteins is the TOM complex. The TOM complex must therefore have the ability to recognize a diverse range of preprotein signals and yet retain specificity to avoid binding preproteins destined for other cellular compartments. This is achieved via the receptor subunits Tom20, Tom22, and Tom70 *(26,30–33)* **(Fig. 3)**. Tom20 acts as the initial docking point for a majority of proteins with an N-terminal classical signal sequence, to which it can bind via the hydrophobic portion of the signal sequence's amphipathic helix *(34)*. Tom22 also binds this class of preproteins, using electrostatic interactions with the basically charged residues of the signal sequence *(35,36)*. Tom22 is more integrally associated with the pore of the translocase and is important for the stability of the TOM complex *(37,38)*. A nucleotide-consuming member of the TOM complex has not been identified, and so it has been hypothesized that a series of preprotein binding sites with increasing affinity across the outer membrane provides the motive force for the transfer of proteins across the outer membrane to the intermembrane space. The amino and carboxy-terminal

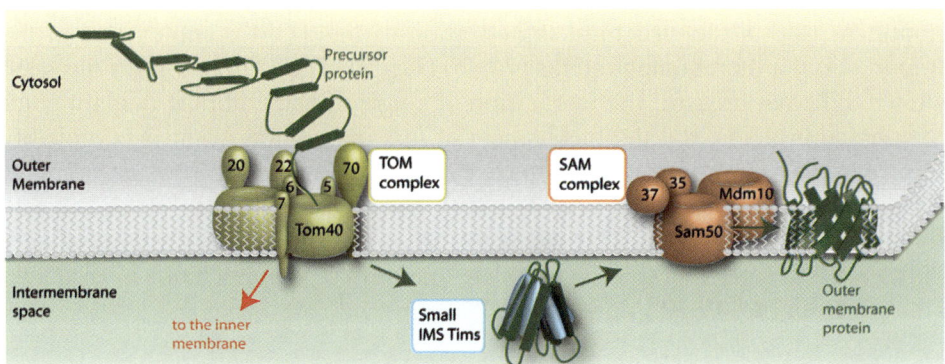

Fig. 3. The import and translocation machineries of the outer membrane. The TOM complex is composed of receptor and core subunits. The receptors Tom70 and Tom20 bind incoming preproteins and pass them to the more integrally associated Tom22, which feeds them into the pore formed by Tom40. The intermembrane space (IMS) domain of Tom22 binds presequences in the IMS. The small TOM proteins Tom5, Tom6, and Tom7 modulate the configuration of the TOM complex. Import of polytopic outer membrane proteins requires shuttling, with the assistance of Tim8–Tim13 and Tim9–Tim10 complexes of the IMS, to a second translocase machinery, the sorting and assembly machinery (SAM) complex. Currently four components of the SAM complex have been identified: Sam50, Sam37, Sam35, and Mdm10. Sam50 and Mdm10 are predicted to form β-barrels. Sam37 and Sam35 are peripheral outer membrane proteins that face the cytosol. The functional mechanism that the SAM utilizes is not yet understood.

acidic domains of Tom22 have been predicted to contribute to this binding chain *(35,36)*. Both Tom20 and Tom22 are important for the import of internal-destined mitochondrial proteins but also for integral outer membrane proteins *(39,40)*.

The largest of the TOM receptor subunits, Tom70, has been shown to have specificity for polytopic carrier proteins destined for the inner mitochondrial membrane *(31)*. Preproteins in this class are very hydrophobic, and it has been shown that these proteins have a high dependence on cytosolic chaperones for their delivery to the outer mitochondrial membrane in an import competent state. Tom70 interacts with a complex consisting of a preprotein and Hsp70 in yeast and additionally Hsp90 in mammals *(41)*. Tom70 most likely then passes proteins to Tom20 before they are passed on to the TOM pore for translocation. A homolog of Tom70, Tom72, has also been described but has not been assigned a function *(42,43)*.

Tom40, a β-barrel protein, is essential and forms the pore of the TOM complex *(44,45)*. The small TOM proteins Tom5, Tom6, and Tom7 play an important role in the structural organization of the TOM complex and in the import of outer membrane proteins *(40,46)* (**Fig. 3**). The TOM complex has been shown to be required for the integration of single transmembrane domain outer membrane proteins, but the mechanism of this integration event is completely unknown *(47)*. Tom5 is involved in passing preproteins to the pore, and it plays a role in the biogenesis of Tom40 itself. The TOM complex forms two and three-ring structures. The TOM channels have a pore diameter of approximately 2 nm, a width sufficient to accommodate one or two peptide chains in α-helical conformation *(45,48,49)*.

4. Integration into the Outer Membrane: The SAM Complex

Polytopic outer membrane proteins such as Tom40 and the outer membrane porins, which possess multiple β-strands that form β-barrel structures, require initial reception and translocation by the TOM complex and then the action of a second outer membrane protein SAM complex *(7,8)*. After translocation of the incoming pre-β-barrel proteins through the TOM channel, the small soluble inter-membrane space chaperones Tim9–Tim10 and Tim8–Tim13 are involved in the transfer of the preproteins to the SAM complex *(50,51)*. The major constituent of the SAM complex is Sam50 *(52–54)*, a homologous protein to the bacterial Omp85 protein that functions to integrate proteins into the outer membrane of Gram-negative bacteria *(55)*. It is therefore likely that the SAM machinery has been conserved from the mitochondria's bacterial ancestry. Other components of the SAM complex include Sam37 and Sam35, two peripheral outer membrane proteins facing the cytosol and Mdm10 required for the import of Tom40 *(7,56–59)* **(Fig. 3)**. The SAM complex is based around the β-barrel protein, Sam50, and structural constraints therefore rule out the possibility that the SAM complex may function in a manner similar to the SEC translocon. The SEC translocon takes advantage of its multimeric α-helical-based channel to allow membrane proteins to diffuse laterally out of the channel and into the membrane *(60,61)*. The mechanism utilized by the SAM complex is completely unknown, but one hypothesis put forward is that it may act as a molecular scaffold in the membrane that can promote the correct folding of β-barrel proteins.

5. Translocation into the Matrix: The TIM23 Complex and the Associated Motor

Once translocated across the outer membrane, preproteins with an N-terminal signal sequence destined for the matrix probably remain bound to the intermembrane space side of the TOM complex until the TIM23 presequence translocase is triggered into action *(6)*. Tim50 is involved in the detection of the incoming presequence preproteins, and it is able to pass the precursors to Tim23 *(62,63)*. In addition, Tim21 transiently binds to the TOM complex and promotes the release of a preprotein from the TOM *(14)*. Tim23 forms the channel of the presequence translocase *(64)*, Another integral inner membrane protein involved in inner membrane sorting, Tim17, remains closely associated to Tim23 *(14)*. Translocation of the N-terminal signal sequence across the inner membrane depends on the electrochemical gradient that exists across the inner membrane (mitochondrial membrane potential, $\Delta\psi$ *(65)*. Ionophores can efficiently block this step of import in vitro. The TIM23 channel is based on α-helices and therefore has the potential to allow proteins to move laterally into the membrane from the complex. In some instances where a preprotein has an N-terminal signal sequence followed by a hydrophobic stop transfer signal, this actually does occur and has been termed "lateral sorting." When a protein requiring sorting is passed to the TIM23 complex, Tim21 remains bound, and therefore the PAM complex is not recruited to the complex *(14)*. Sorted proteins can undergo a series of proteolytic modifications during their import. Some proteins have their N-terminal signal sequence removed by the mitochondrial processing peptidase (MPP) in the matrix, probably while still spanning the membrane in the channel formed by the TIM23 complex. Intermembrane space destined sorted proteins are then released from the membrane after a second cleavage event instigated by the inner membrane protease or the Imp1/Imp2/Som1 complex *(66–68)*. The mitochondrial AAA protease subunits Yta10 and Yta12 and the Pcp1 protease (processing of cytochrome *c* peroxidase) have also been reported to be important for the maturation of preproteins in the sorting pathway *(70,71)*. However the majority of proteins destined for the TIM23 complex are destined for full translocation into the matrix. For these proteins the PAM must be engaged.

The PAM complex acts to fully translocate presequence proteins into the matrix. Tim44 is thought to stabilize the interaction between the TIM23 complex and the mitochondrial Hsp70, Ssc1, and its nucleotide exchange factor, Mge1 *(72–74)*. It is still disputed as to whether the PAM acts to

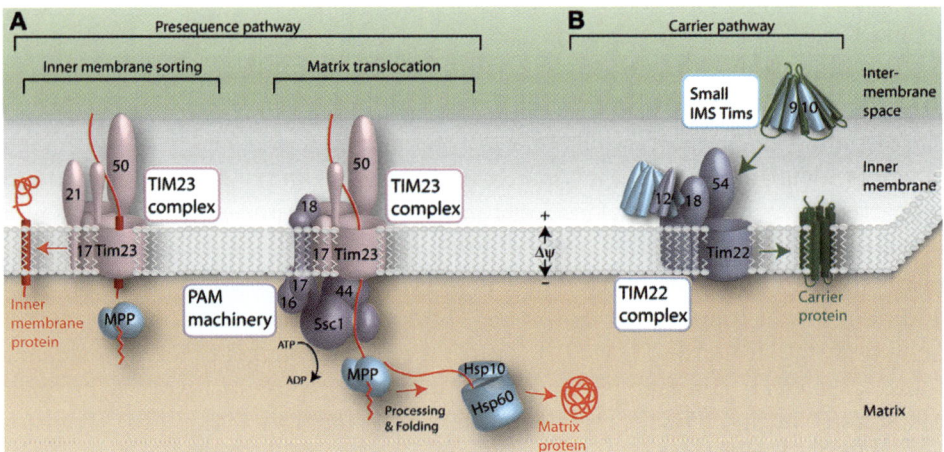

Fig. 4. The translocases of the inner mitochondrial membrane. **(A)** Proteins destined for the mitochondrial matrix are passed from the TOM complex to the TIM23 complex. Tim50 is involved in the initial binding of the preproteins, which it passes to Tim23. Tim23 forms the channel of the complex. Tim17 is closely associated with Tim23. The TIM23 complex switches between two states: a Tim21-bound form, which sorts proteins to the inner membrane, and a PAM-bound form. The PAM is recruited to the TIM23 complex when a protein is destined for matrix translocation. The PAM consists of the Tim44 subunit that recruits Hsp70 (Ssc1) and its nucleotide exchange factor (Mge1) to the translocation site. The integral membrane J protein Pam18 stimulates the activity of Ssc1. Pam16 acts in an antagonistic manner to Pam18. The action exerted by the PAM acts to translocate proteins into the matrix. **(B)** Polytopic inner membrane proteins are translocated through the TOM channel and rapidly bound by the intermembrane space soluble chaperone complexes, Tim9–Tim10 and Tim8–Tim13. These chaperones transport the hydrophobic preproteins to the inner membrane. At the inner membrane these proteins are bound and translocated into the inner membrane by the TIM22 complex. Tim22 forms a twin pore translocase; however, the roles of the peripherally associated Tim12 protein and the integral inner membrane proteins Tim18 and Tim54 are unknown.

translocate proteins using a trapping mechanism or physical pulling mechanism. The trapping model would involve Ssc1 release from the PAM after binding the incoming precursor and therefore allowing diffusion into the matrix and allowing a new Ssc1 molecule to bind closer to the C-terminus, and a repetition of the cycle would result in net inward movement. The pulling model relies on Tim44 to act as a fulcrum for Ssc1 so that after a conformational change (e.g., after ATP binding or hydrolysis) a physical force is exerted to pull proteins into the matrix *(75)*. It is conceivable that both mechanisms play a role in vivo. The ATPase activity of Ssc1 is stimulated by the membrane-bound J-protein Pam18 and can also be stimulated, when overexpressed in a Pam18 deletion background strain, by Mdj2 *(76–79)*. Pam16 can act as an antagonist for this reaction, probably for regulatory purposes *(80)*. A new component Pam17 that stabilizes the interaction between Pam18 and Pam16 has also been recently identified *(81)*. After translocation into the matrix MPP cleaves off the N-terminal signal sequences from the preproteins. Some preproteins require the removal of a further octapeptide by the mitochondrial intermediate peptidase (Mip1/Oct1) *(69)*. The proteins are folded to their native conformation with the assistance of molecular chaperones such as Ssc1, Ssq1, and Chaperonins 60 and 10 *(82)* **(Fig. 4)**.

6. Insertion of Polytopic Inner Membrane Proteins: The TIM22 Pathway

Inner membrane carrier proteins are a highly abundant class of proteins and may represent up to 5% of the total mitochondrial protein mass. After translocation across the outer membrane, these polytopic inner membrane proteins without an N-terminal cleavable signal sequence diverge in their import path from those with cleavable signals sequences *(83)*. After protrusion into the intermembrane space, the soluble small TIM proteins of the intermembrane space rapidly bind the polytopic preproteins *(84)*. The Tim9–Tim10 complex acts not only to shuttle these hydrophobic multimembrane-spanning proteins to the inner membrane, but also to prevent their aggregation *(85–91)*. At the inner membrane the preproteins are handed over to the TIM22 complex *(92)* **(Fig. 4)**. The TIM22 complex is made up of four identified components: Tim18, Tim22, Tim54 and the peripherally associated Tim12. Tim12 may provide a structural link between the soluble intermembrane space small Tim proteins, Tim9 and Tim10, which can also be found stably associated with the TIM22 complex *(85)*. Tim54 is the largest of the TIM22 complex components, and its large intermembrane space domain may bind precursor proteins in a receptor-like fashion *(93)*. Tim22 shares sequence homology with Tim23 and also forms a twin pore α-helical based channel. The gating action of the channel is triggered by the cryptic signals within the carrier proteins *(94)*. The function of Tim18 is unknown *(95)*.

7. Biogenesis of Intermembrane Space Proteins: The MIA Pathway

Very recent findings have revealed a novel import pathway specific for the import and assembly of mitochondrial intermembrane space proteins— the MIA pathway *(19,96,97)*. A consensus pattern for substrate proteins of the MIA pathway is beginning to emerge; substrate proteins contain cysteine repetition motifs. These motifs are not only important for the tertiary folding or the binding of metal cofactors for these proteins, but also to form stable intermediates with constituents of the MIA machinery during their import and assembly *(19,98)*. Mia40 was the first identified component of the MIA import pathway. It was shown that the in vitro synthesized and imported preproteins of the small Tim proteins, Cox17 and Cox19, can form stable intermediates with Mia40 via disulfide bonds and can be visualized by Blue Native polyacrylamide gel electrophoresis (BN-PAGE) *(19)*. The flavin-linked sulfhydryl oxidase Erv1 and the protein Hot13 were also recently shown to be important in the MIA pathway *(98–101)*. The mechanism engaged by the MIA pathway is still unclear and will probably be the focus of intense research in various laboratories in the coming years.

8. Concluding Remarks

Our understanding of the machineries involved in mitochondrial biogenesis has improved tremendously in the last 20 yr. However, many of the underlying mechanisms of protein import remain poorly understood. For example, many components of the SAM and MIA machineries have been identified, but their roles are largely unknown. Although many laboratories are beginning to utilize mammalian systems to elucidate some of the mechanisms, the yeast and fungal model organisms will remain the principal organisms for mitochondrial biogenesis research for years to come.

Acknowledgments

We thank Nils Wiedemann for critical comments on the chapter. Work in the authors' laboratory is supported by the Deutsche Forschungsgemeinschaft, the Sonderforschungsbereich 388, the Gottfried Wilhelm Leibniz Program, the Max Planck Research Award, the Bundesministerium für Bildung und Forschung, and the Fonds der Chemischen Industrie. K.G. is a recipient of an Alexander von Humboldt Foundation research fellowship.

References

1. Green, D. R. (2005) Apoptotic pathways: ten minutes to dead. *Cell* **121**, 671–674.
2. Beal, M. F. (2005) Mitochondria take center stage in aging and neurodegeneration. *Ann. Neurol.* **58**, 495–505.
3. Neupert, W. (1997) Protein import into mitochondria. *Annu Rev Biochem.* **66**, 863–917.
4. Dekker, P. J., Ryan, M. T., Brix, J., Müller, H., Hönlinger, A., and Pfanner, N. (1998) Preprotein translocase of the outer mitochondrial membrane: molecular dissection and assembly of the general import pore complex. *Mol. Cell Biol.* **18**, 6515–6524.
5. Taylor, R. D. and Pfanner, N. (2004) The protein import and assembly machinery of the mitochondrial outer membrane. *Biochim. Biophys. Acta* **1658**, 37–43.
6. Gabriel, K., Egan, B., and Lithgow, T. (2003) Tom40, the import channel of the mitochondrial outer membrane, plays an active role in sorting imported proteins. *EMBO J.* **22**, 2380–2386.
7. Wiedemann, N., Kozjak, V., Chacinska, A., et al. (2003) Machinery for protein sorting and assembly in the mitochondrial outer membrane. *Nature* **424**, 565–571.
8. Pfanner, N., Wiedemann, N., Meisinger, C., and Lithgow T. (2004) Assembling the mitochondrial outer membrane. *Nat. Struct. Mol. Biol.* **11**, 1044–1048.
9. Wiedemann, N., Frazier, A. E., and Pfanner, N. (2004) The protein import machinery of mitochondria. *J. Biol. Chem.* **279**, 14473–14476.
10. Jensen, R. E. and Dunn, C. D. (2002) Protein import into and across the mitochondrial inner membrane: role of the TIM23 and TIM22 translocons. *Biochim. Biophys. Acta* **1592**, 25–34.
11. Chacinska, A., Rehling, P., Guiard, B., et al. (2003) Mitochondrial translocation contact sites: separation of dynamic and stabilizing elements in formation of a TOM-TIM-preprotein supercomplex. *EMBO J.* **22**, 5370–5381.
12. Endo, T., Yamamoto, H., and Esaki, M. (2003) Functional cooperation and separation of translocators in protein import into mitochondria, the double-membrane bounded organelles. *J. Cell Sci.* **116**, 3259–3267.
13. Koehler, C. M. (2004) New developments in mitochondrial assembly. *Annu Rev Cell Dev Biol.* **20**, 309–335.
14. Chacinska, A., Lind, M., Frazier, A. E., et al. (2005) Mitochondrial presequence translocase: switching between TOM tethering and motor recruitment involves Tim21 and Tim17. *Cell* **120**, 817–829.
15. Oka, T. and Mihara, K. (2005) A railroad switch in mitochondrial protein import. *Mol. Cell* **18**, 145–146.
16. Geissler, A., Rassow, J., Pfanner, N., and Voos, W. (2001) Mitochondrial import driving forces: enhanced trapping by matrix Hsp70 stimulates translocation and reduces the membrane potential dependence of loosely folded preproteins. *Mol. Cell Biol.* **21**, 7097–7104.

17. Truscott, K. N., Voos, W., Frazier, A. E., et al. (2003) A J-protein is an essential subunit of the presequence translocase-associated protein import motor of mitochondria. *J. Cell Biol.* **163**, 707–713.
18. Rehling, P., Brandner, K., and Pfanner, N. (2004) Mitochondrial import and the twin-pore translocase. *Nat. Rev. Mol. Cell Biol.* **5**, 519–530.
19. Chacinska, A., Pfannschmidt, S., Wiedemann, N., et al. (2004) Essential role of Mia40 in import and assembly of mitochondrial intermembrane space proteins. *EMBO J.* **23**, 3735–3746.
20. Roise, D. and Schatz, G. (1988) Mitochondrial presequences. *J. Biol. Chem.* **263**, 4509–4511.
21. von Heijne, G., Steppuhn, J., and Herrmann, R.G. (1989) Domain structure of mitochondrial and chloroplast targeting peptides. *Eur. J. Biochem.* **180**, 535–545.
22. van Loon, A. P. and Schatz, G. (1987) Transport of proteins to the mitochondrial intermembrane space: the 'sorting' domain of the cytochrome c_1 presequence is a stop-transfer sequence specific for the mitochondrial inner membrane. *EMBO J.* **6**, 2441–2448.
23. Gärtner, F., Bömer, U., Guiard, B. and Pfanner, N. (1995) The sorting signal of cytochrome b_2 promotes early divergence from the general mitochondrial import pathway and restricts the unfoldase activity of matrix Hsp70. *EMBO J.* **14**, 6043–6057.
24. Bömer, U., Meijer, M., Guiard, B., Dietmeier, K., Pfanner, N., and Rassow, J. (1997) The sorting route of cytochrome b_2 branches from the general mitochondrial import pathway at the preprotein translocase of the inner membrane. *J. Biol. Chem.* **272**, 30439–30446.
25. Herrmann, J. M. and Hell, K. (2005) Chopped, trapped or tacked—protein translocation into the IMS of mitochondria. *Trends Biochem. Sci.* **30**, 205–211.
26. Brix, J., Rüdiger, S., Bukau, B., Schneider-Mergener, J., and Pfanner, N. (1999) Distribution of binding sequences for the mitochondrial import receptors Tom20, Tom22, and Tom70 in a presequence-carrying preprotein and a non-cleavable preprotein. *J. Biol. Chem.* **274**, 16522–16530.
27. Egan, B., Beilharz, T., George, R., et al. (1999) Targeting of tail-anchored proteins to yeast mitochondria in vivo. *FEBS Lett.* **451**, 243–248.
28. Beddoe, T. and Lithgow, T. (2002) Delivery of nascent polypeptides to the mitochondrial surface. *Biochim Biophys Acta* **1592**, 35–39.
29. Marc, P., Margeot, A., Devaux, F., Blugeon, C., Corral-Debrinski, M., and Jacq, C. (2002) Genome-wide analysis of mRNAs targeted to yeast mitochondria. *EMBO Rep.* **3**, 159–164.
30. Söllner, T., Griffiths, G., Pfaller, R., Pfanner, N., and Neupert, W. (1989) MOM19, an import receptor for mitochondrial precursor proteins. *Cell* **59**, 1061–1070.

31. Söllner, T., Pfaller, R., Griffiths, G., Pfanner, N., and Neupert W. (1990) A mitochondrial import receptor for the ADP/ATP carrier. *Cell* **62**, 107–115.
32. Hines, V., Brandt, A., Griffiths, G., Horstmann, H., Brütsch, H., and Schatz, G. (1990) Protein import into yeast mitochondria is accelerated by the outer membrane protein MAS70. *EMBO J.* **9**, 3191–3200.
33. Lithgow, T., Junne, T., Suda, K., Gratzer, S., and Schatz G. (1994) The mitochondrial outer membrane protein Mas22p is essential for protein import and viability of yeast. *Proc. Natl. Acad. Sci. USA* **91**, 11973–11977.
34. Abe, Y., Shodai, T., Muto, T., et al. (2000) Structural basis of presequence recognition by the mitochondrial protein import receptor Tom20. *Cell* **100**, 551–560.
35. Bolliger, L., Junne, T., Schatz, G., and Lithgow, T. (1995) Acidic receptor domains on both sides of the outer membrane mediate translocation of precursor proteins into yeast mitochondria. *EMBO J.* **14**, 6318–6326.
36. Hönlinger, A., Kübrich, M., Moczko, M., et al. (1995) The mitochondrial receptor complex: Mom22 is essential for cell viability and directly interacts with preproteins. *Mol. Cell. Biol.* **15**, 3382–3389.
37. van Wilpe, S., Ryan, M. T., Hill, K., et al. (1999) Tom22 is a multifunctional organizer of the mitochondrial preprotein translocase. *Nature* **401**, 485–489.
38. Model, K., Prinz, T., Ruiz, T., et al. (2002) Protein translocase of the outer mitochondrial membrane: role of import receptors in the structural organization of the TOM complex. *J. Mol. Biol.* **316**, 657–666.
39. Keil, P., Weinzierl, A., Kiebler, M., Dietmeier, K., Söllner, T., and Pfanner, N. (1993) Biogenesis of the mitochondrial receptor complex. Two receptors are required for binding of MOM38 to the outer membrane surface. *J. Biol. Chem.* **268**, 19177–19180.
40. Krimmer, T., Rapaport, D., Ryan, M. T., et al. (2001) Biogenesis of porin of the outer mitochondrial membrane involves an import pathway via receptors and the general import pore of the TOM complex. *J. Cell. Biol.* **152**, 289–300.
41. Young, J. C., Hoogenraad, N. J., and Hartl, F. U. (2003) Molecular chaperones Hsp90 and Hsp70 deliver preproteins to the mitochondrial import receptor Tom70. *Cell* **112**, 41–50.
42. Bömer. U., Pfanner, N. and Dietmeier, K. (1996) Identification of a third yeast mitochondrial Tom protein with tetratrico peptide repeats. *FEBS Lett.* **382**, 153–158.
43. Schlossmann, J., Lill, R., Neupert, W., and Court, D. A. (1996) Tom71, a novel homologue of the mitochondrial preprotein receptor Tom70. *J. Biol. Chem.* **271**, 17890–17895.
44. Baker, K. P., Schaniel, A., Vestweber, D., and Schatz, G. (1990) A yeast mitochondrial outer membrane protein essential for protein import and cell viability. *Nature* **348**, 605–609.
45. Hill, K., Model, K., Ryan, M.,T., et al. (1998) Tom40 forms the hydrophilic channel of the mitochondrial import pore for preproteins. *Nature* **395**, 516–521.

46. Model, K., Meisinger, C., Prinz, T., et al. (2001) Multistep assembly of the protein import channel of the mitochondrial outer membrane. *Nat. Struct. Biol.* **8**, 361–370.

47. Rapaport, D. (2005) How does the TOM complex mediate insertion of precursor proteins into the mitochondrial outer membrane? *J. Cell. Biol.* **171**, 419–423.

48. Künkele, K.,P., Heins, S., Dembowski, M., et al. (1998) The preprotein translocation channel of the outer membrane of mitochondria. *Cell* **93**, 1009–1019.

49. Schwartz, M. P. and Matouschek, A. (1999) The dimensions of the protein import channels in the outer and inner mitochondrial membranes. *Proc. Natl. Acad. Sci. USA* **96**, 13086–13090.

50. Hoppins, S. C. and Nargang, F. E. (2004) The Tim8-Tim13 complex of *Neurospora crassa* functions in the assembly of proteins into both mitochondrial membranes. *J. Biol. Chem.* **279**, 12396–12405.

51. Wiedemann, N., Truscott, K.,N., Pfannschmidt, S., Guiard, B., Meisinger, C., and Pfanner, N. (2004) Biogenesis of the protein import channel Tom40 of the mitochondrial outer membrane: intermembrane space components are involved in an early stage of the assembly pathway. *J. Biol. Chem.* **279**, 18188–18194.

52. Kozjak, V., Wiedemann, N., Milenkovic, D., et al. (2003) An essential role of Sam50 in the protein sorting and assembly machinery of the mitochondrial outer membrane. *J. Biol. Chem.* **278**, 48520–48523.

53. Paschen, S. A., Waizenegger, T., Stan, T., et al. (2003) Evolutionary conservation of biogenesis of β-barrel membrane proteins. *Nature* **426**, 862–866.

54. Gentle, I., Gabriel, K., Beech, P., Waller, R., and Lithgow, T. (2004) The Omp85 family of proteins is essential for outer membrane biogenesis in mitochondria and bacteria. *J. Cell Biol.* **164**, 19–24.

55. Voulhoux, R., Bos, M. P., Geurtsen, J., Mols, M., and Tommassen, J. (2003) Role of a highly conserved bacterial protein in outer membrane protein assembly. *Science* **299**, 262–265.

56. Milenkovic, D., Kozjak, V., Wiedemann, N., et al. (2004) Sam35 of the mitochondrial protein sorting and assembly machinery is a peripheral outer membrane protein essential for cell viability. *J. Biol. Chem.* **279**, 22781–22785.

57. Waizenegger, T., Habib, S. J., Lech, M., et al. (2004) Tob38, a novel essential component in the biogenesis of β-barrel proteins of mitochondria. *EMBO Rep.* **5**, 704–709.

58. Ishikawa, D., Yamamoto, H., Tamura, Y., Moritoh, K., and Endo, T. (2004) Two novel proteins in the mitochondrial outer membrane mediate β-barrel protein assembly. *J. Cell Biol.* **166**, 621–627.

59. Meisinger, C., Rissler, M., Chacinska, A., et al. (2004) The mitochondrial morphology protein Mdm10 functions in assembly of the preprotein translocase of the outer membrane. *Dev. Cell* **7**, 61–71.

60. Johnson, A. E. and van Waes, M. A. (1999) The translocon: a dynamic gateway at the ER membrane. *Annu. Rev. Cell Dev. Biol.* **15**, 799–842.

61. Sadlish, H., Pitonzo, D., Johnson, A. E., and Skach, W. R. (2005) Sequential triage of transmembrane segments by Sec61α during biogenesis of a native multispanning membrane protein. *Nat. Struct. Mol. Biol.* **12**, 870–878.

62. Geissler, A., Chacinska, A., Truscott, K. N., et al. (2002) The mitochondrial presequence translocase: an essential role of Tim50 in directing preproteins to the import channel. *Cell* **111**, 507–518.

63. Yamamoto, H., Esaki, M., Kanamori, T., Tamura, Y., Nishikawa, S., and Endo, T. (2002) Tim50 is a subunit of the TIM23 complex that links protein translocation across the outer and inner mitochondrial membranes. *Cell* **111**, 519–528.

64. Truscott, K. N., Kovermann, P., Geissler, A., et al. (2001) A presequence- and voltage-sensitive channel of the mitochondrial preprotein translocase formed by Tim23. *Nat. Struct. Biol.* **8**, 1074–1082.

65. Geissler, A., Krimmer, T., Börner, U., Guiard, B., Rassow, J., and Pfanner, N. (2000) Membrane potential-driven protein import into mitochondria. The sorting sequence of cytochrome b_2 modulates the $\Delta\psi$-dependence of translocation of the matrix-targeting sequence. *Mol. Biol. Cell* **11**, 3977–3391.

66. Schneider, A., Behrens, M., Scherer, P., Pratje, E., Michaelis, G., and Schatz, G. (1991) Inner membrane protease I, an enzyme mediating intramitochondrial protein sorting in yeast. *EMBO J.* **10**, 247–254.

67. Nunnari, J., Fox, T. D., and Walter, P. (1993) A mitochondrial protease with two catalytic subunits of nonoverlapping specificities. *Science* **262**, 1997–2004.

68. Esser, K, Pratje, E., and Michaelis, G. (1996) SOM 1, a small new gene required for mitochondrial inner membrane peptidase function in Saccharomyces cerevisiae. *Mol. Gen. Genet.* **252**, 437–445.

69. Arlt, H., Tauer, R., Feldmann, H., Neupert, W., and Langer, T. (1996) The YTA10-12 complex, an AAA protease with chaperone-like activity in the inner membrane of mitochondria. *Cell* **85**, 875–885.

70. Esser, K., Tursun, B., Ingenhoven, M., Michaelis, G., and Pratje, E. (2002) A novel two-step mechanism for removal of a mitochondrial signal sequence involves the mAAA complex and the putative rhomboid protease Pcp1. *J. Mol. Biol.* **323**, 835–843.

71. Schneider, H. C., Berthold, J., Bauer, M. F., et al. (1994) Mitochondrial Hsp70/MIM44 complex facilitates protein import. *Nature* **371**, 768–774.

72. Rassow, J., Maarse, A. C., Krainer, E., et al. (1994) Mitochondrial protein import: biochemical and genetic evidence for interaction of matrix hsp70 and the inner membrane protein MIM44. *J. Cell Biol.* **127**, 1547–1556.

73. Kronidou, N. G., Oppliger, W., Bolliger, L., et al. (1994) Dynamic interaction between Isp45 and mitochondrial Hsp70 in the protein import system of the yeast mitochondrial inner membrane. *Proc. Natl. Acad. Sci. USA* **91**, 12818–12822.

74. Strub, A., Lim, J. H., Pfanner, N., and Voos, W. (2000) The mitochondrial protein import motor. *Biol. Chem.* **381**, 943–949.

75. D'Silva, P. D., Schilke, B., Walter, W., Andrew, A., and Craig, E. A. (2003) J protein cochaperone of the mitochondrial inner membrane required for protein import into the mitochondrial matrix. *Proc. Natl. Acad. Sci. USA* **100**, 13839–13844.

76. Truscott, K. N., Voos, W., Frazier, A. E., et al. (2003) A J-protein is an essential subunit of the presequence translocase-associated protein import motor of mitochondria. *J. Cell Biol.* **163**, 707–713

77. Mokranjac, D., Sichting, M., Neupert, W. and Hell, K. (2003) Tim14, a novel key component of the import motor of the TIM23 protein translocase of mitochondria. *EMBO J.* **22**, 4945–4956.

78. Mokranjac, D., Sichting, M., Popov-Celeketic, D., Berg, A., Hell, K. and Neupert, W. (2005) The import motor of the yeast mitochondrial TIM23 preprotein translocase contains two different J proteins, Tim14 and Mdj2. *J. Biol. Chem.* **280**, 31608–31614.

79. Li, Y., Dudek, J., Guiard, B., Pfanner, N., Rehling, P., and Voos, W. (2004) The presequence translocase-associated protein import motor of mitochondria. Pam16 functions in an antagonistic manner to Pam18. *J. Biol. Chem.* **279**, 38047–38054.

80. van der Laan, M., Chacinska, A., Lind, M., et al. (2005) Pam17 is required for architecture and translocation activity of the mitochondrial protein import motor. *Mol. Cell Biol.* **25**, 7449–7458.

81. Isaya, G., Miklos, D., and Rollins, R.A. (1994) MIP1, a new yeast gene homologous to the rat mitochondrial intermediate peptidase gene, is required for oxidative metabolism in Saccharomyces cerevisiae. *Mol. Cell Biol.* **14**, 5603–5616.

82. Voos, W. and Röttgers, K. (2002) Molecular chaperones as essential mediators of mitochondrial biogenesis. *Biochim. Biophys. Acta* **1592**, 51–62.

83. Pfanner, N. and Neupert, W. (1987) Distinct steps in the import of ADP/ATP carrier into mitochondria. *J. Biol. Chem.* **262**, 7528–7536.

84. Truscott, K. N., Wiedemann, N., Rehling, P., et al. (2002) Mitochondrial import of the ADP/ATP carrier: the essential TIM complex of the intermembrane space is required for precursor release from the TOM complex. *Mol. Cell Biol.* **22**, 7780–7789.

85. Sirrenberg, C., Endres, M., Fölsch, H., Stuart, R.A., Neupert, W., and Brunner, M. (1998) Carrier protein import into mitochondria mediated by the intermembrane proteins Tim10/Mrs11 and Tim12/Mrs5. *Nature* **391**, 912–915.

86. Koehler, C. M., Merchant, S., Oppliger, W., et al. (1998) Tim9p, an essential partner subunit of Tim10p for the import of mitochondrial carrier proteins. *EMBO J.* **17**, 6477–6486.

87. Adam, A., Endres, M., Sirrenberg, C., Lottspeich, F., Neupert, W., and Brunner, M. (1999) Tim9, a new component of the TIM22.54 translocase in mitochondria. *EMBO J.* **18**, 313–319.

88. Paschen, S. A., Rothbauer, U., Kaldi, K., Bauer, M. F., Neupert, W., and Brunner, M. (2000) The role of the TIM8-13 complex in the import of Tim23 into mitochondria. *EMBO J.* **19**, 6392–6400.

89. Curran S. P., Leuenberger, D., Oppliger, W. and Koehler, C. M. (2002) The Tim9p-Tim10p complex binds to the transmembrane domains of the ADP/ATP carrier. *EMBO J.* **21**, 942–953.

90. Webb, C. T., Gormann, M. A., Lazarou, M., Ryan, M. T., and Gulbis, J. M. (2006) Crystal structure of the mitochondrial chaperone TIM910 reveals a six-bladed α-propeller. *Mol. Cell* **21**, 123–133.

91. Wiedemann, N., Pfanner, N., and Chacinska, A. (2006) Chaperoning through the Mitochondrial Intermembrane space. *Mol. Cell* **21,** 145–148.

92. Rehling, P., Brandner, K., and Pfanner, N. (2004) Mitochondrial import and the twin-pore translocase. *Nat. Rev. Mol. Cell Biol.* **5**, 519–530.

93. Kerscher, O., Holder, J., Srinivasan, M., Leung, R. S., and Jensen, R. E. (1997) The Tim54p-Tim22p complex mediates insertion of proteins into the mitochondrial inner membrane. *J. Cell Biol.* **139**, 1663–1675.

94. Kovermann, P., Truscott, K. N., Guiard, B., et al. (2002) Tim22, the essential core of the mitochondrial protein insertion complex, forms a voltage-activated and signal-gated channel. *Mol. Cell* **9**, 363–373.

95. Kerscher, O., Sepuri, N. B. and Jensen, R. E. (2000) Tim18p is a new component of the Tim54p-Tim22p translocon in the mitochondrial inner membrane. *Mol. Biol. Cell* **11**, 103–116.

96. Naoé, M., Ohwa, Y., Ishikawa, D., et al. (2004) Identification of Tim40 that mediates protein sorting to the mitochondrial intermembrane space. *J. Biol. Chem.* **279**, 47815–47821.

97. Terziyska, N., Lutz, T., Kozany, C., et al. (2005) Mia40, a novel factor for protein import into the intermembrane space of mitochondria is able to bind metal ions. *FEBS Lett.* **579**, 179–184.

98. Mesecke, N., Terziyska, N., Kozany, C., et al. (2005) A disulfide relay system in the intermembrane space of mitochondria that mediates protein import. *Cell* **121**, 1059–1069.

99. Curran, S. P., Leuenberger, D., Leverich, E. P., Hwang, D. K., Beverly, K. N., and Koehler, C. M. (2004) The role of Hot13p and redox chemistry in the mitochondrial TIM22 import pathway. *J. Biol. Chem.* **279**, 43744–43751.

100. Rissler, M., Wiedemann, N., Pfannschmidt, S., et al. (2005) The essential mitochondrial protein Erv1 cooperates with Mia40 in biogenesis of intermembrane space proteins. *J. Mol. Biol.* **353**, 485–492.

101. Allen, S., Balabanidou, V., Sideris, D. P., Lisowsky, T., and Tokatlidis, K. (2005) Erv1 mediates the Mia40-dependent protein import pathway and provides a functional link to the respiratory chain by shuttling electrons to cytochrome *c*. *J. Mol. Biol.* **353**, 937–944.

8

A Genetic Method to Identify Mitochondrial Proteins in Living Mammalian Cells

Takeaki Ozawa and Yoshio Umezawa

Summary

Mitochondria play pivotal roles in the metabolism and physiology of eukaryotic cells. The composition of mitochondria varies in different cell types, and therefore it is crucial to know the set of proteins that constitutes the organelle in each cell type. The identification of mitochondrial proteins has been furthered by biochemical methods, but half of mitochondrial proteins still remain unidentified. We have developed a genetic method to identify mitochondrial proteins with reconstitution of split enhanced green fluorescent protein (EGFP) by protein splicing. cDNA are randomly fused to the N-terminal half of EGFP and are introduced into cells expressing the C-terminal EGFP in the mitochondrial matrix. If the cDNA encodes a protein that is targeted into mitochondria, full-length EGFP is reconstituted in the mitochondrial matrix by protein splicing. The fluorescent cells are collected by fluorescence-activated cell sorting and the cDNA are identified by DNA sequencing. This method provides a means to map proteins distributed within mitochondria of different mammalian cells.

Key Words: Green fluorescent protein; intein, protein splicing; mitochondria; cDNA library; fluorescence-activated cell sorting.

1. Introduction

Mitochondria play important roles in energy production, apoptosis, and the metabolism of amino acids and lipids. These specialized functions are tightly bound to distinct compartmentalization of mitochondrial proteins. Identification of the sets of proteins is therefore an essential step determining mitochondrial function *(1,2)*. The mitochondrial genome encodes only a dozen mitochondrial proteins (e.g., 8 in yeast, 11 in mouse, 13 in human); most of the mitochondrial proteins are encoded in the nuclear genome, synthesized in the

From: *Methods in Molecular Biology, Vol. 390: Protein Targeting Protocols: Second Edition*
Edited by: M. van der Giezen © Humana Press Inc., Totowa, NJ

cytosol, and targeted to one of the four mitochondrial compartments: the outer membrane, the inner membrane, the intermembrane space, and the mitochondrial matrix. The targeting and sorting to some of the mitochondrial subcompartments is achieved by N-terminal-cleavable presequences of precursor proteins *(3)*.

Historically, comprehensive identification of mitochondrial proteins was first aimed at localizing the proteome in yeast cells. About 60% of all yeast open reading frames (ORFs) were tagged with an epitope, and the cellular localization of each fusion protein was then determined by immunofluorescence microscopy, using antibodies specific to the antigen *(4)*. Using this method, 332 mitochondrial proteins were identified. In the same way, each of the known 6234 ORFs in yeast was tagged at the 3' end with GFP, and the subcellular localization of the fusion was determined in the yeast strains by fluorescence microscopy *(5)*. About 70% of the analyzed proteins showed fluorescence, from which 526 mitochondrial proteins were identified.

In addition to the systematic identification of protein localizations, biochemical methods with mass spectroscopic analysis gave more complete sets of mitochondrial proteins *(6,7)*. Highly purified mitochondria isolated from a target organism are extracted with a detergent, and proteins are partially separated by sucrose gradient centrifugation, electrophoresis, and liquid chromatography. Mitochondrial proteins are identified by mass spectroscopy combined with bioinformatic analysis. Thus, more than 600 mitochondrial proteins are now identified from mitochondria of yeast, mouse, and human, but at least half of the mitochondrial proteins may remain unknown in mammalian cells. Therefore, for the elucidation of a complete set of mitochondrial proteins, several approaches are required for their complementary use *(8)*.

We have developed a genetic method to identify mitochondrial proteins in mammalian cells from large-scale cDNA libraries *(9)*. The basic concept of the genetic method is based on the reconstitution of split enhanced GFP by protein splicing with a DnaE intein (**Fig. 1**). Protein splicing is a reaction in which an intein excises out from a peptide precursor, after which the flanking polypeptides are linked by a peptide bond *(10)*. A tandem fusion protein is used, which contains a mitochondrial targeting signal (MTS) and C-terminal fragments of DnaE and EGFP, which localizes in the mitochondrial matrix in mammalian cells. cDNA libraries generated from mRNAs are genetically fused to the sequences encoding the N-terminal halves of EGFP and DnaE (**Fig. 2**). The cDNAs are converted into retrovirus libraries and used to inject the mammalian cells. If test proteins expressed from cDNA libraries contain a functional MTS, the fusion products translocate into the mitochondrial matrix

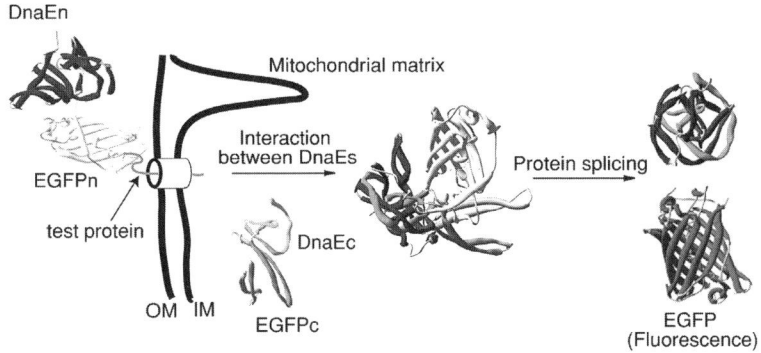

Fig. 1. Schematic showing how enhanced green fluorescent protein (EGFP) is formed by protein splicing of DnaEs when a test protein is localized in the mitochondrial matrix. C-terminal EGFP is connected with C-terminal DnaE and MTS, which is predominantly localized in the mitochondrial matrix. A test protein is connected with N-terminal halves of EGFP and DnaE, which is expressed in the cytosol. When the test protein targets into the mitochondria, the N-terminal DnaE interacts with C-terminal one, resulting in protein splicing. The N- and C-terminal EGFPs are connected together by a peptide bond, and the reconstituted EGFP recovers its fluorescence. EGFPn, N-terminal half (1–157 aa) of EGFP; EGFPc, C-terminal half (158–238 aa) of EGFP; DnaEn and DnaEc, N- and C-terminal DnaEs. OM and IM, mitochondrial outer and inner membranes, respectively.

and bring the N- and C-terminal halves of DnaEs close enough to fold correctly, thereby initiating protein splicing to link the concomitant EGFP halves with a peptide bond. The cells including reconstituted EGFP are screened rapidly and collected by fluorescence-activated cell sorting (FACS). From each clone thus collected, cDNA is retrieved by PCR and its sequence is analyzed. This method is powerful to identify a number of undiscovered genes encoding mitochondria. The method can be used in any laboratory with cDNA library construction capability and FACS equipment.

2. Materials

2.1. Plasmids and Cells

1. Plasmids: pMX-MTS/DEc(Neo), pMX-Mito/LIB-MTS, pMX-Mito/LIB-CaM, pMX-Mito/LIB (*see* **Note 1**) (**Fig. 3**). Retrovirus packaging cell line, PlatE cells (*see* **Note 2**) *(11)*.

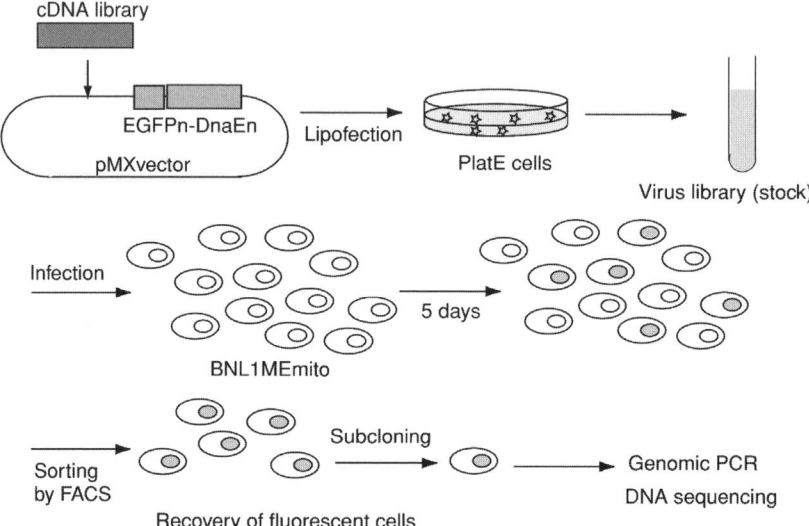

Fig. 2. Strategy for identifying the proteins targeted into mitochondrial matrix from cDNA libraries. cDNAs are connected with the cDNAs of EGFPn and DnaEn and transfected into PlatE cells. After converting the library to retroviruses, cultured mammalian cells including EGFPc-DnaEc (BNL1MEmito cells) are infected with the retrovirus libraries with 20% infection efficiency. Fluorescent cells are collected by FACS on 48-well plates. cDNA integrated in the genome are extracted by PCR, and their sequences are identified by DNA sequencing.

2.2. Cell Culture, Infection, and Transfection

1. Dulbecco's modified Eagle's medium (DMEM) (Gibco/BRL, Bethesda, MD) supplemented with 10% fetal bovine serum (FBS; Gibco/BRL).
2. Solution of trypsin (0.25%) (Gibco/BRL).
3. Geneticin liquid (50 mg/ml) (Invitrogen, Carlsbad, CA).
4. OPTI-MEM medium (Gibco/BRL).
5. Lipofectamine 2000 (Invitrogen).
6. Phosphate-buffered saline (Sigma, St. Louis, MO).
7. Polybrene (hexadimethrine bromide; Sigma) dissolved at 10 μg/ml in sterilized water. Store in aliquot at −30°C.

2.3. Construction of the cDNA Library

1. FastTrack kit (Invitrogen).
2. SuperScript Choice System (Invitrogen).
3. SuperScript III RT (Invitrogen).
4. *Bst*XI adaptor (Invitrogen) dissolved in DEPC-treated water (1 mg/mL).

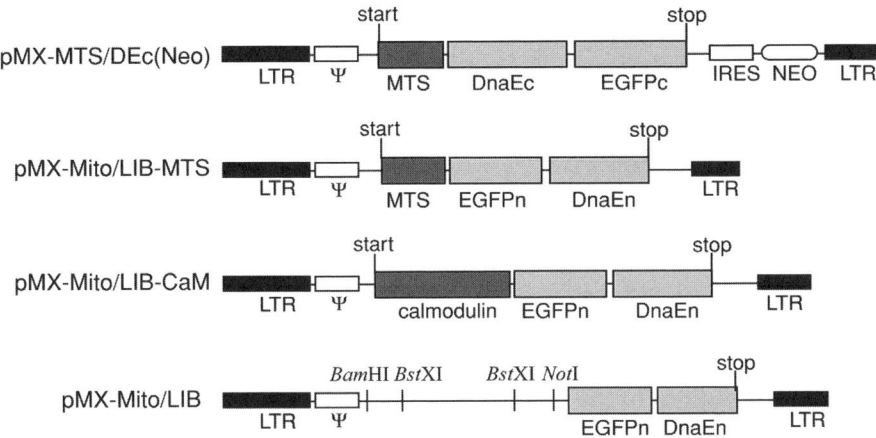

Fig. 3. Schematic structures of the plasmids. MTS, mitochondrial-targeting signal derived from subunit VIII of cytochrome *c* oxidase; EGFPn, N-terminal half (1–157 aa) of EGFP; EGFPc, C-terminal half (158–238 aa) of EGFP; DnaEn and DnaEc, N- and C-terminal DnaEs; IRES, internal ribosome entry site; NEO, neomycin resistance; LTR, long terminal repeat; Ψ, retrovirus-packaging signal.

5. cDNA Size fractionation columns (Invitrogen) equilibrated with TEN buffer at room temperature before use.
6. Phenol:chloroform:isoamyl alcohol (25:24:1) (Sigma).
7. 7.5 M Ammonium acetate (NH$_4$OAc).
8. 70% (v/v) Ethanol stored at −30°C.
9. TE buffer: 10 mM Tris-HCl (pH7.5), 1 mM ethylenediamine tetraacetic acid (EDTA).
10. TEN buffer: 10 mM Tris-HCl (pH7.5), 0.1 mM EDTA, 25 mM NaCl.
11. MAX Efficiency DH10B Competent Cells (Invitrogen).
12. Quiagen Maxi Prep Kit (Qiagen, Hilden, Germany).

2.4. Sorting and Identification of cDNA

1. Cell strainer, 40 μm nylon (BD Falcon, Bedford, MA).
2. Wizard Genomic DNA Purification Kit (Promega, Madison, WI).
3. LA Taq polymerase (Takara, Shiga, Japan).
4. BigDye Terminator Cycle Sequencing Kits (Applied Biosystems, Foster City, CA).
5. PCR primers for cDNA extraction; first nested primers AGGACCTTACACAGTC-CTGCTGACC (forward) and GCCCTCGCCGGACACGCTGAACTTG (reverse), and second nested primers CCGCCCTCAAAGTAGACGGCATCGCAGC (forward) and CGCCGTCCAGCTCGACCAGGAT (reverse).
6. 310 or 3100 Genetic Analyzer (Applied Biosystems).

2.5. Antibodies for Western Blot

1. Mouse monoclonal anti-GFP antibody (Roche Applied Science, Mannheim, Germany). Alkaline phosphatase-conjugated secondary antibody (Jackson, West Grove, PA).
2. Wizard Genomic DNA Purification Kit (Promega).
3. LA Taq polymerase (Takara).

3. Methods

Before screening the mitochondrial proteins from cDNA libraries, a stable cell line harboring C-terminal fragments of DnaE and EGFP in mitochondrial matrix must be generated (*see* **Subheading 3.1**); it will take about 1 mo to generate the cell line. In the meantime, a cDNA library and a stock solution of retrovirus library should be prepared (*see* **Subheadings 3.2–3.4**).

3.1. Preparation of Cells that Include C-Terminal Half of EGFP in the Mitochondrial Matrix

1. Seed BNL1ME cells (1×10^6 cells) onto a 6-cm dish 1 d before the transfection, and incubate the cells at 37°C in a CO_2 incubator.
2. Prepare the following solutions in 1.5-mL tubes: solution A, 2 μg of pMX-MTS/DEc(Neo) in 100 μL OPTI-MEM medium; solution B, 5 μL of Lipofectamine 2000 in 100 μL OPTI-MEM medium. Incubate the mixtures at room temperature for 5 min.
3. Add solution B to solution A, vortex gently, and incubate at room temperature for 20 min.
4. During the incubation, replace the medium in the 6-cm dish into fresh DMEM medium containing 10% FCS.
5. Add the mixture of solution A and solution B to a 6-cm dish.
6. Incubate the cells on the dishes for 2 d at 37°C in a CO_2 incubator.
7. Remove the medium, including the transfection mixture, and add DMEM medium containing 0.5 mg/mL geneticin and 10% of FCS.
8. Incubate the cells for 5 d at 37°C in a CO_2 incubator.
9. Strip the cells with 500 μL of the trypsin solution, dilute the suspension with DMEM medium containing 0.5 mg/mL geneticin and 10% FCS, and then seed the cells onto 10-cm dishes. Adjust 20–50 cells per 10-cm dish.
10. Incubate the cells until the colony of the cells can be observed.
11. Pick up 20 colonies with a pipet tip and again proliferate each cloned cell up to confluent in 6-cm dishes.
12. Examine the expression of the C-terminal EGFP connected with DnaE by Western blot with the GFP antibody (*see* **Note 3**) and select a cell line that expresses the protein in the mitochondria (*see* **Note 4**). This cell line is hereafter referred to as BNL1MEmito cells.

3.2. cDNA Library Construction

1. Poly(A) + RNA was prepared using a FastTrack kit according to the manufacturer's protocol.
2. Add $0.1\,\mu g$ of random hexamers to a sterilized 1.5-mL microcentrifuge tube.
3. Add $5\,\mu g$ of mRNA to the tube and adjust the total volume to $8\,\mu L$ using DEPC-treated water.
4. Heat the mixture at 70°C for 3 min, and quickly chill it on ice.
5. Add $4\,\mu L$ of 5X first strand buffer, $2\,\mu L$ of $0.1\,M$ dithiothreitol, and $1\,\mu L$ of $10\,mM$ dNTP mix.
6. Mix the contents by gentle vortex, and incubate the tube at 37°C for 2 min to equilibrate the temperature.
7. Add $5\,\mu L$ of Superscript III RT (total volume of the reaction mixture is $20\,\mu L$) (*see* **Note 5**), mix gently, and incubate the reaction mixture at 50°C for 1 h.
8. Place the tube on ice to terminate the reaction.
9. Keep the tube on ice and add $93\,\mu L$ of DEPC water, $30\,\mu L$ of 5X second strand buffer, $3\,\mu L$ of $10\,mM$ dNTPmix, $1\,\mu L$ of *E. coli* DNA ligase ($10\,units/\mu L$), $4\,\mu L$ of *E. coli* DNA polymerase I ($10\,units/\mu L$), and $1\,\mu L$ of RNase H ($2\,units/\mu L$) in that order.
10. Vortex the tube and incubate to complete the reaction for 2 h at 16°C.
11. Add $2\,\mu L$ of T4 DNA polymerase ($5\,units/\mu L$), and continue the incbation at 16°C for 5 min.
12. Place the tube on ice and add $10\,\mu L$ of $0.5\,M$ EDTA.
13. Add $150\,\mu L$ of phenol:chloroform:isoamyl alcohol (25:24:1), vortex thoroughly, and centrifuge at room temperature for 5 min at $14,000g$.
14. Vortex the mixture thoroughly, and centrifuge at room temperature for 30 min at $14,000g$.
15. Remove the supernatant and wash the pellet with 0.5 mL of 70% ethanol (−20°C).
16. Centrifuge for 5 min at $14,000g$ and remove the supernatant carefully.
17. Dry the cDNA to evapolate residual ethanol.

3.3. cDNAs Insertion into pMX-ER/LIB Vector

1. Dissolve the cDNA pellet in $12\,\mu L$ of DEPC water, and add $2\,\mu L$ of *Bst*XI adaptor, $4\,\mu L$ of 5X T4 DNA ligase buffer, $2\,\mu L$ of T4 DNA ligase ($1\,unit/\mu L$).
2. Mix gently, and incubate the reaction mixture at 16°C for more than 24 h.
3. Heat the reaction at 70°C for 10 min to inactivate the ligase, and chill on ice.
4. Add $30\,\mu L$ of DEPC water and $50\,\mu L$ of phenol:chloroform:isoamyl alcohol (25:24:1), vortex thoroughly, and centrifuge at room temperature for 5 min at $14,000g$.
5. Remove carefully $45\,\mu L$ of the upper aqueous layer and transfer it to a fresh 1.5-mL tube.
6. Add $5\,\mu L$ of $7.5\,M$ NH$_4$OAc, $1\,\mu g$ of yeast tRNA, and $150\,\mu L$ of ethanol (−30°C).
7. Vortex the mixture, and centrifuge for 20 min at $14,000\,g$ (4°C).

8. Remove the supernatant carefully, and wash the pellet with 0.25 mL of 70% ethanol.
9. Centrifuge for 2 min at 14,000g, and remove the supernatant.
10. Dry the cDNA for 10 min to completely remove residual ethanol.
11. Dissolve the cDNA in 100 μL of TEN buffer, and apply it to the top of the equilibrated cDNA size fractionation column, and discard the effluent.
12. Add 100 μL of TEN buffer to the column top, and discard the effluent.
13. Repeat **step 12** once.
14. Add 100 μL of TEN buffer and collect the effluent into a 1.5-mL microcentrifuge tube.
15. Repeat **step 14** and add 150 μL of TEN buffer and collect the effluent into the 1.5-mL microcentrifuge tube that now contains 350 μL of the adaptor-ligated cDNA solution.
16. Add 35 μL of 7.5 M NH4OAc, 1 μg of yeast, and 700 μL of ethanol (−30°C) into the tube.
17. Repeat **steps 7–10** for cDNA precipitation.
18. Add 40 μL of TE buffer to the cDNA pellet.
19. Separate the cDNAs through a 0.8% SeaPlaque gel. Cut out the gel fragment between 0.6 and 10 kbp, and extract cDNA fragments using a QiaexII according to the manufacturer's protocol.
20. Resuspend the size-fractionated cDNAs in 10 μL of a TE buffer.
21. To ligate the cDNAs with the pMX-Mito/LIB vector, prepare the following solution; 2 μL of the cDNAs, 4 μL of a 5X T4 DNA ligase buffer, 1 μL of the pMX-Mito/LIB vector (100 ng), 1 μL of T4 DNA ligase, and 12 μL of water.
22. Incubate the solution at 16°C for more than 16 h.
23. Heat the reaction mixture at 70°C for 10 min to inactivate the ligase.
24. Add 30 μL of distilled water, 5 μL of 3 M NaOAc, 1 μg yeast tRNA, and 60 μL of ethanol (−30°C).
25. Repeat **steps 7–10** for cDNA precipitation.
26. Resuspend the pellet in 12 μL of a TE buffer (*see* **Note 6**).

3.4. Production of Retrovirus Stock

1. Seed PlatE cells (1×10^6 cells) onto a 10-cm dish 24 h before the transfection, and incubate the cells at 37°C in a CO_2 incubator.
2. Prepare the following solutions in 1.5-mL tubes: solution A, 5 μg of library DNA in 200 μL OPTI-MEM medium; solution B, 18 μL of Lipofectamine 2000 in 200 μL OPTI-MEM medium. Incubate the mixtures at room temperature for 5 min.
3. Add solution B to solution A, vortex gently, and incubate at room temperature for 20 min.
4. During the incubation, replace the medium in the 10-cm dish with DMEM containing 10% of FCS.
5. Add the mixture of solution A and solution B in the 10-cm dish.

6. Incubate the cells on the dishes for 12 h at 37°C in a CO_2 incubator.
7. Remove the medium including the transfection mixture, and add 5 mL of fresh DMEM containing 10% of FCS.
8. Incubate the cells for 48 h at 37°C in a CO_2 incubator, and then collect the supernatant in a 1.5-mL tube (*see* **Note 7**), which is used for infection of target cells.

3.5. Infection of Recombinant Retrovirus and Collection of Fluorescent Cells

1. Seed BNL1MEmito cells (1×10^6 cells) onto a 6-cm dish 24 h before the infection and incubate at 37°C in a CO_2 incubator.
2. Add 5 μL of the polybrene solution, and then incubate the cells at 37°C for 10 min in a CO_2 incubator.
3. Add 20 μL of the virus stock prepared in **Subheading 3.3.**, and then incubate the cells at 37°C for 36 hours in a CO_2 incubator.
4. Move the 6-cm dish in another CO_2 incubator set at 30°C, and incubate for 12 h.
5. Strip off the cells with 500 μL of a trypsin solution, collect the cells by centrifugation for 2 min at 800g, and resuspend the cells in 1 mL of phosphate-buffered saline (PBS) solution (*see* **Note 8**). To remove aggregated cells, pass the cell suspension through a cell strainer (*see* **Note 9**) and collect the cells in a round-bottom tube.
6. The cells are subjected to cell analysis with the FACS (*see* **Note 10**). A region that includes fluorescent cells is determined (see region L in **Fig. 4**).
7. Collect the fluorescent cells in a round-bottom tube filled with a DMEM medium including 10% FCS (*see* **Note 11**). Immediately spread the cells on 6-cm dishes and incubate the cells at 37°C for 2 d in a CO_2 incubator.
8. Repeat the **steps 4–6**. Collect the fluorescent cells on a 48-well microtiter plate (*see* **Note 12**), and incubate the cells at 37°C in a CO_2 incubator until confluent.

3.6. Sequencing of Genome-Integrated cDNA that Encode Mitochondria-Targeting Proteins

Genomic DNA was extracted with a Wizard Genomic DNA Purification Kit according to the manufacturer's protocol.

1. 100 ng of each genomic DNA is subjected to the nested PCR. The PCR was run for 30 cycles (30 s at 98°C for denaturation, 30 s at 58°C for annealing, and 4 min at 72°C for extension) using LA Taq polymerase (*see* **Note 13**).
2. The resulting fragments are sequenced using a BigDye Terminator Cycle Sequencing Kit and analyzed by a genetic analyzer.
3. Compare the obtained cDNA sequence to the databases at the National Center for Biotechnology Information (NCBI) using BLASTn (http://www.ncbi.nlm.nih.gov/BLAST/) (*see* **Note 14**).

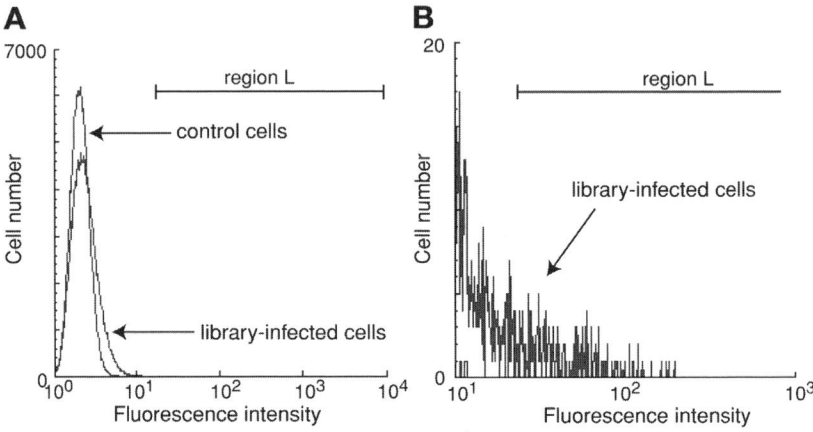

Fig. 4. FACS profiles of BNL1MEmito cells infected with retrovirus cDNA libraries.
(**A**) BNL1MEmito cells were infected with the cDNA retrovirus libraries with an
infection efficiency of 20%. Five days after incubation, the cells were stripped off
and sorted by FACS (library-infected cells). (**B**) Enlarged FACS profiles in region.
Uninfected cells were inserted to show the background fluorescence (control cells).

Notes

1. The plasmids described herein can be obtained from our laboratory
   (umezawa@chem.s.u-tokyo.ac.jp).
2. The PlatE cell line can be obtained from Dr. Toshio Kitamura (The Institute of
   Medical Science, The University of Tokyo, Japan).
3. We have found that the GFP antibody (Roche) is excellent for blotting the
   C-terminal half of EGFP. The other monoclonal-GFP antibodies (Clonetech or
   Takara Bio Inc., Tokyo, Japan) are also available for the blotting.
4. For the following library screening to work accurately, it is required that two conditions be met: (1) cells that include proteins with mitochondrial targeting signals
   must be separated from the ones without mitochondrial targeting signals, and (2)
   the fluorescence intensity of the EGFP reconstituted in mitochondria must be
   strong enough to be detected by FACS analysis. To verify these conditions, pMX-
   Mito/LIB-MTS and pMX-Mito/LIB-CaM are used. According to **Subheading 3.4.**,
   the plasmids are converted into retroviruses, and then the retroviruses are infected
   into a selected cell line that includes C-terminal EGFP. The fluorescence intensities of the cells are examined by FACS (*see* **Subheading 3.4.**), and a cell line
   that meets the criteria is selected.
5. It is also possible to use SuperScript II Reverse Transcriptase; however, SuperScript
   III Reverse Transcriptase has several advantages in the synthesis of first-strand
   cDNAs at temperatures of 50°C, providing increased specificity, higher yields of
   cDNAs, and longer cDNA products.

6. In this step, it is very important to completely dry the pellet. Inclusion of residual amounts of ethanol will greatly decrease the transformation efficiency. To avoid this, do not use more than 10 μL of QiaexII beads in one tube in DNA extraction.
7. The retrovirus library can be stocked for 6 mo at −80°C. Because only 20–50 μL of the retrovirus is used for the subsequent infection procedure, divide the retrovirus-library solution into small portions and stock them at −80°C.
8. Because residual trypsin solution affects the viability of the cells, it is important to remove the trypsin solution completely. If necessary, repeat the washing of the cells with a PBS buffer.
9. The size of mesh of a cell strainer depends on the types of the cells. To prevent the cells from sticking in a flow cell of the FACS, it is much better to use a smaller mesh size.
10. The FACS is generally equipped with collection modes—one for higher collection efficiency or for higher purity. In the first sorting procedure, the FACS is adjusted to set a mode of higher collection efficiency in order to collect as many fluorescent cells as possible. In contrast, a mode of high purity is used in the second sorting procedure for the purpose of accurate identification of the mitochondrial-targeting proteins.
11. Prevent the collection tube from contamination with antibiotics. Antibiotics affect the cells when sorting with FACS, thereby killing the collected cells.
12. If the FACS you use is not equipped with a system for collecting cells onto microtiter plates, it is possible to collect the cells in a round-bottom tube supplemented with a DMEM medium including 10% FCS. The cells are cloned by limiting dilution, and their genomic DNAs are collected.
13. Any Taq polymerase can be used.
14. cDNAs identified according to this scheme may be in reverse orientation, noncoding sequences, or those encoding those proteins lacking an amino terminus. Such false-positive cDNAs can be eliminated by using databases such as FANTOM DB (http://fantom3.gsc.riken.jp/). To identify cDNAs for which bioinformatic data are not deposited in the public databases, full-length cDNA are extracted from cDNA libraries, and ORFs are identified.

Acknowledgments

This work was supported by grants from Core Research for Evolutional Science and Technology (CREST) of Japan Science and Technology (JST) and the Ministry of Education, Science, and Culture, Japan.

References

1. Reichert, A. S. and Neupert, W. (2004) Mitochondriomics or what makes us breathe. *Trends Genet.* **20**, 555–562.
2. Jensen, R. E., Dunn, C. D., Youngman, M. J., and Sesaki, H. (2004) Mitochondrial building blocks. *Trends Cell Biol.* **14**, 215–218.

3. Truscott, K. N., Brandner, K., and Pfanner, N. (2003) Mechanisms of protein import into mitochondria. *Curr. Biol.* **13**, R326–337.
4. Kumar, A., Agarwal, S., Heyman, J. A., et al. (2002) Subcellular localization of the yeast proteome. *Genes Dev.* **16**, 707–719.
5. Huh, W. K., Falvo, J. V., Gerke, L. C., et al. (2003) Global analysis of protein localization in budding yeast. *Nature* **425**, 686–691.
6. Taylor, S. W., Fahy, E., Zhang, B., et al. (2003) Characterization of the human heart mitochondrial proteome. *Nat. Biotechnol.* **21**, 281–286.
7. Mootha, V. K., Bunkenborg, J., Olsen, J. V., et al. (2003) Integrated analysis of protein composition, tissue diversity, and gene regulation in mouse mitochondria. *Cell* **115**, 629–640.
8. Westermann, B. and Neupert, W. (2003) 'Omics' of the mitochondrion. *Nat. Biotechnol.* **21**, 239–240.
9. Ozawa, T., Sako, Y., Sato, M., Kitamura, T., and Umezawa, Y. (2003) A genetic approach to identifying mitochondrial proteins. *Nat. Biotechnol.* **21**, 287–293.
10. Wu, H., Hu, Z., and Liu, X. Q. (1998) Protein trans-splicing by a split intein encoded in a split DnaE gene of Synechocystis sp. PCC6803. *Proc. Natl. Acad. Sci. USA* **95**, 9226–9231.
11. Morita, S., Kojima, T., and Kitamura, T. (2000) Plat-E: an efficient and stable system for transient packaging of retroviruses. *Gene Ther.* **7**, 1063–1066.

9

In Vitro and In Vivo Methods to Study Protein Import Into Plant Mitochondria

Shashi Bhushan, Pavel F. Pavlov, Charlotta Rudhe, and Elzbieta Glaser

Summary

Plant mitochondria contain about 1000 proteins, 90–99% of which in different plant species are nuclear encoded, synthesized on cytosolic polyribosomes, and imported into the organelle. Most of the nuclear-encoded proteins are synthesized as precursors containing an N-terminal extension called a presequence or targeting peptide that directs the protein to the mitochondria. Here we describe in vitro and in vivo methods to study mitochondrial protein import in plants. In vitro synthesized precursor proteins can be imported in vitro into isolated mitochondria (single organelle import). However, missorting of chloroplast precursors in vitro into isolated mitochondria has been observed. A novel dual import system for simultaneous import of proteins into isolated mitochondria and chloroplasts followed by reisolation of the organelles is superior over the single import system as it abolishes the mistargeting. Precursor proteins can also be imported into the mitochondria in vivo using an intact cellular system. In vivo approaches include import of transiently expressed fusion constructs containing a presequence or a full-length precursor protein fused to a reporter gene, most commonly the green fluorescence protein (GFP) in protoplasts or in an *Agrobacterium*-mediated system in intact tobacco leaves.

Key Words: Protein import; mitochondria; chloroplasts; precursor; targeting peptide; presequence; GFP fusion; protoplasts; *Agrobacterium*; transient expression.

1. Introduction

The total proteome of mitochondria has been estimated to contain about 1000 proteins *(1)*. Despite the fact that mitochondria contain their own genome, most of the mitochondrial proteins are encoded in the nucleus. Mitochondrial genomes in different plant species encode only 3–67 proteins *(2)*, which

From: *Methods in Molecular Biology, Vol. 390: Protein Targeting Protocols: Second Edition*
Edited by: M. van der Giezen © Humana Press Inc., Totowa, NJ

are synthesized within the organelle. All the other mitochondrial proteins (several hundred) are synthesized on cytosolic polyribosomes and have to be imported into the mitochondria. Most of the nuclear-encoded proteins are synthesized as precursor proteins with an N-terminal extension called presequence or targeting peptide that directs precursor proteins to the correct organelle. The mitochondrial targeting peptides in plants are about 30–40 amino acid residues long, show no sequence similarity, but are generally enriched in positively charged, hydroxylated, and hydrophobic amino acid residues and form amphiphilic α-helices in membrane environment *(3)*. Cytosolic molecular chaperones maintain an import competent conformation of the newly synthesized precursor proteins and facilitate organellar protein import *(4)*. Mitochondrial protein import is conventionally viewed as a posttranslational process, but several recent observations support the co-translational mitochondrial import of some proteins *(5)*. Presequences are recognized by import receptors on the mitochondrial outer membrane, and precursors are imported into the organelle through oligomeric translocase complexes, TOM and TIM, located on the outer and inner mitochondrial membrane, respectively *(4)*. After translocation into the mitochondrial matrix, presequences are cleaved off by the mitochondrial processing peptidase (MPP) that in plants is an integral part of the cytochrome bc_1 complex of the respiratory chain *(6)*. The mature proteins find their final destination within the mitochondrion and assemble with partner proteins either spontaneously or upon action of molecular chaperones to form functional protein complexes. The cleaved presequences, potentially harmful to biological membranes, are degraded inside the organelles by a novel presequence peptide-degrading metalloendopeptidase, PreP *(7– 10)*, a mitochondrial peptidasome *(11)*.

The targeting of proteins to mitochondria can be studied using both in vitro and in vivo approaches. In the former system, in vitro synthesized radiolabeled precursor proteins are incubated with isolated organelles followed by "shaving"of the organelles with proteases in order to remove nonimported precursor protein and detect the imported product protected inside the organelle. In the in vivo approaches fusion constructs of targeting peptides or precursor proteins coupled to passenger proteins are transiently expressed in protoplasts or in intact leaves. Passenger proteins are selected so that they can be easily detected either enzymatically or by fluorescence microscopy. Both in vitro and in vivo approaches have several advantages as well as some limitations. In vitro investigations allow studies of individual components and kinetics of the import process, but the main limitation of this system is the lack of intact cellular components, such as cytosolic molecular chaperones and competing organelles. In vivo approaches, on the other hand, concern procedures that do contain the intact cellular system, but there are limitations arising from the fact that the investigated proteins are passenger proteins instead of native mature proteins and that the fusion proteins are overexpressed in cells at high levels. Additional complication of the organellar protein import studies in plants arises from the fact that an increasing number of proteins are encoded by a single gene, expressed as a single precursor protein and dually imported into both mitochondria and chloroplasts *(9,12–14)*. This demands studying the specificity of the organellar protein import, and therefore, often it is advantageous to use both types of approaches, including a novel recently developed in vitro dual import system *(15)*.

Mitochondrial protein import was demonstrated to be highly organelle specific using in vivo studies in transgenic plants *(16–18)*. Also, early in vitro import studies using a homologous organelle system from spinach leaves or soybean in combination with mitochondrial and chloroplastic precursor proteins *(19–21)* showed that import is organelle specific. However, whereas missorting of mitochondrial proteins into chloroplasts has not been reported, several more recent in vitro import studies reported mistargeting of chloroplast precursor proteins into mitochondria. The missorting has been observed with several precursors, including the PsaF subunit of photosystem I *(22)*, triose-3-phosphoglycerate phosphate translocator *(23)*, plastocyanin, small subunit of Rubisco, the 33 kDa protein of photosystem II *(24–26)*. It is thus evident that incorrect targeting, which is not seen in vivo, may occur in vitro when an intact cellular system is missing. In order to overcome some of the limitations of the in vitro import system and to ensure the correct specificity of targeting, we have developed a novel in vitro dual import system for simultaneous targeting of precursor proteins into mitochondria and chloroplasts *(15)*. Isolated organelles are mixed for simultaneous incubation with precursor proteins and reisolated after import. This allows the determination of the targeting specificity into either organelle in the presence of the competing organelle. This also allows the use of authentic precursors so that the role of the mature protein in import can be assessed. Furthermore, it has been shown that mistargeting of chloroplast precursors into mitochondria can be abolished.

Here we describe three methodologies to study protein import into plant mitochondria: a single organelle in vitro mitochondrial import, a dual import system to study mitochondrial import in the presence of chloroplasts (*see* **Fig. 1**), and in vivo import studies of chimeric precursors containing green fluorescent protein (GFP) as passenger protein into protoplasts and tobacco leaves.

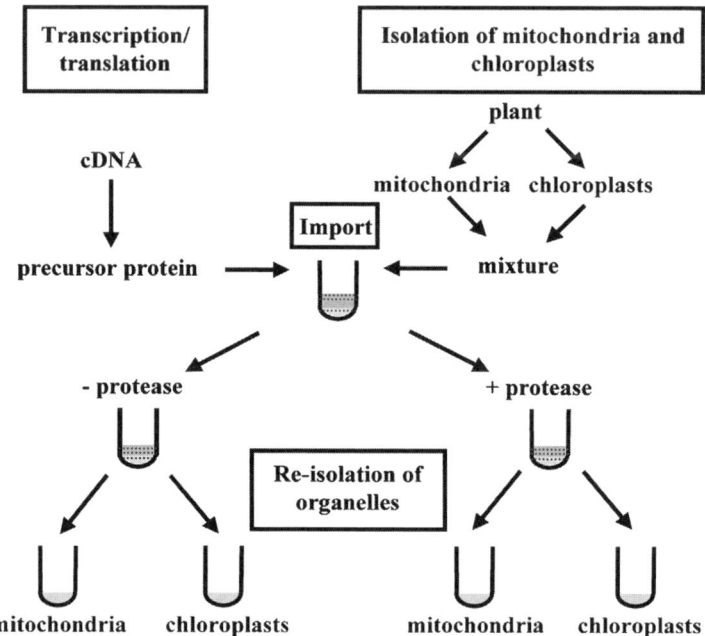

Fig. 1. Overview of the procedure of the dual import assay. Chloroplasts and mitochondria are purified separately, mixed, and incubated with a precursor protein under conditions that support import into both organelles. Immediately after import the sample is divided into two aliquots, and one is treated with protease. After protease treatment the chloroplasts and mitochondria are repurified using a 4% Percoll gradient.

2. Materials

2.1. A Single Organelle In Vitro Mitochondrial Import

2.1.1. Biological Material

1. Potatoes (*Solanum tuberosum* cv. King Edward) were bought in local shop.
2. Spinach (*Spinacia oleracea* L. cv. Medania) was grown hydroponically for 6 wk under artificial light at 25°C with a light period of 10 h.

2.1.1.1. PREPARATION OF POTATO TUBER MITOCHONDRIA

1. Grinding medium: $0.6\,M$ mannitol, $40\,mM$ MOPS-KOH pH 7.5, $10\,mM$ EDTA, $8\,\mu m$ cysteine, and 0.4% bovine serum albumin (BSA).
2. 2× Wash medium: $0.8\,M$ mannitol, $10\,mM$ MOPS-KOH pH 7.2, and 0.2% BSA.
3. Percoll solution: 20 mL of 2× wash medium mixed with 12 mL of Percoll and 8 mL of dH_2O.

2.1.1.2. Preparation of Spinach Leaf Mitochondria

1. Grinding medium: $0.3\,M$ sucrose, $50\,mM$ MOPS-KOH, pH 7.8, $5\,mM\,MgCl_2$, $2\,mM\,EDTA$, $4\,mM$ cysteine, 0.6% polyvinylpyrrolidone (MW 40,000), 0.2% BSA
2. $2.5\times$ Wash medium: $0.75\,M$ mannitol, $25\,mM$ MOPS-KOH, pH 7.2, 0.25% BSA.
3. Percoll solution: $4\,mL$ $2.5\times$ wash medium mixed with $6\,mL$ of Percoll to obtain 60% Percoll solution; $4\,mL$ $2.5\times$ wash medium mixed with $4\,mL$ Percoll and $3\,mL$ dH_2O to obtain 30% Percoll solution; $4\,mL$ $2.5\times$ wash medium mixed with $2\,mL$ Percoll and $4\,mL$ dH_2O to obtain 20% Percoll solution; apply 60, 30, and 20% Percoll solutions stepwise into a centrifuge tube at a ratio of 1:3:1.

2.1.2. In Vitro Synthesis of Radiolabeled Precursor Proteins

1. Coupled in vitro transcription/translation reticulocyte lysate system was obtained from Promega.
2. [^{35}S]-Methionine (specific activity > 1000 Ci/mmol) and RNase inhibitor (GE Healthcare, Uppsala, Sweden).
3. Plasmids encoding proteins of interest were purified from *E. coli* night cultures and isolated using QIAprep® spin mini-prep kit (Quiagen) according to manufacturer's instructions.

2.1.3. Protein Import Into Potato Tuber and Spinach Leaf Mitochondria

1. Import medium: $0.25\,M$ mannitol, $10\,mM$ MOPS-KOH, pH 7.2, $50\,mM$ KCl, $1\,mM$ succinate, $1\,mM$ methionine, $2\,mM\,MgCl_2$, $2\,mM$ KPi, $1\,mM$ ATP, $10\,\mu M$ ADP, 0.1% BSA (make up fresh prior to use).
2. $10\,mg/mL$ Proteinase K.
3. 0.1 M PMSF.
4. $10\,\mu M$ Valinomycin in ethanol.

2.1.4. Sodium Docecyl Sulfate (SDS) Electrophoresis/Phosphoimaging

1. The samples were analyzed by SDS-PAGE using 12% polyacrylamide gels in the presence of $4\,M$ urea.
2. Separating buffer ($4\times$): $1.5\,M$ Tris-HCl, pH 8.8, 0.4% SDS. Store at room temperature.
3. Stacking buffer ($4\times$): $0.5\,M$ Tris-HCl pH 6.8, 0.4% SDS. Store at room temperature.
4. Loading buffer ($2\times$): $1\,M$ Tris-HCl pH 6.8, 0.8% SDS, 20% glycerol, $400\,mM$ β-mercaptoethanol, 0.04% bromophenol blue.
5. 30% Acrylamide (AA) solution (27.3 AA, 2.7% bis-acrylamide) (Hintze AB, Lidingö, Sweden).
6. *N, N, N, N′*-Tetramethylethylene diamine (TEMED).

7. Ammonium persulfate (APS): 10% solution in water, freeze immediately in small aliquots at −20°C.
8. Butanol.
9. Running buffer (5×): 125 mM Tris, 960 mM glycine, 0.5% SDS.
10. Fix solution: 50% methanol, 5% acetic acid.
11. BAS-MP phosphoimaging plates (FujiFilm).
12. FLA-3000 phosphoimaging scanner (FujiFilm).

2.2. Dual In Vitro Import to Mitochondria and Chloroplasts

2.2.1. Biological Material

1. Spinach (*Spinach oleracea* L.cv. Medania).

2.2.2. Preparation of Organelles

2.2.2.1. Preparation of Spinach Mitochondria

As in single in vitro import (*see* **Subheading 2.1.1.2.**).

2.2.2.2. Preparation of Spinach Chloroplasts

1. 2× Grinding medium: 100 mM HEPES, pH 7.3, 660 mM sorbitol, 0.2% (w/v) BSA, 2 mM MgCl$_2$, 2 mM MnCl$_2$, 4 mM EDTA, 60 mg/L ascorbic acid (add the ascorbic acid just prior to use).
2. Import buffer: 50 mM HEPES, pH 8.0, 330 mM sorbitol.
3. Percoll gradient: 25 mL Percoll, 25 mL 2× grinding medium.

2.2.3. In Vitro Transcription/Translation

As in single in vitro import (*see* **Subheading 2.1.2.**).

2.2.4. Dual Import Into Spinach Mitochondria and Spinach Chloroplasts

1. 2× Import medium: 0.6 M sucrose, 30 mM HEPES, pH 7.4, 10 mM KH$_2$PO$_4$, 1% BSA.
2. Import master mix: 2× import medium, 1 M MgCl$_2$, 100 mM methionine, 100 mM ATP, 100 mM GTP, 100 mM ADP, 0.5 M succinate, 0.5 M DTT, 1 M potassium acetate, 1 M NaHCO$_3$ (make up fresh prior to use).
3. 5 mM CaCl$_2$, 2 mg/mL thermolysin, 0.5 M EDTA.
4. Percoll gradient: Percoll, 2× import medium.

2.2.5. SDS Electrophoresis

As in single in vitro import (*see* **Subheading 2.1.4.**).

2.3. *In Vivo Import Using Fusions With GFP*

2.3.1. *Transient Expression of GFP Fusion Constructs in Tobacco Protoplasts (Fig. 2A)*

2.3.1.1. BIOLOGICAL MATERIAL

1. Four- to 6-wk-old plants of *Nicotiana tabacum* cv SRI grown at 23°C under normal conditions.

2.3.1.2. PREPARATION OF PROTOPLASTS

1. Digestion medium (must be freshly prepared): 0.5 M sucrose, 0.125% (w/v) macerozyme R-10 (Onozuka; Yakult Pharmaceutical, Tokyo, Japan), 0.2% (w/v) cellulase R-10 (Onozuka; Yakult Pharmaceutical), 5 mM CaCl$_2$, 0.1% BSA, pH 5.5 adjust with HCl, sterilize by filtration.
2. Flotation medium: 0.6 M sucrose, 0.15% MES-NaOH pH 6.0, 15 mM CaCl$_2$, autoclave.
3. Washing medium: 154 mM NaCl, 125 mM CaCl$_2$, 5 mM KCl, 5 mM glucose, sterilize by filtration.
4. Electroporation medium: 4 mM CaCl$_2$, 80 mM KCl, 8% mannitol, 2 mM NaH$_2$PO$_4$, pH 7.0, 0.1% BSA, sterilize by filtration.
5. K3M medium: 10× diluted macroelements K3, 1000× diluted microelements K3, 500× diluted Morel vitamins, 100 μM Fe-EDTA, 0.5 M glucose, 100 mg/L thiamine, pH 5.7 adjust with KOH.
6. Macroelements K3 (200 mL of 10x): 1.2 g NH$_4$NO$_3$, 3.8 g KNO$_3$, 1.2 g CaCl$_2$ · 2H$_2$O, 0.6 g MgSO$_4$ · 7H$_2$O, 0.34 g KH$_2$PO$_4$, 0.6 g KCl, autoclave.
7. Microelements K3 (100 mL of 1000×): 0.62 g H$_3$BO$_3$, 2.23 g MnSO$_4$ · H$_2$O, 0.86 g ZnSO$_4$ · 7H$_2$O, 0.025 g Na$_2$MoO$_4$ · 2H$_2$O, 2.5 mg CuSO$_4$ · 5H$_2$O, 2.5 mg *CoCl$_2$* · 6H$_2$O, autoclave.
8. Morel vitamins (100 mL of 500×): 16.85 mg thiamine-HCl, 0.9 g myoinositol, 10.25 mg pyridoxine-HCl, 23.8 mg Ca^{2+}-pantothenate, 0.122 mg biotin, 6.15 mg nicotinic acid, sterilize by filtration.
9. Fluorescein diacetate (FDA 20x) 5 mg/mL in acetone (keep at −20°C).

2.3.1.3. ELECTROPORATION OF PROTOPLASTS

Gene Pulser Transfection apparatus and 0.4-cm-gap cuvettes (Bio-Rad, Hercules, CA).

2.3.1.4. CONFOCAL MICROSCOPY

1. Confocal microscopy was performed using a Bio-Rad MRC-1024 laser scanning confocal imaging system.
2. Mitotracker Red CM-H2Xros (Molecular Probes, Eugene, OR) 10 μM in dimethyl sulfoxide (DMSO).

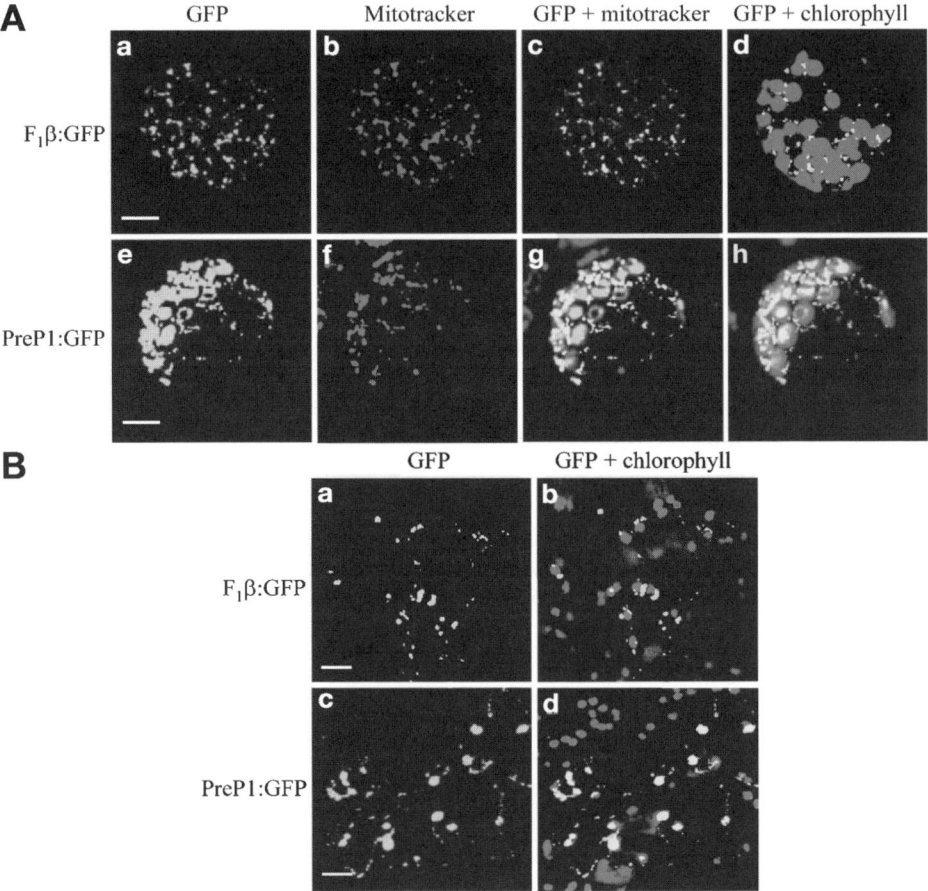

Fig. 2. In vivo organellar targeting of the $F_1\beta$:GFP and PreP1:GFP constructs. (A) Transient expression of the $F_1\beta$:GFP and PreP1:GFP fusion constructs into *Nicotiana tabacum* protoplasts. GFP column shows the signal detected in the green channel (a), mitotracker column shows the signal detected in the red channel (b), GFP + mitotracker corresponds to the merging of the two previous columns, in which yellow represents the superposition of green and red (c), GFP + chlorophyll corresponds to the merging of the green channel and the chlorophyll signal detected in the far red channel (d). Bars = 10μm. It is visualized that the $F_1\beta$:GFP construct has been imported only into mitochondria, whereas the PreP1:GFP constructs has been imported both into mitochondria and chloroplasts. (B) *Agrobacterium*-mediated transient expression of $F_1\beta$:GFP and PreP1:GFP into *Nicotiana tabacum* leaves. GFP column shows the signal detected in the green channel (a), GFP + chlorophyll corresponds to the merging of the green channel and the chlorophyll

2.3.2. Transient Expression Into Tobacco Leaves (Fig. 2B)

2.3.2.1. BIOLOGICAL MATERIAL

1. Four- to 6-wk-old plants of *Nicotiana tabacum* grown at 23°C under normal conditions.
2. *Agrobacterium tumefaciens* competent cells.

2.3.2.2. CLONING

1. Suitable binary vector (pBi) containing GFP construct in a eukaryotic expression cassette under a strong promoter such as 35S promoter *(27)*.

2.3.2.3. PREPARATION OF ELECTROCOMPETENT CELLS OF AGROBACTERIUM

1. 2YT plates: Add 2% agar in 2YT medium and autoclave.
2. 2YT medium: Bacto tryptone 16 g/L, Bacto yeast extract 10 g/L, NaCl 5 g/L. Adjust the pH to 7.0 with 5 M NaOH and adjust the volume to 1 L with deionized water. Sterilize by autoclaving.
3. Antibiotics: kanamycin, gentamycin, rifampicin.
4. 50 mM HEPES buffer: NaCl 16.4 g/l, HEPES 11.9 g/l, Na_2HPO_4 2.1 g/l. Adjust the pH to 7.05 with NaOH and adjust the volume to 1 L with deionized water. Sterilize by autoclaving.

2.3.2.4. TRANSFORMATION OF ELECTROCOMPETENT CELLS

1. YEB medium: 0.5% (w/v) beef extract, 0.1% (w/v) yeast extract, 0.5% (w/v) peptone, 0.5% (w/v) sucrose, 2 mM $MgSO_4$.
2. Acetosyringone (Sigma). Make 200 mM stock in DMSO, aliquot and store at −20°C.
3. Bio-Rad gene pulser.
4. 0.2-cm Bio-Rad electroporation cuvettes.

2.3.2.5. INFILTRATION OF THE AGROBACTERIUM SUSPENSION INTO TOBACCO LEAVES

Infiltration medium: 50 mM MES pH 5.6, 0.5% (w/v) glucose, 2 mM Na_3PO_4, 100 μM acetosyringone (add from stock prior to use).

Fig. 2. *(Continued)* signal detected in the far-red channel (b). Bars = 10 μm. It can be seen that the $F_1β$:GFP construct has been imported only into mitochondria, whereas the PreP1:GFP construct has been imported both into mitochondria and chloroplasts.

2.3.2.6. MICROSCOPY

Fluorescence microscope capable of detecting GFP fluorescence or confocal microscope for better quality of pictures.

3. Methods

3.1. A Single Organelle In Vitro Mitochondrial Import

3.1.1. Preparation of Potato Tuber Mitochondria

1. Grind potato tubers (1 kg) in a Moulinex juice centrifuge in 400 mL of grinding medium (*see* **Notes 1** and **2**).
2. Adjust pH to 7.0–7.5.
3. Filter the slurry through four layers of nylon net (60-μm mesh).
4. Sediment starch and cell wall fragments at 4000 g for 5 min.
5. Recentrifuge the supernatant at 9000 g for 10 min. Resuspend the mitochondrial pellet in 2.5 mL of 1× wash medium.
6. Apply crude mitochondrial suspension onto 40 mL of gradient containing 28% Percoll medium and centrifuge at 32,000 g for 45 min.
7. After centrifugation, collect mitochondrial band positioned close to the bottom of the tube, dilute 10 times in wash medium.
8. Centrifuge at 9000 g for 15 min. Dilute mitochondrial pellet in fresh wash medium to a final protein concentration of 30 mg/mL. Keep mitochondria on ice until use.

3.1.2. Preparation of Spinach Leaf Mitochondria

1. Homogenize depetiolated spinach leaves (300 g) using Waring Blendor in 300 mL of grinding medium 2 × 3 s at high speed (*see* **Note 1**).
2. Filter the slurry through four layers of nylon net (60-μm mesh).
3. Sediment chloroplasts, starch, and cell wall fragments at 5000 g for 10 min.
4. Re-centrifuge the supernatant at 9000 g for 15 min.
5. Resuspend the mitochondrial pellet in small volume of 1× wash medium and load them on top of the same medium containing step gradient of Percoll: 2 mL of 60%, 4 mL of 28%, and 2 mL of 21% for type Ti-21 rotor tubes.
6. Centrifuge at 30,000 g 50 min. Collect the yellowish band containing mitochondria, dilute 10 times with 1× wash medium.
7. Centrifuge at 12,000g for 10 min. Dilute the mitochondrial pellet in fresh wash medium to a final protein concentration of 30 mg/mL. Keep mitochondria on ice until use.

3.1.3. In Vitro Synthesis of Mitochondrial Precursor Proteins

1. All precursor proteins were synthesized with the TNT-coupled reticulocyte lysate system (Promega, SDS Biosciences, Uppsala, Sweden). For the coupled

transcription/translation reaction, mix 1 μg of plasmid with components of the TNT-coupled reticulocyte lysate system, [^{35}S]-methionine, and incubate for 1.5 h at 30°C according to manufacturer's instructions.

3.1.4. In Vitro Mitochondrial Import (28)

1. Import experiments were carried out using isolated potato tuber or spinach leaf mitochondria (200 μg protein) resuspended in 200 μL of import medium. To dissipate mitochondrial membrane potential, add 0.1 μ*M* valinomycin to import mixture.
2. Import reaction is started by addition of 2–10 μL of radiolabeled precursor protein (*see* **Note 3**), and the mixture is incubated for 30 min at 15°C using potato mitochondria or 20 min at 25°C using spinach mitochondria.
3. Stop reaction by placing tubes on ice. Split samples into two, leave one sample on ice, treat the second sample with 10 μg/mL of proteinase K for 20 min at 4°C. Stop the proteolytic reaction by addition of PMSF to 1 m*M*. Reisolate mitochondria by centrifugation in Eppendorf microcentrifuge for 5 min at 15,000 *g*, 4°C, and solubilize mitochondrial pellet with SDS loading buffer (*see* **Fig. 3A**).

3.1.5. SDS Electrophoresis and Image Processing

1. Denature the samples for 5 min at 95°C in 40 μL of the SDS loading buffer and analyze them by SDS-PAGE in the presence of 4 *M* urea.
2. To prepare 12% polyacrylamide gel mix 6.5 mL of 4× separation buffer, 10.5 mL of 30% acrylamide/bis solution, 9 mL H$_2$O, 6.3 g urea, and 20 μl of TEMED. Custom-made glass plates (150 × 150) with 1.5-mm spacers were used for gel formation. AA polymerization begins after addition of 75 μL of APS and solution is immediately applied into gel chamber and overlaid with 1 mL of butanol to achieve even gel formation (*see* **Note 4**).
3. After polymerization of separating gel (polymerization takes approx 30 min, can be visualized as solution/gel partitioning), top solution containing butanol is removed and stacking solution is applied. Stacking gel (4% AA) is prepared by mixing 6.5 mL of 4× stacking buffer, 5 mL of 30% acrylamide/bis solution, 13.5 mL H$_2$O, 6 g urea, 20 μL of TEMED, and 100 μL of APS. The comb is immediately inserted and stacking gel is polymerized for 30 min at room temperature. After removal of comb, pockets of 80 μL loading volume are formed.
4. Glass plates containing polyacrylamide gels are fixed in custom-made electrophoresis apparatus, and upper and lower chambers are filled with 500 mL of SDS running buffer.
5. Apply samples (40 μL) into each well and perform gel electrophoresis overnight at 10 mA constant current at 4°C.
6. After completion of electrophoresis when the bromophenol blue dye front is seen at the bottom of the gel, take the gels out of the glass plates, soak in fix solution for 30 min, and dry using vacuum gel dryer at 70°C for 1 h.

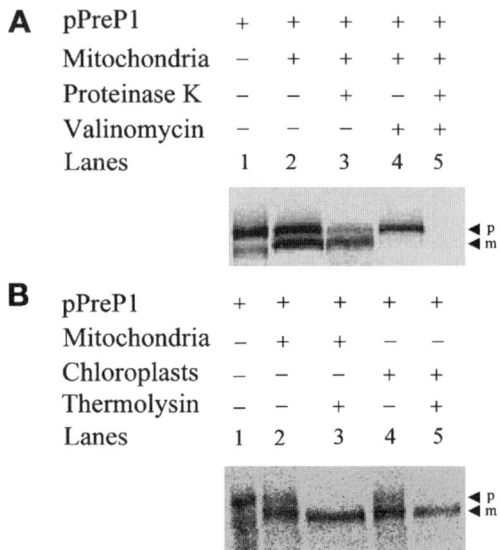

Fig. 3. In vitro import of the presequence protease precursor pPrePl into isolated spinach mitochondria and chloroplasts. (**A**) Single organelle import of pPrePl into spinach leaf mitochondria. Lane 1, precursor protein alone. Lane 2, precursor protein incubated with spinach mitochondria. Lane 3, as lane 2 but with the addition of proteinase K. Lanes 4 and 5, as lanes 2 and 3 but with the addition of valinomycin. p stands for precursor and m for the mature form of the imported protein. (**B**) Dual import of pPrePl simultaneously into spinach mitochondria and chloroplasts. Mitochondria and chloroplasts were isolated separately, mixed, and incubated with the precursor. After import the organelles were treated with thermolysin and reisolated. PrePl is a protein that is dually targeted to both mitochondria and chloroplasts. Lane 1, precursor protein alone. Lane 2, precursor protein import into spinach mitochondria. Lane 3, as lane 2 but with the addition of thermolysin. Lane 4, precursor protein import into spinach chloroplasts. Lane 5, as lane 4 but with the addition of thermolysin. P stands for precursor and m for the mature form of the imported protein.

7. Subsequently, dried gels are exposed overnight to BAS-MP phosphoimaging plates and scanned using a Fujix BAS 1000 MacBAS Bio-imaging Analyzer system.

3.2. Dual In Vitro Import to Mitochondria and Chloroplasts

3.2.1. Biological Material

Approximately 400 g of spinach leaves. Spinach (*Spinacia oleracea* L. cv. Medania) was grown hydroponically for 6 wk under artificial light at 25°C with a light period of 10 h.

3.2.2. Preparation of Organelles

3.2.2.1. PREPARATION OF MITOCHONDRIA

As in single in vitro import (*see* **Subheading 3.1.2.**).

3.2.2.2. PREPARATION OF CHLOROPLASTS

1. Before starting the isolation two 50% Percoll gradients are prepared containing equal parts of Percoll and 2× grinding medium (25 mL Percoll and 25 mL medium) and centrifuged for 30 min at 39, 000 g with the brakes turned off. Keep the gradients on ice until use.
2. Harvest approximately 100 g of spinach leaves from the dark period. Homogenize the tissue in 300 mL of 1× grinding medium using a blender with five short bursts. Filter the homogenate through one layer of chilled and wet Miracloth and transfer into 8 × 50 mL centrifugation tubes and centrifuge for 5 min at 2000 g (*see* **Notes** 1 and **5**).
3. Discard the supernatant and resuspend the pellet with fine hair brush. Gently overlay the crude chloroplasts on the gradients. Centrifuge for 10 min at 12, 100 g with the brakes turned off.
4. Aspirate upper green band (broken chloroplasts and thylakoids) and collect lower (lighter green) using a Pasteur pipet and transfer to a clean centrifugation tube. Add three volumes of import medium and centrifuge for 4 min at 2000 g.
5. Discard the supernatant and resuspend the pellet in a small volume of wash medium. Keep chloroplasts on ice and in dark until use.

3.2.3. In Vitro Transcription/Translation

As in single in vitro import (*see* **Subheading 3.1.3.**).

3.2.4. Dual Import Into Spinach Mitochondria and Chloroplasts *(15)* *(see Note 6)*

1. Measure the concentration of isolated organelles and adjust mitochondria to 10 mg/mL and chloroplasts to 1 mg/mL chlorophyll.
2. Add 65 µL master mix to two (for each precursor protein) prechilled round-bottom Falcon tubes (Eppendorf tubes can be used). Add precursor and incubate for 20 min with gently agitation with a light source.
3. Place the reaction on ice and divide into two aliquots. Supplement one aliquot with 0.1 mM $CaCl_2$ (1 µL of a 5 mM stock) and 120 µg/mL thermolysin (3 µL of a 2 mg/mL stock) and incubate for 30 min on ice. Inhibit the thermolysin activity with 10 mM EDTA (1.25 µL of a 0.5 M stock).
4. Carefully load each reaction on a 4% Percoll gradient prepoured into a 400 µL elongated microfuge tube (Eppendorf tubes can be used) and centrifuge for 30 s at 4000 g. Collect the fractions separately—chloroplasts at the bottom of the gradient

as a pellet and mitochondria at the top. Wash the fractions in 1 mL of import medium. Recover the organelles by centrifugation for 2 min at 830 g for chloroplasts and 2 min maximal speed for mitochondria (*see* **Note 7**).

3.2.5. SDS Electrophoresis

As in single in vitro import (*see* **Subheading 3.1.5.** and **Fig. 3B**).

3.3. In Vivo Import Using Fusions With GFP

3.3.1. Transient Expression of GFP Fusion Constructs in Tobacco Protoplasts

3.3.1.1. CLONING

Use standard cloning techniques to clone the full length or targeting sequence of the protein of interest fused to GFP in an appropriate plasmid containing an eukaryotic expression cassette *(28)* (*see* **Note 8**).

3.3.1.2. PREPARATION OF PROTOPLASTS

1. Abrade the lower face of 30 tobacco leaves with sandpaper No. 1200 A and place them in a sterile Petri dish containing 10 mL digestion medium. The lower (abraded) face is in contact with liquid. Seal the dishes with Parafilm.
2. Leave to digest for 15 h at 25°C in the dark without shaking. To free the protoplasts into the liquid medium, carefully shake the dishes for 30 min (*see* **Note 9**).
3. Remove the undigested pieces of leaf and add 4 mL of flotation medium.
4. Filter the protoplast suspension in a 50 mL Falcon tube through a 64-μm nylon filter.
5. Centrifuge at 110 g for 7 min in the swinging rotor. Transfer the protoplasts (floating band) into another tube (take approx 2.5 mL).
6. Dilute the protoplasts three times with dilution medium. Centrifuge at 110 g for 7 min in the swinging rotor, remove supernatant with vacuum.
7. Wash the protoplasts with 40 mL of electroporation medium and collect by centrifugation at 110 g for 7 min. Resuspend the protoplasts in 20 mL of electroporation medium.
8. Centrifuge at 110 g for 7 min in the swinging rotor, remove supernatant with vacuum, resuspend the protoplasts in electroporation medium to obtain suspension of $1-2 \times 10^6$/mL.

3.3.1.3. DETERMINATION OF PROTOPLAST VIABILITY USING FDA

1. Dilute 5 μL of 20× FDA stock solution in 95 μL of electroporation medium. Pour 10 μL of diluted FDA onto a microscopic slide.
2. Transfer 10 μL of protoplast suspension into the diluted FDA solution.

3. Observe the protoplasts under UV light (375–425 nm) without anything covering it. Count the green (viable) and red (dead) protoplasts and estimate the percentage of dead protoplasts. A normal preparation should contain between 10 and 15% dead protoplasts.

3.3.1.4. ELECTROPORATION OF PROTOPLASTS

1. Set 30 μg of plasmid DNA in a sterile Eppendorf tube.
2. Add 800 μL of well-homogenized protoplasts ($1-2 \times 10^6$/mL in electroporation medium).
3. Mix by slowly inverting the tube and incubate for 15 min at room temperature.
4. Transfer 800 μL of mixture into an electroporation cuvette (0.4 cm). The protoplasts were then electroporated at 250 μF capacitance and voltage of 0.32 kV. Time const = 5 (*see* **Notes 10** and **11**).
5. Leave the protoplasts for 20 min.
6. Transfer into 15-mL Falcon tubes. Add 6 mL of K3M medium.
7. Incubate the tubes in the dark at 25°C for 24 h.

3.3.1.5. CONFOCAL MICROSCOPY ANALYSIS OF THE GFP FUSION EXPRESSION

1. Confocal microscopy was performed using a Bio-Rad MRC-1024 laser scanning confocal imaging system. For GFP detection, excitation was at 488 nm and detection between 506 and 538 nm. Chloroplast autofluorescence was detected between 664 and 696 nm with an excitation at 488 nm. For staining of mitochondria protoplasts, suspension was incubated with 0.1 μ*M* of Mitotracker Red CM-H2Xros for 15 min. Mitotracker Red CM-H2Xros fluorescence was detected between 589 and 621 nm with an excitation at 568 nm (*see* **Fig. 2A**).

3.3.2. Transient Expression Into Tobacco Leaves

3.3.2.1. CLONING

Use standard cloning techniques to clone the full length or targeting sequence of the protein of your interest fused to GFP in an appropriate eukaryotic expression cassette into the *Agrobacterium* pBi plasmid.

3.3.2.2. PREPARATION OF ELECTROCOMPETENT CELLS OF AGROBACTERIUM

1. Streak out *Agrobacterium* on 2YT plates containing rifampicin (50 μg/mL) and gentamycin (20 μg/mL) and grow for 2–3 d at 28°C.
2. Inoculate a single colony into 10 mL of 2YT medium containing the rifampicin (50 μg/mL) and gentamycin (20 μg/mL) antibiotics and grow at 28°C with shaking for 2 d.
3. Inoculate 100 mL of fresh 2YT medium containing the rifampicin and gentamycin antibiotics with 1 mL of preculture and grow (8–10 h) at 28°C until the OD_{600} reaches 0.4–0.6.

4. Place the culture on ice for 15–30 min.
5. Centrifuge the culture at 8000 g for 10 min at 4°C.
6. Pour off supernatant and dissolve the pellet in 20 mL ice-cold 1 mM HEPES, pH 7.0.
7. Centrifuge again and dissolve the pellet in 20 mL ice cold 1 mM HEPES, pH 7.0, 10% glycerol.
8. Centrifuge again and dissolve the pellet in 10 mL ice cold 1 mM HEPES, pH 7.0, 10% glycerol.
9. Aliquot the cells in 50-µL batches and freeze immediately in liquid nitrogen and store at −80°C.

3.3.2.3. Transformation of Electrocompetent Agrobacterium Cells With Plasmid DNA by Electroporation

1. Thaw electrocompetent cells on ice.
2. Incubate 50 µL of cells with 1 µL of plasmid DNA for 2 min on ice. The plasmid DNA may be in either distilled water or a low-salt buffer such as TE buffer. High-salt buffers should be avoided.
3. Transfer the cell–DNA mixture to a chilled 0.2-cm Bio-Rad electroporation cuvette. Tap the mixture to the bottom of the cuvette.
4. Set the Bio-Rad Gene Pulser apparatus to the 25 µF capacitor and the Pulser Controller Unit to 200 Ω.
5. Transfer the cuvette to a chilled Bio-Rad Gene Pulser slide.
6. Set voltage to 2.5 kV.
7. Apply a single pulse and immediately add 1 mL of 2YT broth (room temperature) to the cuvette and gently resuspend the cells.
8. Transfer the cell suspension to a tube and incubate the culture at 28°C with shaking for 2 h.
9. Plate the suspension on rifampicin, gentamycin, kanamycin plate and incubate the plates at 28°C. Transformed colonies will be visible after 2–3 d of incubation at 28°C.

3.3.2.4. Infiltration of the Agrobacterium Suspension Into Tobacco Leaves

1. Culture the 2 mL of *Agrobacterium* (transformed with your construct) in YEB medium containing 100 µL/mL kanamycin, 10 µL/mL gentamycin, and 50 µL/mL of rifampicin at 28°C with shaking to stationary phase (24–48 h).
2. Take 1 mL of the culture, pellet (6000 g for 2 min), and wash twice in infiltration medium then resuspend at OD_{600} of 0.1–0.3.
3. Inject bacterial suspension into the abaxial epidermis of healthy plant leaves from a 1-mL plastic syringe by pressing the nozzle against the leaf surface

(*see* **Notes 12** and **13**). Spread of liquid entering the leaf via stomata can be visualized by a darkening of the leaf. Mark boundaries of the infiltrated area with an indelible pen (*see* **Note 14**).

4. Incubate plants for 2–3 d at 20–25°C under normal growing conditions.

3.3.2.5. MICROSCOPY

Check for expression by analyzing a piece of the infiltrated area under fluorescence microscope, capable of detecting GFP fluorescence. The abaxial surface of the leaf should be facing the incoming light (*see* **Note 15** and **Fig.** 2B).

Notes

1. As a general rule, try to keep operating temperature during isolation of mitochondria and chloroplasts close to 0°C and minimize time between isolation of organelles and import of precursor proteins. Precool buffers, centrifuge, centrifuge tubes, pipets, and beakers before use.
2. The yield and import competence of potato tuber mitochondria can depend on the quality of potatoes.
3. Thaw precursor protein on ice just prior to use.
4. Unpolymerized acrylamide is neurotoxic, and care should be taken to avoid exposure.
5. Be gentle when resuspending chloroplasts to avoid breakage.
6. Prepare the 4% Percoll gradients before starting the import procedure, and keep on ice until use.
7. If mitochondria do not pellet when recovered, extend the centrifugation time to 10 min. The mitochondrial pellet can be quite loose, so remove supernatant as soon as the centrifugation finished.
8. All the manipulations should be performed under the sterile laminar flow.
9. Protoplasts are very fragile, should be handled with care; for example, use trimmed pipet tips to avoid damage.
10. Transient expression of a GFP fusion construct can also be performed using polyethylene glycol-mediated transformation of a plasmid into protoplasts *(29)*.
11. Transient expression of GFP fusion constructs has also been achieved in soybean suspension cells using biolistic transformation *(30)*.
12. Avoid the bottom two leaves and the top two unfolded leaves for infiltration because of the variability in infection rate.
13. Transient expression of GFP constructs in tobacco leaves can also be performed by biolistic transformation *(30,31)*.
14. Plant species so far tested are tobacco, tomato, potato, and petunia, with expression obtained in all cases.

15. Expression of GFP varies with the infiltrated concentration of *Agrobacterium,* which offers an opportunity to control GFP expression levels.

Acknowledgments

The authors would like to thank Prof. J. Whelan and Dr. O. Chew for a fruitful collaboration on the development of the dual import system and Prof. M. Boutry and Dr. B. Lefebvre for instructions and cooperation using in vivo import procedures. This work was supported by a grant from The Swedish Research Council to EG.

References

1. Truscott, K. N., Brandner, K., and Pfanner, N. (2003) Mechanisms of protein import into mitochondria. *Curr. Biol.* **13**, R326–337.
2. Gray, M. W., Burger, G., and Lang, B. F. (1999) Mitochondrial evolution. *Scienc,* **283**, 1476–1481.
3. Zhang, X. P. and Glaser, E. (2002) Interaction of plant mitochondrial and chloroplast signal peptides with Hsp70 molecular chaperone. *Trends Plant Sci.* **7**, 14–21.
4. Glaser, E. and Soll, J. (2004) Targeting signals and import machinery of plastids and plant mitochondria. In *Molecular Biology and Biotechnology of Plant Organelles: Chloroplasts and Mitochondria* (Daniell, H. and Chase, C., eds.) Springer, Dordrecht, The Netherlands, pp. 385–418.
5. Marc, P., Margeot, A., Devaux, F., Blugeon, C., Corral-Debrinski, M., and Jacq, C. (2002). Genome-wide analysis of mRNAs targeted to yeast mitochondria. *EMBO Rep.* **3**, 159–164.
6. Glaser, E. and Dessi, P. (1999) Integration of the mitochondrial processing peptidase into the bc1 complex of the respiratory chain in plants. *J. Bioenerg. Biomembr.* **31**, 259–274.
7. Ståhl, A. Moberg, P., Ytterberg, J., et al. (2002) Isolation and identification of a novel mitochondrial metalloprotease (PreP) that degrades targeting presequences. *J. Biol. Chem.* **277**, 41931–41939.
8. Moberg, P., Ståhl, A., Bhushan, S., et al. (2003) Characterization of a novel zinc metalloprotease involved in degrading signal peptides in mitochondria and chloroplasts. *Plant J.* **36**, 616–628.
9. Bushan, S., Lefebvre, B., Ståhl, A., Boutry, M., and Glaser, E. (2003) Dual targeting and function of a protease in mitochondria and chloroplasts. *EMBO Rep.* **4**, 1073–1078.
10. Ståhl, A., Nilsson, S., Lundberg, P., Bhushan, S., et al. (2005) Two novel targeting peptide degrading proteases, PrePs, in mitochondria and chloroplasts, so similar and still different. *J. Mol. Biol.* **349**, 847–860.

11. Johnson, K. A., Bhushan, S., Ståhl, A., et al. (2006) The closed structure of presequence protease PreP forms a unique 10,000 Å^3 chamber for proteolysis. *EMBO J.* **25**, 1977–1986.

12. Creissen, G., Reynolds, H., Xue, Y., and Mullineaux, P. (1995) Simultaneous targeting of pea glutathione reductase and of a bacterial fusion protein to chloroplasts and mitochondria in transgenic tobacco. *Plant J.* **8**, 167–175.

13. Small, I., Wintz, H., Akashi, K., and Mireau, H. (1998) Two birds with one stone: genes that encode products targeted to two or more compartments. *Plant Mol. Biol.* **38**, 265–277.

14. Hedtke, B., Borner, T., and Weihe, A. (2000) One RNA polymerase serving two genomes. *EMBO Rep.* **1**, 435–440.

15. Rudhe, C., Chew, O., Whelan, J., and Glaser, E. (2002) A novel in vitro system for simultaneous import of precursor proteins into chloroplast and mitochondria. *Plant J.* **30**, 213–220.

16. Boutry, M., Nagy, F., Poulsen, C., Aoyagi, K., and Chua, N.H. (1987) Targeting of bacterial chloramphenicol acetyltransferase to mitochondria in transgenic plants. *Nature* **328**, 340–342.

17. Schmitz, U. K. and Lonsdale, D. M. (1989) A yeast mitochondrial presequence functions as a signal for targeting to plant mitochondria in-vivo. *Plant Cell* **1**, 783–791.

18. Silva Filho, M. d. C., Wieers, M.-C., Flugge, U.-I., Chaumont, F., and Boutry, M. (1997) Different in vitro and in vivo targeting properties of the transit peptide of a chloroplast envelope inner membrane protein. *J. Biol. Chem.* **272**, 15264–15269.

19. Whelan, J., Knorpp, C., and Glaser, E. (1990) Sorting of precursor proteins between isolated spinach leaf mitochondria and chloroplasts. *Plant Mol. Biol.* **14**, 977–982.

20. Glaser, E., Sjoling, S., Tanudji, M., and Whelan, J. (1998) Mitochondrial protein import in plants. *Plant Mol. Biol.* **38**, 311–338.

21. Soll, J. and Tien, R. (1998) Protein translocation into and across the chloroplastic envelope membranes. *Plant Mol. Biol.* **38**, 191–207.

22. Hugosson, M., Nurani, G., Glaser, E. and Franzen, L.G. (1995) Peculiar properties of the PsaF photosystem I protein from the green alga *Chlamydomonas reinhardtii*: presequence independent import of the PsaF protein into both chloroplasts and mitochondria. *Plant Mol. Biol.* **28**, 525–535.

23. Brink, S., Flugge, U. I., Chaumont, F., et al. (1994) Preproteins of chloroplast envelope inner membrane contain targeting information for receptor-dependent import into fungal mitochondria. *J. Biol. Chem.* **269**, 16478–16485.

24. von Stedingk, E. (1999) Sorting and import of plant mitochondrial precursors. PhD thesis, Stockholm University, Stockholm.

25. Lister, R., Chew, O., Lee, M., and Whelan, J. (2001) *Arabidopsis thaliana* ferrochelatase-I and -II are not imported into *Arabidopsis* mitochondria. *FEBS Lett.* **506**, 291–295.

26. Cleary, S. P., Tan, F.-C., Nakrieko, K.-A., et al. (2002) Isolated plant mitochondria import chloroplast precursor proteins in vitro with the same efficiency as chloroplasts. *J. Biol. Chem.* **277**, 5562–5569.

27. Duby, G., Oufattole, M., and Boutry, M. (2001) Hydrophobic residues within the predicted N-terminal amphiphilic alpha-helix of a plant mitochondrial targeting presequence play a major role in in vivo import. *Plant J.* **27**, 539–549.

28. Von Stedingk, E., Pavlov, P. F., Grinkevich, V.A., and Glaser, E. (1999) The precursor of $F_1\beta$ subunit of the ATP synthase is covalently modified upon binding to plant mitochondria. *Plant Mol. Biol.* **41**, 505–515.

29. Datta, K. and Datta, S.K. (1999) Transformation of rice via PEG-mediated DNA uptake into protoplasts. *Meth. Mol. Biol.* **111**, 335–347.

30. Chew, O., Rudhe, C., Glaser, E., and Whelan, J. (2003) Characterization of the targeting signal of dual-targeted pea glutathione reductase. *Plant Mol. Biol.* **53**, 341–356.

31. Lukaszewicz, M., Jerouville, B. and Boutry, M. (1998) Signs of translational regulation within the transcript leader of a plant plasma membrane H(+)-ATPase gene. *Plant J.* **14**, 413–423.

10

Protein Targeting to Mitochondria of *Saccharomyces cerevisiae* and *Neurospora crassa*

In Vitro and In Vivo Studies

Panagiotis Papatheodorou, Grażyna Domańska, and Joachim Rassow

Summary

Most studies on the biogenesis of mitochondrial proteins have been carried out using fungal mitochondria as a model system. In particular, baker's yeast, *Saccharomyces cerevisiae,* combines several experimental advantages, allowing both genetic and biochemical approaches and thus a combination of investigations in vivo and in vitro. However, the red bread mold *Neurospora crassa* has also been an important research tool. Isolated mitochondria can be used from both organisms for import experiments in a reconstituted system, using radiolabeled precursor proteins synthesized in reticulocyte lysate or purified preproteins. Assays are available for studies on the import pathways and localization of mitochondrial proteins and for the characterization of the components of the protein import machinery.

Key Words: Mitochondria; yeast; in vitro assay; protein targeting; protein translocation.

1. Introduction

Our current knowledge of mitochondrial protein targeting emerged primarily from extensive research work using isolated fungal organelles, in particular from *Saccharomyces cerevisiae* or *Neurospora crassa*. Isolated mitochondria from these organisms show a high efficiency of protein uptake in vitro for at least 2 h. If necessary, yeast mitochondria can be frozen for long periods without loss of activity. Moreover, yeast is accessible to powerful methods of molecular genetics. Mutants are available for nearly all components of the mitochondrial protein import machinery. In this chapter we describe the

From: *Methods in Molecular Biology, Vol. 390: Protein Targeting Protocols: Second Edition*
Edited by: M. van der Giezen © Humana Press Inc., Totowa, NJ

isolation of mitochondria from *S. cerevisiae* and *N. crassa*, including an introduction to the application of the standard in vitro system for protein import and assembly *in organello*. Eventually, we include hints regarding the use of yeast as a heterologous expression system to elucidate mitochondrial targeting in intact cells.

The uptake of nuclear-encoded proteins from the cytosol is mainly mediated by the TOM complex (the protein translocase of the mitochondrial outer membrane) and by two independent TIM complexes (translocases of the inner membrane). Additional complexes specifically mediate the insertion of some proteins into the outer membrane, transfer of proteins across the intermembrane space, or transport into the matrix compartment *(1–4)*. The TOM and TIM complexes are highly conserved in all eukaryotes *(5,6)*. Fungal mitochondria therefore provide a suitable model system not only in studies on general features of mitochondrial protein import, but also in the characterization of mitochondrial proteins from mammalian origin. Most mammalian preproteins can efficiently be imported into yeast or *Neurospora* mitochondria. Tests using organelles from different origin can be combined to address different questions. In this case, the advantage of yeast will be the accessibility of genetic methods, together with the unique amount of data on all aspects of the mitochondrial protein import machinery.

2. Materials

If not otherwise stated, prepare all solutions in distilled H_2O.

2.1. Isolation of Mitochondria from Saccharomyces cerevisiae

1. YPG medium: 1% (w/v) yeast extract, 2% (w/v) peptone, 3% (v/v) glycerol; prior to the addition of glycerol, adjust to pH 4.8–5.0 with concentrated HCl; autoclave 1.5 L in a 5-L flask at 121°C for 20 min.
2. Tris/Dithiothreitol (DTT) buffer: 100 mM Tris-H_2SO_4 pH 9.4, 10 mM DTT (add DTT from a freshly made 1 M stock solution).
3. 2.4 M sorbitol: store at 4°C; used to prepare the sorbitol solutions in **steps 4** and **6**.
4. Sorbitol/KH_2PO_4 buffer: 1.2 M sorbitol, 20 mM potassium phosphate (add potassium phosphate from a 100 mM stock solution, pH 7.4), store at 4°C.
5. Zymolyase 20T from *Arthrobacter luteus* (200 U/g; e.g., Seikagaku Corporation): store at 4°C (*see* **Note 2**).
6. Homogenization buffer: 0.6 M sorbitol, 10 mM Tris-HCl (add Tris-HCl from a 1 M stock solution, pH 7.4), 1 mM EDTA (add EDTA from a 250 mM stock solution, pH 8), 0.5% bovine serum albumin (BSA) (fatty acid-free; e.g., Sigma,

9003T); adjust pH after solubilization of the BSA to 7.4, should be prepared just before use, cool on ice.

7. 100 m*M* Phenylmethanesulfonyl fluoride (PMSF) in isopropanol (caution: very toxic): unstable in water, should be added just prior to use, freshly made.

8. SEM buffer: 250 m*M* sucrose, 1 m*M* EDTA, 10 m*M* MOPS; adjust to pH 7.2 with KOH (store at $-20°C$, stable for months).

9. Liquid nitrogen and vials for freezing samples.

10. Dounce homogenizer.

2.2. Isolation of Mitochondria from Neurospora crassa

1. Vogel's minimal medium: 8.4 m*M* Na$_3$ citrate × 2 H$_2$O, 36.7 m*M* KH$_2$PO$_4$, 25 m*M* NH$_4$NO$_3$, 0.8 m*M* MgSO$_4$, 0.68 m*M* CaCl$_2$, 1 n*M* biotin, 23.7 µ*M* citric acid, 12.1 µ*M* Fe(NH$_4$)$_2$(SO$_4$)$_2$, 17.3 µ*M* ZnSO$_4$ × 7 H$_2$O, 1.5 µ*M* CuSO$_4$, 0.2 µ*M* MnSO$_4$, 0.8 µ*M* H$_3$BO$_3$, 0.2 µ*M* Na$_2$MoO$_4$; sterilize by filtration; detailed protocol for growth of *Neurospora crassa* (*see* **ref. 23**).

2. 50% (w/v) Sucrose (autoclave): should be added to a final concentration of 2% (v/v) to Vogel's minimal medium.

3. Filter paper (e.g., Schleicher and Schüll, 595) and Büchner funnel.

4. Mortar and pestle and sterile quartz sand (e.g., Riedel de Haen).

5. PMSF and SEM (*see* **Subheadings 2.1.7.** and **2.1.8.**).

2.3. Assessment of the Mitochondrial Membrane Potential

1. Phosphate buffer (for the assessment of the mitochondrial membrane potential): 0.6 *M* sorbitol, 0.1% (w/v) BSA, 10 m*M* MgCl$_2$, 0.5 m*M* EDTA, 20 m*M* potassium phosphate; adjust to pH 7.2.

2. 3, 3'-Dipropylthiadicarbocyanine iodide [DiSC$_3$(5)] (e.g., Molecular Probes): prepare a 2 m*M* solution in dimethyl sulfoxide (DMSO) and 1 *M* KCN (caution: toxic).

2.4. Protein Import Into Isolated Yeast Mitochondria

1. T$_N$T$^®$-coupled reticulocyte lysate system (Promega) and 10 µCi/µL ^{35}S-methionine (e.g., ICN Biomedical Research Products).

2. Import buffer: 10 m*M* MOPS, 250 m*M* sucrose, 80 m*M* KCl, 5 m*M* MgCl$_2$, 2 m*M* KH$_2$PO$_4$, 3% BSA (essentially fatty acid-free, e.g., Sigma 9003T); adjust pH to 7.2 with KOH, stable for years at $-20°C$.

3. 0.2 *M* ATP (adjust with KOH to pH 7.2; store at $-20°C$)

4. 100 m*M* NADH, freshly made in H$_2$O.

5. 1 m*M* valinomycin (e.g., Sigma) in ethanol: stable for years at $-20°C$, K$^+$-Ionophor, disrupting the membrane potential, very toxic.

6. Radioactive ink: mix 1-2 µL ^{35}S-methionine with 1 mL black ink.

7. Proteinase K (e.g., Sigma): 1 mg/mL in distilled water; freshly made; store in ice just prior to use.

8. PMSF and SEM (*see* **Subheading 2.1.**, **steps 7** and **8**).

2.5. In Vivo Mitochondrial Localization of Expressed Proteins in Yeast

1. Synthetic minimal medium for yeast lacking uracil: 0.67% (w/v) yeast nitrogen base (without amino acids, e.g., Becton Dickinson, Difco™), 0.01% ([w/v] adenine, arginine, cysteine, leucine, lysine, threonine, tryptophan), 0.005% ([w/v] aspartic acid, histidine, isoleucine, methionine, phenylalanine, proline, serine, thyrosine, valine). Dissolve reagents in 90 mL deionized water (if preparing medium containing galactose) or 80 mL deionized water (if preparing medium containing raffinose). For making plates, add 2% (w/v) agar to the medium after dissolving of the reagents. Autoclave at 121°C for 20 min prior to the addition of 10 mL filter-sterilized 20% (w/v) galactose or 20 mL filter-sterilized 10% (w/v) raffinose.
2. 0.1 mM MitoTracker® Orange CMTMRos (Molecular Probes, M-7510) in DMSO; prepare from a 1 mM stock solution of MitoTracker Orange CMTMRos in DMSO.
3. Polylysinated object slides (e.g., Sigma, P0425).
4. PBS (phosphate-buffered saline): 150 mM NaCl, 20 mM sodium phosphate, pH 7.4.

3. Methods
3.1. Isolation of Mitochondria from S. cerevisiae

Yeast can grow on fermentable as well as on nonfermentable media. Glycerol is commonly used as a suitable nonfermentable carbon source. It forces the cells to respire and thus to keep their mitochondria growing. Therefore, glycerol media are preferentially used to grow yeast for the subsequent isolation of mitochondria. In the presence of glucose, yeast is able to obtain ATP independently of mitochondria, exclusively by fermentation, but the mitochondria continue to be essential for growth because several metabolic pathways are dependent on mitochondria enzymes. In a strict sense, mitochondria seem to be essential organelles primarily in the formation of iron–sulfur clusters *(7)*. Mitochondria can therefore be isolated after growth on glucose, and the mitochondria can be used in protein import assays.

Usually, glycerol-containing YPG medium is the medium of choice. The isolation of mitochondria from a yeast culture proceeds in three steps. First, the cell wall is digested by zymolyase; next, the cells are broken up by osmotic shock; and finally, the mitochondria are isolated by differential centrifugation. The entire procedure requires about 5–8 h.

1. Prepare a starter culture by inoculating 50 mL YPG medium with the desired yeast strain (*see* **Note 1**) and incubate the flask under shaking at 30°C until OD_{600} reaches 1–3.
2. Inoculate a further flask containing the desired volume of YPG medium with the starter culture to an OD_{600} of 0.1.

3. Incubate the flask at 30°C under vigorous shaking for 15–18 h until OD_{600} of culture reaches 1–2.
4. Pellet the cells by centrifugation for 5 min at 1600g at room temperature.
5. Wash the cells by resuspension in distilled water, and repeat centrifugation as indicated in **step 4**.
6. Determine the weight of the resulting cell pellet, and resuspend the cells in 1 mL buffer Tris/DTT per 0.5 g wet weight of cells.
7. Incubate the cell suspension for 10 min at 30°C under gentle shaking.
8. Pellet the cells by centrifugation for 5 min at 2800g at 4°C.
9. Resuspend the cells in 1.2 M sorbitol and repeat centrifugation as indicated in **step 8**.
10. Resuspend the cells in 1 mL buffer sorbitol/KH_2PO_4 per 0.15 g wet weight of cells.
11. Supplement the cell suspension with 1–2 mg zymolyase per g wet weight of cells and incubate for 30–45 min at 30°C under gentle shaking (*see* **Note 2**).
12. Monitor the conversion of yeast cells into spheroplasts by dilution of 10 μL of cell suspension in 1 mL distilled water. Diluted suspension becomes clear after several seconds. If this is not the case, repeat **step 11**.
13. Pellet the spheroplasts by centrifugation for 5 min at 2800g at 4°C.
14. Wash the spheroplasts twice by resuspension in 1.2 M sorbitol and centrifugation as indicated in **step 13**.
15. Resuspend spheroplasts in 1 mL cold homogenization buffer per 0.15 g wet weight of cells and add PMSF from a 100 mM stock solution to a final concentration of 1 mM.
16. Carry out 10–20 strokes in a tight-fitting Dounce homogenizer cooled on ice to open spheroplasts.
17. Centrifuge the homogenate for 5 min at 4°C, first at 1100g and then at 1900g. Discard the pelleted cell debris.
18. Isolate mitochondria from supernatant by centrifugation for 10 min at 12,100g at 4°C.
19. Resuspend mitochondria in 5–10 mL buffer SEM + 1 mM PMSF and centrifuge the suspension for 5 min at 1900 rpm at 4°C.
20. Discard the pelleted remaining cell debris and isolate mitochondria from supernatant by centrifugation for 10 min at 12,100g at 4°C.
21. Resuspend mitochondria in 10 μL SEM per 1000 mL original yeast culture and adjust the protein concentration to 5–10 mg/mL based on Bradford protein assay *(8)*.
22. Freeze mitochondria in aliquots in liquid nitrogen. Store mitochondria at −70°C.

Starting with a culture of 10 L, about 10–80 g yeast cells are obtained. The yield of the procedure is usually 0.1–0.5 mg mitochondrial protein per g cells (wet weight).

3.2. Isolation of Mitochondria from N. crassa

Whereas disruption of the cell wall in yeast is achieved by enzymatic digestion, in the case of *N. crassa* a mechanical procedure was established for that purpose. Cell wall disruption is carried out by controlled grinding of the hyphae with quartz sand. Afterwards, mitochondria can easily be obtained by differential centrifugation. In comparison to yeast, the cultivation of *N. crassa* is more complicated and requires additional equipment. However, once the hyphae are obtained, the isolation of mitochondria can be completed within 1–2 h. It is recommended to use only freshly isolated mitochondria for protein import assays.

1. Inoculate the cell culture for the isolation of mitochondria with 2×10^9 conidia (*see* **Note 1**) per L Vogel's medium supplemented with 2% (w/v) sucrose.
2. Incubate cell culture for 15 h at 25 °C under bright illumination and vigorous aeration via a glass tube introducing sterile compressed air.
3. Lay a filter paper in a Büchner funnel and isolate hyphae from cell culture by suction filtration. Peel off the hyphae from filter paper and determine the wet weight.
4. Stir the hyphae into a slurry consisting of 1.5 g sterile quartz sand per g and 2 mL buffer SEM per g wet weight of hyphae, supplemented with 2 mM PMSF.
5. Grind down hyphae for 1–2 min at 4°C using a mortar and pestle.
6. Remove sand and cell walls twice by centrifugation for 5 min at 2000g at 4°C.
7. Isolate mitochondria from cell homogenate by centrifugation of the combined supernatants for 12 min at 17,000g at 4°C.
8. Wash mitochondria by resuspension in buffer SEM supplemented with 2 mM PMSF and centrifugation as indicated in **step 7**.
9. Resuspend mitochondria in buffer SEM and adjust protein concentration to 5 mg/mL based on Bradford protein assay.

3.3. Assessment of the Mitochondrial Membrane Potential

Import of mitochondrial outer membrane proteins is usually completely independent of the mitochondrial membrane potential. In contrast, transport of proteins into or across the mitochondrial inner membrane strictly requires an intact membrane potential. The membrane potential can be assessed by using the fluorescence dye DiSC$_3$(5) (3, 3'-dipropylthiadicarbocyanine iodide). The positively charged dye accumulates in mitochondria and the fluorescence is quenched by an unknown mechanism. After dissipation of the membrane potential, the dye is released again. Therefore, the difference in fluorescence before and after treatment of mitochondria with an uncoupling reagent, such as cyanide or valinomycin, can be used to monitor the mitochondrial membrane potential. An example for a typical result of a fluorescence measurement of the mitochondrial membrane potential is shown in **Fig. 1**.

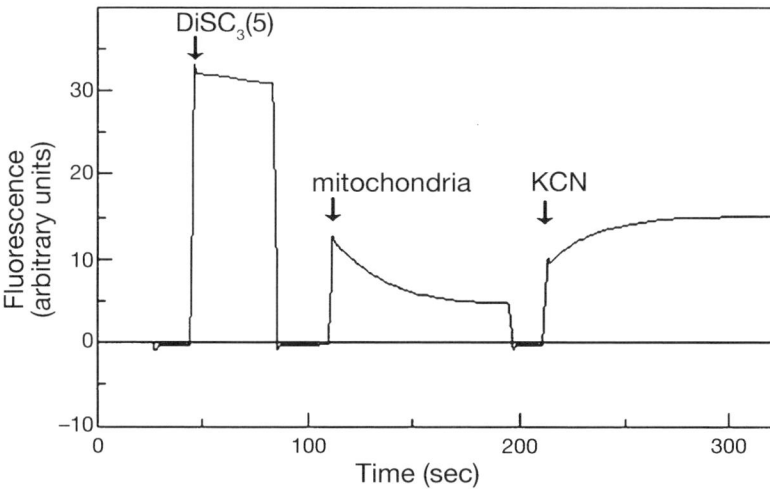

Fig. 1. Assessment of the mitochondrial membrane potential. $DiSC_3(5)$ fluorescence is decreased by quenching after the addition of mitochondria. Dissipation of the membrane potential by the addition of KCN leads to the release of $DiSC_3(5)$ and increase of fluorescence.

1. Add 3 mL phosphate buffer (*see* **Subheading 2.3.**) to a cuvette and determine fluorescence in a fluorescence spectrophotometer (excitation at 622 nm and emission at 670 nm) at 25°C.
2. Complement with 3 µL of a 2 mM $DiSC_3(5)$ solution.
3. Add 20 µL mitochondria from a 10 mg/mL suspension and mix immediately. The uptake of the dye by the mitochondria should lead to a significant decrease in fluorescence.
4. Wait for at least 2 min and add 3 µL of a 1 m*M* KCN solution to dissipate the membrane potential. The release of the dye results in an increase of fluorescence, correlating with the mitochondrial membrane potential (*see* **Note 3**).

3.4. Protein Import Into Isolated Yeast Mitochondria

Mitochondrial protein import can simply be reconstituted in an in vitro experiment by incubating isolated yeast mitochondria with radiolabeled precursor proteins from reticulocyte lysate or with isolated precursor proteins. Subsequent analysis of the import reaction serves to determine the mitochondrial localization of the imported precursor protein. By introducing special conditions (e.g., dissipation or reduction of the membrane potential *[9]*, variation of internal or external nucleotide levels *[10]*) or by using mitochondria derived from mutant strains with specific defects in components of the translocation machinery, it is possible to study the biogenesis of an imported protein in more detail.

3.4.1. Synthesis of Radiolabeled Precursor Proteins in Reticulocyte Lysate

A precursor protein can be synthesized in an in vitro protein synthesis system, based on a plasmid, containing a RNA polymerase promotor (e.g., SP6 or T7, in front of the protein coding sequence of interest), and by coupled transcription and translation in a commercially available reticulocyte lysate. The reaction is carried out in a single tube for 60–90 min at 30°C. It should be noted that the concentrations of the proteins that are synthesized during this procedure are very low (about 1–10 pmol/mL). They are generally not detectable by immunoblotting. For that reason, [^{35}S]methionine is usually included to radiolabel the newly synthesized protein for subsequent detection. The following protocol corresponds to the manufacturer's recommendations for a volume of 50 μL of the TnT-coupled reticulocyte lysate system (Promega). Other reticulocyte lysates are similarly suitable:

1. Prepare an Eppendorf tube containing 1 μg plasmid DNA diluted in 18 μL nuclease-free water (*see* **Note 4**).
2. Supplement the solution with 25 μL rabbit reticulocyte lysate, 2 μL reaction buffer, 1 μL amino acid mixture (w/o methionine), 2 μL [^{35}S]methionine ({GT}1000 Ci/mmol at 10 mCi/mL), 1 μL ribonuclease inhibitor (RNasin$^{®}$), and 1 μL RNA polymerase (SP6 or T7).
3. Incubate the reaction mixture for 90 min at 30°C without shaking.
4. Store the reaction mixture at −80°C.

3.4.2. Standard Protocol for Import of Proteins in Isolated Yeast Mitochondria

Most mitochondrial preproteins can be imported using standard conditions: The samples have volumes of 50 or 100 μL, containing 25–50 μg mitochondrial protein and 1–10 μL reticulocyte lysate. The buffer contains 10 mM MOPS pH 7.4 and 3% (w/v) BSA. To protect the mitochondria from osmotic swelling, 250 mM sucrose is included. Optimal salt concentrations are achieved with 80 mM KCl. To support the respiratory chain and the membrane potential, 1 mM NADH is added. In some cases the efficiency of protein import is improved by addition of 0.5 mM ATP, 20 mM potassium phosphate, or 5 mM MgCl$_2$. Preproteins that are prone to aggregation can sometimes be stabilized by raising the content of reticulocyte lysate in the samples to 25%.

Following expression in *Escherichia coli*, purified precursor proteins can be imported using similar protocols *(11)*. Successful import into mitochondria is usually demonstrated by treatment with proteinase K (20–100 μg/mL, 10 min at 0°C). The fraction of imported precursor proteins (5–50%) should be protected against degradation. Binding of some proteins to the outer surface of the mitochondria can be demonstrated by testing for the formation of specific proteolytic fragments *(12)*. Membrane potential-dependent steps of translocation can be blocked by addition of valinomycin. The result of a typical standard experiment is shown in **Fig. 2**. The experiment was carried out according to the following protocol:

1. Thaw suspension of isolated yeast mitochondria on ice (*see* **Note 5**).
2. Prepare master mix containing 415 μL import buffer (*see* **Note 6**), 10 μL 0.1 *M* NADH, 5μL 0.2 *M* ATP, and 30μL suspension of mitochondria (10 mg/mL).
3. Take aliquots of 94 μL from the mixture and add 2 μL ethanol to samples 1–3, or 2 μL 100 μ*M* valinomycin (in ethanol) to sample 4, respectively, in four separate tubes.
4. Incubate samples for 3 min at 25°C.
5. Initiate import reactions by adding 4 μL reticulocyte lysate (containing radiolabeled precursor protein) to each of the samples (*see* **Note 7**). Mix and incubate at 25°C.
6. After 2, 5, and 20 min, respectively, terminate the import reactions by cooling the samples on ice.

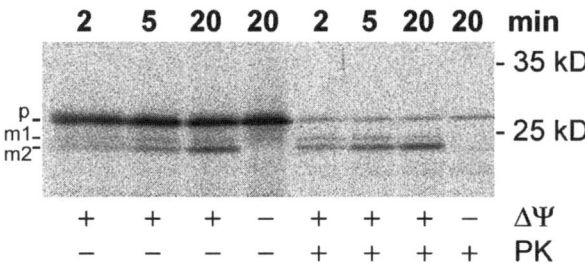

Fig. 2. In vitro import of the Rieske iron–sulfur protein into isolated yeast mitochondria. The precursor (p) is imported in a time-dependent manner into the mitochondrial matrix, followed by a two-step processing to the premature form (m1) and mature form (m2) *(18)*. Import is inhibited in the absence of a membrane potential $(-\Delta\Psi)$. Unspecifically bound precursor (p) shows neither a dependence on mitochondrial membrane potential nor a legible increase in import over time and remains accessible to externally added proteinase K. A minor fraction of the precursor accumulates in the intermembrane space.

7. Split all samples into halves and transfer into new tubes.
8. Add 1 μL proteinase K (1 mg/mL, f.c. 20 μg/mL) to one half of samples, mix (vortex) and incubate for 10 min on ice.
9. Add 0.5 μL PMSF (200 mM in ethanol, final concentration 2 mM) to all samples, mix (vortex) and incubate for 5 min on ice.
10. Centrifuge all samples in a table centrifuge (10 min, 16,000g, 4°C) and remove the supernatant (*see* **Note 8**).
11. Wash mitochondrial pellet by resuspending in 100 μL SEM buffer and repeat centrifugation as indicated in **step 10**.
12. Discard the supernatant and resuspend the mitochondrial pellet in 20 μL sample buffer.
13. Heat the samples for 5 min at 95°C. Separate proteins by SDS-PAGE. Stain, destain, and dry the SDS polyacrylamide gel. Expose on phosphorimager screen (*see* **Note 9**).

Techniques to further characterize the import pathways or the localization of mitochondrial proteins are summarized in **Table 1**.

3.5. In Vivo Mitochondrial Localization of Expressed Proteins in Yeast

S. cerevisiae can also be used for in vivo mitochondrial targeting studies by choosing a vector for inducible expression of recombinant proteins. The desired precursor protein can be fused N- or C-terminally to a GFP sequence or an epitope tag sequence. GFP fluorescence allows direct monitoring of the intracellular localization of the precursor protein in yeast by fluorescence microscopy. Epitope tags can be used for the detection of expressed proteins when antibodies recognizing the protein of interest are not available. For heterologous protein expression in yeast, the pYES2 vector (Invitrogen) can be used, which allows easy cloning and selection of transformants by uracil prototrophy on yeast host strains with an *ura3* genotype. pYES2-based protein expression in yeast can be induced from a *GAL1* promoter by the addition of galactose to the medium. After the induction, the intracellular localization of GFP-tagged proteins can be directly monitored in living cells by confocal fluorescence microscopy. Mitochondrial localization can be shown by colocalization of GFP fluorescence with signals of mitochondria stained by a specific dye (MitoTracker, Molecular Probes; *see* **Note 10**). Alternatively, the intracellular localization of epitope-tagged proteins can be analyzed by lysis of the cells and subsequent fractionation of the different cell compartments. Heterologous expression of a GFP moiety C-terminally fused to a mitochondrial targeting sequence is shown in **Fig. 3**. The protocol can similarly be applied with other preproteins:

Table 1
Characterization of Mitochondrial Proteins

Application	Ref.
Generation of translocation intermediates	*(20–22)*
Analysis of the import reaction and localization of imported proteins	*(23–26)*
Chemical crosslinking of imported proteins with components of the mitochondrial translocation machinery	*(27–29)*
Analysis of protein folding after import into mitochondria	*(30,31)*
Analysis of mitochondrial protein complexes by blue native polyacrylamide gel electrophoresis (BN-PAGE)	*(32–35)*
Analysis of mitochondrial protein complexes by coimmunoprecipitation	*(36,37)*
Monitoring the association of imported proteins with mitochondrial membranes by carbonate extraction	*(38)*
Additional protocols for isolation of fungal mitochondria	*(23,32,39–41)*

1. Inoculate 5 mL synthetic minimal medium for yeast lacking uracil and containing 2% raffinose with yeast transformants harbouring the pYES2 construct and incubate under shaking until culture OD_{600} reaches 1–3 (*see* **Note 11**).
2. Centrifuge x mL culture volume in a table centrifuge at room temperature for 5 min at 1500g, according to the formula: x (mL) $= 0.5/OD_{reached}$ * 2 mL.
3. Resuspend the cell pellet in 2 mL synthetic minimal medium for yeast lacking uracil and containing 2% galactose to induce protein expression and incubate for 4 h at 30°C under gentle shaking.
4. Add 1 μL MitoTracker Orange CMTMRos from a 0.1 mM working solution in DMSO (*see* **Note 12**) to the medium (final concentration 50 nM) and continue incubation for 30 min at 30°C under gentle shaking.
5. Collect the cells by centrifugation in a table centrifuge at room temperature for 5 min at 1500g and resuspend in 2 mL buffer PBS.
6. Transfer 10 μL of the cell suspension to a polylysinated object slide.
7. Allow the cells to sediment on the surface of the slide by incubation for 15 min at room temperature in the dark (reduces fluorescent fading).
8. View cells by (confocal) fluorescence microscopy using a ×100 oil immersion objective (spectral characteristics of GFP and different MitoTracker dyes; *see* **Table 2**).

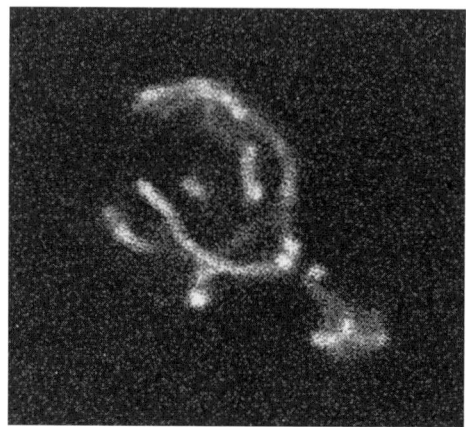

Fig. 3. In vivo mitochondrial localization in *Saccharomyces cerevisiae*. Green fluorescent protein (GFP) was fused C-terminally to the mitochondrial targeting sequence of the Map toxin of enteropathogenic *E. coli* (EPEC) and fusion protein expression induced for 4 h prior to visualization by confocal fluorescence microscopy. The GFP moiety localizes at subcellular, worm-like structures, which represent the yeast mitochondrial network. MitoTracker™ dye can be used to confirm mitochondrial localization *(19)*.

Notes

1. The cultivation of *S. cerevisiae* and *N. crassa* are described in detail in **refs. *13*** and ***14***.
2. Successful formation of spheroplasts, as well as intactness of mitochondria after lysis of the spheroplasts, may depend on the source of the zymolyase enzyme.

Table 2
Absorption and Emission Maxima of GFP and Different MitoTracker® Dyes

Fluorophore	Absorption (nm)	Emission (nm)
GFP	475	509
EGFP	488, 498	507, 516
MitoTracker Orange CMTMRos	551[a]	576[a]
MitoTracker Red CMXRos	578[a]	599[a]
MitoTracker Red 580	581[a]	644[a]
MitoTracker Deep Red 633	644[a]	665[a]

GFP, green fluorescent protein; EGFP, enhanced GFP.
[a]Determined in methanol; may vary somewhat in cellular environments

High concentrations of the enzyme and prolonged incubation times may damage the mitochondria.

3. This assay serves to compare the relative membrane potential of different preparations of isolated mitochondria. It does not show the absolute values of the membrane potential.

4. Synthesis of precursor proteins in reticulocyte lysate is also possible using linearized vector. As an interesting alternative, the protein coding sequence can be amplified by PCR with a 5'-oligonucleotide containing an SP6 or T7 promotor. Purified PCR product ($0.5\,\mu g$), resuspended in RNase-free H_2O, can be introduced instead of plasmid DNA to the in vitro transcription and translation system *(15)*.

5. Thawed mitochondria should be used immediately (within 2 h) and should not be frozen again.

6. For some precursor proteins, higher salt concentrations in the import buffer (up to $250\,mM$ KCl) may stimulate in vitro import efficiency.

7. It is recommendable to initiate and terminate the import reaction between several samples at regular intervals of 30 s to ensure the correct observance of incubation times. Alternatively, the import reaction can be performed in a larger volume and samples can be removed at desired time points for subsequent termination of import by incubation on ice.

8. Pellets of PMSF-treated mitochondria may be unstable. The supernatant should be removed carefully.

9. Molecular weight markers can be marked by dots of radioactive ink. The dots can then be covered with a suitable tape to protect the phosphorimager screen. Instead of using a phosphorimager screen, the dried gel can be placed in a light-tight X-ray film holder and covered with a sheet of X-ray film. In this case, the sensitivity compared to phosphoimager screens is about tenfold lower.

10. Mitotracker dyes are mitochondrion-selective dyes that accumulate only in mitochondria with an intact membrane potential.

11. Transcription from the *GAL1* promoter is repressed in the presence of glucose. Raffinose does not repress or induce transcription *(16)*. Therefore, more rapid induction of the *GAL1* promoter by galactose can be achieved by growing cells maintained in raffinose when compared to cells maintained in glucose. For proper initiation of translation of genes inserted into the pYES2 vector, a Kozak translation initiation sequence should be included as part of the ATG initiation codon (example of a Kozak consensus sequence: ANN-ATGG) *(17)*.

12. Prepare a stock solution of MitoTracker dye by dissolving the lyophilized reagent to a final concentration of $1\,mM$ in DMSO. It is important for the stability of the dye to avoid repeated freezing and thawing. From our own experience, lyophilized reagent and stock solution can be stored at $-20°C$ for at least 1 yr.

Acknowledgments

We authors thank E. Dian and Dr. C. Motz for discussion. This work was supported by the Deutsche Forschungsgemeinschaft and by the Sonder-forschungsbereich 495.

References

1. Herrmann, J. M. and Neupert, W. (2003) Protein insertion into the inner membrane of mitochondria. *IUBMB Life* **55**, 219–225.
2. Neupert, W. and Brunner, M. (2002) The protein import motor of mitochondria. *Nat. Rev. Mol. Cell Biol.* **3**, 555–565.
3. Pfanner, N., Wiedemann, N., Meisinger, C., and Lithgow, T. (2004) Assembling the mitochondrial outer membrane. *Nat. Struct. Mol. Biol.* **11**, 1044–1048.
4. Rehling, P., Brandner, K., and Pfanner, N. (2004) Mitochondrial import and the twin-pore translocase. *Nat. Rev. Mol. Cell Biol.* **5**, 519–530.
5. Rassow, J. and Pfanner, N. (2000) The protein import machinery of the mitochondrial membranes. *Traffic* **1**, 457–464.
6. Likic, V. A., Perry, A., Hulett, J., et al. (2005) Patterns that define the four domains conserved in known and novel isoforms of the protein import receptor Tom20. *J. Mol. Biol.* **347**, 81–93.
7. Lill, R. and Kispal, G. (2000) Maturation of cellular Fe-S proteins: an essential function of mitochondria. *Trends Biochem. Sci.* **25**, 352–356.
8. Bradford, M. M. (1976) A rapid and sensitive method for the quantitation of microgram quantities of protein utilizing the principle of protein-dye binding. *Anal. Biochem.* **72**, 248–254.
9. Geissler, A., Krimmer, T., Bömer, U., Guiard, B., Rassow, J., and Pfanner, N. (2000) Membrane potential-driven protein import into mitochondria: the sorting sequence of cytochrome b_2 modulates the $\Delta\Psi$-dependence of translocation of the matrix-targeted sequence. *Mol. Biol. Cell.* **11**, 3977–3991.
10. Glick, B. S. (1995) Pathways and energetics of mitochondrial protein import in *Saccharomyces cerevisiae. Methods Enzymol.* **260**, 224–271.
11. Becker, K., Guiard, B., Rassow, J., Söllner, T., and Pfanner, N. (1992) Targeting of a chemically pure preprotein to mitochondria does not require the addition of a cytosolic signal recognition factor. *J. Biol. Chem.* **267**, 5637–5643.
12. Motz, C., Martin, H., Krimmer, T., and Rassow, J. (2002) Bcl-2 and porin follow different pathways of TOM-dependent insertion into the mitochondrial outer membrane. *J. Mol. Biol.* **323**, 729-738.
13. Davenport, R. R. (1980) An introduction to yeasts and yeast-like organisms, in *Biology and Activities of Yeasts* (Skinner, F. A., Passmore, S. M., and Davenport, R. R., eds.), Academic Press, London, pp. 1–27.

14. Weiss, H., von Jagow, G., Klingenberg, M., and Bücher, T. (1970) Characterization of Neurospora crassa mitochondria prepared with a grind-mill. *Eur. J. Biochem.* **14**, 75-82.

15. Ryan, M. T., Voos, W., and Pfanner, N. (2001) Assaying protein import into mitochondria. *Methods Cell Biol.* **65**, 189–215.

16. West, R. W. J., Yocum, R. R., and Ptashne, M. (1984) Use of *lacZ* fusions to delimit regulatory elements of the inducible divergent *GAL1-GAL10* promoter in *Saccharomyces cerevisiae. Mol. Cell. Biol.* **4**, 1985–1998.

17. Kozak, M. (1987) An analysis of 5'-noncoding sequences from 699 vertebrate messenger RNAs. *Nuc. Acids Res.* **15**, 8125–8148.

18. Fu, W., Japa, S., and Beattie, D. S. (1990) Import of the iron-sulfur protein of the cytochrome b/c_1 complex into yeast mitochondria. *J. Biol. Chem.* **265**, 16541–16547.

19. Papatheodorou, P., Domanska, G., Öxle, M., et al. (2006) The enteropathogenic *Escherichia coli* (EPEC) Map effector is imported into the mitochondrial matrix by the TOM/Hsp70 system and alters organelle morphology. *Cell. Microbiol.* **8**, 677–689.

20. Söllner, T., Rassow, J., and Pfanner, N. (1991) Analysis of mitochondrial protein import using translocation intermediates and specific antibodies, in (Tartakoff, A. M., ed.), Academic Press, London, pp. 245–358.

21. Cyr, D. M., Ungermann, C., and Neupert, W. (1995) Analysis of mitochondrial protein import pathway in *Saccharomyces cerevisiae* with translocation intermediates, in (Attardi, G. M. and Chomyn, A., eds.), Academic Press, London, pp. 241–252.

22. Alconada, A., Gärtner, F., Hönlinger, A., Kübrich, M., and Pfanner, N. (1995) Mitochondrial receptor complex from *Neurospora crassa* and *Saccharomyces cerevisiae*, in (Attardi, G. M. and Chomyn, A., eds.), Academic Press, London, pp. 263–286.

23. Wienhues, U., Koll, H., Becker, K., Guiard, B., and Hartl, F. U. (1992) Protein targeting to mitochondria, in *Protein targeting: a practical approach* (Magee, A. I. and Wileman, T., eds.), 135–159. IRL Press, Oxford.

24. Wiedemann, N., Pfanner, N., and Rehling, P. (2006) Import of precursor proteins into isolated yeast mitochondria, in (Xiao, w., ed.), Humana Press Inc., Totowa, NJ, pp. 373–383.

25. Ryan, M. T., Voos, W., and Pfanner, N. (2001). Assaying protein import into mitochondria. *Methods Cell Biol.* **65**, 189–215.

26. Martin, H., Eckerskorn, C., Gärtner, F., Rassow, J., Lottspeich, F., and Pfanner, N. (1998) The yeast mitochondrial intermembrane space: purification and analysis of two distinct fractions. *Anal. Biochem.* **265**, 123–128.

27. Tartakoff, A. M. (ed.) (1991) Academic Press, London.

28. Söllner, T., Rassow, J., Wiedmann, M, et al. (1992) Mapping of the protein import machinery in the mitochondrial outer membrane by crosslinking of translocation intermediates. *Nature* **355**, 84–87.

29. Blom, J., Kübrich, M., Rassow, J., et al. (1993) The essential yeast protein MIM44 (encoded by MPI1) is involved in an early step of preprotein translocation across the mitochondrial inner membrane. *Mol. Cell. Biol.* **13**, 7364–7371.
30. Rospert, S., Looser, R., Dubaquié, Y., Matouschek, A., Glick, B. S., and Schatz, G. (1996) Hsp60-independent protein folding in the matrix of yeast mitochondria. *EMBO J.* **15**, 764–774.
31. Rassow, J., Mohrs, K., Koidl, S., Barthelmess, I. B., Pfanner, N., and Tropschug, M. (1995) Cyclophilin 20 is involved in mitochondrial protein folding in cooperation with molecular chaperones Hsp70 and Hsp60. *Mol. Cell. Biol.* **15**, 2654–2662.
32. Rassow, J. (1999) Protein folding and import into organelles, in *Post-translational Processing: A Practical Approach* (Higgins, S. J. and Hames, B. D., eds.), IRL Press, Oxford, UK, pp. 43–94
33. Schägger, H., and von Jaggow, G. (1991) Blue native electrophoresis for isolation of membrane protein complexes in enzymatically active form. *Anal. Biochem.* **199**, 223–231.
34. Dekker, P. J. T., Martin, F., Maarse, A. C., et al. (1997) The Tim core complex defines the number of mitochondrial translocation contact sites and can hold arrested preproteins in the absence of matrix Hsp70-Tim44. *EMBO J.* **16**, 5408–5419.
35. Wiedemann, N., Kozjak, V., Chacinska, A., et al. (2003) Machinery for protein sorting and assembly in the mitochondrial outer membrane. *Nature* **424**, 565–571.
36. Model, K., Meisinger, C., Prinz, T., et al. (2001) Multistep assembly of the protein import channel of the mitochondrial outer membrane. *Nat. Struct. Biol.* **8**, 361–370.
37. Kolodziej, P. A. and Young, R. A. (1991) Epitope tagging and protein surveillance, in (Guthrie, C. and Fink, G. R., eds.), Academic Press, London, pp. 508–519.
38. Fujiki, Y., Hubbard, A. L., Fowler, S., and Lazarow, P. B. (1982) Isolation of intracellular membranes by means of sodium carbonate treatment: application to endoplasmic reticulum. *J. Cell. Biol.* **93**, 97–102.
39. Daum, G., Böhni, P.C., and Schatz, G. (1982) Import of proteins into mitochondria. cytochrome b_2 and cytochrome c peroxidase are located in the intermembrane space of yeast mitochondria. *J. Biol. Chem.* **257**, 13028–13033.
40. Hartl, F. U., Ostermann, J., Guiard, B., and Neupert, W. (1987) Successive translocation into and out of the mitochondrial matrix: targeting of proteins to the intermembrane space by a bipartite signal peptide. *Cell* **51**, 1027–1037.
41. Meisinger, C., Pfanner, N., and Truscott, K. N. (2006) Isolation of yeast mitochondria, in (Xiao, W., ed.,), Humana Press Inc., Totowa, NJ, pp. 33–39.

11

A Mitosome Purification Protocol Based on Percoll Density Gradients and Its Use in Validating the Mitosomal Nature of *Entamoeba histolytica* Mitochondrial Hsp70

Jorge Tovar, Siân S. E. Cox, and Mark van der Giezen

Summary

Mitochondria are indispensable for aerobic respiration, but many microbial eukaryotes have lost this function through reductive evolution. Their modified mitochondria are known as hydrogenosomes or mitosomes depending on whether or not they produce molecular hydrogen. The intestinal parasite *Entamoeba histolytica* contains mitosomes whose role in cellular metabolism is unclear. Only three proteins have been shown thus far to reside in these organelles: the molecular chaperones Hsp10 and Hsp60 and an unusual ADP/ATP carrier. Here we describe the isolation of *E. histolytica* mitosomes by cellular fractionation and density gradient centrifugation and show that the mitochondrial-type chaperone Hsp70 is also housed in *Entamoeba* mitosomes.

Key Words: *Entamoeba histolytica*; mitosomes; mitochondria; differential centrifugation; density gradient.

1. Introduction

Mitochondria are essential organelles for eukaryotes because of their role in aerobic respiration and the subsequent energy yield. However, not all environments are capable of supporting this oxygen-requiring process. Examples are swamps, marine sediments, and the body cavities of animals. Fermentation is the dominant mode of energy production for eukaryotes

From: *Methods in Molecular Biology, Vol. 390: Protein Targeting Protocols: Second Edition*
Edited by: M. van der Giezen © Humana Press Inc., Totowa, NJ

inhabiting such habitats. Indeed, mitochondria were thought to be absent in these organisms (1). However, over the last decade or so, in contrast to what had been published before, genes encoding mitochondrial proteins and even highly modified mitochondria have been discovered in these eukaryotes (reviewed in **ref.** 2). These modified mitochondria are known as mitosomes or hydrogenosomes depending on whether or not they produce molecular hydrogen.

The human intestinal parasite *Entamoeba histolytica* is an example of a mitosome-containing organism; in fact, mitosomes were first described in this species (3). *Entamoeba* was regarded as a typical example of a "primitive" eukaryote because of the lack of mitochondria and peroxisomes and its supposedly simple cellular structure (4). However, the discovery of genes encoding proteins normally located to mitochondria in other organisms in *Entamoeba* changed this primitive status. Some 10 years ago, Clark and Roger described the discovery of the two amoebal genes encoding heat-shock protein Hsp60 and pyridine nucleotide transhydrogenase (5). Both proteins seemed to contain mitochondrial-like targeting signals at their amino-terminus and phylogenetic analysis of Hsp60 clearly demonstrated its mitochondrial ancestry. Subsequent work confirmed the mitochondrial nature of the Hsp60-targeting signal showing that its removal caused an accumulation of the protein in the cytosol while the use of a *bona fide* mitochondrial targeting signal restored its mitosomal localization (3).

Although the existence of several other mitochondrial proteins has recently been documented by the *Entamoeba* genome project (6,7) only three proteins have so far been localized to *Entamoeba* mitosomes: the original chaperonin Hsp60 (3), its co-chaperonin Hsp10 (8), and an unusual member of the mitochondrial carrier family (9). One of the problems in understanding the roles of these mitochondrially related organelles is the paucity of information relating to the actual organellar proteome, mainly caused by the inability to obtain sufficient quantities of organelles for more thorough proteomic analysis. Thus we have developed a mitosome isolation protocol based on various methods used to isolate mitochondria. Here we describe this protocol in sufficient detail to enable the isolation of large numbers of mitosomes. These highly enriched mitosomal preparations should allow further research into these organelles in a bid to understand the role they may play in overall cellular metabolism. As an example we demonstrate for the first time that the previously reported mitochondrial heat-shock protein Hsp70 (10) is indeed localized in the mitosomal fraction, further demonstrating the mitochondrial nature of these enigmatic organelles.

2. Materials

2.1. Organism

1. *Entamoeba histolytica* cells (trophozoites) strain HM-1:IMSS Clone 9 (ATCC 50528) (*see* **Note 1**).
2. LYI-S-2 axenic culture medium, pH 6.8 *(11)* supplemented with 2% Vitamin Mix 18 (Biofluids Inc., Rockville, MD) and 10% adult bovine serum (Sera Laboratories International Ltd, Horsted Keynes, UK) (*see* **Notes 2 *and* 3**).
3. Borosilicate glass culture tubes, $16 \times 125\,mm$ (Fisher Scientific).
4. 225-cm^2 Cell culture flasks, rectangular, angled neck, polystyrene, TC-treated with plug seal cap (Corning Life Sciences).

2.2. Cellular Fractionation

1. $1\,M$ Dithiothreitol (DTT): dissolve $0.154\,g$ of DTT in $1\,mL$ of H_2O (*see* **Note 4**).
2. $1\,M$ MOPS-HCl, pH 7.2: add $23.12\,g$ MOPS to $80\,mL$ of H_2O and adjust the pH to 7.2 using concentrated HCl, adjust to a final volume of $100\,mL$ using H_2O.
3. $2.5\,M$ sucrose/$100\,mM$ MOPS pH 7.2: add $85.58\,g$ sucrose to $10\,mL$ of $1\,M$ MOPS-HCl pH 7.2 and subsequently add enough H_2O to make up to $100\,mL$.
4. 10x PBS: to $800\,mL$ of H_2O add $80\,g$ NaCl, $2\,g$ KCl, $11.5\,g$ Na_2HPO_4, and $2\,g$ KH_2PO_4, sterilize by autoclaving (*see* **Note 5**).
5. 1x PBS: take $100\,mL$ of 10x PBS and add $900\,mL$ of H_2O.
6. SMDI ($0.25\,M$ sucrose, $10\,mM$ MOPS-HCl pH 7.2, $10\,mM$ DTT, inhibitors): take $5\,mL$ of $2.5\,M$ sucrose/$100\,mM$ MOPS-HCl pH 7.2 (*see* **Subheading 2.2., step 3**), add $0.5\,mL$ $1\,M$ DTT and 2 Complete inhibitor tablets (Roche), add H_2O up to $50\,mL$ and make sure the tablets are completely dissolved (*see* **Note 6**). Make fresh.
7. SMDI-Percoll 90% ($30\,mL$): take $27\,mL$ 100% Percoll (Sigma), add $3\,mL$ $2.5\,M$ sucrose/$100\,mM$ MOPS-HCl pH 7.2, $300\,\mu L$ $1\,M$ DTT and 2 Complete inhibitor tablets (Roche), make sure the tablets are completely dissolved.
8. Some liquid nitrogen.
9. Dome-top sealable centrifuge tubes, $16 \times 76\,mm$ (Quick-Seal, Beckman Coulter).
10. Heat-sealer/tube topper (Beckman Coulter).

2.3. Polyacrylamide Gel Electrophoresis

1. Monomer solution: 30.8% T, 2.7% C_{bis} (Biorad) (*see* **Note 7**).
2. 4x Running buffer ($1.5\,M$ Tris-HCl, pH 8.8): dissolve $36.3\,g$ Tris (MW 121.1) in $150\,mL$ water, adjust pH to 8.8 using HCl, and top up volume to $200\,mL$.
3. 4x Stacking gel buffer ($0.5\,M$ Tris-HCl, pH 6.8): dissolve $3.0\,g$ Tris (MW 121.1) in $40\,mL$ of water, adjust pH to 6.8 using HCl, and top up to $50\,mL$.
4. 10% Sodium dodecyl sulfate (SDS): dissolve $10\,g$ SDS in $100\,mL$ of water.
5. 10% Ammonium persulfate (APS; Sigma): dissolve $0.1\,g$ APS in $1.0\,mL$ of water (*see* **Note 8**).

6. 2x Treatment buffer (0.125 M Tris-HCl, 4% SDS, 20% v/v glycerol, 0.2 M DTT, 0.02% bromophenol blue, pH 6.8): combine the following, 2.5 mL 4x stacking gel buffer, 4.0 mL 10% SDS, 2.0 mL glycerol, 2.0 mg bromophenol blue, 0.31 g DTT, top up to 10 mL using water, store 0.5 mL aliquots at −20°C.

7. 10x Tank buffer (0.25 M Tris, 1.92 M glycine, 1% SDS): dissolve 30.28 g Tris (MW 121.1) in 500 mL of water, add 144.13 g glycine, stir until dissolved, add 10 g SDS, top up to 1 L. It is not necessary to check the pH.

8. 1x Tank buffer (0.025 M Tris, 0.192 M glycine, 0.1% SDS, pH 8.3): add 100 mL 10x tank buffer to 900 mL of water; there is no need to check the pH.

9. Tetramethylethylene diamine (TEMED; Sigma).

10. Protein molecular weight standards: Trail mix (Novagen).

11. Mini-gel apparatus (*see* **Note 9**).

2.4. Western Transfer Blotting

1. Modified Towbin transfer buffer (25 mM Tris, 192 mM glycine, 10% methanol): dissolve 3.0 g Tris (mol. weight 121.1) into 700 mL of water, add 14.4 g glycine and dissolve, add 100 mL methanol, top up to 1 L using water. The pH should be about 8.3. Put in an ice bucket before use to chill.

2. Transfer unit, either a tank transfer or a semi-dry transfer will suffice; we use a semi-dry system.

3. Blotter paper (e.g., Whatmann).

4. Membrane: nitrocellulose.

5. Ponceau S (0.1% (w/v) Ponceau S, 5% acetic acid): add 1.0 g Ponceau S to 900 mL of water, add 50 mL acetic acid, and top up to 1 L.

2.5. Immunodetection

1. 5x TBST: dissolve 12.11 g Tris in 700 mL of water, add 82.32 g NaCl, dissolve, add 5 mL Tween-20, top up to 1 L with water, adjust the pH to 8.0.

2. 1x TBST: add 200 mL 5x TBST to 800 mL water and mix.

3. Skimmed milk powder (from a supermarket will suffice).

4. 1x TBST-milk: add 2–4% skimmed milk powder (w/v) to 1x TBST.

5. Chemiluminescence or chromogenic detection kit.

6. Primary antibodies: anti-*Entamoeba histolytica* mitosomal Cpn60 *(3)* and anti-*Neocallimastix patriciarum* hydrogenosomal Hsp70 *(12)*.

7. Secondary antibodies (*see* **Note 10**).

3. Methods

3.1. Cellular Fractionation and Percoll Gradient Centrifugation

1. Subculture *E. histolytica* trophozoites in screw-capped borosilicate glass culture tubes containing 13 mL LYI-S-2 axenic culture media and incubate at 35–37°C.

2. When cells have reached a density of approx 2×10^5 cells/mL (usually 2–3 d following subculture), pour entire contents of one tube into a sterile flask containing 1 L of complete LYI-S-2 axenic culture media. Mix carefully and distribute into four 225 cm^2 flasks and grow at 35–37°C until cell density reaches approx 2×10^5 cells/ml (usually 3–4 d following transfer).

3. Chill the *Entamoeba* culture flasks on ice for 20 min and rock the flasks gently every 5 min to detach the cells from the walls (*see* **Note 11**).

4. Transfer the medium containing the cells to large centrifuge tubes and collect the cells by centrifugation at 800g for 10 min at 4°C (*see* **Note 12**).

5. Wash the cell pellet gently with 20 mL 1x PBS and centrifuge at 800g for 10 min at 4°C.

6. Resuspend the cells gently in 10 mL SMDI.

7. Break the cells by freeze-thaw cycles; 3 min in liquid nitrogen (*see* **Note 13**) followed by 3 min at 37°C; repeat three times. Ensure that cell membranes have ruptured by light microscopy.

8. Keep some of the total cell lysate for protein work ($\sim$ 200 µL) and freeze at −20°C.

9. To remove cell debris and nuclei, centrifuge at 700g for 10 min at 4°C.

10. Keep 0.5–1 mL of the supernatant (cell-free extract) and freeze at −20°C.

11. Transfer the remaining supernatant (which contains cytosol and the organellar fraction) to a centrifuge tube (*see* **Note 14**) and chill on ice. This supernatant will be used in **step 13**. The pellet (which contains the nuclei) needs to be resuspended in 10 mL SMDI.

12. To concentrate the nuclei, centrifuge the pellet at 700g for 10 min at 4°C. Resuspend the nuclear pellet in 3 mL SMDI and freeze at −20°C.

13. Now centrifuge the supernatant from **step 11,** which contains the cytosol and organellar pellet in order to separate these; 30,000g for 1 h at 4°C (*see* **Note 15**).

14. Keep the supernatant (cytosol) and freeze at −20°C for later use.

15. Carefully wash the organellar pellet twice using 5 mL SMDI.

16. Resuspend the organellar pellet in 1 mL SMDI if you want to stop (freeze at −20°C), otherwise, resuspend the pellet in 7 mL SMDI.

17. Add 7 mL 90% Percoll in SMDI to give 45% Percoll.

18. Using a needle and syringe, carefully load the Percoll/organelle mix into a dome-top centrifuge tube and seal by applying heat to the inlet using the tube topper. Before sealing, ensure that any air bubbles are removed.

19. Centrifuge this in a swing-out rotor to obtain organellar fractions, 68,000g (*see* **Note 16**) for 30 min at 4°C.

20. Collect approx 1 mL fractions (or one band if you can) by puncturing (*see* **Note 17**) the bottom of the tube (work in a cold room) (*see* **Fig. 1**).

21. Dilute the fractions by adding 10 mL SMDI to remove the Percoll. Centrifuge the fractions at 100,000g for 2 h at 4°C.

22. The samples will sit on top of the hard Percoll pellet, resuspend the samples in 100 µL SMDI and analyze on a sodium dodecyl sulfate (SDS) polyacrylamide gel.

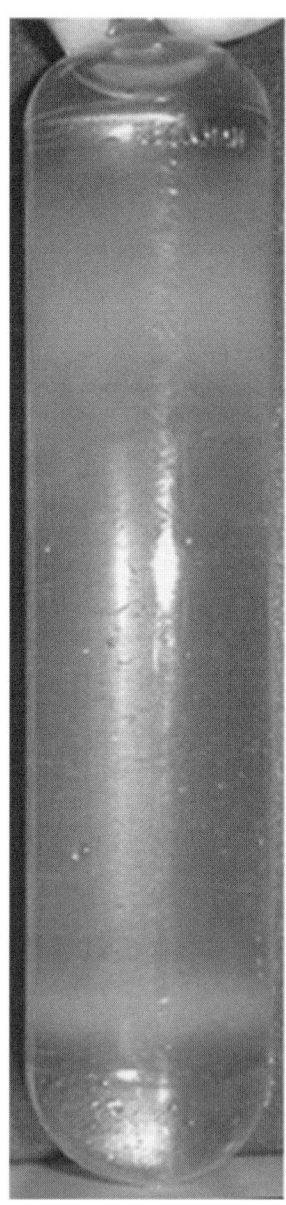

Fig. 1. A typical organellar fraction containing *Entamoeba histolytica* mitosomes (white band near the bottom of the tube). The absence of typical iron-containing proteins found in classic mitochondria is obvious from the color, which would otherwise be red–brown.

3.2. Protein Electrophoresis

1. Clean all glass plates, spacers, combs, and mini-gel system using a mild detergent, rinse with water.
2. Assemble the mini-gel system and casting stand according to your make and model (*see* **Note 9**). Make sure it stands on a flat surface.
3. Prepare the monomer solution for a 10% gel by combining the following in a 15 mL Falcon tube: 2 mL monomer solution, 2.5 mL 4x running gel buffer, 0.1 mL 10% SDS, 3.2 mL water and mix well. Add 50 μL 10% APS and 3.3 μL TEMED and mix well (*see* **Notes 7 *and* 18**).
4. Immediately pour the monomer mix between the glass plates.
5. Add 70% ethanol on top of the monomer mix to prevent exposure to oxygen.
6. Allow the gel to polymerise for 30 min (*see* **Notes 7 *and* 19**).
7. Rinse off the ethanol with water.
8. Prepare the stacking gel solution in a 15 mL Falcon tube: 0.44 mL monomer solution, 0.83 mL 4x stacking gel buffer, 33 μL 10% SDS, 2.03 mL water, mix well (*see* **Notes 7 *and* 20**).
9. Add 16.7 μL 10% APS and 1.7 μL TEMED, mix well (*see* **Note 18**).
10. Immediately pour the monomer solution on top of the polymerized running gel.
11. Insert the comb into the sandwich and avoid trapping air under it. Allow the gel to polymerise for 30 min.
12. Remove gel sandwich from casting stand and install the gel according to the instructions, specific to your exact make and model.
13. Fill the lower and upper buffer chambers with 1x tank buffer.
14. Carefully remove the comb and rinse the wells using 1x tank buffer.
15. Load molecular weight markers into lane 1.
16. Load your fractions in subsequent lanes (*see* **Note 21**).
17. Run the gels using 20 mA constant current per gel. Check that the blue indicator dye actually moves downward in the gel. The run is complete when the blue front reaches the bottom of the gel.
18. Disassemble the gel sandwich and remove the stacking gel from the running gel by cutting the gel with the spacer. Remove the top left corner of the running gel for orientation purposes.
19. Place the running gel in ice-cold transfer buffer and continue with the Western blotting protocol.

3.3. Western Blotting Using a Semi-Dry Blotter

1. Measure the dimensions of the running gel as obtained under **Subheading 3.2., step 18**, and cut six pieces of blotting paper and one piece of nitrocellulose membrane of the same dimensions. Remove the top left corner of the membrane.
2. Wet one piece of blotting paper in ice-cold transfer buffer and transfer to the bottom plate of a semi-dry electroblotter. Smooth out any bubbles by rolling a Pasteur's

pipet over the paper. Repeat with two more pieces of blotting paper, but place these on top of the previous piece of blotting paper in order to make a small stack.

3. Wet the nitrocellulose membrane in the ice-cold transfer buffer and place on top of the wet blotting paper stack.

4. Take out the gel from the transfer buffer and gently lay on top of the stack. Ensure that the cut top left corners of the gel and membrane are aligned. As before, smooth out any trapped air bubbles using a rod or a Pasteur's pipet.

5. Repeat **step 2** using three more pieces of blotting paper, but now place the blotting paper on top of the gel.

6. Place the lid on the electroblotter.

7. Apply $0.8 \, mA/cm^2$ for about 1 h (*see* **Note 22**).

8. Remove the lid from the transfer unit and disassemble the stack.

9. Place the membrane in a tray and cover with Ponceau S for about 1 min. Remove the Ponceau S and rinse the membrane with water until protein bands become visible. Mark the position of the lanes and markers on the membrane using pencil. Proceed to the immunolabelling section (*see* **Note 23**).

3.4. Immunolabeling and Detection Using Chemoluminescence

1. Place the membrane in 50 mL 1x TBST-milk to block nonspecific binding sites on the membrane, place on a rocker, and incubate for 1 h at room temperature or overnight at 4°C.

2. Dilute the primary antibody in 1x TBST-milk (*see* **Note 24**) and incubate the membrane 1 h at room temperature on a rocking platform.

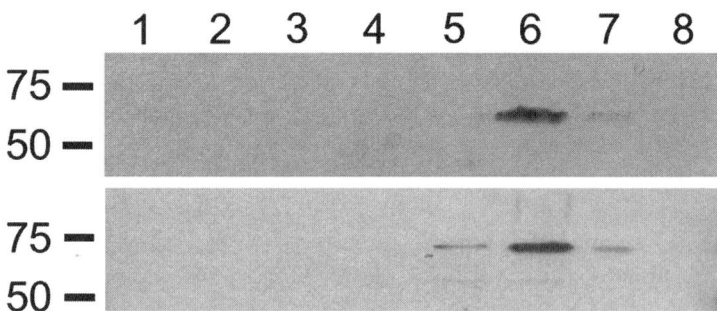

Fig. 2. Distribution of *Entamoeba histolytica* molecular chaperones. *E. histolytica* cells were fractionated by differential and density centrifugation. Fractions were separated on SDS polyacrylamide gels and analyzed using the homologous anti-Cpn60 antiserum *(3)* (**top**) or using a heterologous antiserum raised against the hydrogenosomal Hsp70 from *Neocallimastix patriciarum (12)* (**bottom**). Fractions are indicated above the gels by numbers; high numbers correspond to fractions near the bottom of the centrifuge tube. Molecular weight markers are shown on the left in kDa.

3. Rinse the membrane three times for 5 min each using 1x TBST-milk.
4. Dilute the secondary antibody in 1x TBST-milk (*see* **Note 25**) and incubate the membrane 1 h at room temperature on a rocking platform.
5. Rinse the membrane three times for 5 min each using 1x TBST-milk.
6. Incubate the membrane in the chemiluminescent detection buffer according to the specific kit used.
7. Compare a membrane probed with anti-cpn60 (which will indicate the protein fraction derived from purified mitosomes) and one probed with anti-Hsp70 to establish that both heat-shock proteins localize to the same purified fraction (**Fig. 2**).

Notes

1. *Entamoeba histolytica* is a known pathogen, and suitable containment facilities are therefore needed to be able to cultivate and handle this organism.
2. Make sure that the serum is heat-inactivated because this inactivates proteins of the complement system that are toxic to *Entamoeba*.
3. The serum used should be adult serum and not fetal serum. Fetal serum contains fetuin protein, which is toxic to *Entamoeba*.
4. When mentioning "water," we mean water of Millipore quality.
5. Only add the next salt after the first one has completely dissolved.
6. The Complete tables are a bit tough to dissolve; breaking them up in smaller bits might be helpful.
7. Nonpolymerized acrylamide is a neurotoxin; take extreme care when handling.
8. Normally one should make the 10% APS solution fresh, but it stays good for about 5 d at 4°C.
9. Several good working mini-gel systems are commercially available; we have successfully used systems from Hoefer, Biorad, and Owl, but are confident that other systems would work equally well.
10. Depending on the detection systems used, secondary antibodies (against the host in which the primary antibodies have been raised—rabbit in our case) conjugated to either horseradish peroxidase or alkaline phosphatase need to be used. Several detection systems are commercially available, and it is generally a matter of using the kit that happens to be present in the lab (or in the lab next door). When using a chemiluminescence method, images are detected using photographic film, which are a kind of hit-and-miss approach, since the actual development of the signal cannot be directly followed (unless one has a state-of-the-art gel documentation system with a cooled camera). The advantage is that obtained images generally have good contrast and can be easily reproduced for publications. On the other hand, when using a chromogenic substrate that produces a colored substrate on the membrane, one has the advantage of actually being able to follow the development of the precipitate. The disadvantage is that the contrast is not as good as when

using chemiluminescence, which might be a problem when preparing publication-quality images (however, with today's image processing software, one should be able to change the tone and increase the contrast).

11. Work as much as possible on ice because *E. histolytica* does what it says on the tin—it lyses tissue and avidly degrades proteins, including its own, hence the cocktail of protease inhibitors.

12. We use Sorvall GSA rotors, so it would be 2500 rpm.

13. Please take care when handling liquid nitrogen, wear protective clothing, gloves, and goggles (a face shield would be better).

14. We use Sorvall SS-34 tubes.

15. Since we use Sorvall SS-34 tubes, the speed is 16,000 rpm.

16. We use a TY-40 fixed-angle rotor in a Beckmann Coulter centrifuge: 24,500 rpm.

17. The tube can be punctured using an 18-gauge hypodermic needle. Take the needle between the thumb and index finger and try to make a hole by rotating the needle between your fingers as if drilling. Do not collect the drops via the needle, but let the drops escape via the hole. Be careful when handling hypodermic needles, and always dispose of them properly using sharps bins.

18. Once the initiator (APS) and catalyst (TEMED) are added, work fast because the monomer mix will start to polymerize.

19. After about 20 min, one can check whether the gel has polymerized by carefully tilting the gel. The 70% ethanol overlay will maintain a horizontal meniscus, while a polymerized gel will follow the angle of the tilt.

20. To aid in visualizing the wells, we normally use $2 \mu L$ of 0.4% phenol red (w/v in water). This will lightly stain the stacking gel but will not interfere with any aspect of gel electrophoresis.

21. Normally, specified amounts of protein are loaded onto SDS polyacrylamide gels. However, when using fractions obtained after differential centrifugation followed by a density gradient, it is customary to load identical volumes on the gel to represent the actual distribution of protein as obtained from the gradient.

22. The exact time will vary from sample to sample and will need to be determined empirically.

23. There is no need to completely destain the blot; the blocking solution will turn into a pinkish color if Ponceau S is left on the membrane, but this does not affect the outcome of the experiment, and it is solely a matter of personal preference whether one prefers a pinkish or whitish blocking solution.

24. The best working dilution needs to be determined empirically. For new antibodies, starting dilutions of 1:500, 1:1000, and 1:5000 are recommended.

25. As with primary antibodies, the working dilutions need to be discovered empirically. Good quality secondary antibodies can be diluted up to 20,000–50,000 times if needed.

Acknowledgments

We would like to thank Neil Sommerville for parasite cultures. Work in our laboratories is supported by grants from the BBSRC (BB/C507145) to JT and the Wellcome Trust (078566/A/05/Z) to MvdG.

References

1. Cavalier-Smith, T. (1983). A 6 kingdom classification and a unified phylogeny, in *Endocytobiology II* (Schwemmler, W. and Schenk, H. E. A., eds.), De Gruyter, Berlin, pp. 1027–1034.
2. van der Giezen, M. and Tovar, J. (2005). Degenerate mitochondria. *EMBO Rep.* **6**, 525–530.
3. Tovar, J., Fischer, A., and Clark, C.G. (1999). The mitosome, a novel organelle related to mitochondria in the amitochondrial parasite *Entamoeba histolytica*. *Mol. Microbiol.* **32**, 1013–1021.
4. Bakker-Grunwald, T. and Wostmann, C. (1993). *Entamoeba histolytica* as a model for the primitive eukaryotic cell. *Parasitol. Today* **9**, 27–31.
5. Clark, C. G. and Roger, A. J. (1995). Direct evidence for secondary loss of mitochondria in *Entamoeba histolytica*. *Proc. Natl. Acad. Sci. USA* **92**, 6518–6521.
6. Loftus, B., Anderson, I., Davies, R., et al. (2005). The genome of the protist parasite *Entamoeba histolytica*. *Nature* **433**, 865–868.
7. van der Giezen, M., Tovar, J., and Clark, C.G. (2005). Mitochondrion-derived organelles in protists and fungi. *Int. Rev. Cytol.* **244**, 175–225.
8. van der Giezen, M., León-Avila, G., and Tovar, J. (2005). Characterization of chaperonin 10 (Cpn10) from the intestinal human pathogen *Entamoeba histolytica*. *Microbiology* **151**, 3107–3115.
9. Chan, K. W., Slotboom, D. J., Cox, S., et al. (2005). A novel ADP/ATP transporter in the mitosome of the microaerophilic human parasite *Entamoeba histolytica*. *Curr. Biol.* **15**, 737–742.
10. Bakatselou, C., Kidgell, C., and Clark, C.G. (2000). A mitochondrial-type hsp70 gene of *Entamoeba histolytica*. *Mol. Biochem. Parasitol.* **110**, 177–182.
11. Clark, C.G. and Diamond, L.S. (2002). Methods for cultivation of luminal parasitic protists of clinical importance. *Clin. Microbiol. Rev.* **15**, 329–341.
12. van der Giezen, M., Birdsey, G. M., Horner, D. S., et al. (2003). Fungal hydrogenosomes contain mitochondrial heat-shock proteins. *Mol. Biol. Evol.* **20**, 1051–1061.

12

The Chloroplast Protein Import Machinery: A Review

Friederike Hörmann, Jürgen Soll, and Bettina Bölter

Summary

Plastids are a heterogeneous family of organelles found ubiquitously in plant and algal cells. Most prominent are the chloroplasts, which carry out such essential processes as photosynthesis and the biosynthesis of fatty acids as well as of amino acids. As mitochondria, chloroplasts derived from a single endosymbiotic event. They are believed to have evolved from an ancient cyanobacterium, which was engulfed by an early eukaryotic ancestor. During evolution the plastid genome has been greatly reduced and most of the genes have been transferred to the host nucleus. Consequently, more than 98% of all plastid proteins are translated on cytosolic ribosomes. They have to be posttranslationally targeted to and imported into the organelle. Targeting is assisted by cytosolic proteins, which interact with proteins destined for plastids and thereby keep them in an import-competent state. After reaching the target organelle, many proteins have to conquer the barrier of the chloroplast outer and inner envelopes. This process is mediated by complex molecular machines in the outer (Toc complex) and inner (Tic complex) envelope of chloroplasts, respectively. Most proteins destined for compartments inside the chloroplast contain a cleavable N-terminal transit peptide, whereas most of the outer envelope components insert into the membrane without such a targeting peptide.

Key Words: Chloroplasts; protein targeting; in vitro import; translocation machinery; Toc/Tic; localization.

1. What Happens in the Cytosol?

The proteins destined for the chloroplast are synthesized as precursor proteins with a transit sequence that is necessary and sufficient for correct targeting. There are two types of targeting signals. The first class of targeting signal involves internal noncleavable signals, mostly found in outer envelope proteins (OEPs), such as OEP14, OEP16, OEP21, OEP24, Toc34, and Toc159 (*1–6*). Some

From: *Methods in Molecular Biology, Vol. 390: Protein Targeting Protocols: Second Edition*
Edited by: M. van der Giezen © Humana Press Inc., Totowa, NJ

OEPs can insert spontaneously into the outer envelope. Although the process of insertion is not energy dependent, a step in the insertion or more likely the assembly process of Toc34 and OEP14 was stimulated by nucleotides (*7,8*).

With some exceptions, the majority of proteins that are targeted inside chloroplasts are synthesized as precursor proteins with a cleavable N-terminal presequence. Targeting sequences reveal little similarity at the level of primary sequence or length (*9*), but all contain predominantly positively charged and hydroxylated amino acids such as threonine and serine (*10*). In vitro, serine and threonine can be phosphorylated by a cytosolic protein kinase that exclusively phosphorylates chloroplast but not mitochondrial targeting signals (*11*). Phosphorylation of the presequence is important not only for the binding to a cytosolic guidance complex, but also for the interaction with the isolated import receptor Toc34. The lack of a secondary structure (*9*) makes preproteins good candidates for interacting with heat shock proteins. For mitochondrial protein import the interaction of precursor proteins with chaperones is well described (*12*). In chloroplast protein import experiments using preLHCP, purified cytosolic Hsp70 could partially substitute for leaf extract (*13*). No stimulation of import could be observed for soluble stroma proteins such as preFd and preSSU (*14,15*). But May and Soll (*16*) showed an interaction of preSSU as well as preOE23 with an Hsp70 homolog after translation. Nevertheless, the chaperone interaction seems not to occur in an organelle-specific manner (*17*). Thus, other organelle-specific factors are needed. Other than a putative Hsp70-binding, site the presequence contains a kinase phosphorylation motif, as mentioned above. This motif shows strong similarities to phosphopeptide-binding motifs for 14-3-3 proteins, indicating that 14-3-3 proteins might be organelle-specific factors. May and Soll (*16*) presented evidence that cytosolic 14-3-3 proteins interact with phosphorylated preproteins, and together with an Hsp70 isoform and perhaps other components, a guidance complex is formed. This guidance complex is thought to dock the preprotein to the chloroplast surface component prior to translocation. Upon arrival of the precursor protein at the chloroplast surface, translocation is initiated.

2. How to Pass the Outer Envelope?

The core of the translocon in the outer envelope of chloroplasts comprises three proteins: Toc34, Toc75, and Toc159 (*see* **Fig. 1**). They are thought to build a complex of approx 500 kDa in which four copies of Toc34, four copies of Toc75, and one copy of Toc159 are found (*18*).

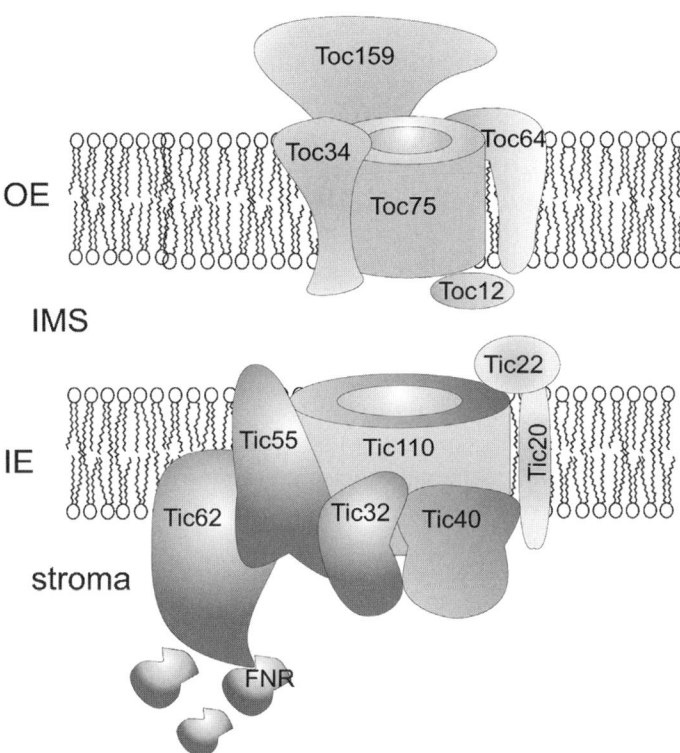

Fig. 1. Model of the chloroplast envelope translocation machinery. The translocon at the outer envelope of chloroplasts (Toc) consists of five components: The receptor proteins Toc159 and Toc34 and the translocation pore Toc75 comprise the "core" complex, whereas Toc64 and Toc12 are more loosely associated. Seven members of the translocon at the inner envelope (Tic) have been identified so far. Tic110 and Tic20 are discussed as being part of the import channel. Tic62, Tic55, and Tic32 exhibit features typical for redox regulated proteins and could therefore function in the regulation of the import process. Tic40 has been shown to bind molecular chaperones. Tic22 forms an intermembrane complex with Toc12 and Toc64.

Toc75 is the most abundant protein of the outer envelope membrane, and it forms the import pore of the Toc translocon. When reconstituted into liposomes, Toc75 forms a cation-selective, voltage-gated channel with a pore width of 15–25 Å. Resistance to high amounts of protease indicates that Toc75 is deeply embedded in the outer envelope membrane *(19)*. Before entering the translocation channel, dephosphorylation of the preprotein must occur because phosphorylated precursors are unable to pass Toc75 *(11)*. Toc75 contains a precursor-binding site distinct from that on the chloroplast surface and does not recognize the mature part of the preprotein. In *Arabidopsis thaliana*, four homologs of Toc75 are found: atToc75-I, atToc75-III, atToc75-IV, and atToc75-V *(20,21)*. Until now, only expression of atToc75-III and atToc75-V could be detected. Because atToc75-III is expressed in all tissues, it is believed to be the general import pore in *A. thaliana*. No gene insertion mutants of atToc75-III have been described so far, indicating that Toc75 function might be essential for chloroplast biogenesis. The strong similarities of atToc75-V to a class of bacterial transport proteins *(22)* suggest that it represents the most ancestral or earliest form of a Toc75-like channel that was further modified during endosymbiosis. It has been proposed that the Toc75-like channel derived from an ancient prokaryotic channel of smaller size and evolved by partial gene duplication in the amino-terminal region of the protein *(23)*. The fairly abundant presence of atToc75-V (approx 5–10% of atToc75-III) indicates that it has retained special functional properties required for the import (or export?) of its substrate(s). Alternatively, atToc75-V could be involved in membrane insertion of ß-barrel proteins, as has been shown for proteins of the Omp85-family, to which Toc75-V belongs.

Toc34 and Toc159 have both been characterized as receptor proteins. They share the presence of a GTP-binding domain, and it has been shown that they function as GTP-dependent precursor protein receptors. For Toc159 a second possible function has been described; namely, it provides the driving force needed for traversing the membrane. Toc159 can be divided into three parts: the A-domain, which is rich in acidic amino acids; the G-domain, which contains the GTP-binding domain; and the M-domain, which is the carboxy-terminal membrane domain. Four homologs of Toc159 can be found in *A. thaliana*: atToc159, atToc132, atToc120, and atToc90 *(24)*. The proteins differ mainly in the size of the A-domain. It is completely absent in atToc90. The expression of atToc159 seems to be essential for proper chloroplast development. The analysis of T-DNA insertion mutants of atToc159 *(ppi2)* led to the conclusion that Toc159 is the receptor for photosynthetic preproteins, while atToc132 and atToc120 seem to have a more redundant function in protein import.

During protein import, Toc34 is in close proximity to the preprotein as could be shown by crosslinking *(25)* and co-immunoprecipitation with αToc34 antibodies *(26)*. Additionally, Toc34 can be connected covalently with the translocation pore Toc75 via a disulfide bridge *(27)*. The close physical proximity ensures efficient translocation initiation and delivery of the preproteins to the translocation channel. Toc34 is anchored by a single transmembrane domain at the carboxy terminus in a way that the amino terminus is facing the cytosol *(28–30)*. A preprotein receptor is regulated both by GTP/GDP-binding and by phosphorylation in an elaborate mechanism: The affinity of the specific interaction of Toc34 with the transit peptide of the preprotein is drastically increased in the GTP-bound state. Bound precursor is released upon hydrolysis of GTP. Either the GDP can be replaced by GTP and the Toc34 is ready for a new round of transit peptide recognition, or Toc34 is phosphorylated and in this way inactivated because the phosphorylated Toc34 can neither recognize preproteins nor bind GTP. Dephosphorylation and at the same time activation of Toc34 is carried out by an ATP-dependent phosphatase *(31)*. To make the story even more complex, Toc34 has a higher affinity for phosphorylated precursors than for nonphosphorylated ones. After dephosphorylation, Toc34 binds GTP and subsequently forms a high-affinity complex with phosphorylated precursors *(32)*. A homo-dimerization *(33)* of Toc34 and a hetero-dimerization with the GTP-binding domain of Toc159 might cause the GDP/GTP exchange of the intrinsic GTPase. On the other hand, the preprotein can act as a GTPase-activating factor for Toc34, which causes a 10- to 50-fold increase in GTPase activity. The resulting GDP–Toc34 precursor complex has a lower affinity for the preprotein than the GTP-bound form and allows the preprotein to dissociate from the receptor and continue its passage through the translocon *(34)*. Dimerization with Toc159 occurs in the GDP form of Toc34. In *A. thaliana* two Toc34 homologs could be identified: atToc33 and atToc34, both of which are expressed in vivo *(35)*. While atToc34 is expressed nearly equally in all tissues at all stages of development, the expression of atToc33 seems to be upregulated in photosynthetic and meristematic tissue *(36)*. Another difference between both proteins is that only atToc33 is regulated by phosphorylation. T-DNA insertion lines of atToc33 *(ppi1)* showed a pale green phenotype and retarded chloroplast development *(35,37)*. In later stages of development, however, plants recovered and were able to grow on soil. This observation is probably a result of the fact that atToc34 can partially take over the function of atToc33. Nevertheless, both proteins show clear preferences for different classes of preproteins. Although biochemical studies of pea chloroplasts have revealed the presence of only one Toc complex so far *(38,39)*, the discovery of the two very similar Toc components atToc33 and atToc34, the small protein family of atToc159, atToc132, and atToc120 as well as two atToc75 isoforms could be an indication for the existence of at least two distinct translocon complexes in *Arabidopsis*. The capacity of plastids to import proteins is regulated developmentally and is maximal during the early days of organ expansion *(40,41)*. The reason for changing import efficiencies could be translocon complexes made out of slightly different subunits. In addition to the core components, there are two auxiliary components: Toc64 and Toc12 *(42,43)*. Toc64 could be crosslinked to several Toc components, and its proposed function is to be the recognition site for chaperone-bound preproteins. Toc64 is an integral membrane protein that seems to be built from two independent modules. One module exhibits homologies to amidases. No amidase activity could be measured either from isolated envelope membranes or overexpressed Toc64 since the amidase function seems to be inactivated by a point mutation in the active site. The second module is a threefold repeated TPR (tetratricopeptide repeat) motif *(44)*. TPR motifs are found in several proteins of protein import machineries. The mitochondrial import receptor Tom70 contains seven TPRs and was proposed to interact preferentially with precursor proteins that require the mitochondrial import stimulating factor *(45)*. Pex5, a protein related to perox-isomal import, also contains seven repeats of the TPR motif *(46)*. TPR motifs are potential domains for protein–protein interaction *(47)*, thus Toc64 might recognize chaperone-bound preproteins (K. Sohrt and J. Soll, unpublished) and bind it with the help of its tetratricopeptide repeats. This binding might even take place before interaction of the precursor with the import receptor Toc159.

There are three homologs of Toc64 in the genome of *A. thaliana*, all of which are expressed in vivo *(21,48)*. AtToc64-I, however, seems to be targeted to mitochondria. Because in plant mitochondria an ortholog to Tom70 has not been found *(49)*, one could speculate that this form of atToc64 represents a substitution for this mitochondrial import receptor *(50)*.

In chloroplasts Toc64 can interact with Toc12, which is located toward the intermembrane space. Toc12 contains a J-domain that is common to a family of Dna-J proteins *(43)*. This J-domain is required for the interaction of those proteins with Hsp70. Becker and co-workers *(43)* showed that Toc12 interacts with an Hsp70 in the intermembrane space, and together with Toc64 and Tic22 they form a so-called intermembrane space complex. This complex might provide assistance in guiding preproteins from the Toc complex to the Tic complex.

How the precursor engages the Tic complex is unknown, but the association of Toc and Tic complexes has been shown *(51,52)* and indicates that a *de novo* formation of joint translocation sites is not absolutely required. The precursor proteins engage both the Toc and the Tic complexes simultaneously during translocation.

3. What About the Inner Envelope?

Unlike the Toc complex, the knowledge about the composition of the Tic complex and the function of the sofar-identified components is rather scarce. Up to now, seven components have been identified: Tic110, Tic62, Tic55, Tic40, Tic32, Tic22, and Tic20 (*see* **Fig. 1**).

Tic110 was the first component to be found. After crosslinking of preproteins, Tic110 and Toc75 could be co-immunoprecipitated with antibodies against preSSU *(53)*, indicating a close proximity of Tic110 to the Toc complex. Additionally, Tic110 is thought to recruit chaperones from the stroma such as cpn60 (Hsp60) or the Hsp100 homolog ClpC *(54,55)*. ClpC is thought to play a role in driving preprotein translocation, while cpn60 is believed to be responsible for folding newly imported proteins. Recently, Heins and co-workers showed by electrophysiological measurements and binding studies with preSSU that Tic110 is at least a central part of the import pore of the Tic complex, forming a cation-selective channel similar to Toc75 *(56)*. There have been other candidates for the import pore of the Tic complex, such as Tic20 or PIRAC (protein import related anion channel) *(57–59)*, but no experimental data to prove these assumptions were presented. In *Arabidopsis* only one homolog of Tic110 is present *(60)*, and T-DNA insertion lines of atTic110 are embryo lethal, indicating a general role for atTic110 in translocation of plastidic precursor proteins across the inner envelope membrane.

Using blue native gel electrophoresis, Tic55 and later Tic62 were found in a core translocation complex together with Tic110 *(61,62)*. Intriguingly, both proteins show features from known redox proteins: Tic55 contains a Rieske-type iron–sulfur cluster and a mononuclear iron-binding site. Usually Rieske iron clusters are known to be involved in electron transfer chains like in photosynthesis and the respiratory chain. The cytochrome b_6f complex, for example, contains such a Rieske-type iron–sulfur cluster. Nevertheless, there is also some evidence that iron–sulfur proteins function as sensors. The SoxR system in *E. coli*, for example, is composed of a protein iron–sulfur center to regulate gene expression. The oxidation state of the FeS center in the redox-sensing SoxR protein controls its own activity as a transcription activator independent of DNA-binding ability. Thus, FeS centers link cellular oxidative stress to the expression of defense genes *(63,64)*. Another clue for the participation of iron–sulfur proteins in plastidic protein translocation came from import studies in which the chloroplasts have been pretreated with diethyl pyrocarbonate (DEPC). DEPC changes histidine residues in iron–sulfur centers. Import into pretreated chloroplasts was inhibited at the level of inner envelope translocation *(61)*, possibly because the function of Tic55 was disturbed. Tic62, on the other hand, shows similarities to dehydrogenases with an $NADP^+$-binding site at its extreme N-terminus. Moreover, it also contains an FNR (ferredoxin-oxidoreductase)-binding motif in its C-terminal region. Tic62 could be co-immunprecipitated with antibodies against Tic110 and Tic55, indicating that this component is a member of the Tic complex *(62)*. Thus, both proteins seem to be *bona fide* components of a regulatory system at the inner envelope. So far, no regulation of protein import over the inner envelope membrane has been described, but it is known that light-dependent signal transduction in plants regulates, for example, gene expression in both the nucleus and the chloroplast (*65,66*). Furthermore, Hirohashi and co-workers (*67*) recently demonstrated that Ferredoxin I (FdI) and Ferredoxin III (FdIII) showed different import patterns into maize chloroplasts under dark and light conditions. FdI was imported equally well by isolated chloroplasts in both light and dark. In contrast, FdIII accumulated in the intermembrane space in the presence of light, whereas in the dark the protein was processed correctly (*68*).

Tic32 was found by screening for interaction partners of the N-terminus of Tic110 (*69*). Interestingly, sequencing of this protein showed that it is a dehydrogenase like Tic62 and that it belongs to the class of short-chain dehydrogenases. T-DNA insertion lines of atTic32 are embryo-lethal, suggesting that Tic32 plays a rather important role during protein import. Alternatively, Tic32 could be responsible for the proper assembly of the Tic complex.

The function of Tic40 is also not clear. It was originally identified as a member of the chloroplast inner membrane/chloroplast outer membrane protein family and was suggested to be involved in preprotein translocation at both membranes. Therefore, it was named Cim/Com44 and later bnToc36 (*Brassica napus* Toc36) (*70,71*). Stahl and co-workers demonstrated by crosslinking and immunoprecipitation that indeed it is a component of the Tic complex (*72*). Tic40 is predicted to be integrated into the inner membrane by a hydrophobic N-terminal domain. In its C-terminus it shows homologies to Hip (heat shock-interacting) proteins, which could point to a chaperone-like function. An *E. coli* mutant defective in SecA could be complemented with a soluble fragment of Tic40, supporting the possible role as a chaperone (*73*). There is a single Tic40 homolog in *Arabidopsis*, and T-DNA insertion mutants of atTic40 revealed pale seedlings that recovered after a while and showed reduced import rates in vitro (*74*).

The two smallest components, Tic22 and Tic20, were identified by label transfer crosslink experiments, giving evidence that both proteins are in close proximity to the preprotein during translocation (*75*). Tic20 is shown to be an integral membrane protein and, as mentioned above, was therefore favored to be part of the translocation pore. Tic22 is thought to be located in the intermembrane space, where it takes over the preprotein from the Toc complex. However, these assumptions lack experimental evidence.

In the *Arabidopsis* genome there are two homologs of Tic20 and Tic22 (*76*). AtTic20-I and atTic20-IV are only 40% similar to each other, atTic22-III and atTic22-VI only 35%. The similarity between the *Arabidopsis* isoforms is rather low, which could hint at slightly different functions of the homologous proteins.

A different function would support the theory of different Tic complexes. Another hint supporting this notion could be in the finding that Tic20 and Tic22 can only be found in a supercomplex together with Tic110 and the Toc complex, but do not associate with the Tic complex without the Toc components. As mentioned above, this is in contrast to Tic55 and Tic62, which both were isolated as a part of a stable Tic complex together with Tic110. To investigate the role of Tic20, the expression of one of the *Arabidopsis* homologs of Tic20 was altered by antisense expression. The plants exhibited chloroplast defects illustrated by pale leaves, reduced accumulation of plastid proteins, and significant growth defects *(77)*.

There is some disagreement about participation of particular components in the Tic complex. One explanation for this discrepancy in data is that there might be two more more Tic complexes. A plurality of translocons has been described in mitochondrial protein import: there are two Tim (translocase in the inner mitochondrial membrane) complexes *(78)*. The Tim23 complex imports matrix proteins with typical N-terminal targeting sequences, and the Tim22 complex is in charge of transporting integral inner membrane proteins. For chloroplastic proteins there is no evidence for a different import pathway for inner membrane and stromal proteins, but it could be possible that chloroplastic proteins are divided into photosynthetic and nonphotosynthetic proteins, which are then imported differently under different light conditions to meet the requirements of the chloroplast.

However, this is a hypothesis and research is needed to prove the existence of different molecular mechanisms of preprotein translocation at the inner chloroplast envelope.

4. Arrival in the Stroma

Similar to Hsp70 in mitochondria, a stromal Hsp100-like protein ClpC is thought to pull precursor proteins across the envelope membranes in an ATP-dependent manner *(79)*. ClpC was found to be permanently associated with the Tic complex, whereas the stromal chaperone cpn60 seems to interact with the Tic complex only in the presence of a translocating precursor *(52)*. Cpn60 is most likely involved in folding and assembly of the translocated protein; the role of Hsp70 and ClpC is less clear *(80)*. After translocation into the stroma, the processing peptidase completes import by the removal of the transit sequence *(81)*. The proteins are then either folded into an active conformation or further sorted toward the thylakoid membrane or lumen. The internal protein-sorting mechanisms of chloroplasts are reviewed elsewhere in detail *(82,83)*.

References

1. Bölter, B., Soll, J., Hill, K., Hemmler, R., and Wagner, R. (1999) A rectifying ATP-regulated solute channel in the chloroplastic outer envelope from pea. *EMBO J.* **18**, 5505–5516.
2. Li, H. M. and Chen, L. J. (1996) Protein targeting and integration signal for the chloroplastic outer envelope membrane. *Plant Cell* **8**, 2117–2126.
3. Pohlmeyer, K., Soll, J., Steinkamp, T., Hinnah, S., and Wagner, R. (1997) Isolation and characterization of an amino acid-selective channel protein present in the chloroplastic outer envelope membrane. *PNAS* **94**, 9504–9509.
4. Pohlmeyer, K., Soll, J., Grimm, R., Hill, K., and Wagner, R. (1998) A high-conductance solute channel in the chloroplastic outer envelope from Pea. *Plant Cell* **10**, 1207–1216.
5. Salomon, M., Fischer, K., Flugge, U. I., and Soll, J. (1990) Sequence analysis and protein import studies of an outer chloroplast envelope polypeptide. *PNAS* **87**, 5778–5782.
6. Seedorf, M., Waegemann, K., and Soll, J. (1995) A constituent of the chloroplast import complex represents a new type of GTP-binding protein. *Plant J.* **7**, 401–411.
7. Salomon, M., Fischer, K., Flugge, U. I., and Soll, J. (1990) Sequence analysis and protein import studies of an outer chloroplast envelope polypeptide. *PNAS* **87**, 5778–5782.
8. Seedorf, M., Waegemann, K., and Soll, J. (1995) A constituent of the chloroplast import complex represents a new type of GTP-binding protein. *Plant J.* **7**, 401–411.
9. von Heijne, G., Steppuhn, J., and Herrmann, R. G. (1989) Domain structure of mitochondrial and chloroplast targeting peptides. *Eur. J. Biochem.* **180**, 535–545.
10. Cline, K. (2000) Gateway to the chloroplast. *Nature* **403**, 148–149.
11. Waegemann, K. and Soll, J. (1996) Phosphorylation of the transit sequence of chloroplast precursor proteins. *J. Biol. Chem.* **271**, 6545–6554.
12. Pfanner, N., Craig, E. A., and Honlinger, A. (1997) Mitochondrial preprotein translocase. *Annu. Rev. Cell Dev. Biol.* **13**, 25–51.
13. Waegemann, K., Paulsen, H., and Soll, J. (1990) Translocation of proteins into chloroplasts requires cytosolic factors to obtain import competence. *FEBS Lett.* **261**, 89–92.
14. Pilon, M., de Boer, A. D., Knols, S. L., et al. (1990) Expression in Escherichia coli and purification of a translocation-competent precursor of the chloroplast protein ferredoxin. *J Biol. Chem.* **265**, 3358–3361.
15. Pilon, M., Rietveld, A. G., Weisbeek, P. J., and de Kruijff, B. (1992) Secondary structure and folding of a functional chloroplast precursor protein. *J. Biol. Chem.* **267**, 19907–19913.
16. May, T. and Soll, J. (2000) 14-3-3 proteins form a guidance complex with chloroplast precursor proteins in plants. *Plant Cell* **12**, 53–64.
17. Ellis, R. J. and van der Vies, S. M. (1991) Molecular chaperones. *Annu. Rev. Biochem.* **60**, 321–347.

18. Schleiff, E., Soll, J., Küchler, M., Kühlbrandt, W., and Harrer, R. (2003) Characterization of the translocon of the outer envelope of chloroplasts. *J. Cell Biol.* **160**, 541–551.
19. Tranel, P. J., Froehlich, J., Goyal, A., and Keegstra, K. (1995) A component of the chloroplastic protein import apparatus is targeted to the outer envelope membrane via a novel pathway. *EMBO J.* **14**, 2436–2446.
20. Eckart, K., Eichacker, L., Sohrt, K., Schleiff, E., Heins, L., and Soll, J. (2002) A Toc75-like protein import channel is abundant in chloroplasts. *EMBO Rep.* **3**, 557–562.
21. Jackson-Constan, D. and Keegstra, K. (2001) Arabidopsis genes encoding components of the chloroplastic protein import apparatus. *Plant Physiol.* **125**, 1567–1576.
22. Schleiff, E. and Soll, J. (2005) Membrane protein insertion: mixing eukaryotic and prokaryotic concepts. *EMBO Rep.* **6**, 1023–1027.
23. Reumann, S. and Keegstra, K. (1999) The endosymbiotic origin of the protein import machinery of chloroplastic envelope membranes. *Trends Plant Sci.* **4**, 302–307.
24. Hiltbrunner, A., Bauer, J., Alvarez-Huerta, M., and Kessler, F. (2001) Protein translocon at the Arabidopsis outer chloroplast membrane. *Biochem. Cell Biol.* **79**, 629–635.
25. Kessler, F., Blobel, G., Patel, H. A., and Schnell, D. J. (1994) Identification of two GTP-binding proteins in the chloroplast protein import machinery. *Science* **266**, 1035–1039.
26. Sveshnikova, N., Soll, J., and Schleiff, E. (2000) Toc34 is a preprotein receptor regulated by GTP and phosphorylation. *PNAS* **97**, 4973–4978.
27. Seedorf, M., Waegemann, K., and Soll, J. (1995) A constituent of the chloroplast import complex represents a new type of GTP-binding protein. *Plant J.* **7**, 401–411.
28. Hirsch, S., Muckel, E., Heemeyer, F., von Heijne, G., and Soll, J. (1994) A receptor component of the chloroplast protein translocation machinery. *Science* **266**, 1989–1992.
29. Kessler, F., Blobel, G., Patel, H. A., and Schnell, D. J. (1994) Identification of two GTP-binding proteins in the chloroplast protein import machinery. *Science* **266**, 1035–1039.
30. Seedorf, M., Waegemann, K., and Soll, J. (1995) A constituent of the chloroplast import complex represents a new type of GTP-binding protein. *Plant J* **7**, 401–411.
31. Sveshnikova, N., Soll, J., and Schleiff, E. (2000) Toc34 is a preprotein receptor regulated by GTP and phosphorylation. *PNAS* **97**, 4973–4978.
32. Sveshnikova, N., Soll, J., and Schleiff, E. (2000) Toc34 is a preprotein receptor regulated by GTP and phosphorylation. *PNAS* **97**, 4973–4978.
33. Sun, Y. J., Forouhar, F., Li Hm, H. M., et al. (2002) Crystal structure of pea Toc34, a novel GTPase of the chloroplast protein translocon. *Nat. Struct. Biol.* **9**, 95–100.

34. Jelic, M., Sveshnikova, N., Motzkus, M., Hörth, P., Soll, J., and Schleiff, E. (2002) The chloroplast import receptor Toc34 functions as preprotein regulated GTPase. *J Biol. Chem.* **383**, 1875–1883.

35. Jarvis, P., Chen, L. J., Li, H., Peto, C. A., Fankhauser, C., and Chory, J. (1998) An Arabidopsis mutant defective in the plastid general protein import apparatus. *Science* **282**, 100–103.

36. Hirsch, S., Muckel, E., Heemeyer, F., von Heijne, G., and Soll, J. (1994) A receptor component of the chloroplast protein translocation machinery. *Science* **266**, 1989–1992.

37. Gutensohn, M., Schulz, B., Nicolay, P., and Flügge, U. I. (2000) Functional analysis of the two Arabidopsis homologues of Toc34, a component of the chloroplast protein import apparatus. *Plant J.* **23**, 771–783.

38. Cline, K. and Henry, R. (1996) Import and routing of nucleus-encoded chloroplast proteins. *Annu. Rev. Cell Dev. Biol.* **12**, 1–26.

39. Fuks, B. and Schnell, D. J. (1997) Mechanism of protein transport across the chloroplast envelope. *Plant Physiol.* **114**, 405–410.

40. Tranel, P. J., Froehlich, J., Goyal, A., and Keegstra, K. (1995) A component of the chloroplastic protein import apparatus is targeted to the outer envelope membrane via a novel pathway. *EMBO J.* **14**, 2436–2446.

41. Dahlin, C. and Cline, K. (1991) Developmental regulation of the plastid protein import apparatus. *Plant Cell* **3**, 1131–1140.

42. Sohrt, K. and Soll, J. (2000) Toc64, a new component of the protein translocon of chloroplasts. *J.Cell Biol.* **148**, 1213–1221.

43. Becker, T., Hritz, J., Vogel, M., Caliebe, A., Bukau, B., Soll, J., and Schleiff, E. (2004) Toc12, a novel subunit of the intermembrane space preprotein translocon of chloroplasts. *Mol. Biol. Cell* **15**, 5130–5144.

44. Sohrt, K. and Soll, J. (2000) Toc64, a new component of the protein translocon of chloroplasts. *J. Cell Biol.* **148**, 1213–1221.

45. Komiya, T., Rospert, S., Schatz, G., and Mihara, K. (1997) Binding of mitochondrial precursor proteins to the cytoplasmic domains of the import receptors Tom70 and Tom20 is determined by cytoplasmic chaperones. *EMBO J.* **16**, 4267–4275.

46. Brocard, C., Kragler, F., Simon, M. M., Schuster, T., and Hartig, A. (1994) The tetratricopeptide repeat-domain of the PAS10 protein of Saccharomyces cerevisiae is essential for binding the peroxisomal targeting signal-SKL. *Biochem. Biophys. Res. Commun.* **204**, 1016–1022.

47. Lamb, J. R., Tugendreich, S., and Hieter, P. (1995) Tetratrico peptide repeat interactions: to TPR or not to TPR? *Trends Biochem. Sci.* **20**, 257–259.

48. Kessler, F., Blobel, G., Patel, H. A., and Schnell, D. J. (1994) Identification of two GTP-binding proteins in the chloroplast protein import machinery. *Science* **266**, 1035–1039.

49. Werhahn, W. and Braun, H. P. (2002) Biochemical dissection of the mitochondrial proteome from Arabidopsis thaliana by three-dimensional gel electrophoresis. *Electrophoresis* **23**, 640–646.

50. Chew, O., Lister, R., Qbadou, S., et al. (2004) A plant outer mitochondrial membrane protein with high amino acid sequence identity to a chloroplast protein import receptor. *FEBS Lett.* **557**, 109–114.

51. Akita, M., Nielsen, E., and Keegstra, K. (1997) Identification of protein transport complexes in the chloroplastic envelope membranes via chemical cross-linking. *J. Cell Biol.* **136**, 983–994.

52. Nielsen, E., Akita, M., Davila-Aponte, J., and Keegstra, K. (1997) Stable association of chloroplastic precursors with protein translocation complexes that contain proteins from both envelope membranes and a stromal Hsp100 molecular chaperone. *EMBO J.* **16**, 935–946.

53. Lübeck, J., Soll, J., Akita, M., Nielsen, E., and Keegstra, K. (1996) Topology of IEP110, a component of the chloroplastic protein import machinery present in the inner envelope membrane. *EMBO J.* **15**, 4230–4238.

54. Kessler, F. and Blobel, G. (1996) Interaction of the protein import and folding machineries of the chloroplast. *PNAS* **93**, 7684–7689.

55. Nielsen, E., Akita, M., Davila-Aponte, J., and Keegstra, K. (1997) Stable association of chloroplastic precursors with protein translocation complexes that contain proteins from both envelope membranes and a stromal Hsp100 molecular chaperone. *EMBO J.* **16**, 935–946.

56. Heins, L., Mehrle, A., Hemmler, R., Wagner, R., Küchler, M., Hörmann, F., Sveshnikov, D., and Soll, J. (2002) The preprotein conducting channel at the inner envelope membrane of plastids. *EMBO J.* **21**, 2616–2625.

57. Dabney-Smith, C., van den Wijngaard, P. W., Treece, Y., Vredenberg, W. J., and Bruce, B. D. (1999) The C terminus of a chloroplast precursor modulates its interaction with the translocation apparatus and PIRAC. *J. Biol. Chem.* **274**, 32351–32359.

58. Kouranov, A., Chen, X., Fuks, B., and Schnell, D. J. (1998) Tic20 and Tic22 are new components of the protein import apparatus at the chloroplast inner envelope membrane. *J. Cell Biol.* **143**, 991–1002.

59. van den Wijngaard, P. W. and Vredenberg, W. J. (1997) A 50-picosiemens anion channel of the chloroplast envelope is involved in chloroplast protein import. *J. Biol. Chem.* **272**, 29430–29433.

60. Jackson-Constan, D. and Keegstra, K. (2001) Arabidopsis genes encoding components of the chloroplastic protein import apparatus. *Plant Physiol.* **125**, 1567–1576.

61. Caliebe, A., Grimm, R., Kaiser, G., Lübeck, J., Soll, J., and Heins, L. (1997) The chloroplastic protein import machinery contains a Rieske-type iron-sulfur cluster and a mononuclear iron-binding protein. *EMBO J.* **16**, 7342–7350.

62. Küchler, M., Decker, S., Hormann, F., Soll, J., and Heins, L. (2002) Protein import into chloroplasts involves redox-regulated proteins. *EMBO J.* **21**, 6136–6145.

63. Ding, H., Hidalgo, E., and Demple, B. (1996) The redox state of the [2Fe-2S] clusters in SoxR protein regulates its activity as a transcription factor. *J. Biol. Chem.* **271**, 33173–33175.
64. Hidalgo, E., Ding, H., and Demple, B. (1997) Redox signal transduction via iron-sulfur clusters in the SoxR transcription activator. *Trends Biochem. Sci.* **22**, 207–210.
65. Fankhauser, C. and Chory, J. (1999) Light receptor kinases in plants! *Curr. Biol.* **9**, R123–R126.
66. Li, H. M., Washburn, T., and Chory, J. (1993) Regulation of gene expression by light. *Curr. Opin. Cell Biol.* **5**, 455–460.
67. Hirohashi, T., Hase, T., and Nakai, M. (2001) Maize non-photosynthetic ferredoxin precursor is mis-sorted to the intermembrane space of chloroplasts in the presence of light. *Plant Physiol.* **125**, 2154–2163.
68. Hirohashi, T., Hase, T., and Nakai, M. (2001) Maize non-photosynthetic ferredoxin precursor is mis-sorted to the intermembrane space of chloroplasts in the presence of light. *Plant Physiol.* **125**, 2154–2163.
69. Hormann, F., Küchler, M., Sveshnikov, D., Oppermann, U., Li, Y., and Soll, J. (2004) Tic32, an essential component in chloroplast biogenesis. *J. Biol. Chem.* **279**, 34756–34762.
70. Ko, K., Budd, D., Wu, C., Seibert, F., Kourtz, L., and Ko, Z. W. (1995) Isolation and characterization of a cDNA clone encoding a member of the Com44/Cim44 envelope components of the chloroplast protein import apparatus. *J. Biol. Chem.* **270**, 28601–28608.
71. Schnell, D. J., Blobel, G., Keegstra, K., Kessler, F., Ko, K., and Soll, J. (1997) A nomenclature for the protein import components of the chloroplast envelope. *Trends Cell Biol.* **7**, 303–304.
72. Stahl, T., Glockmann, C., Soll, J., and Heins, L. (1999) Tic40, a new "old" subunit of the chloroplast protein import translocon. *J. Biol. Chem.* **274**, 37467–37472.
73. Pang, P., Meathrel, K., and Ko, K. (1997) A component of the chloroplast protein import apparatus functions in bacteria. *J. Biol. Chem.* **272**, 25623–25627.
74. Budziszewski, G. J., Lewis, S. P., Glover, L. W., et al. (2001) Arabidopsis genes essential for seedling viability: isolation of insertional mutants and molecular cloning. *Genetics* **159**, 1765–1778.
75. Kouranov, A., Chen, X., Fuks, B., and Schnell, D. J. (1998) Tic20 and Tic22 are new components of the protein import apparatus at the chloroplast inner envelope membrane. *J. Cell Biol.* **143**, 991–1002.
76. Jackson-Constan, D. and Keegstra, K. (2001) Arabidopsis genes encoding components of the chloroplastic protein import apparatus. *Plant Physiol.* **125**, 1567–1576.
77. Chen, X., Smith, M. D., Fitzpatrick, L., and Schnell, D. J. (2002) In vivo analysis of the role of atTic20 in protein import into chloroplasts. *Plant Cell* **14**, 641–654.
78. Koehler, C. M. (2000) Protein translocation pathways of the mitochondrion. *FEBS Lett.* **476**, 27–31.

79. Keegstra, K. and Cline, K. (1999) Protein import and routing systems of chloroplasts. *Plant Cell* **11**, 557–570.

80. Tsugeki, R. and Nishimura, M. (1993) Interaction of homologues of Hsp70 and Cpn60 with ferredoxin-NADP+ reductase upon its import into chloroplasts. *FEBS Lett.* **320**, 198–202.

81. Richter, S. and Lamppa, G. K. (1998) A chloroplast processing enzyme functions as the general stromal processing peptidase. *PNAS* **95**, 7463–7468.

82. Eichacker, L. A. and Henry, R. (2001) Function of a chloroplast SRP in thylakoid protein export. *Biochim. Biophys. Acta* **1541**, 120–134.

83. Robinson, C., Hynds, P. J., Robinson, D., and Mant, A. (1998) Multiple pathways for the targeting of thylakoid proteins in chloroplasts. *Plant Mol. Biol.* **38**, 209–221.

13

Import of Plastid Precursor Proteins Into Pea Chloroplasts

Bettina Bölter and Jürgen Soll

Summary

Plastids are a heterogeneous family of organelles found ubiquitously in plant and algal cells *(1)*. Most prominent are the chloroplasts, which carry out such essential processes as photosynthesis and the biosynthesis of fatty acids as well as of amino acids. As mitochondria, chloroplasts derived from a single endosymbiotic event *(2)*. They are believed to have evolved from an ancient cyanobacterium, which had been engulfed by an early eukaryotic ancestor. During evolution the plastid genome has been greatly reduced, and most of the genes have been transferred to the host nucleus. Consequently, more than 98% of all plastid proteins are translated onto cytosolic ribosomes. They have to be posttranslationally targeted to and imported into the organelle. Targeting is assisted by cytosolic proteins, which interact with proteins destined for plastids and thereby keep them in an import-competent state. After reaching the target organelle, many proteins have to conquer the barrier of the chloroplast outer and inner envelopes. This process is mediated by complex molecular machines in the outer (Toc complex) and inner (Tic complex) envelope of chloroplasts, respectively *(3)*. Most proteins destined for compartments inside the chloroplast contain a cleavable N-terminal transit peptide *(4)*, whereas most of the outer envelope components insert into the membrane without such a targeting peptide *(5)*.

Key Words: Chloroplasts; protein targeting; in vitro import; translocation machinery; Toc/Tic; localization.

1. Introduction

The import process of plastid-designated proteins into isolated chloroplasts is not an easy system to study. It depends not only on the quality of the translation product (including the necessary posttranslational modifications), but especially on the integrity of the isolated organelles.

From: *Methods in Molecular Biology, Vol. 390: Protein Targeting Protocols: Second Edition*
Edited by: M. van der Giezen © Humana Press Inc., Totowa, NJ

The source of the translation product—reticulocyte lysate or wheat germ lysate—can influence the results of an import experiment dramatically. Often the translation in the mammalian system is more efficient and gives higher yields of radioactively labeled protein. However, posttranslational modifications typical for plant systems are not made, and therefore proteins tend to aggregate. Those aggregates are not import competent and simply "stick" to the chloroplast surface. The plant-derived wheat germ lysate results in native-like translated proteins, although often the amount is quite low. For each protein one wishes to study, the best system has to be determined empirically.

Optimal conditions to analyze the import process can be analyzed in some detail in vitro. It is known today that several pathways of protein import into chloroplasts exist. Their typical features, such as energy requirement or dependence on proteinaceous components, can be studied, and successive steps can be dissected using different experimental setups. In this chapter we provide protocols to study the import process of plastid precursor proteins into pea chloroplasts. The methods can be adapted to organelles originating from other species such as spinach or *Arabidopsis*, but we have found that import into pea chloroplasts is the easiest system to handle.

2. Materials

2.1. Plant Growth

1. Seeds from *Pisum sativum* var. Golf, Arnika, or Violetta were purchased from Bayerische Futtersaatbau (Ismaning, Germany).
2. Vermiculite.

2.2. DNA Preparation

1. DH5α cells were used for propagation of plasmid DNA.
2. Macherey-Nagel (Düren, Germany) MIDI Kit.
3. Isopropanol.
4. Ethanol.

2.3. Transcription

1. 5x Transcription buffer (Fermentas, St. Leon-Rot, Germany), stored at −20°C.
2. RNA Polymerase (SP6/T7; Fermentas), stored at −20°C.
3. Dithiothreitol (DTT; Promega, Madison, WI), stored at −80°C.
4. RNase inhibitor (GE Healthcare, Freiburg, Germany), stored at −20°C.
5. Bovine serum albumin (BSA; Fermentas), stored at −20°C.
6. NTPs (Roche, Penzberg, Germany), stored at −80°C.
7. m^7 -Guanosin (5') ppp (5') Guanosin (cap), stored at −80°C.

2.4. Translation

1. Rabbit reticulocyte/wheat germ translation kit (Promega).
2. [S^{35}]-Methionine/cysteine mixture (GE Healthcare).

2.5. Isolation of Intact Chloroplasts

1. Isolation buffer: 330 mM sorbitol, 20 mM MOPS, 13 mM Tris-HCl pH7.6, 3 mM MgCl$_2$, 0.1% (w/v) BSA.
2. Four layers of mull and one layer of gauze (30-μm pore size).
3. Washing buffer: 330 mM sorbitol, 50 mM HEPES/KOH, pH7.6.
4. 40/80% Percoll in 330 mM sorbitol, 50 mM HEPES/KOH, pH7.6.
5. 80% Acetone.
6. Glass cuvettes, spectrophotometer.

2.6. In Vitro Import

1. Import buffer: 330 mM sorbitol, 50 mM HEPES/KOH pH7.6, 3 mM Mgso$_4$, 10 mM methionine, 10 mM cysteine, 20 mM K-gluconate, 10 mM NaHCO$_3$, 2% BSA (w/v); stock solutions: 10 × HMS (3.3 M sorbitol, 500 mM HEPES/KOH pH 7.6, 30 mM MgCl$_2$); 250 mM methionine; 250 mM cysteine; 1 M K-gluconate; 1 M NaHCO$_3$; 5% BSA.
2. ATP (prepare fresh!).
3. Percoll.
4. Washing buffer: 330 mM sorbitol, 50 mM HEPES/KOH, pH7.6.
5. LaemmLi buffer for denaturation: 50 mM Tris-HCl pH6.8, 100 mMβ-mercapto-ethanol, 2% sodium dodecyl sulfate (SDS), 0.1% bromophenol blue, 10% glycerol.

2.7. Thermolysin Treatment

1. Digestion buffer: 330 mM sorbitol, 50 mM HEPES/KOH, pH7.6, 0.5 mM CaCl$_2$.
2. Thermolysin (Roche) (*see* **Note 1**), always prepare fresh!
3. EDTA.
4. Washing buffer: 330 mM sorbitol, 50 mM HEPES/KOH, pH7.6, 5 mM EDTA.

2.8. Subfractionation of Chloroplasts

1. Lysis buffer: 25 mM HEPES/KOH pH 7.6, 1 mM PMSF (prepare fresh in 100% ethanol; this stock solution can be kept for 1 wk at 4°C).
2. 0.465 M, 0.8 M, and 0.996 M sucrose in lysis buffer, prepare fresh.
3. 50% Trichloracetic acid (TCA).

2.9. Extraction of Chloroplasts With Urea

1. Lysis buffer: 25 mM HEPES/KOH pH 7.6, 4–6 M urea.
2. 50% TCA.

2.10. SDS-PAGE

1. Separating gel: 375 mM Tris-HCl pH 8.8, 10–12% acrylamide/bis (30:1).
2. Stacking gel: 125 mM Tris-HCl pH 6.8, 4.8% acrylamide/bis.
3. 10% Ammonium persulfate (APS), stock solution stored at −20°C, working aliquot at 4°C.
4. TEMED.
5. Running buffer: 125 mM Tris-HCl, 960 mM glycine, 0.5% SDS (we use a 10× stock solution).
6. Molecular weight marker: MW-SDS-70 (Sigma, Munich, Germany).
7. Coomassie: 0.18% Coomassie brilliant blue R250 (Sigma), 50% MeOH, 7% acetic acid; Destain: 40% MeOH, 7% acetic acid, 3% glycerol.

2.11. Western Blotting

1. Nitrocellulose, 0.2 μm (Schleicher and Schüll, Dassel, Germany), and 0.37 mm blotting-papers (Macherey-Nagel).
2. Anode buffer 1: 0.3 M Tris, 20% methanol, pH 10.4 (*see* **Note 2**).
3. Anode buffer 2: 25 mM Tris, 20% methanol, pH 10.4.
4. Cathode buffer: 25 mM Tris, 4 mM amino caproic acid, 20% methanol, pH 7.6.
5. Ponceau red: 1% Ponceau red (sigma), 1% HAc.
6. Tris-buffered saline (TBS): prepare 10x stock with 1.5 M NaCl, 1 M Tris-HCl pH 7.5, dilute 100 mL with 900 mL dest. water.
7. Blocking buffer: 0.3% nonfat dry milk, 0.03% BSA in TBS for most primary antibodies. For others 1% BSA, 0.05% Tween 20 in TBS was used.
8. Primary antibodies raised in rabbits against the respective target protein.
9. Secondary antibody: anti-rabbit IgG conjugated to alkaline phosphatase (AP) (Sigma).
10. AP-buffer: 100 mM Tris-HCl pH 9.5, 100 mM NaCl, 50 mM MgCl$_2$.
11. Nitrotetrazolium blue (NBT) and 5-bromo-4-chloro-3-indolyl phosphate (BCIP) (Sigma).

3. Methods

3.1. Plant Growth

1. Pea seeds (approx 300 mL per tray) are imbibed overnight in running tab water.
2. The water is discarded and the pea seeds are distributed evenly into trays with a thick layer of vermiculite. Another layer of vermiculite is added on top and the tray transferred to a climate chamber with 25°C and a 16 h/8 h light/dark cycle of about 120 μmol of photons. Plants are watered generously on a daily basis. Plants are grown for 9–11 d until use.

3.2. DNA Preparation

1. The gene coding for the protein of interest has to be cloned into a vector with a suitable promoter. In our hands best results are obtained with the SP6 RNA-polymerase, but T7 and T3 can also be used.
2. The plasmid should be linearized with a suitable restriction enzyme immediately downstream of the stop codon (*see* **Note 3**).
3. The linearized plasmid is extracted with phenol/chloroform. Use 1 volume of buffered phenol/chloroform, mix vigorously and centrifuge at room temperature for 2–5 min to separate phases. Transfer the aqueous upper phase to a fresh Eppendorf cup, add one volume of chloroform, mix vigorously, and centrifuge at room temperature for 2–5 min to separate phases. Again take the upper phase to a fresh tube and precipitate the DNA by adding 1/10 volume of $3\,M$ NaCl, 2.5 volumes of ethanol p.a. Incubate this for 30 min at $-80°C$ (alternatively overnight at $-20°C$) and centrifuge for 20 min at 4°C at $14,000\,g$. Remove the supernatant and wash the pellet twice with 70% ethanol p.a. After drying the pellet is resuspended in RNase-free water and the concentration is determined via agarose electrophoresis or spectrophotometry.

3.3. Transcription

1. In vitro transcription of linearized plasmids is carried out in a reaction volume of $50\,\mu L$ containing transcription buffer, $10\,mM$ DTT, 100 U RNase inhibitor, 0.05% (w/v) BSA, $0.5\,mM$ ATP, CTP, and UTP, $0.375\,mM$ m^7-guanosine (5') ppp (5') guanosine (cap), 10 U SP6 RNA-polymerase and $2–2.5\,\mu g$ plasmid DNA. The reaction mixture is incubated for 15 min at 37°C to yield RNA with cap at the 5' end. Finally, $1.2\,mM$ GTP is added and the transcription mixture is incubated for another 60 min. mRNA is either used directly for in vitro translation or stored in 10-μL aliquots under liquid N_2 (*see* **Notes 4** and **5**).

3.4. Translation

1. As a first step you will need to test your RNA concerning optimal amounts of salts and RNA itself.
2. We normally try the reticulocyte lysate translation kit first. We start by varying RNA and MgOAc amounts. Prepare a premix for 10 samples consisting of $66\,\mu L$ Retic, $4\,\mu L$ ^{35}S-methionine/cysteine, $2\,\mu L$ amino acid mix without methionine (in case you intend to label cysteines you need to take the aa mix without cysteine!), $2.8\,\mu L$ KCl, $1\,\mu L$ $0.1\,M$ DTT and $2\,\mu L$ RNasin. Of this mixture you need $7.78\,\mu L$ for each test translation. We test 0, 0.2, 0.4, and $0.6\,\mu L$ MgOAc, respectively, combined each with $1\,\mu L$ RNA or $1\,\mu L$ of a 1:10 dilution. Fill up to $10\,\mu L$ with RNAse-free water. RNA is added last and the mixture is carefully centrifuged for a few seconds. It is incubated for 45 min at 30°C, then $1\,\mu L$ is taken out for SDS-PAGE analysis and the rest is shock-frozen in liquid nitrogen. If this does not result in satisfying amounts of radiolabeled protein, you might try changing the DTT concentration and/or the amount of KCl.

In case you want to ensure posttranslational modifications to your protein, you need to use wheat germ lysate for translation. For a typical test translation of 10 samples prepare a premix of $50\,\mu L$ wheat germ lysate, $8\,\mu L$ amino acids without methionine, and $2.5\,\mu L$ of ^{35}S-methionine/cysteine. In this case the KOAc concentration is tested together with different amounts of RNA as described above. The mixture is incubated for 60 min at 25°C and handled as above.

3.5. Isolation of Intact Chloroplasts

1. The first initial step to a successful isolation of import competent plastids is harvesting the pea leaves within a dark period, so either you start early in the morning or place the tray in the dark already in the evening (see Note 6).
2. All procedures are carried out at 4°C. About 200 g of pea leaves are ground in a kitchen blender in 330 mL isolation medium and filtered through four layers of mull and one layer of gauze. The homogenate is centrifuged for 1 min at 1500g, and the pellet is gently resuspended in about 1 mL wash medium with a cut-off pipet tip. Take care to leave the suspension on ice all the time!
3. Intact chloroplasts are reisolated via a discontinuous Percoll gradient of 13 mL 40% and 8 mL 80% (see Note 7) and centrifuged for 5 min at 3000g in a swing-out rotor. Two green bands of chloroplasts can be observed: the bottom band represents the intact chloroplasts.
4. Again use a cut-off pipet tip to collect intact plastids, transfer to a fresh centrifuge tube, and fill it with washing buffer. Centrifuge for 1 min at 1500g, discard supernatant, and resuspend again *carefully* in washing buffer. After centrifugation use $500\,\mu L$ to resuspend the chloroplasts. Keep them on ice in the dark until further use.
5. Five μL of the chloroplast suspension are diluted in 5 mL 80% acetone, mixed well, and chlorophyll concentration is determined by measuring optical density at 645, 663, and 750 nm against the solvent (6). We have a program in the photometer that directly calculates the chlorophyll concentration, but you can easily use this formula: mg chlorophyll/mL $= 8.02 \times (E_{663} - E_{750}) + 20.2 \times (E_{645} - E_{750})$.

3.6. In Vitro Import Into Isolated Chloroplasts

The import process can be subdivided into four different steps: (1) binding of the (pre) protein to the chloroplast surface, (2) translocation across the envelope via the translocation machineries, (3) processing of precursor proteins by the stromal processing peptidase to their mature form, and (4) assembly into functional complexes in the stroma or further sorting to thylakoidal compartments.

One can investigate the individual steps by altering experimental conditions. Binding and translocation can be separated by different amounts of ATP as energy source and by using different temperatures. Binding requires micromolar amounts of ATP and occurs at 4°C, whereas translocation needs 0–$3\,mM$ ATP and 25°C. In case you want to study the ATP dependence of the import process, you need to deplete the internal ATP from your translation product. We usually do this by applying the translation mixture on small columns (e.g., Micro-spin 6 chromatography columns from Biorad). But you still have to keep in mind that the translation kit contains an ATP-regenerating system! In the following we will describe a standard import assay usually done in our lab.

1. Use chloroplasts equivalent to 15 μg of chlorophyll in a final volume of 100 μL. Mix import buffer and translation product (*see* **Note 8**). Never use more than 10% translation product because this would inhibit import. The final concentration of sorbitol, magnesium, and HEPES is achieved by adding 10× HMS depending on the actual amount of chloroplasts you need to add. For a 100 μL sample you calculate (100 − x μL chloroplasts)/10. It is essential to mix well after adding 10× HMS! Start the reaction with chloroplasts and incubate the sample at 25°C for the desired times. Import is linear, depending on the precursor protein, over a range of 2–15 min.
2. Place vials on ice and transfer the reaction on top of a 40% Percoll cushion (same as for chloroplast preparation) and centrifuge for 5 min at 3300g.
3. Remove the supernatant carefully (*see* **Note 9**). It contains broken chloroplasts and unbound precursor.
4. Wash pellet twice with 100 μL washing buffer. Chloroplasts are collected by centrifugation for 1 min at 800g.
5. Either resuspend pellet in Laemmli buffer and heat it for 3 min at 95°C or treat chloroplasts with thermolysin to remove precursor which is bound to the organelle surface.

3.7. Thermolysin Treatment of Chloroplasts After Import

1. Chloroplasts are resuspended in 100 μL digestion buffer. Thermolysin is added to a final concentration of 1–3 μg/μg chlorophyll (*see* **Note 1**). Incubate for 20 min on ice.
2. Stop digestion with 10 m*M* EDTA. Reisolate chloroplasts via centrifugation for 1 min at 800g and wash them once in washing buffer. Finally, the organelles are solubilized in Laemmli buffer and analyzed via SDS-PAGE. An example of a typical import experiment of a radioactively labeled precursor protein into chloroplasts is shown in **Fig. 1**.

3.8. Subfractionation of Chloroplasts After Import

To determine the subcellular localization of the protein of interest, you can fractionate chloroplasts into the different compartments, i.e. inner and outer envelope membranes, thylakoids, and stroma/intermembrane space, respectively.

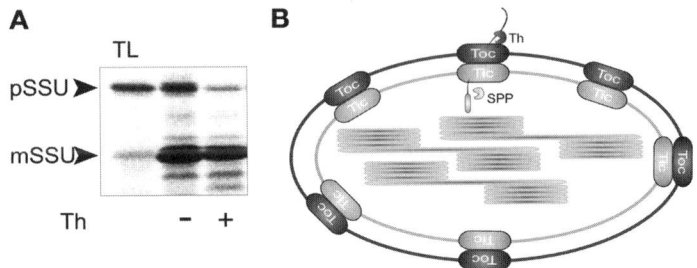

Fig. 1. Import of pSSU into intact chloroplasts. **(A)** In vitro synthesized pSSU
was incubated with isolated chloroplasts for 5 min at 25°C, resulting in the mature
protein (mSSU). Intact chloroplasts were reisolated and treated with thermolysin for
20 min on ice. Lane 1 represents 1/10 of translation product (TL); lane 2 shows import
without thermolysin treatment (Th-). The sample in lane 3 was treated with thermolysin
(Th+). A radiograph is shown. **(B)** Model of the import process of a precursor protein
across chloroplast envelope membranes. The transit peptide (depicted as a small box)
is cleaved off by the stromal processing peptidase (SPP) as soon as it reaches the
stroma. Thermolysin digests only proteins on the chloroplast surface, in this case the
C-terminal part of the precursor, which has not been imported completely.

1. After the last washing step once import is completed, resuspend the organelles in
   lysis buffer (try to bring the concentration to 1 mg/mL, it should be at the most
   2.5 mg/mL) and incubate for 30 min on ice.
2. Meanwhile, prepare a sucrose gradient consisting of $0.998 M$ $0.8 M$, and $0.46 M$
   sucrose, respectively. The volume depends on the volume of lysis buffer. We
   normally make gradients of 4 mL and add 200–400 µL suspension onto them.
3. Centrifuge for 3 h at $137,000g$ at 4°C.
4. Take the top layer from the gradient (represents the soluble compartments) and
   precipitate adding TCA to a final concentration of 15%. Incubate for 15– 20 min on
   ice and centrifuge for 15 min at $15,000g$. The pellet is then solubilized in Laemmli
   buffer (take care to adjust the pH with Tris base so that the buffer is still blue) and
   heated for 3 min at 95°C.
5. The upper band between 0.46 and $0.8 M$ sucrose constitutes the outer envelope.
   Take it off and dilute at least 1:3 in lysis buffer.
6. The lower band between 0.8 and $0.998 M$ sucrose represents the inner envelope
   membrane. Dilute this as in **step 5**.
7. Centrifuge the inner and outer envelope fractions for 1 h at $137,000g$, resuspend
   the pellet in Laemmli buffer, and heat it.
8. In the pellet of the gradient you will find the thylakoids. Resuspend in lysis buffer
   and centrifuge 15 min at $4500g$. Repeat that twice, resuspend the final pellet in
   Laemmli buffer, and also heat it.

3.9. Extraction of Chloroplasts With Urea

Proteins that are found to be in a membrane fraction can be further classified into peripheral and integral membrane proteins. Those that are merely attached to or slightly embedded in the lipid bilayer can be extracted with urea *(7)*, whereas deeply embedded proteins are resistant to extraction.

1. After the completed import reaction chloroplasts are washed once and then resuspended in 50 µL lysis buffer containing 4–6 *M* urea. Incubate on ice for 30 min.
2. Centrifuge for 10 min at 250,000g to separate extractable proteins from integral ones.
3. Remove supernatant and add 1/5 volume of 5× Laemmli buffer. Alternatively, precipitate proteins with TCA as described under **Subheading 3.8., step 4** (*see* **Note 10**).
4. Resuspend pellet in Laemmli buffer and analyze on SDS gel. Do **NOT** heat to 95°C!! If necessary, heat the samples at 65°C for 5 min.

3.10. SDS-PAGE

1. We always use the Hoefer SE-600 system for analyzing import experiments. Minigels are not suitable for this purpose. Our glass plates are usually self-made (the original ones from Hoefer are too expensive) and only need to be etched with 0.1 *M* NaOH before the first use. Afterwards we merely clean them with demineralized water.
2. For two 12.5% 0.75-mm gels mix 10 mL bidest. water, 7.5 mL separating buffer, and 12.5 mL acrylamide/bis (30:1). Add 120 µL 10% APS and 35 µL TEMED, mix well, and pipet between the glass plates with a 20-mL glass pipet, leaving space for the stacking gel. Overlay with water or 2-propanol. The separating gel should polymerize for approx 30–45 min, depending on the room temperature. (In summer you need to be quick with filling the gel between the plates! Precool solutions if necessary.)
3. Pour off water or isopropanol (rinse once with water).
4. Prepare the stacking gel by mixing 5.8 mL water, 2.5 mL buffer, and 1.6 mL acrylamide. Add 150 µL APS and 15 µL TEMED and pipet onto the separating gel. Insert the comb and let in polymerize for approx 30 min.
5. Prepare the running buffer by dilution the 10× stock. We prepare 2 L of 1x buffer. Keep 600 mL for the upper buffer tank and dilute the remaining 1.2 L 1:2 with demineralized water for the lower tank.
6. Carefully remove the comb form the stacking gel, rinse the slots thoroughly with water, and take care to remove residual water completely.
7. Assemble the gel unit, fill the upper buffer tank, and load your samples with a 100 µL Hamilton syringe. Use 10 µL of low molecular weight marker.

8. During the day we run one gel at 27 mA until the blue front has reached the separating gel, at which point we raise the current to 35 mA. Alternatively, run the gel overnight at 5 mA.

9. After the run is completed, eiher (1) stain the gel with Coomassie, destain and water it before applying it to a vacuum dryer—put the dry gel on a film or phosphoimager screen overnight—or (2) proceed with immunoblotting (*see* **Subheading 3.11.**)

3.11. Western Blotting

1. The separated proteins are transferred onto nitrocellulose via a Hoefer semi-dry blot system. We assemble the system on the anode.

2. Prepare 3 mm Whatman filter sheets and nitrocellulose the same size of your gel. You need three sheets for anode buffer $1 + 2$ (A1 + 2), six for the cathode buffer (C).

3. The gel is disassembled, the stacking gel removed, and the gel incubated in A2.

4. Bathe three sheets of filter paper in A1, three in A2, and six in C. First put papers in A1 on the anode plate, use a glass tube for removing air bubbles (this is crucial, so repeat after each step), layer those in A2 on top, then add the nitrocellulose to A2 and put it carefully on top. After this the gel is applied (in the same orientation as you had it in the apparatus) and the filters in buffer C are finally stacked on top. Put the cathode plate in place, position a heavy object (e.g., a full 1-L glass bottle) on the blot apparatus, and apply $0.8 \, mA/cm^2$ for 1 h.

5. When the transfer is completed, rinse the nitrocellulose briefly in bidest. water and then stain it with Ponceau red for 1–5 min.

6. Destain with bidest. water until you see the marker bands clearly. Mark them with a pencil and destain the membrane completely in blocking buffer.

7. Ideally, the blocking buffer is changed three times every 10 min before the primary antibody is added.

8. The dilution of the antibody (we do this mostly in blocking buffer) has to be determined for every individual antibody. We normally use a dilution of 1:1000 (*see* **Note 11**).

9. Incubate with gentle agitation either for 2 h at room temperature or overnight at 4°C.

10. Wash three times with blocking buffer.

11. Apply secondary antibody in a dilution of 1:10,000 and incubate for 1 h at room temperature with gentle agitation.

12. Wash three times with blocking buffer.

13. Rinse the membrane briefly in AP buffer. Add 66 µL NBT and 132 µL BCIP to 10 mL AP buffer, and apply this to the membrane and wait for the appearance of bands. This should, depending on the antibody and protein amount, not take more than a few minutes. In some cases, however, an overnight incubation is necessary.

14. Stop the reaction by transferring the membrane into bidest. water and a little EDTA. Dry the membrane.

Notes

1. We ordered thermolysin from different companies and found that the optimal working concentration must be determined each time you change the supplier. Even different lots from the same supplier might behave differently.
2. The pH of anode buffer 1 and 2 does not need to be adjusted!
3. The linearization of the plasmid results in a "run-off transcription," which is more efficient than just using the terminator provided by the vector.
4. RNA can be freeze/thawed up to four times safely. However, some RNAs are only functional directly after transcription. If the first try with fresh RNA was successful and the following ones are not, consider this the major problem. In such a case one might want to avoid in vitro transcription (which is quite costly and cannot be down scaled to less than half volumes). Try a coupled transcription/translation kit then!
5. In some cases translation efficiency is higher after denaturation of RNA directly before translation. Incubate the RNA for 5 min at 65°C and place it *immediately* on ice. This should solve unwanted secondary structure within the RNA, which might hinder the proper binding and procession of the RNA–polymerase.
6. This reduces the amount of starch synthesized by plastid in the light. The starch granules are so heavy that they break many chloroplasts during centrifugation. Placing the peas in the dark on the previous evening results in a low starch content. However, keeping them in the dark too long interferes with the metabolism of the plants, and chloroplasts tend to be very sensitive under such conditions!
7. Percoll gradients should be prepared in advance. It is best to pipet the 40% solution first and then underlay it with the 80% Percoll (easily done with a glass pipet).
8. In some cases in vitro translated proteins tend to aggregate. Therefore, we centrifuge the sample for 10 min at 50,000g to remove precipitated protein and take the supernatant for the import experiment.
9. The chloroplast pellet is very fluffy and easily disturbed. It is best to leave about 10 μL of supernatant instead of trying to remove it completely.
10. Please be careful when taking off the supernatant—sometimes the pellet is very fluffy!
11. In most cases we work with polyclonal antisera from rabbits. The optimal conditions for each one concerning blocking buffer and dilution must be experimentally determined. For the weaker antibodies, a blocking buffer consisting of 1xTBS, 1% BSA, 0.05% Tween 20 has been found to work well.

References

1. Whatley, J. (1978) A suggested cycle of plastid developmental interrelationships. *New Phytol.* **80**, 489–502.
2. Moreira, D., Le Guyader, H., and Philippe, H. (2000) The origin of red algae and the evolution of chloroplasts. *Nature* **405**, 69–72.

3. Soll, J. and Schleiff, E. (2004) Protein import into chloroplasts. *Nat. Rev. Mol. Cell Biol.* **5**, 198–208.

4. Vothknecht, U. C. and Soll, J. (2002) Chloroplast quest: a journey from the cytosol into the chloroplast and beyond. *Rev. Physiol Biochem. Pharmacol.* **145**, 181–222.

5. Schleiff, E. and Klosgen, R. B. (2001) Without a little help from 'my' friends: direct insertion of proteins into chloroplast membranes? *Biochim. Biophys. Acta* **1541**, 22–33.

6. Arnon, D.J. (1949) Copper enzymes in isolated chloroplasts. Polyphenoloxidase in *Beta vulgaris. Plant Physiol.* **6**, 1–15.

7. Molloy, M. P., Herbert, B. R., Walsh, B. J., et al. (1998) Extraction of membrane proteins by differential solubilization for separation using two-dimensional gel electrophoresis. *Electrophoresis* **19**, 837–844.

14

Protein Targeting in "Secondary" or "Complex" Chloroplasts

Balbir K. Chaal and Beverley R. Green

Summary

All the algae with chlorophyll (Chl) *c* (haptophytes, cryptophytes, and heterokonts such as diatoms) acquired their chloroplasts by secondary endosymbiosis, where a nonphotosynthetic eukaryote host engulfed (or was invaded by) a red alga. This resulted in chloroplasts with four bounding membranes. The outermost membrane (chloroplast endoplasmic reticulum [ER]), is physically continuous with the rough ER, and in some algal species can be seen to have cytoplasmic ribosomes attached to its outer surface. All nuclear-encoded chloroplast proteins have an N-terminal ER targeting sequence, which is cleaved off during transit across this membrane. We know little about how proteins cross the next membrane and engage the import translocons of the envelope membranes. One way to study the targeting of proteins across the inner membranes is to make constructs lacking the ER signal sequence, translate them in vitro, and assay their import into pea chloroplasts.

Key Words: Plastid; complex chloroplast; secondary endosymbiosis; protein import; in vitro transcription-translation; pea chloroplasts; intact chloroplasts.

1. Introduction

A substantial fraction of global carbon fixation is carried out by eukaryotic marine algae distinguished by having chlorophyll *c* rather than chlorophyll *b* as accessory chlorophyll (*1*). They include the haptophytes (coccolithophorids), cryptophytes, diatoms, and a variety of other heterokont algae. The chloroplasts of these algae were obtained from an endosymbiotic red algal ancestor,

From: *Methods in Molecular Biology, Vol. 390: Protein Targeting Protocols: Second Edition*
Edited by: M. van der Giezen © Humana Press Inc., Totowa, NJ

Present Address: School of Biological Sciences, Nanyang Technological University, Singapore 637515

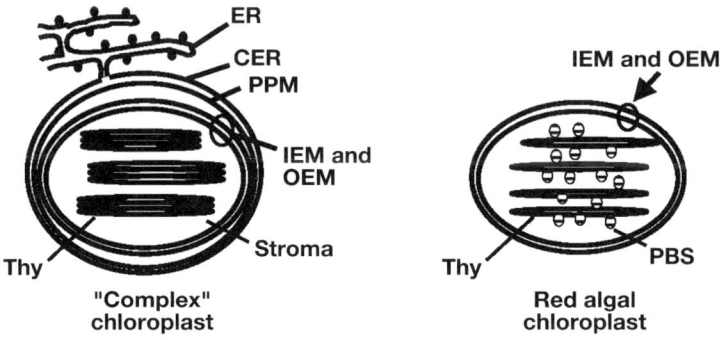

Fig. 1. Comparison of the "complex" chloroplasts of heterokonts, haptophytes, and cryptophytes (algae with Chl *c*) with the primary chloroplasts of red algae. The complex chloroplast is surrounded by four membranes; the outermost is called the chloroplast ER (CER) because it is continuous with the rough endoplasmic reticulum (ER). PPM, periplastidal membrane; OEM, outer envelope membrane; IEM, inner envelope membrane; Thy, thylakoid (photosynthetic membrane); PBS, phycobilisome, a light-harvesting antenna complex located on the cytoplasmic surface of the thylakoid membrane, as in cyanobacteria. Plant and green algal chloroplasts have only the two envelope membranes (IEM and OEM), like those of red algae, but their thylakoids are appressed over part of their length (grana) and they do not have phycobilisomes.

which was engulfed by a nonphotosynthetic eukaryote (*2,3*). As a result, these "secondary" or "complex" chloroplasts are surrounded by four membranes rather than the two that surround a higher plant or red algal chloroplast (**Fig. 1**). The inner two were probably derived from the two envelope membranes of the red algal (primary) plastid, the next membrane (periplastidal membrane) from the red algal plasma membrane, and the outermost membrane from the phagocytic vacuole/endomembrane system of the host. This last membrane is physically continuous with the rough endoplasmic reticulum (ER) and in some algal species can be seen to have cytoplasmic ribosomes attached to its outer surface.

Because a large fraction of chloroplast proteins are nucleus encoded, this means that proteins synthesized on cytoplasmic ribosomes have to be imported across four membranes rather than two to arrive in the chloroplast stroma. It has been shown that all such proteins have an N-terminal ER signal sequence, which is cleaved off during transit across the chloroplast ER (*4,5*). Very little is known about how the proteins are targeted across the next three membranes. We have found that if gene constructs lacking the ER signal sequence are transcribed and translated in vitro, the proteins can be recognized and imported by intact pea chloroplasts, providing a useful system for studying targeting *(6)*.

2. Materials

2.1. DNA Template Preparation

1. QIAquick PCR Purification Kit (QIAGEN, Inc., cat. no. 28104).

2.2. Transcription

1. T3 Cap-Scribe RNA polymerase kit (Roche Diagnostics, cat. no. 1 581 058).
2. RNasin® Ribonuclease Inhibitor (Promega, cat. no. N2111).

2.3. Translation

1. Rabbit Reticulocyte Lysate System, Nuclease Treated (Promega, cat. no. L4960).
2. RNasin Ribonuclease Inhibitor (Promega, cat. no. N2111).
3. Redivue L-[^{35}S]methionine (1000 Ci/mmol, 10 µCi/µL) (Amersham Biosciences, Inc. cat. no. AG1594).

2.4. Pea Seedlings

1. Potting mix (Metro-Mix 290, Grace Horticultural Products).
2. Pea seeds (*Pisum sativum* var. Maestro).

2.5. Preparation of Intact Pea Chloroplasts

1. Sucrose isolation medium (SIM): to make 1 L dissolve 120 g sucrose, 6 g HEPES, and 0.8 g EDTA in 900 mL distilled water. Adjust the pH to 7.6 with NaOH and bring total volume to 1 L with distilled water. Store aliquots of 500 mL at −20°C. The evening before use, store at 4°C.
2. 5x Sorbitol resuspension medium (5x SRM): to make 1 L dissolve 300.6 g sorbitol and 59.6 g HEPES in 600 mL distilled water. Adjust the pH to 8.0 using 10 M KOH and bring total volume to 1 L. Store aliquots of 50 mL at −20°C. The evening before use, store at 4°C.
3. 1x Sorbitol resuspension medium (1x SRM): to make 1 L mix 200 mL 5x SRM with 800 mL distilled water. Store aliquots of 100 mL at −20°C. The evening before use, store at 4°C.
4. 40% Percoll® in import buffer: prepare on day of use. To a beaker (250 mL) add 24 mL distilled water and 12 mL 5x SRM. Mix by using a magnetic stirrer. While still stirring add 24 mL Percoll (Sigma-Aldrich, cat. no. P1644). Divide the 60-mL suspension into two 50-mL centrifuge tubes. Using a weighing balance, make sure that the two centrifuge tubes are of equal weight. Store on ice till ready to use.
5. 80% (v/v) Acetone.

2.6. Import Reaction

1. 100 m*M* MgATP (ATP magnesium salt): dissolve 50.8 mg of MgATP (Sigma-Aldrich, cat. no. A-9187) in 0.5 mL 1x SRM. Adjust the pH to 7.0 with 0.5 *M* NaOH. This can be done by adding 20–30 μL 0.5 *M* NaOH and then pipeting a few microliters onto pH indicator paper. Bring total volume to 1 mL with 1x SRM. Store aliquots of 100 μL at −20°C. Thaw just before use and discard any unused solution.
2. 100 m*M* Methionine: dissolve 14.9 mg of L-methionine (Sigma-Aldrich, cat. no. M9625) in 1 mL 1x SRM. Store aliquots of 100 μL at −20°C. Thaw just before use and discard any unused solution.
3. Thermolysin (4 mg/mL): dissolve 4 mg thermolysin (Sigma-Aldrich, cat. no. P1512) in 1 mL 1x SRM. Store aliquots of 100 μL at −20°C. Thaw just before use and discard any unused solution.
4. 0.5 *M* EDTA in SRM: prepare on day of use. Mix 5 mL of 1 *M* EDTA pH 8.0 with 2 mL 5x SRM. Make final volume to 10 mL with distilled water. Store on ice until used.
5. 40% Percoll, 50 m*M* EDTA in SRM: prepare on day of use. To make 10 mL, add to a beaker (50 mL) 1 mL 0.5 *M* EDTA pH 8.0, 2 mL 5x SRM, and 3 mL distilled water. Mix by using a magnetic stirrer. While still stirring add 4 mL Percoll (Sigma-Aldrich, cat. no. P1644). Aliquot 800 μL into 1.5-mL Eppendorfs. Store on ice until ready to use.

2.7. Fractionation and Preparation for Electrophoresis

1. 10 m*M* HEPES-KOH pH 8.0. Dissolve 1.19 g of HEPES in 400 mL distilled water, adjust pH to 8.0 with KOH, and make the total volume to 500 mL with distilled water. Always store at 4°C or on ice.
2. Protein sample buffer: 8% w/v sodium dodecyl sulfate, 20% v/v glycerol, 0.01% w/v bromophenol blue, and 0.2 *M* Tris-HCl pH 6.8.

2.8. SDS-PAGE and Autoradiography

1. PROTEAN II xi vertical electrophoresis cell (Bio-Rad Laboratories).
2. 1.5 *M* Tris-HCl pH 8.8.
3. Acrylamide/bis-acrylamide stock solution. Dissolve 30 g acrylamide and 0.8 g bisacrylamide in distilled water and bring volume to 100 mL.
4. 10% (w/v) Sodium dodecyl sulfate (SDS): dissolve 10 g sodium dodecyl sulfate in 60 mL distilled water. Bring total volume to 100 mL with distilled water.
5. 10% (w/v) Ammonium persulfate: prepare on day of use. Dissolve 0.1 g ammonium persulfate in 1 mL distilled water.
6. TEMED (*N*, *N*, *N'*, *N'*-tetramethylethylenediamine). Buy a small amount (e.g., 100 mL) and keep in refrigerator. It lasts a long time, but check the color occasionally and discard if it turns brown.

7. Water-saturated isobutanol: mix equal volume of water and isobutanol. Leave to separate into two layers. Use top layer. Store at room temperature.
8. 1 *M* Tris-HCl pH 6.8.
9. β-Mercaptoethanol.
10. Running buffer: to make 2 L dissolve 6.06 g Tris, 28.8 g glycine, and 2 g SDS in 2 L distilled water.
11. Kaleidoscope Pre-stained Standards (Bio-Rad Laboratories, cat. no. 161-0324).
12. Fixing buffer: to make 1 L, mix 500 mL methanol, 400 mL distilled water, and 100 mL acetic acid.
13. X-ray film (Kodak).

3. Methods

The success of an import experiment depends on gently isolated intact chloroplasts from healthy plants. It is essential to keep reagents and equipment as cold as possible, work quickly, and prepare things on the same day as the import experiment where noted. Intact chloroplasts CANNOT be frozen for later use!

3.1. DNA Template Preparation

1. cDNA is cloned into a plasmid with an upstream T3 promoter site (*see* **Notes 1** and **2**).
2. The plasmid is linearized by restriction digest at a site downstream from the cDNA stop codon. Restriction digestion is carried out overnight to assure complete linearization (*see* **Note 3**).
3. The linearized plasmid is purified using the QIAquick PCR Purification Kit (QIAGEN). The DNA is eluted with 30 μL of 10 m*M* Tris-HCl pH 8.0 (sterile) after 5 min incubation at room temperature.

3.2. Transcription

1. DNA template (1 μg) is added to 4 μL CAP Scribe buffer, 1 μL RNasin (40 units/μL), and 2 μL T3 RNA polymerase and made up to a final volume of 20 μL with sterile distilled water.
2. Incubation is carried out for 60 min at 37°C. Transcription products are immediately frozen and stored at −80°C until use.

3.3. Translation in Vitro must be Carried Out on the Same Day as the Import Reaction

1. Transcription product (2 μL) is added to 8 μL sterile distilled water, 0.5 μL amino acid mixture minus methionine, 0.5 μL RNasin (40 units/μL), 12.5 μL rabbit reticulocyte lysate, and 1.5 μL Redivue L-[^{35}S] methionine (1000 Ci/mmol) (*see* **Notes 4** and **5**).

2. Incubation is carried out for 90 min at 30°C.
3. Keep products on ice till ready to use.

3.4. Growing Pea Seedlings

1. Two flat trays (60 cm × 30 cm × 10 cm) are two-thirds filled with potting mix (Metro-Mix 290, Grace Horticultural Products). Pea seeds (*Pisum sativum* var. Maestro) are sown on top, approx 2 cm apart, covered with potting mix, and watered.
2. Trays are kept in a greenhouse for 7–9 d at 18–25°C with supplementary lighting providing photosynthetically active radiation of 125 μmol photons per m^2/s for 16 h per day (*see* **Notes 6** and **7**).
3. The seedlings are watered every second day.

3.5. Preparation of Intact Pea Chloroplasts (6,7)

1. The evening before preparing intact pea chloroplasts, place frozen aliquots of SIM (500 mL), 5x SRM (50 mL), and 1x SRM (100 mL) at 4°C. Also put centrifuge rotors, buckets and tubes, Waring blender, muslin cheesecloth, cotton swabs, beakers, measuring cylinders, Eppendorfs, and pipet tips in a cold room at 4–11°C. Set any centrifuges to be used at 4°C.
2. Before harvesting the pea seedlings, make 40% Percoll in import buffer.
3. Using a clean pair of scissors, cut the pea shoots (*see* **Note 7**) and place them in a 1-L beaker until it is two-thirds full. This should give 40–50 g of material.

All subsequent steps should be carried out in a cold room at 4–11°C. Gloves should be worn at all times. Preparation of intact chloroplasts should take no longer than a total of 2 h.

4. To the pea shoots add 200 mL SIM (invert the bottle several times before using).
5. Transfer the contents to a Waring blender. With the blender set at high, blend three times, each time for 2 s only.
6. Carefully pour the suspension over eight layers of cheesecloth (muslin) covering a 500-mL beaker. Raising the sides of the muslin cloth, gently squeeze any excess liquid through the muslin into the beaker.
7. Divide the filtrate into two 250-mL centrifuge bottles and centrifuge at 2000*g* for 3 min at 4°C. Carefully pour out the liquid without disturbing the dark green pellet.
8. To each centrifuge bottle add 5 mL SIM. Gently resuspend the pellets using a cotton swab (*see* **Note 8**). Add 95 mL SIM to each suspension. Centrifuge at 2000*g* for 3 min at 4°C. Pour off supernatant.
9. Add 2 mL 1x SRM to each pellet, resuspend with cotton swab.
10. Layer each suspension onto a tube containing 30 mL 40% Percoll in import buffer (made the same day), using a plastic tip with the edge cut off (*see* **Note 9**). Centrifuge in a swing-out rotor at 3000*g* with the brake off for 15 min at 4°C.

11. Using a Pasteur pipet, remove all the supernatant, leaving the pellet of intact chloroplasts undisturbed.

12. To remove the Percoll, add 2 mL 1x SRM to each pellet, resuspend with cotton swab. Add 10 mL 1x SRM to each tube. Pool suspension into one tube and centrifuge at $2000g$ for 3 min at 4°C. Gently pour off the liquid and resuspend the pellet in 500 μL 1x SRM using a cotton swab. To determine chlorophyll concentration, add 10 μL of chloroplasts to 990 μL 80% (v/v) acetone. Vortex for 30 s and centrifuge at $12,000g$ for 3 min. The absorbance of the supernatant is measured at 652 nm using 80% (v/v) acetone as a blank. The chlorophyll concentration of the total suspension in mg/mL is estimated by multiplying the absorbance by 2.9.

3.6. Import Reaction (6,7)

There are two parts to this procedure: the import reaction (**steps 1–5**) and protease treatment to remove adsorbed protein that was not imported (**steps 6–10**).

1. Place a 500-mL beaker filled with water in a water bath set at 24°C. The top half of the beaker should not be inserted into the water-bath but exposed to a lamp giving 100 μmole photons per m^2/s (*see* **Note 6**). The temperature of the water inside the beaker should remain at 24°C throughout.

2. In a 15-mL Falcon tube (Fisher Scientific, cat. no. 14-959-53A) put 40 μL MgATP (100 mM stock), 20 μL methionine (100 mM stock), and intact chloroplasts containing 200 μg chlorophyll (using plastic tip with edge cut off). Make the volume up to 380 μL with 1x SRM. Add 20 μL of translated product, made the same day. Mix by stirring with a plastic tip.

3. With the Falcon tube in a polystyrene float, incubate for 40 min in the illuminated beaker. Make sure that the import sample is receiving 100 μmol photons per m^2/s by measuring with a light meter.

4. Stop the import reaction by placing the sample on ice for 5 min. Add 1200 μL 1x SRM. Mix by stirring with a plastic tip.

5. Using a plastic tip with the edge cut off, aliquot 400 μL into another Falcon tube and label as wC. This is the control untreated with thermolysin. Keep on ice.

6. Label the remaining 1200 μL as C. To C add 60 μL thermolysin (4 mg/mL stock) and mix by stirring (*see* **Note 10**). Incubate it on ice for 30 min.

7. Prepare 0.5 M EDTA in SRM. Keep on ice. Add 140 μL to C and mix by stirring. Total volume is now 1400 μL.

8. Divide C into two aliquots and layer each on 800 μL 40% Percoll, 50 mM EDTA in SRM in small tubes that fit the swing-out rotor. Layer the 400 μL wC on 800 μL of the Percoll solution in the same way.

9. Centrifuge in a swing-out rotor at $3000g$ for 5 min at 4°C with brake off. Using a Pasteur pipet remove the supernatant without disturbing the pellet of intact chloroplasts.

10. To each pellet add 800 μL 1x SRM. Invert gently several times. Centrifuge at $2000g$ for 5 min at 4°C. Slowly remove as much as possible of the supernatant.

3.7. Fractionation and Preparation for Electrophoresis

1. Add 30 μL 10 m*M* HEPES-KOH pH 8.0 to the wC sample to break the chloroplasts. Resuspend the pellet by gently pipetting up and down with a plastic tip, trying not to cause any bubbles. Make total volume to 50 μL with 10 m*M* HEPES-KOH 8.0, add 50 μL protein sample buffer, vortex for 5 s, and store at −80°C. This is the washed chloroplast fraction.

2. Add 50 μL 10 m*M* HEPES-KOH pH 8.0 to each of the C samples, resuspend by pipetting, and pool. Make total volume to 150 μL with 10 m*M* HEPES-KOH pH 8.0. Remove a 50-μL aliqout into a 1.5-mL Eppendorf and label as pC. Add 50 μL protein sample buffer, vortex for 5 s, and store at −80°C. This is the protease-treated chloroplast fraction.

3. Centrifuge the remaining 100 μL C sample at $10,000 \times g$ for 10 min at 4°C. Transfer the supernatant into a 1.5-mL Eppendorf and label as Stroma. To the pellet (thylakoids) add 800 μL 10 m*M* HEPES-KOH pH 8.0 and invert several times. Centrifuge both the stroma and thylakoid samples at $10,000 \times g$ for 10 min at 4°C (*see* **Note 11**).

4. From the stroma sample carefully transfer the supernatant (stroma) into a 1.5-mL Eppendorf. Make total volume to 100 μL with 10 m*M* HEPES-KOH pH 8.0. Add 100 μL protein sample buffer, vortex, and store at −80°C. Discard the pellet.

5. Discard the supernatant from the thylakoid sample. Resuspend the thylakoid pellet in 60 μL 10 m*M* HEPES-KOH pH 8.0. Make total volume to 100 μL. Aliquot 50 μL from the thylakoid sample into a 1.5-mL Eppendorf and label as wThy (washed thylakoid fraction). Add 50 μL protein sample buffer, vortex, and store at −80°C.

6. If the imported protein is targeted to the thylakoid membrane itself or to the thylakoid lumen, carry out the following steps. To the remaining thylakoid fraction, add 350 μL 10 m*M* HEPES-KOH pH 8.0 and 20 μL thermolysin (4 mg/mL stock). Mix by stirring and leave on ice for 30 min.

7. To stop the thermolysin reaction, add 840 μL 10 m*M* HEPES-KOH pH 8.0 and 140 μL 0.5 *M* EDTA in SRM. Mix by stirring. Centrifuge at $10,000 \times g$ for 10 min at 4°C and discard the supernatant.

8. Suspend the pellet in 30 μL 10 m*M* HEPES-KOH pH 8.0. Label as pThy (protease-treated thylakoid fraction). Make total volume to 50 μL, add 50 μL protein sample buffer, vortex for 5 s, and store at −80°C.

3.8. SDS-PAGE and Autoradiography

1. Electrophoresis of polypeptides is carried out using the Bio-Rad gel system with vertical slab gels (20×20 cm).

2. The glass plates are washed with detergent, rinsed with distilled water and air-dried. Before assembly the plates are wiped down with 95% ethanol and air-dried.

3. To make a 1.0-mm-thick 15% polyacrylamide resolving gel, mix 11.5 mL water, 12.5 mL 1.5 *M* Tris-HCl pH 8.8, 25 mL 30% acrylamide/bis-acrylamide stock solution, 500 μL 10% (w/v) SDS, 500 μL freshly prepared 10% (w/v) ammonium

persulfate, and 20 μL TEMED. After the gel is poured, overlay with a thin layer of water-saturated isobutanol and leave to polymerize.

4. After 2 h, pour off the water-saturated isobutanol and rinse the top of the gel several times with distilled water.

5. To make the stacking gel, mix 13.6 mL distilled water, 2.5 mL 1 *M* Tris-HCl pH 6.8, 3.4 mL 30% acrylamide/bis-acrylamide stock solution, 200 μL 10% (w/v) SDS, 200 μL freshly prepared 10% (w/v) ammonium persulfate, and 20 μL TEMED. Pour onto the top of the resolving gel, insert the well-forming comb, and leave for 1 h.

6. A sample of the untreated translation reaction is prepared with 1 μL of translated product, 19 μL distilled water, and 20 μL protein sample buffer. Preparation of the other samples was explained in the previous section. Just before loading, a one-tenth volume of β-mercaptoethanol is added to each sample, and the samples are then boiled for 5 min. Load 20 μL of T, 52.5 μL of wC, pC, stroma, wThy, and pThy into individual wells. Return unused samples to −80°C. Include one lane for Kaleidoscope prestained molecular weight markers.

7. Electrophoresis is carried out overnight at 8 mA with running buffer at room temperature.

8. Discard the stacking gel and soak the resolving gel in 500 mL of fixing buffer for 30 min (*see* **Note 12**).

9. Wet a piece of 3MM Whatman paper (20 × 20 cm) in fixing buffer and place the resolving gel on top. Remove any excess buffer with paper towels.

10. Using a gel dryer, dry the gel for 3 h at 80°C.

11. The gel is placed in a film cassette and exposed to X-ray film (Kodak) for 3 d (*see* **Note 13**) at −80°C.

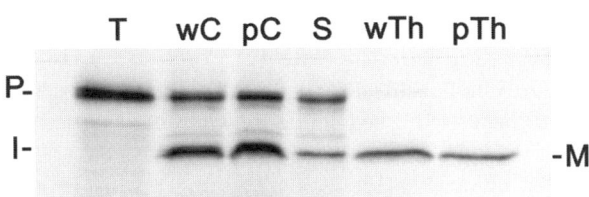

Fig. 2. Import of the haptophyte *Isochrysis galbana* PsbO precursor into pea chloroplasts. The PsbO precursor was incubated with intact pea chloroplasts. After incubation samples were treated with thermolysin, reisolated, washed, lysed, fractionated, and analyzed by SDS-PAGE followed by autoradiography. Translation products (lane T); washed chloroplast fraction (lane wC); protease-treated chloroplast fraction (lane pC); stromal fraction (lane S); washed thylakoid fraction (lane wThy), and protease-treated thylakoid fraction (lane pThy). Precursor, intermediate, and mature forms are denoted P, I, and M, respectively.

12. **Figure 2** shows an example of an import result using the nuclear-encoded thylakoid lumen protein PsbO, also called OEC33 or 33 kDa oxygen-enhancer 1 protein, of the haptophyte alga *Isochrysis galbana* (**8**). Radiolabeled translation products from the *I. galbana* PsbO construct were incubated with isolated intact pea chloroplasts giving two products, labeled as intermediate (I) and mature (M) forms (**Fig. 2**, lane wC). Both forms were protease protected (**Fig. 2**, lane pC). The I form was localized in the stroma and the M form in the thylakoid fraction (**Fig. 2**, lanes S and wThy). The M form remained intact even after the thylakoids were treated with thermolysin, indicating localization within the thylakoid lumen (**Fig. 2**, lane pThy).

Notes

1. The cDNA can also be cloned into a plasmid with an upstream SP6 or T7 promoter site, in which case transcription would be carried out with SP6 or T7 RNA polymerase respectively.
2. The ER signal peptide of the PsbO precursor from the heterokont alga *Heterosigma akashiwo* was shown to inhibit import across higher plant chloroplast envelope membranes (**6**). Therefore, cDNA constructs encoding both the full-length precursor and the precursor without the ER signal peptide should be tested for import.
3. Linearizing the plasmid prevents transcription continuing on after the stop site.
4. Sometimes, the translation product will produce a weak signal or there may be several polypeptides because of incorrect termination of translation. To find the optimum conditions for translation, reactions can be carried out in a total volume of 25 µL with differing amounts of RNA (1, 2, and 4 µL), L-[^{35}S] methionine (1.5 and 3 µL), and 1 M KOAc (0.5 and 1 µL).
5. If the full-length precursor or the mature processed protein does not contain methionine residues, it is possible to use L-[4, 5 − ^{3}H]leucine (120–190 Ci/mmol) (Amersham Biosciences, cat. no. TRK 510) as an alternative. Using leucine will require a different fixing step (*see* **Note 12**).
6. Use a light meter to make sure that the correct amount of light is provided.
7. The ideal time to harvest the pea shoots is when the first two leaves are starting to expand.
8. Resuspending the chloroplast pellet by swirling results in a decrease of intact chloroplast yield.
9. To prevent breakage and lysis of intact chloroplasts, pipetting should always be carried out with a 1000-µL plastic tip (Fisher Scientific, cat. no. S63213) with the edge cut off, leaving an opening at least 5 mm wide.
10. The protease thermolysin will degrade anything that is bound to the outer membrane of the chloroplast but does not degrade any protein stuck in between the two envelope membranes. The protease trypsin is able to penetrate the outer envelope and degrade proteins in the intermembrane space and those bound to the outer surface of the inner envelope but leaves the inner envelope membrane intact unless

incubation is prolonged. To test the effect of trypsin, add $75\,\mu L$ $1.8\,mM$ $CaCl_2$ in SRM and $75\,\mu L$ trypsin in SRM ($2\,mg/mL$ stock) to the $1200\,\mu L$ C sample. Incubate at $20°C$ for $45\,min$. Stop digestion by adding $150\,\mu L$ soybean trypsin inhibitor in SRM ($5\,mg/ml$ stock). Leave on ice for $5\,min$ before proceeding with the Percoll gradient. When using trypsin, EDTA is excluded from the 40% Percoll.

11. The stromal fraction is centrifuged to pellet any contaminating thylakoids. The thylakoid fraction is washed and centrifuged so to remove any stromal contamination.

12. A different method of fixing the gel is used when using the radiolabeled L-$[4, 5 - {}^3H]$ leucine. Soak the resolving gel in 10% acetic acid and 1% glycerol for $40\,min$ followed by soaking in distilled water for $30\,min$. Afterwards, soak the gel in freshly prepared $1\,M$ sodium salicylate for $30\,min$. Continue with **steps 9–11**.

13. The length of time of exposure of the X-ray film has to be determined in each individual case. This normally varies from $3\,d$ to $2\,w$.

Acknowledgments

Our work was supported by a grant from the Natural Sciences and Engineering Research Council of Canada and a Killam Research Fellowship from the Canada Council to BRG.

References

1. Armbrust, E. V., Berges, J. A., Bowler, C., Green, B. R., (2004) The genome of the diatom *Thalassiosira pseudonana*: ecology, evolution and metabolism. *Science* **306**, 79–86.

2. Gibbs, S. P. (1981) The chloroplast endoplasmic reticulum: structure, function, and evolutionary significance. *Int. Rev. Cytol.* **72**, 49–99.

3. Cavalier-Smith, T. (2000) Membrane heredity and early chloroplast evolution. *Trends Plant Sci.* **5**, 174–182.

4. Bhaya, D. and Grossman, A. (1991) Targeting proteins to diatom plastids involves transport through an endoplasmic reticulum. *Mol. Gen. Genet.* **229**, 400–404.

5. Kroth, P. G. (2002) Protein transport into secondary plastids and the evolution of primary and secondary plastids. *Int. Rev. Cytol.* **221**, 191–255.

6. Chaal, B. K and Green, B.R. (2005) Protein import pathways in 'complex' chloroplasts derived from secondary endosymbiosis involving a red algal ancestor. *Plant Mol. Biol.* **57**, 333–342.

7. Mould, R. M. and Gray, J. C. (1998) Preparation of chloroplasts for protein synthesis and protein import, in *Cell Biology: A Laboratory Handbook* (Celis, J. E., ed.), Academic Press, New York, pp. 81–86.

8. Ishida, K. and Green, B. R. (2002) Second- and third-hand chloroplasts in dinoflagellates: Phylogeny of oxygen-evolving enhancer 1 (PsbO) protein reveals replacement of a nuclear-encoded plastid gene by that of a haptophyte tertiary endosymbiont. *Proc. Natl. Acad. Sci. USA* **99**, 9294–9299.

15

Protein Trafficking to the Complex Chloroplasts of *Euglena*

Rostislav Vacula, Silvia Sláviková, and Steven D. Schwartzbach

Summary

Proteins are delivered to *Euglena* chloroplasts using the secretory pathway. We describe analytical methods to study the intracellular trafficking of *Euglena* chloroplast proteins and a method to isolate preparative amounts of intact import competent chloroplasts for biochemical studies. Cells are pulse labeled with ^{35}S-sulfate and chased with unlabeled sulfate allowing the trafficking and posttranslational processing of the labeled protein to be followed. Sucrose gradients are used to separate a ^{35}S-labeled cell lysate into cytoplasmic, endoplasmic reticuum (ER), Golgi apparatus, chloroplast and mitochondrial fractions. Immunoprecipitation of each gradient fraction allows identification of the intracellular compartment containing a specific ^{35}S-labeled protein at different times after synthesis delineating the trafficking pathway. Because sucrose gradients cannot be used to isolate preparative amounts of highly purified chloroplasts for biochemical characterization, a preparative high-yield procedure using Percoll gradients to isolate highly purified import competent chloroplasts is also presented.

Key Words: *Euglena*; chloroplast purification; chloroplast protein trafficking; complex chloroplasts; intracellular localization; pulse chase; secondary plastids; subcellular fractionation.

1. Introduction

Euglena contains complex chloroplasts which are surrounded by three membranes rather than two membranes as found in higher plants and other algae suggesting that they evolved through an endosymbiotic relationship between a eukaryotic host and a photosynthetic eukaryotic endosymbiont (*1*). Precursors targeted to *Euglena* chloroplasts are characterized by having a tripartite presequence composed of an ER targeting signal peptide, a region having the properties of a higher plant chloroplast stromal targeting sequence

From: *Methods in Molecular Biology, Vol. 390: Protein Targeting Protocols: Second Edition*
Edited by: M. van der Giezen © Humana Press Inc., Totowa, NJ

and a stop transfer membrane anchor sequence approx 60 amino acids from the predicted signal peptidase processing site *(2)*. The presequence structure and immunoelectron microscopic localization of chloroplast proteins to the Golgi apparatus *(3)* suggested that the secretory system was used to deliver precursors to the outer chloroplast envelope membrane while passage through the intermediate and innermost membranes utilized an import system similar to that of higher plant chloroplasts.

In vivo studies of the import pathway require methods to identify chloroplast precursor containing intracellular compartments and, more importantly, follow the temporal sequence of precursor transfer between compartments. Organelles have different densities allowing the major organelles in a crude cell lysate to be separated on isopycnic sucrose gradients *(4)*. The identity of the organelles in each membrane fraction can be determined through enzymatic or immunological assays for marker enzymes, enzymes recognized as localized in specific subcellular organelles *(5)*. Co-localization of a protein of interest with a marker enzyme localizes that protein to a specific organelle. Intracellular fractionation studies can thus be used to determine which organelles contain a specific chloroplast precursor protein but they do not provide information regarding the transfer sequence from organelle to organelle.

Determination of the transport sequence requires methods for distinguishing newly synthesized proteins from preexisting proteins and methods to differentiate the proteins made over a specific time span from proteins made prior to and after this time. Pulse labeling with ^{35}S-sulfate provides a convenient method to produce a pool of protein synthesized during the labeling period which can be distinguished by its radioactive content from preexisting proteins *(6)*. Addition of a large excess of unlabeled sulfate, a chase, stops further ^{35}S-sulfate incorporation into newly synthesized protein so that the only radioactive protein in the cell is the protein labeled during the pulse period *(7)*. A pulse-labeling period corresponding to less than the half-life for the transport process produces a pool of radioactive protein that can be followed during a chase as it moves from its site of synthesis to the chloroplast. Combining a pulse chase ^{35}S- sulfate labeling protocol, subcellular fractionation and immunoprecipitation to localize specific proteins within organelles at different times after their synthesis provides a method to study the in vivo trafficking of *Euglena* chloroplast proteins *(5,8)*.

Isopycnic sucrose gradient fractionation of crude cell lysates is an analytical method that is not suited for the rapid preparation of the large quantities of highly purified import competent chloroplasts required for in vitro biochemical studies. A preparative method for *Euglena* chloroplast isolation utilizes a French pressure cell to rapidly prepare cell extracts from a large number of cells. A chloroplast-enriched pellet relatively free of whole cell, nuclear, endomembrane, and mitochondrial contamination is rapidly obtained by differential centrifugation and collection on an 80% Percoll cushion *(9)*. Final purification by sedimentation through 40% Percoll completes the purification providing highly purified intact chloroplasts suitable for in vitro biochemical studies including the characterization of the novel chloroplast import pathway *(10)*.

2. Materials

2.1. Cell Growth

1. Culture of *Euglena gracilis* Klebs var. *bacillaris* Cori.
2. Trace metal mix for approx 2000 L of EM 3.5: grind in a mortar and pestle 52.8 g $ZnSO_4 \bullet 7H_2O$, 24.8 g $MnSO_4 \bullet H_2O$, 28 g $Fe(NH_4)_2(SO_4)_2 \bullet 6H_2O$, 4.8 g $CoSO_4 \bullet 7H_2O$, 0.8 g $CuSO_4 \bullet 5H_2O$, 0.36 g $(NH_4)_6Mo_7O_{24}$, 0.37 g $Na_3VO_4 \bullet 16H_2O$, 1.14 g H_3BO_3. The ground powder is stored in the dark in a tightly sealed bottle at room temperature (*see* **Note 1**).
3. A 10 μg/mL stock solution of vitamin B_{12} is prepared in deionized water and filter-sterilized. The solution is stored at 4°C.
4. A 10 mg/mL stock solution of thiamine-HCl is prepared in deionized water and filter-sterilized. The solution is stored at 4°C.
5. 2X EM 3.5: mix 8.0 g KH_2PO_4, 10.0 g $MgSO_4 \bullet 7H_2O$, 4.0 g $CaCO_3$, 100 g L-glutamic acid, 40 g DL-malic acid, 0.1 g $FeCl_3 \bullet 6H_2O$, 4.0 g $(NH_4)_2HPO_4$, 1.0 g trace metal mix, 20 mg thiamine-HCl (2 mL of a 10 mg/mL stock solution), and 4 μg vitamin B_{12} (0.4 mL of a 10 μg/mL stock solution) with deionized water for a final volume of 10 L. Autoclave and store at room temperature or in a coldroom.
6. EM 3.5: dilute 2X EM 3.5 twofold, placing it in an appropriate size flask and autoclaving (*see* **Note 2**).
7. Low sulfate 2X EM 3.5: prepare the same way as 2X EM 3.5 except the $MgSO_4 \bullet 7H_2O$ is replaced with 8.25 g $MgCl_2 \bullet 6H_2O$.
8. Low sulfate EM 3.5: dilute the double strength media twofold, placing it in an appropriate size flask and autoclaving.
9. Low B_{12} EM 3.5: aseptically add sterile vitamin B_{12} to autoclaved single-strength vitamin B_{12} free EM 3.5 for a final concentration of 50 ng/L. (*see* **Note 3**).
10. Resting media pH 5.0 (RM 5.0): mix 5 g mannitol, 5 mL 1 M KH_2PO_4, and 5 mL 1 M $MgCl_2$ with deionized water for a final volume of 1 L. The media is autoclaved and stored at room temperature.

2.2. Pulse Chase Labeling with ^{35}S-Sulfate

1. Carrier-free sulfate in aqueous solution or $H_2^{35}SO_4$ (*see* **Note 4**).
2. 0.6 M K_2SO_4.

2.3. Isolation and Purification of Cellular Organelles

2.3.1. Separation and Fractionation of Euglena Subcellular Organelles by Isopycnic Sucrose Gradient Centrifugation

1. Grinding buffer: 0.4 M sucrose, 25 mM Tris-HCl, pH 7.4, 1 mM EDTA.
2. Protease inhibitor stock solutions (Roche Applied Science, Indianapolis, IN): antipain, 1 mg/mL in deionized water; chymostatin, 10 mg/mL in dimethyl sulfoxide (DMSO); pepstatin, 10 mg/mL in methanol; E-64; 1 mg/mL in 50% ethanol; PMSF, 87 mg/mL (0.5 M) in isopropanol; aprotinin, 1 mg/mL in 0.1 M Tris-HCl, pH 7.5; leupeptin, 1 mg/mL in deionized water. Prepare 1 mL of each solution and store at −20°C for no more than 1mo. DMSO is readily absorbed through the skin and the protease inhibitors are toxic. Care should be taken to avoid inhalation and skin contact.
3. Grinding buffer containing protease inhibitors is prepared immediately before use by adding 10 µL antipain, E-64, PMSF, aprotinin, leupeptin and pepstatin stock solutions and 100 µL chymostatin stock solution to 10 mL grinding buffer for a final concentration of: 0.4 M sucrose, 25 mM Tris-HCl, pH 7.4, 1 mM EDTA, 1 µg/mL antipain, 100 µg/mL chymostatin, 10 µg/mL pepstatin, 1 µg/mL E-64, 0.5 mM PMSF, 1 µg/mL aprotinin, 1 µg/mL leupeptin. (*see* **Note 5**).
4. 20% (w/w), 25% (w/w), 50% (w/w) and 55% (w/w) Sucrose prepared in 25 mM HEPES-KOH, pH 7.4, 1 mM EDTA.
5. 80% Acetone.
6. 20% Trichloracetic acid (TCA): mix 500 g TCA with 227 mL of deionized water for a 100% TCA solution. Mix one volume 100% TCA with four volumes deionized water to produce a 20% solution.
7. Sodium dodecyl sulfate (SDS) buffer: 60 mM Tris-HCl, pH 8.6, 2% SDS.
8. 212–300 µm Acid-washed glass beads (Sigma, St. Louis, MO).
9. Dual conical tissue grinder pestle for 15 mL conical tubes (Fisher Scientific, Pittsburgh, PA).
10. 15-mL Linear gradient maker (Hoefer, San Francisco, CA)
11. Ultra-Clear 14 × 89 mm ultracentrifuge tube (Beckman Coulter, Fullerton, CA)
12. SW 41 Ti rotor (Beckman Coulter).
13. Isco 640 Gradient Fractionator (Teledyne Isco, Lincoln, NE)

2.3.2. Purification of Intact Biologically Functional Chloroplasts

1. Gauze.
2. 5X breaking buffer: 100 mM HEPES-KOH, pH 7.4, 1.65 M sorbitol, 2.5 mM EDTA prepared with sterile deionized water.
3. Prepare breaking buffer by diluting 5X breaking buffer fivefold with sterile deionized water. Add BSA (Sigma) for a final concentration of 0.1% (W/V) just prior to use.
4. Plant cell extract protease inhibitor cocktail (Sigma).

5. Breaking buffer containing protease inhibitors is prepared immediately before use by adding $5\,\mu L/mL$ plant cell extract protease inhibitor to breaking buffer (*see* **Note 5**).

6. French Pressure cell (Thermo Electron Corporation, Waltham, MA).

7. PBF Percoll: dissolve 5.1 g polyethylene glycol (PEG) 3350 (Sigma), 1.7 g BSA (Sigma), 1.7 g Ficoll PM 400 (Sigma) in sterile Percoll (GE Healthcare Life Sciences Corp, Piscataway, NJ) for a final volume of 170 mL mixing the solution with a sterile magnetic stirbar. Store at 4°C.

8. 80% PBF Percoll: mix one volume 5X breaking buffer with four volumes PBF Percoll.

9. 40% PBF Percoll: mix one volume 5X breaking buffer with two volumes PBF Percoll and two volumes sterile deionized water.

10. Small fine paintbrush for resuspending chloroplast pellets.

11. Kimble high-strength round-bottom 30-mL glass centrifuge tubes (VWR Scientific, West Chester, PA).

2.4. Protein Immunoprecipitation

1. SDS buffer: $60\,mM$ Tris – HCl, pH 8.6, 2% SDS.

2. RIPA buffer: 1% Nonidet P40 substitute (USB, Cleveland, OH), 1% sodium deoxycholate, $5\,mM$ NaCl, $5\,mM$ Tris-HCl, pH 7.0. Store at 4°C.

3. 1 M NaCl RIPA buffer: 1% Nonidet P40 substitute (USB), 1% sodium deoxycholate, $1M$ NaCl, $5\,mM$ Tris-HCl, pH 7.0. Store at 4°C.

4. Protein A Sepharose CL-4B (GE Healthcare Life Sciences Corp) is prepared by swelling 0.25 g of protein A Sepharose overnight in 10 mL of RIPA buffer at 4°C. The RIPA buffer is removed and the protein A Sepharose gel is resuspended in 1 mL of fresh RIPA buffer. The protein A Sepharose should be prepared immediately before use and transferred using wide-bore pipets.

5. 3X Sample loading buffer: $180\,mM$ Tris-HCl, pH 6.8, 30% glycerol, 6% (W/V) SDS, 5% 2-mercaptoethanol, 0.0003% (W/V) bromophenol blue.

2.5. SDS Gel Electrophoresis

1. Acrylamide/bis-acrylamide stock solution (37.5/12.6% C): mix 30 g acrylamide (Bio-Rad, Hercules, CA) and 0.8 g (Bio-Rad) bis-acrylamide with deionized water for a final volume of 100 mL. Store at 4°C in the dark for up to 3 wk. Unpolymerized acrylamide is a neurotoxin and care should be taken to avoid inhalation and skin contact.

2. Separating gel solution: 1.5 M Tris-HCl, pH 8.8, 0.4% (W/V) SDS. Add SDS after adjusting the pH.

3. Stacking gel solution: 0.5 M Tris-HCl, pH 6.8, 0.4% (W/V) SDS. Add SDS after adjusting the pH.

4. Ammonium persulfate: prepare a 10% (W/V) ammonium persulfate solution (Bio-Rad) in sterile deionized water immediately before use.

5. N, N, N', N'-Tetramethylethylenediamine (TEMED) (Bio-Rad). Store at room temperature.
6. 50% (w/w) Sucrose prepared in water.
7. Prepare water-saturated isobutanol by mixing equal amounts of deionized water and isobutanol, shaking well and allowing the two phases to separate. The top layer is isobutanol. Store indefinitely at room temperature.
8. SDS-PAGE running buffer: 25 mM Tris, 192 mM glycine, 0.1% (W/V) SDS. Store at room temperature.
9. 3X Sample loading buffer: 180 mM Tris − HCl, pH 6.8, 30% glycerol, 6% (W/V) SDS, 0.003% (W/V) bromophenol blue, 5% β-mercaptoethanol. Store in aliquots at −20°C.
10. Electrophoresis apparatus: 16-cm Protean II xi Cell apparatus (Bio-Rad), 1-mm spacers, 1-mm Teflon combs with 15 teeth.
11. 30 mL Linear gradient maker (Hoefer).
12. Gel loading micropipet tips (Genesee Scientific Corporation, San Diego, CA).

2.6. Visualization and Quantitation of ^{35}S-Labeled Proteins Separated by SDS Gel Electrophoresis by Fluorography and Phosphorimaging

1. Whatman 3 MM filter paper (Fisher Scientific).
2. Saran wrap.
3. Mylar (polyethylene terephthalate) 0.001 in. thick (Fralock Corporation, Canoga Park, CA).
4. 1 M Sodium salicylate. Sodium salicylate is light sensitive. Keep tightly closed and store in a cool dry place.
5. Glow Writer Autoradiography Pen (Diversified Biotech, Boston, MA).
6. Kodak X-Omat AR X-ray film (GE Healthcare Life Sciences Corp) (*see* **Note 6**).
7. Mounted general purpose (GP) 20 × 25 cm storage phosphor screen; (GP) screen, (GE Healthcare Life Sciences).
8. Exposure cassette (GE Healthcare Life Sciences) for 20 × 25 cm storage phosphor screen.
9. X-ray film cassette (GE Healthcare Life Sciences).
10. Chemicals to develop X-ray film.

3. Methods
3.1. Cell Growth

1. Maintain a *Euglena* stock culture in a 250-mL flask containing 50 mL of EM 3.5 in the dark or in the light $(30 \mu mol/s/m^2)$ without shaking at 26°C and transfer weekly by adding a few drops of culture to 50 mL EM 3.5.
2. Maintain a low-B$_{12}$ *Euglena* stock culture in a 250-mL flask containing 50 mL of low-B$_{12}$ EM 3.5 in the dark or in the light $(30 \mu mol/s/m^2)$ without shaking at 26°C and transfer weekly by adding a few drops of culture to 50 mL of low-B$_{12}$ EM 3.5. Every 3 wk start a new low-B$_{12}$ stock using cells grown on EM 3.5.

3. For cells to be pulse-labeled with ^{35}S-sulfate during light-induced chloroplast development, inoculate 5–10 mL of a 7-d-old stock culture grown on EM 3.5 into 1 L of low-sulfate EM 3.5 in a 2-L flask. Grow in the dark for 4 d at 26°C on an orbital shaker at 150 rpm. After 4-d growth, harvest the cells aseptically in the dark by centrifugation at room temperature for 2 min at 1000g (2500 rpm) in a Sorvall GSA rotor. Gently resuspend the cell pellet in 500 mL of sterile RM 5.0 and recover the 1 L of cells by centrifugation at room temperature for 2 min at 1000g. Resuspend the cells in 500 mL of fresh resting medium and incubate the cells in the dark at 26°C on an orbital shaker at 150 rpm. After 3 d, cell division has ceased and cells can be transferred to an orbital shaker at 150 rpm in the light (30 μmol/s/m^2) at 26°C to initiate chloroplast development (*see* **Note 7**).
4. For cells to be used for chloroplast isolation, inoculate 5–10 mL of a 7-d-old stock culture grown on low-B$_{12}$ EM 3.5 into 1 L of low-B$_{12}$ EM 3.5 in a 2-L flask. Between 6 and 7 d after inoculation, the culture should go from a pale to a deep green indicative of cessation of cell division as a result of vitamin B$_{12}$ deficiency. Cells should be used for chloroplast isolation within 2 d after the color shift (*see* **Note 8**).

3.2. Pulse Chase Labeling with ^{35}S-Sulfate

1. Aseptically transfer 20 mL of low-sulfate-grown 3-d resting cells exposed to light for 24 h into a 125-mL Erlenmeyer flask and initiate the pulse by addition of 600 μCi/mL carrier-free ^{35}S-sulfate. Incubate in the light for 10 min (*see* **Note 9**).
2. Initiate the chase by addition of 4 mL 0.6 M K$_2$SO$_4$ for a final concentration 0.1 M K$_2$SO$_4$. (*see* **Note 10**).
3. At the end of the pulse and appropriate times after the start of the chase, a 10-mL sample is transferred to a plastic 15-mL conical centrifuge tube and the cells are harvested at room temperature by centrifugation for 1 min at maximum speed (approx 7000 rpm) in a tabletop centrifuge. A Pasteur pipet is used to completely remove the supernatant.
4. Resuspend the sample in the appropriate buffer for subsequent analysis (*see* **Note 11**).

3.3. Isolation and Purification of Cellular Organelles

3.3.1. Separation and Fractionation of Subcellular Organelles from ^{35}S-Labeled Cells by Isopycnic Sucrose Gradient Centrifugation

1. All operations are performed at 0–4°C. The cell pellet from 10 mL of ^{35}S-labeled cells in a 15-mL conical centrifuge tube is resuspended in 0.5 mL of grinding buffer containing protease inhibitors, 1 g of glass beads are added to the 15 mL conical centrifuge tube and the cells are disrupted by grinding with a glass pestle for 2.5 min (*see* **Note 12**).

2. Allow the beads to settle and remove the homogenate from the settled beads. Wash the beads three times with 0.3 mL of grinding buffer containing protease inhibitors and combine the homogenates in a clean 15-mL conical centrifuge tube.
3. The combined homogenates is clarified by centrifugation at 145g (1000 rpm) for 2 min at 4°C in a Sorvall SA-600 rotor.
4. The clarified homogenate is loaded onto a sucrose gradient consisting of a 2 mL 20% (w/w) sucrose step on top of an 8-mL 25–50% (w/w) linear sucrose gradient formed over a 0.5-mL cushion of 55% (w/w) sucrose. The gradients are prepared by adding 0.5 mL 55% (w/w) sucrose to the bottom of an Ultra-Clear ultracentrifuge tube for the SW 41 TI rotor. The 25–50% (w/w) sucrose gradient is prepared and overlayed onto the 55% (w/w) sucrose cushion using a 15-mL linear gradient maker and a peristaltic pump set to deliver the sucrose solution at a rate of 2 mL/min. Load 4 mL of 25% (w/w) sucrose into the back chamber of the gradient maker. Open the stopcock connecting the front and back chambers and let a minimal amount of sucrose flow into the front chamber to displace the air in the connecting tube and close the stopcock. Place a stirring bar in the front chamber and add 4 mL of 50% (w/w) sucrose. Turn the magnetic stirrer on, open the stopcock connecting the two chambers, open the outlet stopcock from the gradient maker and start the peristaltic pump using a pump rate of 2 mL/min. Check that the stirring rate is sufficiently rapid to mix the two sucrose solutions before they leave the gradient maker. The gradient should flow out of the gradient maker down the side of the centrifuge tube overlayering the 55% (w/w) sucrose in the bottom of the tube. When the gradient is completed it is overlayed with a 2-mL 20% (w/w) sucrose step.
5. The gradients are centrifuged at 100,000g for 3 h at 4°C in a SW 41 TI rotor.
6. A gradient fractionator is used to collect 30 0.4-mL fractions from the top of the gradient.
7. Chlorophyll is measured to localize broken (less dense) and intact chloroplasts on the gradient. Remove 25 μL of each fraction, mix with 1 mL of 80% acetone, incubate in the dark for 15 min at room temperature and clarify by centrifugation in a microfuge for 2 min. Determine relative chlorophyll amounts using a spectrofluorometer with an excitation wavelength of 435 nm and emission wavelength of 670 nm (*see* **Note 13**).
8. Proteins are concentrated for immunoprecipitation or electrophoresis by TCA precipitation. Add 0.4 mL 20% TCA to each gradient fraction, incubate for 1 h on ice and centrifuge for 10 min at maximum speed in a microfuge. Remove the supernatant and solubilize the pellet by resuspending in 110 μL of SDS buffer and boiling for 2 min. Samples can be stored at −20°C. Prior to immunoprecipitation or SDS gel electrophoresis, resolubilize the samples by boiling for 2 min.

3.3.2. Purification of Intact Biologically Functional Chloroplasts

1. All operations are performed at 0–4°C. *Euglena* grown for 6 d on low-vitamin-B$_{12}$ media are filtered through two layers of gauze and harvested by centrifugation at

4°C for 3 min at 1000*g* (2500 rpm) in a Sorvall GSA rotor. Cells are washed twice with breaking buffer, 50 mL/L cells, and recovered by centrifugation for 3 min at 1000*g* (2500 rpm in a Sorvall GSA rotor (*see* **Note 14**).

2. The washed cells are resuspended in breaking buffer, 27 mL/L cells. Cells are broken by passage through an ice-cold French pressure cell at 1500 psi. The pressate is collected directly into 12 mL/L cells of breaking buffer containing protease inhibitors and mixed well with the aid of a sterile 10-mL plastic pipet (*see* **Note 15**).

3. The diluted pressate is centrifuged twice for 2 min at 145*g* (1000 rpm) in a Sorvall SA600 rotor to remove unbroken cells and debris. Care must be taken not to disturb the loose dark green cell pellet when pouring off the chloroplast-containing supernatant (*see* **Note 16**).

4. Forty mL of chloroplast supernatant is layered onto a 5-mL 80% PBF Percoll cushion in 50-mL polycarbonate centrifuge tubes. The tubes are centrifuged for 6 min at 2600*g* (4000 rpm) in a Sorvall HB-6 swinging bucket rotor. Most of the paramylum sediments through the Percoll cushion while broken and intact chloroplasts form a dark green band on top of the cushion. Discard approx 15 mL of the clear yellow-green upper layer. Transfer the remaining chloroplast containing supernatant and Percoll cushion to a new centrifuge tube being careful not to disturb the paramylum pellet and mix well (*see* **Note 17**).

5. Chloroplasts are pelleted by centrifugation for 3 min at 1500*g* (3000 rpm) in a Sorvall HB-6 swinging bucket rotor. Discard the supernatant, add 0.5–1 mL of breaking buffer to the centrifuge tube and use a fine sterile paint brush to gently resuspend the loose chloroplast pellet taking care not to disturb the tightly packed white paramylum pellet under the chloroplasts. Repeat the resuspension using small buffer volumes as many times as necessary until most of the chloroplasts have been resuspended. Start resuspending the pellet at its outer edge, working towards the central area where the parmylum is found.

6. Add breaking buffer containing protease inhibitors to the isolated chloroplasts from each centrifuge tube for a final volume of 10 mL, mix well and layer the chloroplast suspension over a 5-mL 40% PBF Percoll cushion in a glass 30-mL centrifuge tube. Recover the intact chloroplasts by centrifugation for 2 min at 4400*g* (5200 rpm) in a Sorvall HB-6 swinging bucket rotor. Intact chloroplasts form a dark green pellet below the 40% Percoll cushion while broken chloroplasts remain on top of the Percoll cushion. Resuspend the chloroplasts in a small volume of breaking buffer and keep on ice for further analysis.

7. The chloroplast concentration is normally expressed as the chlorophyll concentration which is determined spectrophotometrically by the method of Arnon *(11)* as described in **ref. 12**. A chloroplast aliquot (typically 5 µL) is diluted to 1 mL with water and made to 80% acetone by addition of 4 mL of acetone. The solution is mixed and clarified by centrifugation in a tabletop centrifuge and the A_{652} is determined. The chlorophyll concentration in the aliquot is calculated using the equation:

$$\text{Chlorophyll (mg/mL)} = 5 \times A_{652} \times 0.02899 / \text{sample volume}$$

The sample volume is the volume of the chloroplast suspension in mL added to the acetone. The chloroplast suspension is adjusted to a final concentration of approx 2 mg chlorophyll/mL and stored on ice in the dark for subsequent analysis.

3.4. Protein Immunoprecipitation

1. All operations are performed at 0–4°C and samples are mixed by constant shaking provided by a vortex mixer having a foam insert for microfuge tubes.
2. Protein samples in SDS buffer are resolubilized by boiling for 2 min prior to immunoprecipitation. A 100-μL aliquot of each resolubilized gradient fraction is mixed with 1350 μL of RIPA buffer. Samples are preabsorbed with 30 μL of protein A Sepharose by incubation for 20 min with constant shaking. The protein A Sepharose is removed by centrifugation for 2 min in a microfuge at maximum speed and the preabsorbed protein supernatant is transferred to a new tube. Preclearing the lysate will reduce nonspecific binding of proteins to the protein A Sepharose when it is used to recover the antigen–antibody complex reducing background.
3. The preabsorbed protein sample is mixed with 2 μL of rabbit polyclonal antibody at an appropriate dilution and incubated for 30 min with constant shaking followed by incubation with 50 μL protein A Sepharose for 30 min with constant shaking (*see* **Note 18**).
4. The antigen–antibody–protein A Sepharose complex is recovered by centrifugation for 2 min in a microfuge at maximum speed. Remove as much of the supernatant as possible without disturbing the pellet.
5. The antigen–antibody–protein A Sepharose pellet is washed three times by resuspension with 1 *M* NaCl RIPA buffer and once by resuspension with RIPA buffer followed by centrifugation at maximum speed in a microfuge. Care must be taken not to disturb the antigen–antibody–protein A Sepharose pellet when removing the wash solution.
6. The antigen–antibody complexes in the final washed pellet are eluted by resuspending the pellet in 25 μL of 3X sample loading buffer and incubating at 37°C for 15 min. The protein A Sepharose is removed by centrifugation at room temperature for 2 min in a microfuge at maximum speed. The antigen-containing supernatant is removed being careful not to disturb the Sepharose bead pellet and transferred to a new tube. The recovered protein is stored at −20°C for analysis by SDS-PAGE.

3.5. SDS-PAGE

1. This method utilizes a Bio-Rad 16-cm Protean II xi Gel Apparatus and 1-mm-thick gels. It is easily adapted to other types of electrophoresis chambers and gel sizes. The glass plates must be handled carefully to avoid chips, cracks and scratches. The plates are cleaned immediately after use in a liquid detergent such as Alconox and rinsed extensively with deionized water. Plates are stored upright in a plastic stand. Prior to use, plates are cleaned with 95% ethanol and air-dried.

2. Assemble the glass plates and spacers locking the assembled glass sandwich to the casting stand.
3. The percentage gel used must be chosen to provide maximum separation in the molecular weight range of the proteins being studied with 10–12% gels being the most commonly used. The two *Euglena* chloroplast proteins most extensively studied are synthesized as polyprotein precursors with the precursor molecular weight approximately eight times the weight of the mature protein (*7,13*). Studies of precursor processing utilize 8–12% linear gradient gels in order to obtain maximum resolution of the large precursors and the small mature proteins on a single gel. An 8–12% gradient gel is prepared using a 30-mL linear gradient maker and a peristaltic pump set to deliver the gel mixture at a rate of 2 mL/min. The 8% gel solution is prepared in the back chamber of the gradient maker by adding 2.49 mL separating gel solution, 4.82 mL water, 2.66 mL acrylamide/bis-acrylamide and mixing well with a disposable pipet. Open the stopcock connecting the front and back chambers and let a minimal amount of acrylamide solution flow into the front chamber to displace the air in the connecting tube and close the stopcock. Place a stirring bar into the front chamber of the gradient maker and prepare the 12% gel solution in the front chamber by adding 2.49 mL separating gel solution, 3.98 mL acrylamide/bis-acrylamide, 3.49 mL of 50% (w/w) sucrose in water and mixing well with a disposable pipet. Rapidly add 33.6 µL 10% ammonium persulfate and 43.8 µL 10% TEMED into the back chamber and 33.6 µL 10% ammonium persulfate and 21.9 µL 10% TEMED into the front chamber and immediately mix using a disposable pipet. Turn the magnetic stirrer on, open the stopcock connecting the two chambers, open the outlet stopcock from the gradient maker and start the peristaltic pump using a pump rate of 2 mL/min. Check that the stirring rate is sufficiently rapid to mix the two gel solutions before they leave the gradient maker. The gradient should flow out of the gradient maker down the side of one of the glass plates. When all of the gel solution has been transferred to the gel apparatus, gently overlay the gel with water-saturated isobutanol using a disposable pipet to slowly drip the isobutanol down the side of one of the glass plates. Allow the gel to polymerize for 2 h (*see* **Note 19**).
4. After the gel polymerizes, the isobutanol is poured off and the top of the separating gel is rinsed twice with deionized water. The sample comb is positioned at the top of the gel at a slight angle. The stacking gel solution is prepared in a 25-mL sidearm flask by mixing 3 mL deionized water, 1.25 mL stacking gel solution, and 0.65 mL acrylamide/bis-acrylamide. Degass the solution under vacuum for 10–15 min, add 25 µL freshly prepared 10% ammonium persulfate, 5 µL TEMED and immediately pour the stacking gel into the glass sandwich using a 5-mL plastic pipet until all the teeth of the comb are covered. Align the comb in the glass sandwich being careful not to introduce air bubbles into the gel and complete filling the gel. Let the gel polymerize for 30–45 min.

5. Once the stacking gel has polymerized, carefully remove the comb pulling straight up being careful not to tear the wells. Wash the wells thoroughly with sterile deionized water using a squeeze bottle. Fill the wells with SDS-PAGE running buffer.

6. Dilute each 25-μL immunoprecipitated gradient fraction on ice in a capped microfuge tube with 20 μL SDS buffer, mix and denature the samples by incubation at 95°C for 3 min.

7. Attach the gel sandwich to the upper buffer chamber. Fill the lower buffer chamber with SDS-PAGE running buffer and place the gel sandwich upper buffer chamber assembly into the lower buffer chamber. If excessive air bubbles are present at the bottom of the gel sandwich, remove them using a bent Pasteur pipet to squirt buffer across the bottom of the gel sandwich. Fill the upper buffer chamber with SDS-PAGE running buffer and check for leaks. Using a gel loading micropipet tip, carefully load 40 μL of denatured sample into each well loading the first 15 gradient fractions onto one gel and the last 15 fractions onto the second gel.

8. Complete assembly of the gel unit and connect the power supply. Run the gel at 5 mA/gel constant current until the bromophenol blue tracking dye enters the separating gel and then raise the current to 20 mA/gel for a 4- to 5-h run. Mix the buffer in the lower chamber with a magnetic stirrer during the run. Run the gel until the bromophenol blue tracking dye reaches the bottom of the gel. Gels can be run overnight at 40 V constant voltage.

9. At the end of the run, the gel sandwich is removed from the apparatus and placed on the laboratory bench. Wearing gloves, the clamps are removed and one of the spacers is pushed out to the side of the plate. The spacer is used to gently pry the plates open. The stacking gel is cut off and a diagonal cut is made on the left side of the gel for orientation. The gel is separated from the plate by floating into deionized water.

3.6. Visualization and Quantitation of ^{35}S-Labeled Proteins Separated by SDS Gel Electrophoresis by Fluorography and Phosphorimaging

1. Gently shake the gel for 10 min in 100 mL of sterile deionized water.

2. Transfer the gel to a dry dish. Place a precut piece of Whatman 3 MM paper that is slightly larger than the gel over the gel making sure no air bubbles are trapped under the gel. Lift the gel/filter paper out of the dish and place gel-side down onto a smooth sheet of 0.001-in. mylar if it will be analyzed with the PhosphorImager or saran wrap if it will only be used for fluorography. Wrap the gel ensuring that the surface is completely smooth.

3. The wrapped gel is placed gel-side up onto a large piece of filter paper in a gel dryer (Bio-Rad). Dry the gel for 2 h at 80°C under vacuum. The gel is dry when the surface of the gel is the same temperature as the dryer surface.

4. For fluorography, use tape to mount the dried wrapped gel on a used piece of X-ray film oriented gel-side up and place the mounted gel in the X-ray cassette. Mark

three pieces of tape with the glow writer autoradiography pen and place on the plastic wrap at the top and bottom edge of the gel, providing orientation points for aligning the gel and exposed film. In the darkroom, make a diagonal cut on the left side of the X-ray film for orientation, place the film onto the gel, close the cassette, and expose at -80°C for the appropriate time.

5. After an appropriate exposure time, remove the cassette from the −80°C freezer, allow it to reach room temperature, remove the X-ray film in a darkroom and develop. Fluorographs of a pulse chase subcellular fractionation experiment are shown in **Fig. 1**.

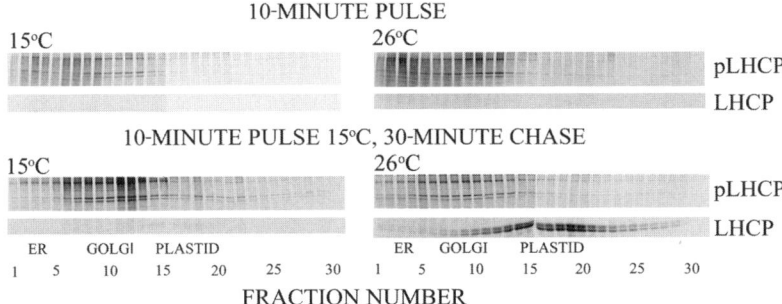

Fig. 1. Fluorograph showing pLHCPII and LHCPII intracellular localization in cells growing at 15°C and in cells growing at 15°C and transferred to 26°C for 30 min. Dark-grown resting *Euglena* exposed to light for 24 h were incubated at 15 or 26°C for 2 h and pulse-labeled at 15 or 26°C for 10 min with ^{35}S-sulfate (top). At the end of the pulse, cells pulse-labeled at 15°C were chased for 30 min at 15 or 26°C by addition of unlabeled sulfate (bottom). Cell-free extracts were prepared at the end of the pulse and chase. Organelles were separated by isopycnic sucrose gradient centrifugation, pLHCPII was immunoprecipitated from each gradient fraction, the immunoprecipitates were analyzed on SDS gels, and the immunoprecipitated proteins were visualized by fluorography using unflashed X-ray film. The regions of the fluorographs containing pLHCPII and LHCPII are presented. pLHCPII is similarly distributed between the ER, fractions 2–5, and Golgi apparatus, fractions 6–14, after a 10-min pulse at both 15 and 26°C, while mature LHCPII is undetectable in any of the gradient fractions (top). In cells pulse labeled at 15°C and chased at 15 or 26°C for 30 min (bottom), pLHCPII is predominately in the Golgi fractions with lesser amounts in the ER or in fractions 14–22 containing broken and intact chloroplasts. Mature LHCPII is found only in the broken and intact chloroplast fraction of cells chased at 26°C being undetectable in the cells chased at 15°C demonstrating that incubation at 15°C blocks transport of pLHCPII from the Golgi apparatus to the chloroplast where it is proteolytically processed to mature LHCPII. (Data from Zhiwei Fang and S. D. Schwartzbach, unpublished.)

6. Radioactivity in individual bands can be quantitated by scanning the fluorograph
   with a densitometer if preflashed film was used or analyzed with a PhosphorImager
   (*see* **Note 20**).

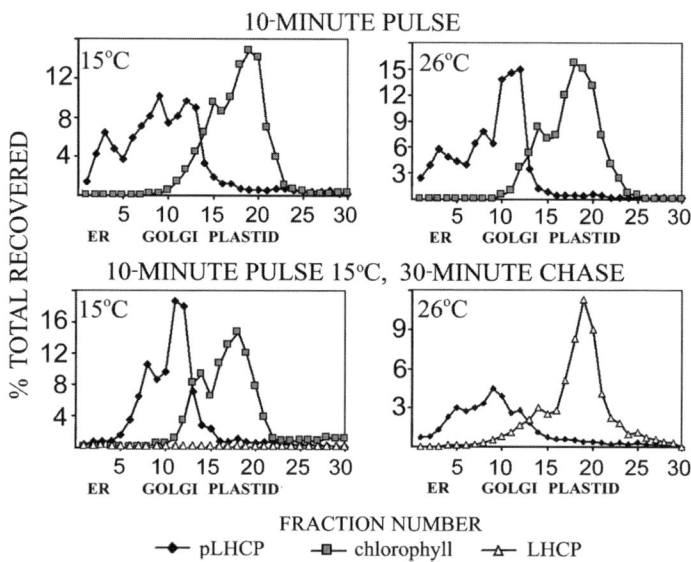

Fig. 2. Quantitative analysis showing pLHCPII and LHCPII intracellular localization
in cells growing at 15°C and in cells growing at 15°C and transferred to 26°C for 30 min.
The SDS gels presented in **Fig. 1** were scanned with a PhosphorImager. pLHCPII
levels are defined as the sum of the image intensity for the 207-, 161-, 122-, and 106-
kDa pLHCPIIs and LHCPII levels are defined as the sum of the image intensity for
the 26- and 27-kDa LHCPIIs. To allow direct comparisons between gradients loaded
with differing amounts of ^{35}S-labeled protein, the amount of pLHCPII and LHCPII
in each fraction is plotted as a percent of the total immunoprecipitate (pLHCPII and
LHCPII) recovered from the gradient. Note the discrepancy between what appears
in the fluorograph (**Fig. 1**, bottom right) to be significant amounts of pLHCPII in
chloroplast fractions 14–22 and the negligible amounts found by quantitative analysis
with the phosporimager (**Fig. 2**, bottom right). This and other discrepancies between
the visual and PhosphorImager analysis result from the nonquantitative response of the
unflashed film. The low levels of pLHCPII in the chloroplast fractions produce film
darkening proportional to their amount as their levels do not saturate the film response
while the large amount of pLHCPII in the Golgi fractions saturate the film response
resulting in a large underestimation of the actual amount of radioactive protein present
in these fractions. (Modified from **Fig. 1** [10] with permission.)

7. To analyze the gel using a PhosphorImager, place the dried wrapped gel into the storage phosphor cassette oriented gel-side up. Orient the gel against the ruler on the side of the cassette (*see* **Note 21**).

8. Erase the storage phosphor screen. Remove the screen from the eraser avoiding exposure to direct light and place it on top of the gel in the cassette. Close the cassette and expose for an appropriate time (overnight) at room temperature.

9. After an appropriate exposure, unlock the cassette and carefully but quickly remove the screen from the gel minimizing exposure of the screen to direct light. Scan the screen and quantitate the amounts of protein in the bands of interest. The SDS gels used for the fluorographs in **Fig. 1** were scanned with a PhosphorImager and the amounts of the precursor to the light harvesting chlorophyll a/b binding protein of photosystem II (pLHCP) and LHCP were quantified (**Fig. 2**). Because it is difficult to know how many counts are loaded on each gradient, the amount of LHCPII and pLHCPII in each gradient fraction is plotted as the percent of total immunoprecipitate (LHCPII and pLHCPII) recovered from the gradient permitting a direct comparison between gradients.

Notes

1. The chemicals should be ground in the order given. The color of the final powder has varied from batch to batch being either light blue or reddish-brown. Mixes of either color support normal cell growth.

2. Media is normally prepared in 20-L carboys. After autoclaving, a precipitate may appear which can be redissolved by shaking the warm carboy. Media can be stored at room temperature for 1–2 wk or in a cold room for 1 mo before becoming contaminated as a result of frequently opening the carboy to dispense media into individual flasks.

3. Vitamin B_{12} is heat sensitive and its aseptic addition after autoclaving ensures reproducible yields of vitamin B_{12}-deficient cells.

4. We have used carrier-free aqueous solutions of $^{35}SO_4$ and $H_2^{35}SO_4$ interchangeably. The choice of isotope form and supplier is dictated by which supplier will offer the lowest price for either a single purchase or a contract purchase. Because of the short half-life of ^{35}S, experiments are planned to utilize the entire shipment within 2 wk to avoid buildup of toxic radioactive decay products.

5. The plant cell protease inhibitor cocktail at a concentration of $5\,\mu L/mL$ buffer can be used interchangeably with the individual protease inhibitor stock solutions.

6. When densitometry will be used for quantitative analysis of fluorographs, preflashed X-ray film should be used to maintain a linear relationship between radioactivity and film density. Film is preflashed in a darkroom by exposure to a photographic flash unit in a light tight box with its output reduced and diffused by thin filter paper to achieve a film density of approximately $0.1–0.2\ A_{540}$ above background. The unit is calibrated by placing the flash unit approx 2 ft above a sheet of X-ray film. Cardboard is placed over the film so that only a thin strip is exposed

allowing multiple exposure conditions for a single film sheet. The film is flashed, the cardboard is moved to expose a new section of film and the flash is repeated. The A_{540} of the strips receiving different exposures are determined relative to an unexposed strip of film in a spectrophotometer. By varying the number of flashes and the distance of the flash unit from the film, the number of flashes at a specific distance required for a uniform film density of $0.1–0.2A_{540}$ above background can be determined. A flash unit (Sensitize preflash unit; GE Healthcare Life Sciences) specifically designed for pre-flashing X-ray film is commercially available.

7. Cells for chloroplast development are grown in the dark under dim green safelights. Centrifuge bottles are placed in the rotor in the dark and the rotor is sealed and then carried to the centrifuge. Care must be taken to maintain sterile conditions during centrifugation and resuspension because cultures are easily contaminated with yeast and bacteria. In 3-d resting cells, chloroplast development occurs over a 72-h period without a net change in total cell protein. Photosynthetic proteins such as LHCPII (*6,14*) are generally synthesized at their maximum rate 24 h after light exposure whereas for mitochondrial proteins such as fumarase maximal synthesis is seen 8–12 h after light exposure (*6,15*). Preliminary experiments should be performed to determine the time after light exposure when the protein to be studied is synthesized at its maximum rate.

8. The storage carbohydrate paramylum sediments faster than intact chloroplasts disrupting chloroplasts during isolation and reducing chloroplast yield. Vitamin B_{12} deficiency allows chloroplast replication to continue in the absence of cell division depleting cellular paramylum reserves. Isolation of chloroplasts from vitamin B_{12}-deficient cells optimizes the yield of intact chloroplasts by producing cells containing a high chloroplast number and low paramylum content. For large-scale chloroplast isolation, cells can be grown in 20-L carboys aerated with air and illuminated by banks of fluorescent lights at the carboy surface. To prepare cells for isolation of chloroplasts at different times after the start of light-induced chloroplast development, grow cultures in the dark on low-vitamin-B_{12} medium, harvest the cells after 4 d of growth and transfer to resting medium as described for the low-sulfur cells used for pulse labeling.

9. Each subcellular fractionation experiment requires approx 10 mL of ^{35}S-sulfate-labeled cells. Experiments are normally done in pairs so that 20 mL of cells are pulse labeled providing sufficient material for two time points. The $t_{1/2}$ for pLHCPII transport is 20 min while the $t_{1/2}$ for the transport of the small subunit of ribulose-bis-phosphate carboxylase/oxygenase to the chloroplast is only 10 min. After a 10-min pulse, all of the pLHCPII remains in the ER while for pSSU a 5-min pulse is needed if all of the material is to be recovered in the ER (*5,8*). Thus, the duration of the pulse must be determined empirically for each protein to ensure that the pulse duration is long enough to accumulate sufficient ^{35}S-labeled protein for easy detection and short enough so that the majority of the labeled protein remains within a single subcellular compartment which is presumably its site of synthesis.

10. $MgSO_4$ cannot be used for the chase because $0.1\,M$ Mg^{2+} will cause membranes to aggregate, making subsequent organelle fractionation impossible. If pulse-chase experiments are to be done with cells on a different medium, preliminary experiments should be performed to assess the effectiveness of the chase and the concentration of unlabeled sulfate adjusted accordingly.

11. For studies of the kinetics of protein synthesis and precursor processing, cells can be labeled with ^{35}S-sulfate, 20 µCi/mL (specific activity 200 Ci/mM adjusted with $MgSO_4$), resuspended directly in SDS buffer (60 mM Tris-HCl, pH 8.6, 2% SDS), heated at 90°C for 10 min and stored at -20°C for subsequent immunoprecipitation *(7)*.

12. To scale up the procedure for intracellular localization of specific proteins by enzymatic or immunological methods, harvest 250 mL cells, resuspend them in 1 mL grinding buffer/g cells, and grind them in a small mortar with 2.5 g glass beads per mL of suspension. By maintaining a constant ratio of g cells/mL grinding buffer/mL glass beads, the procedure can be scaled up to any culture volume.

13. The location of specific organelles on the sucrose gradient is determined by using Western blotting, immunoprecipitation, or enzyme assays to identify gradient fractions containing marker proteins whose intracellular localization is known. Once the position of specific organelles in the sucrose gradient is known, chlorophyll provides a convenient internal marker for comparing organelle locations on different gradients and for inferring the location of other organelles based on their position in the gradient relative to broken and intact chloroplasts. The dense chlorophyll peak localizes intact chloroplasts, and the less dense chlorophyll peak localizes broken chloroplasts *(5,8)*.

14. One liter of vitamin-B$_{12}$-deficient cells yields 4–6 g cells and purified chloroplasts equivalent to 1.8–2.5 mg chlorophyll. The procedure can be scaled up, but the culture volume:buffer ratio must be maintained constant to obtain good chloroplast yields.

15. The French press is assembled and precooled by covering it with ice. The pressate is collected into breaking buffer with protease inhibitors contained in an appropriate-sized flask surrounded by ice. The pressure used for breakage depends on the cell-to-buffer ratio, so it is essential that this ratio remain constant. The breaking pressure represents a tradeoff between obtaining maximum cell breakage while minimizing chloroplast rupture so that only a fraction of the cells are broken. It is advisable to monitor breakage with a light microscope.

16. The centrifugation should be timed with a stopwatch as longer centrifugation leads to increased chloroplast loss. It is best to pour the supernatant out of the tube over the cell pellet. Just as the dark green layer reaches the edge of the tube, stop pouring. Although this layer contains a large number of chloroplasts, it also has an unacceptable number of whole cells. A microscope should be used to monitor the chloroplast-containing supernatant for whole cell contamination. If large numbers of cells are found after the second centrifugation, the centrifugation should be repeated until whole cell contamination is reduced to a low level.

17. As much as 30 mL of the yellow-green supernatant can be removed to reduce the total volume for the next centrifugation step. Removing more than this amount will result in the diluted Percoll concentration being too high preventing the chloroplasts from being pelleted efficiently. Percoll must be kept well below 40% at this stage.

18. Each antibody will have a different titer and preliminary experiments must be run to determine the amount of antibody needed. Prepare a cell extract for immunoprecipitation by resuspending 2 mL pulse-labeled cells in 100 µL SDS buffer and heating at 90°C for 2 min. Immunoprecipitate 20 µL cell extract with increasing amounts of antibody determining the saturating amount of antibody required to immunoprecipitate the labeled antigen. Direct counts of the immunoprecipitate are highly inaccurate and the immunoprecipitate should be analyzed by SDS gel electrophoresis.

19. When pouring a gradient gel, the gel solutions are not degassed.

20. The PhosphorImager is at least 10 times more sensitive than fluorography and has a linear response range of at least 4 orders of magnitude while the linear response range of pre-flashed X-ray film is at best 2 orders of magnitude. We routinely do a fluorograph without preflashed film first because the fluorographs produce a better hard copy of the results than can be obtained from the computer printouts from the phosphorimager. After fluorography, the gel is exposed to the PhosphorImager for quantitative analysis.

21. Extreme care must be exercised in handling the storage phosphor screens. Common causes of screen damage include compression lines from ballpoint pens, contamination with long-lived radioisotopes resulting from failure to properly wrap the dried gel and warping or bending of the backing plates from improper storage and from being dropped.

Acknowledgments

This work was supported by National Science Foundation Grant MCB-0080345.

References

1. Gibbs, S. P. (1978) The chloroplasts of *Euglena* may have evolved from symbiotic green algae. *Can. J. Bot.* **56**, 2883–2889.

2. van Dooren, G. G., Schwartzbach, S. D., Osafune, T., and McFadden, G. I. (2001) Translocation of proteins across the multiple membranes of complex plastids. *Biochim. Biophys. Acta* **1541**, 34–53.

3. Osafune, T., Sumida, S., Schiff, J. A., and Hase, E. (1991) Immunolocalization of LHCPII apoprotein in the Golgi during light-induced chloroplast development in non-dividing *Euglena* cells. *J. Electron Microsc.* **40**, 41–47.

4. Dockerty, A. and Merrett, M. J. (1979) Isolation and enzymatic characterization of *Euglena* proplastids. *Plant Physiol.* **63**, 468–473.

5. Sulli, C. and Schwartzbach, S. D. (1995) The polyprotein precursor to the *Euglena* light harvesting chlorophyll a/b-binding protein is transported to the Golgi apparatus prior to chloroplast import and polyprotein processing. *J. Biol. Chem.* **270**, 13084–13090.

6. Monroy, A. F., McCarthy, S. A., and Schwartzbach, S. D. (1987) Evidence for translational regulation of chloroplast and mitochondrial biogenesis in *Euglena*. *Plant Sci.* **51**, 61–76.

7. Rikin, A. and Schwartzbach, S. D. (1988) Extremely large and slowly processed precursors to the *Euglena* light harvesting chlorophyll a/b binding proteins of photosystem II. *Proc. Natl. Acad. Sci. USA* **85**, 5117–5121.

8. Sulli, C. and Schwartzbach, S. D. (1996) A soluble protein is imported into *Euglena* chloroplasts as a membrane-bound precursor. *Plant Cell* **8**, 43–53.

9. Suzuki, E., Tsuzuki, M., and Miyachi, S. (1987) Photosynthetic characteristics of chloroplasts isolated from *Euglena gracilis* Z grown photoautotrophically. *Plant Cell Physiol.* **28**, 1377–1388.

10. Slavikova, S., Vacula, R., Fang, Z., Ehara, T., Osafune, T., and Schwartzbach, S. D. (2005) Homologous and heterologous reconstitution of Golgi to chloroplast transport and protein import into the complex chloroplasts of *Euglena*. *J. Cell Sci.* **118**, 1651-1661.

11. Arnon, D. (1949) Copper enzymes in isolated chloroplasts. Polyphenoloxidase in *Beta vulgaris*. *Plant Physiol.* **24**, 1–15.

12. Perry, S. E., Li, H. M., and Keegstra, K. (1991) In vitro reconstitution of protein transport into chloroplasts. *Methods Cell Biol.* **34**, 327–344.

13. Chan, R. L., Keller, M., Canaday, J., Weil, J. H., and Imbault, P. (1990) Eight small subunits of *Euglena* ribulose 1-5 bisphosphate carboxylase/oxygenase are translated from a large mRNA as a polyprotein. *EMBO J.* **9**, 333–338.

14. Rikin, A. and Schwartzbach, S. D. (1989) Regulation by light and ethanol of the synthesis of the light harvesting chlorophyll a/b binding protein of photosystem II in Euglena. *Planta* **178**, 76–83.

15. Rikin, A. and Schwartzbach, S. (1989) Translational regulation of the synthesis of *Euglena* fumarase by light and ethanol. *Plant Physiol.* **90**, 63–69.

16

Fluorescent Protein Fusions for Protein Localization in Plants

John Runions, Chis Hawes, and Smita Kurup

Summary

Protein localization in living plant cells is commonly studied using fluorescent protein fusions. Stable transformation of plant cells requires the use of binary vectors, which are larger and not as amenable to genetic manipulation as animal cell transfection vectors. Binary vectors containing fluorescent protein fusion constructs are prepared using standard molecular biological techniques. Fusion genes as well as promoters and selection markers are stably incorporated into the plant cell genome via *Agrobacterium*-mediated transfer. Presented here are a series of protocols that detail binary vector construction, bacterial transformation, and a rapid transient assay technique that can be used to evaluate fusion protein fluorescence in leaves.

Key Words: Fluorescent protein; green fluorescent protein; binary vector; *Agrobacterium*; *Nicotiana*; tobacco; *Arabidopsis*.

1. Introduction

Fusion of fluorescent proteins to proteins of interest has become the method of choice for studying subcellular localizations in living tissues *(1–3)*. The techniques required to study protein localization by this method are common regardless of the organelle under study. For this reason, the technique is useful not only for studying known proteins but also for determining the subcellular localization of previously uncharacterized proteins. Expression of fusion proteins is successful for more than 80% of the constructs that we have tested. Several of the common uses of this technique include:

From: *Methods in Molecular Biology, Vol. 390: Protein Targeting Protocols: Second Edition*
Edited by: M. van der Giezen © Humana Press Inc., Totowa, NJ

1. The study of a particular organelle: In this case, the protein localized to the organelle under study may not be of interest in itself except that its expression results in the organelle becoming fluorescent *(2–6)*.
2. Characterization of unknown proteins: Functional genomics and proteomics aim to characterize proteins and the study of protein subcellular localization is a useful tool in this respect. Here, full-length coding sequences of DNA (open reading frames) discovered by sequencing or cDNA derived from expression screens are fused to the fluorescent protein DNA *(7–10)*.
3. Characterization of protein motifs: Fusion of short protein domains that may convey targeting information upon fluorescent proteins is useful in two respects—either as a motif discovery technique or to add the known functionality of the motif to the fluorescent protein, i.e., membrane anchoring via a transmembrane domain *(11,12)*.

Transformation of plant cells is slightly more complicated than transformation of animal cells. Plant cells will not take up a transformation vector directly via transfection. Rather, for transformation of *Arabidopsis*, tobacco, tomato, wheat, rice, etc., the bacterium *Agrobacterium tumefaciens* is employed because of its ability to introduce DNA into the plant cell (for further information, *see* **refs. *13–15***). In nature, *Agrobacterium* species are "gall-forming," they inject a region of their own DNA (the transfer-, or T-DNA) into the plant stem. T-DNA, which originates in the tumor-inducing (Ti) plasmid of *A. tumefaciens,* is incorporated into the plant genome by homologous recombination and results in tissue proliferation when the virulence (vir) genes are expressed. For plant transformation, the vir genes of the T-DNA are replaced with a genetic construct containing at least the protein to be expressed and a selectable marker gene and perhaps promoter genes and other DNA as required. These modified vectors are known as "binary vectors" because they work with a smaller plasmid called a helper plasmid to effect plant cell transformation. The strains of *Agrobacterium* commonly used for transformation contain the smaller helper plasmid and must be transformed with the larger T-DNA-containing vector *(16)*.

There is often no requirement to construct binary vectors. Many are available from the public community and commercial suppliers and can be easily obtained (**Table 1** and **Fig. 1**). They come with a variety of different T-DNA regions depending on the requirements. The most commonly used for fluorescent protein fusions contain a 35S CaMV promoter that is generally thought to drive transcription in virtually all plant tissues with some exceptions, e.g., early embryonic stages. Immediately downstream of the promoter is a multiple-cloning region for insertion of your gene or DNA segment of interest. Many binary vectors now include the fluorescent protein gene so that the fusion gene can simply be cloned as a 5′ or 3′ in-frame fusion. Many different selection

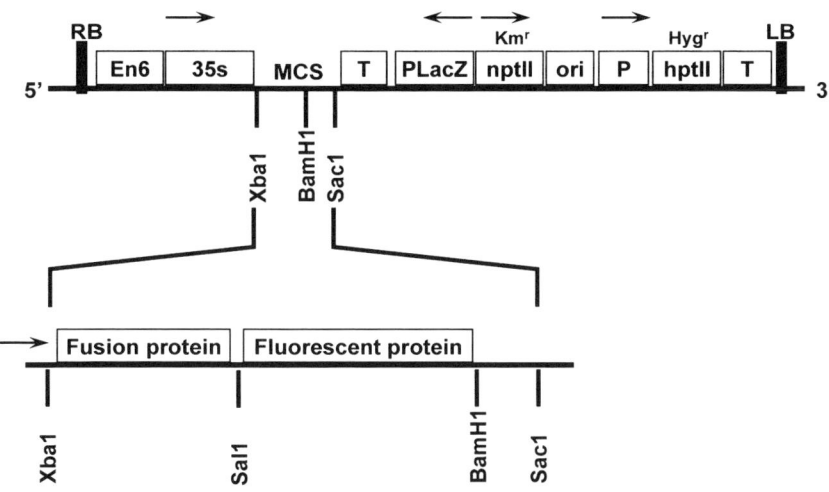

Fig. 1. Schematic of the pVKH18En6 binary vector used as an example in this chapter. Only the T-DNA region of the vector is shown. The gene of interest (fusion protein) is cloned as an in-frame fusion with a fluorescent protein using *Xba*I and *Sal*I restriction sites. RB, right border; En6 35s, enhanced 35s promoter; MCS, multicloning site with unique restriction sites shown; T, *Nos* terminator; PlacZ, *LacZ* promoter; *nptII*, kanamycin resistance gene; ori, *ColE1* origin of replication; P, *Nos* promoter; *hptII*, hygromycin resistance gene.

marker genes are available each with either bacterial or plant promoters. In certain cases, the selection marker will differ between bacteria and plants, e.g., in bacteria the vector might convey resistance to kanamycin, while once stably transformed into plants the T-DNA might convey resistance to hygromycin.

Binary vectors are large (10–20 kb) and more difficult to work with than the smaller vectors used in animal cell transfection. For this reason, it is best to carry out the intermediate cloning steps in smaller plasmid vectors. In addition, it is easier to maintain and manipulate binary vectors in *E. coli. Agrobacterium* are slow growing compared to *E. coli*, they contain recombinase genes that can alter a vector sequence, and binary vectors are maintained only at low copy number. Maintenance of a binary vector in a RecA$^-$ strain of *E. coli* (e.g., DH5α or XL1-blue) results in reduced likelihood of recombination and yields far more DNA from mini and midi preps. Once the binary vector is constructed, it is transferred to an *Agrobacterium* strain for plant transformation.

Inserting your gene of interest into a binary vector can be done with traditional cloning techniques involving restriction digestion and ligation or with

Table 1
Binary Vectors[a]

Vector	Source (ref.)	Details
pVKH18En6	Available from John Runions, Oxford Brookes University *(26)*	Enhanced 35s vector for C- and N-terminal fusions to FPs
pEGAD	Available though ABRC *(7)*	C-term fusion to GFP, Basta selection, 35S promoter
pBIN 35S-mGFP5-er	Jim Haseloff, University of Cambridge, UK	Based on pBI121, can be used to for C- and N-term fusion to GFP, kanamycin selection, 35S promoter
pGREEN	http://www.pgreen.ac.uk *(27)*	Optimized for plant transformation, can be used with a range of promoters and selectable markers
pCAMBIA series	CAMBIA, http://www.cambia.org	Derived from pPZP vectors *(28)*, some IP limitations, 35S promoter, GFP, GUS as reporter genes
pBI121	Clonetech, Mountain View, CA	Can be modified for GFP fusions, derivative of pBIN19 *(29)*
pART27	*(30)*	Can be modified for GFP fusions
pMDC series	Available though ABRC *(31)*	Gateway™ assisted cloning, C- and N-term fusions to GFP

[a] Most are useful for traditional cloning of fluorescent protein fusions. Newer vectors, e.g., the pMDC series, have been adapted for cloning by recombination-based methods. (*See* **ref. 17** for review.)

recombination-based protocols like the Gateway™ system from Invitrogen *(17)*. Here we provide a step-by-step guide for cloning of a fluorescent protein fusion construct in a binary vector and transiently expressing it in tobacco leaves. This technique requires some basic knowledge of molecular biology techniques, e.g.,

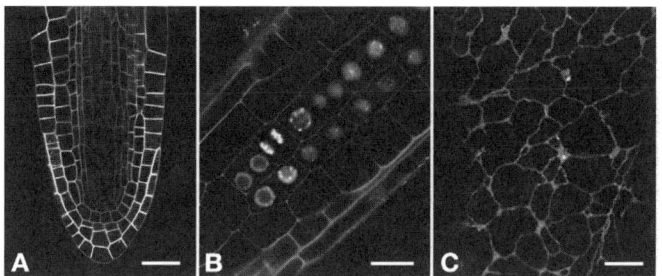

Fig. 2. Live tissue examples of fluorescent protein fusions that have been targeted to different subcellular locations using the techniques described in this chapter. (a) Green fluorescent protein (GFP) targeted to the plasma membrane in a developing lateral root of *Arabidopsis* using the vector pBIB 35s- EGFP-LTI6b *(32)*. LTI6b is a plasma membrane protein of unknown function. Scale bar = 50 μm. (b) Yellow fluorescent protein targeted to the nuclei of an *Arabidopsis* main root using the vector pBIN 35s-H2B-YFP *(33)*. H2B is a histone protein that binds chomatin. Scale bar = 10 μm. (c) GFP localized to the endoplasmic reticulum using the transient expression technique described in this chapter. The vector used is pVKH18En6 35s-GFP-HDEL. HDEL is not a full gene sequence but simply a 12-base-pair-long sequence of DNA that confers targeting specificity *(34)*. Scale bar = 5 μm.

PCR primer design, restriction digestion, etc. These basic molecular biology techniques do not differ from those used in animal cell biology research and are detailed in other publications *(18–21)* and in the literature supplied with the required enzymes and chemicals. Many of the cloning steps utilized by this procedure only require the use of DNA preparation "kits" supplied commercially. For these steps we will list the kit that we use and leave it to the reader to follow the instructions that come with the kit.

Because this chapter is mainly concerned with fluorescent protein fusions to your gene of interest with respect to studying subcellular localization, we also present a protocol for the transient expression system in tobacco leaves. In our experience, this transient system works extremely well as a rapid method for checking constructs and in cell biological research and is also amenable to imaging. Other rapid and simple transient assay techniques have been reviewed *(16)*. If stable transformants are required, other well-documented methods, such as floral dip in *Arabidopsis*, may be used as well *(22, 23)*.

The simplest approach to construction of a fluorescent protein fusion is to select a binary vector that already contains a fluorescent protein. Otherwise, clone first one gene and then repeat the procedure using the second gene so that the gene of interest and the fluorescent protein gene are added sequentially.

1.1. Cloning Your Gene of Interest

1. Procure DNA for your gene of interest and a suitable binary vector.
2. Amplify your gene by PCR, if necessary, to incorporate restriction sites for insertion into the binary vector.
3. Perform restriction digestion on both the PCR product (your gene) and the binary vector to produce complementary ends.
4. Ligate the digested gene and vector.

1.2. Bacterial Transformation (E. coli and Agrobacterium Protocols)

1. Transform *E. coli* with the ligated vector, which should now contain your gene of interest.
2. Pick colonies of transformed *E. coli* and grow cultures for mini preps.
3. Repeat **Subheading 1.1., step 1** to **Subheading 1.2., step 2** for the fluorescent protein gene using the mini prep DNA from **Subheading 1.2., step 2** as the binary vector (only if you have used a binary vector that did not initially carry a fluorescent protein gene).
4. Transform *Agrobacterium* with the binary vector.
5. Pick colonies of the transformed *Agrobacterium* and grow cultures.
6. Prepare glycerol stocks of each culture.

1.3. Plant Transformation to Check Fluorescent Protein Expression and Targeting

1. Transiently transform tobacco leaves to check that the new construct works, i.e., that there is fluorescent protein expression, and to evaluate the expression pattern.
2. Check mini prep DNA from **Subheading 1.2., step 2** by sequencing to ensure that the gene of interest has been correctly cloned.

The procedure from **Subheading 1.1., step 2** to **Subheading 1.3., step 1** can be accomplished in 10 d if all goes well.

2. Materials

2.1. Cloning Your Gene of Interest

1. DNA for gene of interest.
2. Binary vector.
3. PCR primers.
4. Taq polymerase, PCR reaction buffer, dNTPs from commercial supplier (e.g., New England Biolabs or Promega).
5. DNA cleaning kit, e.g., Wizard® SV Gel and PCR Clean-Up System from Promega.

6. Restriction enzymes and appropriate buffers from commercial supplier (e.g., New England Biolabs or Promega).
7. Ligase enzyme and appropriate buffer from commercial supplier (e.g., New England Biolabs or Promega).

2.2. Preparation of Heat-Shock Competent Bacterial Cells (CaCl₂ Method)

1. Sterile 100 mM $CaCl_2$.
2. Sterile 100% glycerol.
3. Bacterial strains: *E. coli* DH5α, *Agrobacterium tumefaciens* GV3101 (with pMP90Ti helper plasmid).

2.3. Bacterial Transformation (E. coli and Agrobacterium Protocols)

1. Transformation competent cells of a RecA⁻ *E. coli* strain such as DH5α.
2. *E. coli* culture medium such as LB (liquid and solidified with agar).
3. Mini prep kit, e.g., Wizard Plus SV Mini Preps from Promega.
4. Transformation competent cells of an *Agrobacterium* strain such as GV3101.
5. Small amount of liquid nitrogen.
6. *Agrobacterium* culture medium such as LB or YEB (liquid and solidified with agar).
7. Sterile 80% glycerol.

2.4. Plant Transformation to Check Fluorescent Protein Expression and Targeting

1. Plants of *Nicotiana tabacum*. Plants should be about 5–6 wk old. Grown at 21°C, 14 h light, 10 h dark.
2. 1-mL Disposable syringes (without needles).
3. Syringe needle.
4. Infiltration medium (50 mL): 250 mg D-glucose, 5 mL MES stock solution, 5 mL $Na_3PO_4 \cdot 12H_2O$ stock solution, 5 μL 1 M acetosyringone stock solution, make up to 50 mL with dH_2O.
5. MES stock solution: 500 mM MES (4.88 g in 50 mL) made with dH_2O.
6. Na_3PO_4 stock solution: 20 mM $Na_3PO_4 \cdot 12H_2O$ (trisodium orthophosphate) (0.38 g in 50 mL) made with dH_2O.
7. Acetosyringone stock solution: 1 M acetosyringone (3',5'-dimethoxy-4'-hydroxy acetophenone) (0.196 g in 1 mL) made with dimethyl sulfoxide (DMSO). Aliquot this into small amounts (one use) and store at −20°C. (Acetosyringone at 1 M in DMSO dissolves better than when made up in ethanol.)

3. Methods

All of these steps need to be carried out in a lab that is equipped for basic molecular biology with access to a PCR cycler, flow hood, incubators (28 and

37°C), autoclave, and a plant growth facility (greenhouse or growth chamber). In addition, you will be required to satisfy your local health and safety requirements for working with genetically modified organisms prior to beginning.

Basic molecular biology reactions (PCR, restriction digestion, ligation, mini prep) have not been included in detail except where this protocol differs from those in common usage.

Thoughout this section we will use as an example the cloning of a gene of interest into the binary vector pVKH18En6 so that it is in frame and fused to green fluorescent protein (GFP).

Subheadings 3.3. and **3.4.** detail the heat-shock transformation procedure for *E. coli* and *Agrobacterium*, respectively. It is best to have the heat-shock competent bacteria prepared by the protocols described in **Subheading 3.2.** before starting the cloning procedure.

3.1. Cloning Your Gene of Interest

1. DNA for your gene of interest and the vectors that you will need for this procedure will be available from a large number of sources (*see* **Note 1**).
2. Binary vectors are available from many labs (*see* **Table 1**). These are used for traditional, restriction digestion and ligation-based cloning or for recombination-based cloning as required. Our example will follow the traditional cloning technique as most departments are equipped to do this. **Fig. 1** is a schematic of the binary vector pVKH18En6 and the cloning that we will do in this example.
3. Design PCR forward and reverse primers that incorporate the restriction sites you require for cloning into the binary vector you have chosen. In our example we need to add an *Xba*I restriction site at the 5′ end of our gene sequence and a *Sal*I site at the 3′ end (*see* **Note 2**).
4. Follow a standard PCR protocol as per the Taq polymerase supplier's instructions. We find that a 10- to 50 − μL reaction volume is sufficient for most cloning procedures (*see* **Note 3**).
5. Purify the PCR amplified DNA from primers, dNTPs, etc. using a commercially supplied kit, e.g., Wizard SV Gel and PCR Clean-Up System from Promega.
6. Follow a standard restriction digestion protocol as per the enzyme supplier's instructions. A 10- to 50-μL reaction volume is sufficient for most cloning procedures. In our example, *Xba*I and *Sal*I supplied by New England Biolabs can be used in a double restriction digest if buffer number 3 and bovine serum albumin are added.
7. Gel-purify the digested binary vector using a commercially supplied kit, e.g., Wizard SV Gel and PCR Clean-Up System (*see* **Notes 4** and **5**).
8. Purify the PCR amplified/restriction digested gene using a commercially supplied kit, e.g., Wizard SV Gel and PCR Clean-Up System from Promega.

9. Use of standard ligation protocols works well at this stage. We prefer the 2x Rapid Ligation system supplied by Promega because it reduces the time required for this step from overnight to 30–60 min (*see* **Note 6**).

3.2. Preparation of Heat-Shock Competent Bacterial Cells (CaCl$_2$ Method)

1. Streak bacteria on a LB agar plate with appropriate antibiotics. Grow at 29°C (*Agrobacterium*) or 37°C (*E. coli*) overnight or longer, as required. Agrobacteria have a much longer doubling time than *E. coli*. Various other media formulations are used for *Agrobacterium* culture as well, e.g., YEB, but LB seems to work well and does away with the requirement of maintaining two different media.
2. Inoculate 5 mL of LB medium plus antibiotics with well-isolated colonies. Grow at 29°C (*Agrobacterium*) or 37°C (*E. coli*) overnight.
3. The next morning, inoculate 200 mL LB with the 5 mL that have been growing overnight and grow these at the appropriate temperature with shaking.
4. Place sterile 100 mM CaCl$_2$ in refrigerator or freezer to cool to 4°C.
5. When the culture reaches an OD$_{600}$ of 0.4–0.6 (2–5 h), harvest the culture by centrifugation at 4°C for 10 min at 3600 rpm. Split the culture into, e.g., 4× 50 mL for centrifugation. Pour off the supernatant carefully and drain well.
6. While ensuring that bacteria remain chilled, add 1 mL of cold (4°C) 100 mM CaCl$_2$. Resuspend the cells gently by swirling. If it is difficult to obtain complete suspension, they can be gently pipetted up and down in the chilled pipet. After a smooth suspension has been obtained, add an additional 19 mL of cold 100 mM CaCl$_2$ (final volume of 20 mL). The cells are very fragile at this stage. Do not at any point vortex these cells, and be gentle at all stages when handling them.
7. Combine the suspended cells into 2× 40 mL volumes and incubate them on ice for 30 min.
8. Centrifuge the cells again and drain the pellet well. Resuspend cells in 1mL of ice cold 85 mM CaCl$_2$ and 15% glycerol (i.e., 850 μL 100 mM CaCl$_2$ + 150 μL sterile 100% glycerol).
9. Immediately aliquot the cell suspension in 50-μL portions into sterile prelabeled microfuge tubes and freeze in liquid nitrogen. This protocol will make more than 40 × 50 μL aliquots because of the combined volume of cells and resuspension solution i.e., it can yield 60–70 aliquots. Store at −80°C.

3.3. Transformation of E. coli

Transform your DNA into *E. coli*. The simplest way to transform bacteria is by heat shock (*see* **Note 7**). Use a RecA$^-$ strain of *E. coli* such as DH5α.

1. Add 25 ng DNA to 50 μL of thawed, heat-shock competent bacteria in a 1.5-mL microfuge tube.
2. Incubate on ice for 30 min.

3. Heat shock in a water bath at 42°C for 45 s.
4. Place back onto ice for 1–2 min.
5. Add 500 μL of LB medium.
6. Incubate at 37°C for 30–60 min.
7. Centrifuge at 5000 rpm at room temperature for 3 min.
8. Discard supernatant.
9. Resuspend pellet in 150 μL LB.
10. Spread on LB plates that contain the appropriate antibiotics for selection, e.g., kanamycin.
11. Incubate plates at 37°C overnight.
12. Pick single colonies and grow in 10 mL LB overnight cultures. Prepare mini prep DNA using a kit from a commercial supplier, e.g., Wizard Mini Prep kit from Promega.
13. Transform the mini prep binary vector into *Agrobacterium* for plant transformation.
14. Repeat these steps as required using the newly transformed and purified vector as starting material if you wish to insert other DNA sequences (*see* **Notes 8** and **9**).

3.4. Transformation of Agrobacterium (24)

You must first make heat-shock competent *Agrobacterium* (**Subheading 3.2.**). Use *Agrobacterium tumefaciens* strain GV3101 (with pMP90 Ti plasmid).

1. Add 25 ng of the mini prep DNA from **step 6** to 50 μL of frozen, heat-shock competent bacteria in a 1.5-mL microfuge tube. It is usually sufficient to just add 5 μL of miniprep DNA to the frozen bacteria.
2. Let cells thaw on ice for 10–15 min.
3. Freeze tubes in liquid nitrogen.
4. Heat shock in a water bath at 37°C for 5 min directly from the liquid nitrogen.
5. Add 1 mL of LB medium.
6. Incubate at 29°C for 2–4 h (*see* **Note 10**).
7. Centrifuge at 5000 rpm at room temperature for 3 min.
8. Discard supernatant.
9. Resuspend pellet in 150 μL LB medium.
10. Spread on LB plates that contain the appropriate antibiotics for selection, e.g., gentamycin, rifampicin, and kanamycin.
11. Incubate plates at 29°C for 36–48 h.
12. Pick individual colonies of the transformed *Agrobacterium* and grow them in 10 mL LB culture overnight at 29°C. These cultures will be used for two things: (1) to prepare glycerol stocks of bacteria containing your binary vector for future use (**Subheading 3.4., step 13**) and (2) to transform tobacco leaves to evaluate constructs (**Subheading 3.3.**).

13. To make glycerol stocks, add 0.25 mL of sterile 80% glycerol to 1 mL of *Agrobacterium* culture from **step 9**. Vortex briefly and freeze at −80°C (*see* **Note 11**).

3.5. Plant Transformation to Check Fluorescent Protein Expression and Targeting

1. This is a transient assay to check whether the new binary vector gives fluorescent protein expression. (*see* **Note 12**).
2. Use *Nicotiana tabacum* plants (*see* **Note 13**). Plants should be about 5–6 wk old, grown at 21°C, 14 h light, 10 h dark.
3. See the **Subheading 2.** for stock solution and infiltration solution recipes. Make up infiltration medium freshly (the day of infiltrating).
4. To prepare bacteria, place 5-mL aliquots from overnight suspension culture (29°C) (media + appropriate antibiotic) into centrifuge tubes.
5. Spin for 5 min at 4000 rpm at room temperature.
6. Remove supernatant.
7. Add 1 mL infiltration medium at room temperature and resuspend pellet.
8. Again, spin for 5 min at 4000 rpm, remove supernatant, and resuspend pellet in 1 mL infiltration medium (*see* **Note 14**).
9. Make a dilution using a bit of the resuspension and measure OD_{600} (*see* **Note 15**).
10. Dilute the bacteria with infiltration medium until the desired OD_{600} is reached. When testing new constructs, start with an OD_{600} of 0.1 (*see* **Note 16**).
11. For leaf infiltration, select leaves that are one-half to two-thirds expanded near the top of the plant. Mark the upper leaf epidermis with a permanent black marker to indicate the leaf sectors that you are going to inject the bacteria into. Use a syringe needle to make needle marks to inject into on the lower leaf epidermis.
12. Take up resuspended bacterial cells in a 1-mL sterile syringe (no needle). Place the syringe tip against the underside of a young leaf on a needle mark and press the plunger down while exerting gentle pressure against the other side of the leaf with your finger. You should see liquid diffusing thoughout the mesophyllar air space (*see* **Note 17**).
13. Repeat in different areas on different leaves on at least two different plants (*see* **Note 18**).
14. Place the plant back into growth conditions for at least 30 h to allow transformation and gene expression to occur. This growth period depends on the construct and protein longevity, so look early, but most commonly we find that 36–48 h are required before expression can be observed (*see* **Note 19**).
15. To observe fluorescent protein expression, cut out small pieces (approx 1 cm^2) from the transformed region of the leaf and mount them in water, lower epidermis facing upward, on a microscope slide under a cover slip. Follow standard epifluorescence or confocal microscopy procedures.

16. It will be a good idea, and probably required, that you check the new binary vector construct by sequencing the gene insertion region.

Notes

1. Generally, since the *Arabidopsis* genome has been sequenced, most genes (cDNA) should be straightforward to acquire. Many other species have fairly detailed genomic information available on the web. Options for obtaining DNA include:

    a. Write to a laboratory that already has the gene and have published their work with it.

    b. Search for your gene of interest in online databases and order the appropriate clones. Good sources include TAIR and ABRC (http://www.arabidopsis.org) and RIKEN (http://www.brc.riken.jp/lab/epd/Eng/). In particular, a gene supplied as cDNA does not include introns and will be shorter and easier to work with.

    c. PCR-amplify your gene directly from genomic DNA (full length including introns).

    d. It is feasible to search for DNA similar or homologous to *Arabidopsis* genes within the genomic DNA of another species via PCR.

2. The binary vector pVKH18En6 usually already contains a fluorescent protein fusion construct. In this example, the sialyl transferase signal anchor sequence (STtmd) is fused to mGFP5 *(25)* and targets GFP to the Golgi bodies. We will replace STtmd with the gene of interest, so PCR primers need to add the *Xba*I/*Sal*I restriction sites as well as modify the 5′ end of the gene to include a start codon (ATG) and modify the 3′ end to remove a stop codon (TGA, TAA, or TAG), if necessary, so that translation continues in frame though the gene into the GFP. GFP, in this case, contains the stop codon. Suitable primers in this example would be:
   Forward 5′- gctt/ctagaggtatg (+9 codons homologous to gene sequence).
   *Xba*I site added (underlined and cut site indicated) and start codon ATG added.
   Reverse 5′- tgag/tcgacgat (+10 codons homologous to gene sequence).
   *Sal*I site added (underlined and cut site indicated) and stop codon (TGA, TAA, TAG) removed if required.

3. PCR amplification of genes from plants is no different from PCR amplification of any other genes. Use a small amount of template (this depends on template source, genomic DNA, or plasmid) and ensure that buffers, primers, dNTPs, water, etc. remain contamination free by using filtered pipet tips in all steps. At the end of this step, and at the end of restriction digestion reactions, DNA purifications, etc., it is advisable to check for products of the expected length on agarose gels.

4. In our example, both the PCR amplified gene and the binary vector are digested in the same way, but subsequent handling to purify the reaction products differs. In the case of the PCR amplified gene, the small end fragments will not be retained on the column during the purification process using a commercially supplied kit,

and there is no danger of religation, which would interfere with the next step. The binary vector will need to be extracted and purified from an agarose gel if a fragment has been removed to prevent the fragment from religating into the vector in the next step. When gel-purifying binary vector DNA, care needs to be taken. Binary vectors are large (10–20 kb) and so should be run on a low percentage (0.6–0.7%) TAE agarose gel. Use a large gel tank with large well volume (50–75 μL) and run the gel for a long time (> 1 h) at relatively low voltage. Use UV light to observe and image the vector band sparingly because it can damage the DNA. Do not vortex the tube containing the vector DNA at any step of the procedure because the DNA will shear. Use warmed water (70°C) to elute vector DNA from the purification column.

5. It is useful to dephosphorylate restriction digested vectors when doing "nondirectional" cloning, i.e., when the vector is cut with a single enzyme and could potentially recircularize. In this case the vector has been restriction digested with two enzymes and will not recircularize if digestion has been performed to completion.

6. As in all ligation reactions, the best results occur when the gene is added to the reaction in equivalent molar concentration to the vector. In our example, 2–3 μL of vector and 5–10 μL of insert would probably be sufficient. Try various combinations of vector:insert ratio in different ligation reactions. Always run a control ligation that includes vector but no insert as a check on complete vector restriction digestion.

7. Transformation of bacteria by electroporation is more efficient but is a lot more equipment and time costly. For the heat-shock procedure, the preparation of bacterial cells is accomplished more quickly (**Subheading 3.2.**), and because electrical conductivity is not an issue, the ligation reaction can be directly transformed into the cells without first being purified.

8. If you wish to insert other DNA sequences into the transformed and purified vector, it will be best to check the vector first by PCR or restriction digestion to ensure that the first DNA sequence has been cloned properly.

9. In our example there is no further cloning to do—the gene of interest should now be cloned adjacent to the GFP ready for expression.

10. There is much debate about the time required for bacteria to incorporate a transforming plasmid so that they become resistant to a selection antibiotic. *E. coli* become antibiotic resistant within minutes of being transformed. Agrobacteria, however, have a much longer generation time and are generally left for 2 h posttransformation before being plated on antibiotics. In any case, if the bacteria are left too long in culture media without selection, there is a danger of overestimating the efficiency of the transformation procedure because of cell doubling.

11. Glycerol stocks are invaluable because they store your binary vectors and eliminate the need to transform bacteria each time that you want to transform plants. Glycerol stocks can be made from *E. coli* as well and are the best way to store vectors that you might want to increase your stocks of via mini or larger preps. *Agrobacterium*

is not a suitable host for use in DNA prep procedures because it holds vectors at very low copy number relative to *E. coli*.

12. This transient assay is the fastest way to evaluate fluorescent protein fusion expression. Stably transformed plants can be constructed at a later stage if so desired by floral dipping *(22)* or by tissue culture of transiently transformed tissues. We recommend that you transform some leaf sectors with a known binary vector at the same time as you are doing transformation with an unevaluated vector as a control of the transformation process.

13. We use any of *Nicotiana tabacum* cv Petite Havana, *N. benthamiana*, or *N. cleavlandii*. Other species and varieties of tobacco will, no doubt, work as well.

14. At least two rounds of centrifugation and resuspension are required to ensure that all traces of antibiotic, which would otherwise kill the plant cells, are removed from the *Agrobacterium*.

15. You will have to determine what dilution to use to get a good spectrophotometer reading. We use dilutions of 10–50 times depending on the concentration of cells in culture.

16. Optimum concentration of cells for transient expression will vary according to construct, we recommend trying a few different concentrations until you get the expression level you want. An OD of 0.1 is a good place to start with a new construct. Once you know it works, use a lower OD such as 0.03. Here is a handy formula: to make 5 mL (5000 μL) of cells at an OD of 0.1 when the initial OD is 0.9:

(Final conc. × final vol.)/initial conc. = volume of cells in diluted infiltration medium

Or

$$(0.1\,OD_{600} \times 5000\,\mu L)/0.9\,OD_{600} = 555.56\,\mu L$$

so use 556 μL of original bacterial resuspension and make the volume up to 5 mL with infiltration solution.

17. Wear protective clothing, eyewear, and gloves. The *Agrobacterium* tends to spray around a bit until you become competent with the procedure.

18. It is important to inject bacteria into different leaves on different plants because sometimes one leaf or plant will just work better than another. Overall, the process is very robust if your plants are in good condition.

19. Some constructs will yield no expression of a fluorescent protein. Do not be fooled by the low-level autofluorescence that occurs in most wavelength ranges. Look at a known fluorescent protein construct to get an idea of how bright they really are. If you do not see expression, it might be because the protein fusion is lethal to the plant cells when expressed at the high levels the 35S promoter produces. In these circumstances, try looking earlier than 48 h postinfiltration. You may observe a

low level of real fluorescence before the lethal effects take hold. In our hands, more than 80% of fusions work and do not appear to have an adverse effect on the cells.

References

1. Chiu, W., Niwa, Y., Zeng, W., Hirano, T., Kobayashi, H., and Sheen, J. (1996) Engineered GFP as a vital reporter in plants. *Curr. Biol.* **6**, 325–330.
2. Brandizzi, F., Irons, S., Johanson, J., Kotzer, A., and Neumann, U. (2004) GFP is the way to glow: bioimaging of the plant endomembrane system. *J. Microsc.* **214**,138–158.
3. Dixit, R., Cyr, R., and Gilroy, S. (2006) Using intrisically fluorescent proteins for plant cell imaging. *Plant J.* **45**, 599–615.
4. Niedz, R. P., Sussman, M. R., and Satterlee, J. S. (1995) Green fluorescent protein—an in-vivo reporter of plant gene- expression. *Plant Cell Rep.* **14**, 403–406.
5. Grebenok, R. J., Pierson, E., Lambert, G. M., et al. (1997) Green fluorescent protein fusions for efficient characterization of nuclear targeting. *Plant J.* **11**, 573–586.
6. Haseloff, J., Siemering, K. R., Prasher, D. C., and Hodge, S. (1997) Removal of a cryptic intron and subcellular localisation of green fluorescent protein are required to mark transgenic *Arabidopsis* plants brightly. *Proc. Natl. Acad. Sci. USA* **94**, 2122–2127.
7. Cutler, S. R., Ehhardt, D. W., Griffitts, J. S., and Somerville, C. R. (2000) Random GFP::cDNA fusions enable visualization of subcellular structures in cells of Arabidopsis at a high frequency. *Proc. Natl Acad. Sci. USA* **97**, 3718–3723.
8. Escobar, N. M., Haupt, S., Thow, G., Boevink, P., Chapman, S., and Oparka, K. (2003) High-thoughput viral expression of cDNA-green fluorescent protein fusions reveals novel subcellular addresses and identifies unique proteins that interact with plasmodesmata. *Plant Cell* **15**, 1507–1523.
9. Tian, G.W., Mohanty, A., Chary, S.N., et al. (2004) High-thoughput fluorescent tagging of full-length Arabidopsis gene products in planta. *Plant Physiol.* **135**, 25–38.
10. Koroleva, O. A., TomLinson, M. L., Leader, D., Shaw, P., and Doonan, J. H. (2005) High-thoughput protein localization in Arabidopsis using *Agrobacterium*-mediated transient expression of GFP-ORF fusions. *Plant J.* **41**, 162–174.
11. Ambard-Bretteville, F., Small, I., Grandjean, O., and Colas des Francs-Smal, C. (2003) Discrete mutations in the presequence of potato formate dehydrogenase inhibit the in vivo targeting of GFP fusions into mitochondria. *Biochem. Biophys. Res. Commun.* **311**, 966–971.
12. Mackenzie, S. A. (2005) Plant organellar protein targeting: a traffic plan still under construction. *Trends Cell Biol.* **15**, 548–554.
13. Draper, J., Scott, R., Armitage, P., and Walden, R. (1988) *Plant Genetic Transformation and Gene Expression, A Laboratory Manual*, Blackwell Scientific.

14. Gelvin, S.B. and Schilperoort, R. (eds.) (1995) *Plant Molecular Biology Manual*, 2nd ed., Kluwer Academic Publishers, Dordrecht.

15. Veluthambi, K., Gupta, A. K., and Sharma, A. (2003) The current status of plant transformation technologies. *Curr. Sci.* **84**, 368–380.

16. Wroblewski, T., Tomczak, A., and Michelmore, R. (2005) Optimization of *Agrobacterium*-mediated transient assays of gene expression in lettuce, tomato and *Arabidopsis. Plant Biotechnol. J.* **3**, 259–273.

17. Earley, K. W., Haag, J. R., Pontes, O., et al. (2006) Gateway-compatible vectors for plant functional genomics and proteomics. *Plant J.* **45**, 616–629.

18. Old, R.W. and Primrose, S. B. (1994) *Principles of Gene Manipulation: An Introduction to Genetic Engineering*, 5th ed. Blackwell Scientific.

19. Brown, T. A. (ed.) (1998) *Molecular Biology Labfax*. Academic Press Inc., London.

20. Brent, R., Kingston, R. E., Moore, D. D., Seidman, J. E., Smith J. A., and Struhl, K. (1999) *Short Protocols in Molecular Biology*, 4th ed., John Wiley, New York.

21. Sambrook, J. and Russell, D.W. (2001) *Molecular Cloning: A Laboratory Manual*. Cold Spring Harbor, New York.

22. Clough, S. J. and Bent, A. F. (1998) Floral dip: a simplified method for *Agrobacterium*-mediated transformation of *Arabidopsis thaliana. Plant J.* **16**, 735–743.

23. Martinez-Zapater, J. M. and Salinas, J. (eds.) (1998) *Arabidopsis Protocols*. Humana Press, Totowa, NJ.

24. An, G., Ebert, P. R., Mitra, A., and Ha, S. B. (1988) Binary Vectors, in *Plant Molecular Biology Manua.* (Gelvin, S. B. and Schilperoort, R. A., eds.), Kluwer Academic Publishers, Dordrecht, The Netherlands, A3,pp. 1–19

25. Haseloff, J. and Siemering, K. (1998) The uses of GFP in plants, in *Green Fluorescent Protein: Properties, Applications and Protocols* (Chalfie, M. and Kain, S., eds.) John Wiley, New York.

26. Batoko, H., Zheng, H.-Q., Hawes, C., and Moore, I. (2000) A Rab1 GTPase is required for transport between the endoplasmic reticulum and Golgi apparatus and for normal Golgi movement in plants. *Plant Cell* **12**, 2201–2218.

27. Hellens, R. P., Edwards, E. A., Leyland, N. R., Bean, S., and Mullineaux, P. M. (2000) pGreen: a versatile and flexible binary Ti Vector for *Agrobacterium*-mediated plant transformation. *Plant Mol. Biol.* **42**, 819–832.

28. Hajdukiewicz, P., Svab, Z. and Maliga, P. (1994) The small versatile pPZP family of *Agrobacterium* binary vectors for plant transformation. *Plant Mol. Biol.* **25**, 989–994.

29. Bevan, M. (1984) Binary *Agrobacterium* vectors for plant transformation. *Nucl. Acids Res.* **12**, 8711–8721.

30. Gleave, A.P. (1992) A versatile binary vector system with a T-DNA organisational structure conducive to efficient integration of cloned DNA into the plant genome. *Plant Mol. Biol.* **20**, 1203–1207.

31. Curtis, M. D. and Grossniklaus, U. (2003) A gateway cloning vector set for high-thoughput functional analysis of genes in planta. *Plant Physiol.* 133, 462–469.

32. Kurup, S., Runions, J., Köhler, U., Laplaze, L., Hodge, S., and Haseloff, J. (2005) Marking cell lineages in living tissues. *Plant J.* **42**, 444–453.

33. Boisnard-Lorig, C., Colon-Carmona, A., Bauch, M., et al. (2001) Dynamic analyses of the expression of the HISTONE::YFP fusion protein in Arabidopsis show that syncytial endosperm is divided in mitotic domains. *Plant Cell*, **13**, 495–509.

34. Runions, J., Thorsten, B., Kühner, S., and Hawes, C. (2005) Photoactivation of GFP reveals protein dynamics within the endoplasmic reticulum membrane, *J Exp. Bot.* **57**, 43–50.

17

Genetic Transformation
A Tool to Study Protein Targeting in Diatoms

Peter G. Kroth

Summary

Diatoms are unicellular photoautotrophic eukaryotes that play an important role in ecology by fixing large amounts of CO_2 in the oceans. Because they evolved by secondary endocytobiosis—a process of uptake of a eukaryotic alga into another eukaryotic cell—they have a rather unusual cell biology and genetic constitution. Because the preparation of organelles is rather difficult as a result of the cytosolic structures, genetic transformation and expression of preproteins fused to green fluorescent protein (GFP) became one of the major tools to analyze subcellular localization of proteins in diatoms. Meanwhile several groups successfully attempted to develop genetic transformation protocols for diatoms. These methods are based on "biolistic" DNA delivery via a particle gun and allow the introduction and expression of foreign genes in the algae. Here a protocol for the genetic transformation of the diatom *Phaeodactylum tricornutum* is described as well as the subsequent characterization of the transformants.

Key Words: Diatom; transformation; screening; targeting; microscopy.

1. Introduction

Diatoms are single-celled, sometimes colonial organisms belonging to the division Bacillariophycea. Almost all of them are photoautotrophic and can be found in nearly any aquatic and even in some terrestrial habitats *(1)*. Although they are a phylogenetically rather young group—they first appeared about 180 million years ago—there is a huge diversity of estimated some 10,000–100,000 different species in about 250 genera *(2)*. The frustules (extracellular silica cell walls) are formed by two valves that fit together like Petri dishes, are often highly ornamented, and show species-specific structures. The ability

From: *Methods in Molecular Biology, Vol. 390: Protein Targeting Protocols: Second Edition*
Edited by: M. van der Giezen © Humana Press Inc., Totowa, NJ

to genetically define these structures makes diatoms highly interesting for nanotechnological applications *(3)*. Diatoms also play a significant role in ecology. About half of the global annual net primary production in the oceans is to the result of phytoplankton, which is dominated by diatoms *(4)*.

Another peculiar aspect of diatoms is their evolution by secondary endocytobiosis. In this process, eukaryotic algae have been taken up by host cells and were transformed into plastids. Because of redundancy, nearly all of the cytoplasmic structures of the endosymbiotic algae have vanished—usually including the nuclear genome—while the plastids have been preserved; because of their ability to perform photosynthesis they were probably highly attractive for the host cell *(5)*. Diatom plastids are also termed "complex" plastids, because in comparison to land plant plastids they possess two additional surrounding membranes probably originating from the phagotrophic membrane of the host and the plasma membrane of the endosymbiont. Because of gene transfer processes during secondary endocytobiosis, diatom nuclei may possess genes of both primary and secondary host cells as well as of the plastid. The recent sequencing projects exploring the genome of the diatoms *Thalassiosira pseudonana (6)* and *Phaeodactylum tricornutum (7)* now allow us to investigate the genes in all of these processes, while genetic transformation enables the overexpression and modification of diatom genes in vivo. Meanwhile, methods are available for different diatoms *(8–12)* that differ mainly in the use of different promoters and reporter genes but not in the method itself. Because *Phaeodactylum* is mainly utilized for genetic transformation, the respective protocol will be described here. It is also shown how the green fluorescent protein (GFP) may serve as marker protein for subcellular localization of fusion proteins in diatoms *(13,14)*.

2. Materials

2.1. Cell Culture

1. *Phaeodactylum tricornutum*, strain UTEX646 (available at UTEX Culture Collection of Algae, University of Texas, Austin, TX, http://www.bio.utexas. edu/research/utex/). Store on agar plates as described here at 22°C with continuous illumination. Replate on fresh plates all 4 wk.
2. f/2 culture medium *(15)* at a concentration of 50% sea water, dissolve 16.6 g of sea salt (Tropic Marin™, Dr. Biener, Wartemberg) in distilled water or filtered sea water (diluted to 50% with distilled water), adjust pH to 7.0, if necessary. Autoclave and before use add 0.5 mL f/2 vitamin solution, 1 mL 2 M Tris-HCl (ultrapure, pH 8), 1 mL NaNO$_3$ (75 g/L), 1 mL NaH$_2$PO$_4 \cdot$H$_2$O (5 g/L), 1 mL trace metals solution

by first mixing all added solutions and filtering them through a sterile 0.2-μm filter directly into the culture medium. Store at 4°C.

For the f/2 vitamin solution, first prepare stock solutions: vitamin B_{12}: 1 mg/mL, biotin: 1 mg/10 mL. Sterilize by filtering through 0.2-μm disposable filter and freeze aliquots. Prepare the f/2 vitamin solution by dissolving 20 mg thiamine-HCl in 50 mL of water, add 100 μL of the vitamin B_{12} stock solution and 1 mL of the biotin stock solution, fill up with sterile water to 100 mL. Store at 4°C.

For the trace metal solution, first prepare stock solutions: Cu $SO_4 \cdot 5\ H_2O$ (1 g/100 mL), $ZnSO_4 \cdot 7\ H_2O$ (2.2 g/100 mL), $CoCl_2 \cdot 6\ H_2O$ (1 g/100mL), $MnCl_2 \cdot 4\ H_2O$ (1.8 g/100 mL), $NaMoO_4 \cdot 2\ H_2O$ (0.63 g/100 mL). Then dissolve 0.436 g Na_2-EDTA and 0.315 g $FeCl_3 \cdot 6\ H_2O$ in water, add 100 μL of each stock solution, fill up to 100 mL. Store at 4°C.

3. Similarly prepare f/2 plates by adding 2% Bacto-Agar (Becton, Dickinson and Co., Heidelberg, Germany) before autoclaving and add metals and vitamins after cooling to 55°C, mix carefully without introducing bubbles, and pour plates. If the plates are needed for selection, add Zeocin (Invitrogen, Karlsruhe, Germany) to a final concentration of 75 μg/L before pouring. The plates can be kept at 4–8°C for 3–4 wk in the dark.

2.2. Constructs

1. Prepare pPha-T1 vector *(9)* (Genbank acc. no. AF219942) from *E. coli* cells according to standard plasmid preparation protocols. You can use various restriction sites to clone the gene of interest into the vector. pPha-T1 or related vectors with other promoters are not commercially available, but can be obtained by the author or by other groups *(9,16,17)*. To analyze protein targeting you can decide either to fuse the DNA fragment encoding the respective N-terminal presequence to the GFP gene or to fuse the gene encoding the whole preprotein. This can be helpful if the mature protein also contains some targeting information, like for instance membrane anchors in membrane proteins (see **Notes 1** and **2**).

2.3. Biolistic Transformation

1. Tungsten particles. Weigh 60 mg of particles (M10 size, BioRad, Hercules, CA) into a sterile 1.5-mL reaction tube, add 1mL ethanol (of high purity) and vortex for 3–5 min. Spin down in a microcentrifuge at maximum speed for 1 min. Discard supernatant and add 1 mL of distilled water and repeat resuspension and centrifugation two times. Then resuspend particles in 1 mL of water and prepare aliquots of 50 μL in fresh tubes. Avoid sedimentation of the particles during this step by frequent vortexing. Store the aliquots at −20°C for further use.
2. 0.1 *M* Spermidin (Sigma-Aldrich, culture grade), store 200-μL aliquots at −20°C.
3. 2.5 *M* $CaCl_2$, store aliquots of 1 mL at −20°C.
4. Grow *Phaeodactylum* cells (*see* **Notes 3–5**) by resuspending some cells from a plate in 50 mL f/2 media in a 200-mL Erlenmeyer flask with a sterile top that allows gas

exchange for up to 7 d on a shaker at 22°C at a light intensity of 50 μmol photons per m^2/s until the culture is brownish. Spin down the cells at room temperature at 3000*g* for 10 min, and dissolve in a small amount of culture medium.

5. Thoma counting chamber and microscope for counting the cells. Estimate the cell density and adjust to a density of 1×10^9 cells per mL with culture medium. Carefully resuspend the cells and place 100 μL of the suspension in the center of an f/2 agar plate (without antibiotic) and spread the cells to a circle with about 3 cm in diameter. Put the plate with an open lid into the sterile chamber (*see* **Note 6**).

2.4. Particle Bombardment

1. BioRad PDS-1000/He Particle Delivery System.
2. Helium supply with a minimum pressure of 1500 psi.
3. Rupture discs 1350 psi (BioRad).
4. Stopping screens (BioRad).
5. Isopropanol.
6. Ethanol.
7. Laminar flow hood or sterile bench that is large enough for the biolistic device.
8. Vacuum pump.

2.5. Fluorescence Analysis

1. Epifluorescence microscope with a digital camera or confocal microscope for fluorescence analysis. Filter sets should be optimized for GFP fluorescence, but they should clearly separate the GFP fluorescence from the chlorophyll autofluorescence. Narrow-band GFP filter sets are available for this, but because they decrease the GFP signal substantially, they are often not included in standard epifluorescence filter sets.
2. Microscope slides and cover slips.
3. Immersion oil (for fluorescence microscopy) for stronger magnification.

2.6. Western Blot Analysis

1. French Press.
2. Isolation buffer: 50 m*M* Tris-HCl (pH 7), 1 m*M* phenylmethylsulfonyl fluoride (PMSF).
2. Kit for determining the protein concentrations (according to Bradford or others).
3. 5X sample buffer: 125 m*M* Tris-HCl (pH 8.0), 10% (w/v) socium dodecyl sulfate (SDS), 25% (v/v) glycerol, 25% (v/v) β-mercaptoethanol, 0.025% (w/v) bromophenol blue.
3. Stacking gel: acrylamide (30% (w/v)), 200 m*M* Tris/PO$_4$ (pH 6.7), 0.5% (w/v) SDS, 0.1% (v/v) TEMED, 0.15% (w/v) ammonium peroxodisulfate.
4. Separating gel: acrylamide (30% (w/v)), 1.875 *M* Tris-HCl (pH 8.9), 0.5% (w/v) SDS, 0.05% (v/v) TEMED, 0.07% (w/v) ammonium peroxodisulfate.
5. Running buffer: 0.25 *M* Tris-HCl, 1.92 *M* glycine, 0.5% (w/v) SDS.

6. Phosphate-buffered saline (PBS): $137 \, mM$ NaCl, $2.7 \, mM$ KCl, $8 \, mM$ Na$_2$HPO$_4$, $1.8 \, mM$ KH$_2$PO$_4$, 0.02% (v/v) Tween 20, pH 7.4. For blocking membranes include 5% (w/v) milk powder.
7. Primary antiserum: anti-GFP (Molecular Probes, Leiden, the Netherlands, or Becton, Dickinson and Co).
8. Secondary antiserum: antigoat IgG conjugated to horseradish peroxidase (Roche, Mannheim, Germany).
9. BM Chemoluminescence Kit (Roche) or any other supplier.
10. Nitrocellulose membrane.
11. Ponceau-S: 0.5% (v/v) in 1% (v/v) acetic acid.

3. Methods

The biolistic procedure of genetic transformation of diatoms is based on the partial damage of cells by biolistic bombardment with tungsten particles that are loaded with DNA. This way the DNA may enter the cells and might be transported into the nucleus, where it may integrate randomly into the nuclear genome. It is assumed that it is important that the cells are only slightly damaged, so that they can recover and seal the membranes again. As the transformation vector results in expression of a resistance protein, selection may occur by plating the cells on selective media containing antibiotics. Various vectors that use different antibiotic resistances and different promoters to drive the expression of the gene of interest are meanwhile available for diatom transformation.

3.1. Preparation of Particles for Biolistic Bombardment

1. Thaw and resuspend thoroughly an aliquot of the tungsten particles.
2. Pipet the following solutions into a 1.5-mL reaction tube: $50 \, \mu L$ resuspended tungsten particles, $5 \, \mu g$ plasmid DNA (in 5–$10 \, \mu L$ water), $50 \, \mu L$ $2.5 \, M$ CaCl$_2$, $20 \, \mu L$ $0.1 \, M$ spermidine. Vortex carefully, then let the particles sediment for 10 min. Additionally, you may also centrifuge the particles down very carefully (3 s at maximum $2000g$). Remove supernatant, then add $250 \, \mu L$ pure ethanol and vortex again.
3. Again, let the particles sediment for 3–5 min, then remove supernatant.
4. Add $50 \, \mu L$ ethanol to the pellet and resuspend. Particles should be used within the next one hour.

3.2. Preparation of Biolistic Device for Bombardment

1. Place biolistic device in the hood or on the sterile bench. Wipe the device on the outside and in the chamber with ethanol. Connect to helium supply and vacuum pump according to instructions.

2. Dip a rupture disc shortly in isopropanol and install the assembly within the device as described by the manufacturer.
3. For cleaning the helium lines, fire one shot without cells and particles. Again, wipe interior with ethanol.
4. Place macrocarriers into macrocarrier holders (metal rings), place them for a short time in ethanol, and allow them to dry completely on a sterile surface on the working bench.
5. Vortex the particles thoroughly and place 10 μL in the middle of a macrocarrier and let them dry. Repeat for all DNA/particle mixtures you want to transform that day. During this step the ventilation of the working bench should be switched off to prevent clogging of particles because of vibrations.
6. After the ethanol has evaporated, install macrocarrier together with a stopping screen and a wet isopropanol-soaked rupture disc into the assembly, fetch a prepared agar plate with the diatoms, remove the lid, and place the plate in the respective distance onto the tray in the chamber. Close door.
7. Deflate air in the chamber by pressing the vacuum switch until you have a vacuum of 25 psi, then press "Hold." Press the helium button and wait until the rupture disc bursts and the tungsten particles are accelerated onto the diatom cells.
8. Release the vaccum, take out the agar plate, and close the lid.
9. Repeat the shooting steps until all prepared macrocarriers are used up.
10. Seal the agar plates with parafilm and place the agar plates with the cells in the culture room for 24 h to recover and to start expression of the *sh ble* resistance protein (against Zeocin) (*see* **Note 7**).

3.3. Replating and Selection

1. Take agar plates from the culture room and wash the cells off the plate by repeatedly pipetting 500–1000 μL of culture medium on the spot of algae, holding the plate on a slightly tilted position. You may also use a sterile spatula to remove cells sticking to the surface of the agar plate. Remove the medium together with the cells and transfer into a reaction tube.
2. Take four fresh agar plates containing Zeocin and pipet one-fourth of the resuspended diatoms onto each plate. Spread the cells carefully with a Drigalski spatula or any other sterile device for plating bacteria (*see* **Notes 8** and **9**).
3. Seal the plates with parafilm to avoid dessication and incubate them for 2–3 wk in the culture chamber at 22°C at 50 μmol photons per m^2/s (*see* **Note 10**). You may put several plates on each other, but make sure the light intensity is enough, otherwise the resistance protein will not be expressed sufficiently and hence also the transformants will eventually die (*see* **Note 11**).
4. After 2–4 wk (*see* **Notes 12** and **13**) sharp-edged, dark brownish colonies should be visible against an empty background of dead cells (*see* **Notes 14** and **15**). Number these colonies and transfer cells from these colonies on fresh plates containing Zeocin. Cells that survive replating definitely express the resistance protein and in many cases also the respective reporter gene (*see* **Note 16**).

3.4. Fluorescence Microscopy

1. For fluorescence analysis scratch some cells from an agar plate and incubate them in a small volume of f/2 medium in the light on a shaker. You can either use reaction tubes or 8-, 16-, or 96-well reaction plates, depending on the amount of cells you need or the number of strains you want to analyze. You can also use cells directly from plates, but depending on the growth phase, GFP fluorescence can be very variable. After growth for 1 or 2 d, take the cells for microscopy. If you need a denser cell suspension you may also carefully spin them down and resuspend them in a smaller volume.
2. You can observe the cells by light microscopy, by epifluorescence microscopy, or by confocal laser scanning microscopy. Fluorescence of GFP is usually unstable under UV light, and therefore it is helpful first to identify cells or groups of cells by transmitted light microscopy and then to switch to fluorescence excitation. Include observations of wild-type cells to know the level of background fluorescence, especially from the chlorophylls. Depending on your imaging system you can take pictures of the transmitted light view and of the GFP—and red chlorophyll—fluorescence separately and combine the images afterwards (**Fig. 1**).

3.5. Western Blot Analysis

A good indicator for successful protein targeting into plastids is the cleavage of the complete presequence after plastid import. To visualize this process you may label the protein with antisera against the GFP domain of the fusion protein and measure the respective size on an SDS gel. Because the size differences can be rather small (1.5–10 kDa), it is reasonable to achieve a good separation by using high percentage polyacrylamide gels on large gel systems. As a standard for mature GFP you can order overexpressed GFP from Molecular Probes or other suppliers or use a *Phaeodactylum* transformant that expresses GFP without any presequence.

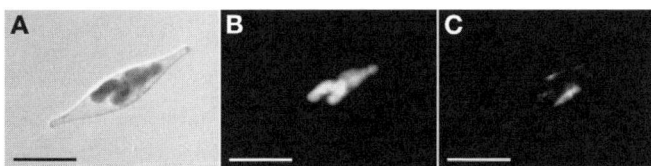

Fig. 1. Transformant of *Phaeodactylum tricornutum* expressing green fluorescent protein (GFP) within the cytosol. (**A**) Transmitted light microscopy. The plastids are dark brownish. (**B**) Chlorophyll autofluorescence within the plastids. (**C**) GFP fluorescence within the cytosol. Note that GFP is not seen within the areas occupied by the plastids. Other GFP-free areas within the cytoplasm are the result of vacuoles. Bars = 10 μm. (The image was taken by Ansgar Gruber, University of Konstanz.)

1. Grow the cells in 25 mL f/2 medium in 100-mL Erlenmeyer flasks with a sterile top that allows gas exchange. A dense culture will be spun down at 4°C for 15 min at 5000g and the pellet is dissolved in 2 mL of isolation medium. The cells can be broken in a French Press (or similar systems) in a precooled cell at a pressure of 110 MPa with a flow rate of 2 drops/min. Spin down the cell lysate at 14, 000g at 4°C for 30 min and transfer the supernatant into a fresh tube. Use one of the standard assays for protein determination.
2. Add 5x SDS sample buffer to the lysate.
3. Prepare a 1.5-mm-thick 15% SDS gel by mixing 32 mL separating gel buffer with 40 mL acrylamide/bis-acrylamide solution, 7.7 mL water, 260 μL ammonium peroxodisulfate solution, and 40 μL TEMED. Pour gel, leave space for the stacking gel, and overlay with ethanol. The gel should polymerize in 30–60 min, depending on the size.
4. Prepare a stacking gel by mixing 6 mL stacking gel buffer with 5 mL acrylamide/bis-acrylamide solution, 18.8 mL water, 200 μL ammonium peroxodisulfate solution, and 30 μL TEMED. Rinse ethanol from gel, rinse with water and pour stacking gel quickly. Insert comb.
5. After polymerization remove comb, complete the assembly of the gel unit, and connect to power supply. Load samples and let the gel run either overnight at low voltage or during the day depending on the size of the gel.
6. After the run, remove gel from unit and eventually cut off lanes with markers. Transfer the proteins from the gel onto a nitrocellulose (or PVDF) membrane using either a semi-dry blotting apparatus or a blotting tank according to the recommendations of the manufacturer.
7. After the blot is complete stain the membrane with a 1% solution of Ponceau Red (only for nitrocellulose) to see whether transfer is complete or whether air bubbles might have been trapped. If you are not using prestained markers, label the marker bands with a ball pen. Wash off the Ponceau Red by rinsing the membrane with water.
8. Label the bands with primary and secondary antisera according to the recommendation of the manufacturer.

Notes

1. Because of the codon usage of *Phaeodactylum,* only the enhanced GFP version optimized for the human codon usage is expressed sufficiently in *Phaeodactylum.* This has to be taken into consideration especially for expression of foreign genes in *Phaeodactylum.* Several codons are present in a range of 5–10%, and this can drastically reduce the expression of the respective gene. A codon usage table to identify rare codons in *Phaeodactylum* is available in *(18)*.
2. Cotransformation of two different constructs may be obtained simply by mixing the DNA of two different constructs before bombardment. Both vectors may carry

the same resistance gene (e.g., *sh ble* against Zeocin). In about 50% of the obtained transformants both reporter genes will be expressed simultaneously.

3. Make sure that the diatom cultures you use for transformation are sterile. After bombardment the cells will be kept on plates at about room temperature for several weeks. Because Zeocin is not stable in the light, bacterial growth may be obtained after some time.

4. Keep stock cultures of *Phaeodactylum* on plates at low light and replate every 4 wk. If you encounter bacterial infections, you can plate the cells on plates containing 50 µg/mL kanamycin. At this concentration diatom growth is not affected, while bacteria usually die. Fungal infection of the stock strains is more severe, as there are no known treatments that affect fungi only. The only way to get rid of fungi is to replate the cells repeatedly.

5. The stock cells used for transformation should be selected to be fusiform and nonmotile. From time to time they tend to revert to oval cell shape and motility. In this case cells should be plated in low density to select single cells as stares for a new stock culture.

6. It has been reported recently by Bowler et al. (personal communication) and confirmed in our lab that plating the cells on agar plates for bombardment 1 d in advance may increase the transformation rates more than 10-fold compared to cells directly plated before transformation. A possible reason for this effect is the division of the cells on the plate, putting the cells in a state in which membrane regeneration might be more easily feasible.

7. If replating the cells after transformation, make sure that you are not plating too many cells on a single plate; otherwise you might obtain false positives after antibiotic screening.

8. After bombardment do not leave the cells on the agar plates without antiobiotics for longer than 24 h.

9. Always replate colonies obtained from Zeocin screening after transformation on fresh plates with the antibiotic to remove false positives.

10. For some approaches it might be helpful to reduce the intracellular GFP concentration. As the fcp promoter is light-driven, keeping the cells in the dark for several days results in a complete degradation of GFP and a new expression after several hours of light. *Phaeodactylum* easily survives longer dark periods of up to weeks.

11. If you do not obtain transformants, check the light spectra of the light bulbs in your culture room to match the requirements of the fcp promoters (for an action spectrum, *see* **ref. (8)**).

12. You may keep stocks and transformants on plate for up to 4 wk until replating is necessary. Transformants will be stable for generations without selection pressures, however, from time to time we replate them on Zeocin to avoid strain loss. It turned out to be easiest to keep only important transformants on plate until the end of the experiment and to store the DNA to be able to transform cells again if necessary.

13. If a plate should dry for any reason, it is sometimes possible to dissolve "dried" colonies by mixing them with 10–20 μL f/2 medium followed by replating on a fresh agar plate.

14. Because integration of foreign DNA occurs randomly, you will obtain a set of transformants with rather low and rather high expression of the gene of interest. This can be very helpful when you are interested in expression of a gene that in high concentration might be lethal to the cell.

15. If colonies obtained from transformation give a hard and leathery appearance, they are usually false positives that do not recover after replating on antibiotic.

16. Transformations rates can vary strongly from one experiment to the next.

Acknowledgements

The author would like to thank Ansgar Gruber and Doris Ballert for careful reading of the manuscript and helpful suggestions. This work was supported by the German Research Foundation (DFG, project Kr 1661/3-1), the European Community (MARGENES, contract QLRT-2001-01226), and the University of Konstanz.

References

1. Lee, R. E. (1989) Phycology, 2nd ed., Cambridge University Press, Cambridge, UK.

2. Norton, T. A., Melkonian, M., and Andersen, R. A. (1996) Algal biodiversity. *Phycologia* **35**, 308–326.

3. Drum, R. W. and Gordon, R. (2003) Star Trek replicators and diatom nanotechnology. *Trends Biotechnol.* **21**, 325–328.

4. Falkowski, P. G., Katz, M. E., Knoll, A. H., et al. (2004) The evolution of modern eukaryotic phytoplankton. *Science* **305**, 354–360.

5. Delwiche, C. F. and Palmer, J. D. (1997) The origin of plastids and their spread via secondary symbiosis. *Plant Syst. Evol.* **11**, 53–86.

6. Armbrust, E. V., Berges, J. A., Bowler, C., et al. (2004) The genome of the diatom *Thalassiosira pseudonana*: ecology, evolution, and metabolism. *Science* **306**, 79–86.

7. Maheswari, U., Montsant, A., Goll, J., et al. (2005) The Diatom EST Database. *Nucleic Acids Res.* **33**, D344–D347.

8. Falciatore, A., Casotti, R., Leblanc, C., Abrescia, C., and Bowler, C. (1999) Transformation of nonselectable reporter genes in marine diatoms. *Marine Biotechnology* **1**, 239-251.

9. Zaslavskaia, L. A., Lippmeier, J. C., Kroth, P. G., Grossman, A. R., and Apt, K. E. (2000) Transformation of the diatom *Phaeodactylum tricornutum* (Bacillariophyceae) with a variety of selectable marker and reporter genes. *J. Phycol.* **36**, 379–386.

10. Poulsen, N. and Kröger, N. (2005) A new molecular tool for transgenic diatoms: control of mRNA and protein biosynthesis by an inducible promoter-terminator cassette. *FEBS J.* **272**, 3413–3423.

11. Dunahay, T. G., Jarvis, E. E., and Roessler, P. G. (1995) Genetic transformation of the diatoms *Cyclotella cryptica* and *Navicula saprophila*. *J. Phycol.* **31**, 1004–1012.

12. Poulsen, N., Chesley, P. M., and Kröger, N. (2006) Molecular genetic manipulation of the diatom *Thalassiosira pseudonana* (Bacillariophyceae). *J. Phycol.* **42**, 1059–1065.

13. Kilian, O. and Kroth, P. G. (2005) Identification and characterization of a new conserved motif within the presequence of proteins targeted into complex diatom plastids. *Plant J.* **41**, 175–183.

14. Tanaka, Y., Nakatsuma, D., Harada, H., Ishida, M., and Matsuda, Y. (2005) Localization of soluble β-carbonic anhydrase in the marine diatom *Phaeodactylum tricornutum*. Sorting to the chloroplast and cluster formation on the girdle lamellae. *Plant Physiol.* **138**, 207-217.

15. Guillard, R. R. L. (1975) Culture of phytoplankton for feeding marine planktonic diatoms, in *Culture of Marine Invertebrate Animals* (Smith, W. L. and Chanley, M. H., eds.), Plenum Press, New York, pp. 26–60.

16. Poulsen, N. and Kröger, N. (2005) A new molecular tool for transgenic diatoms: control of mRNA and protein biosynthesis by an inducible promoter-terminator cassette. *FEBS J.* **272**, 3413–3423.

17. Scala, S., Carels, N., Falciatore, A., Chiusano, M. L., and Bowler, C. (2002) Genome properties of the diatom *Phaeodactylum tricornutum*. *Plant Physiol.* **129**, 993–1002.

18. Montsant, A., Jabbari, K., Maheswari, U., and Bowler, C. (2005) Comparative genomics of the pennate diatom *Phaeodactylum tricornutum*. *Plant Physiol.* **137**, 500–513.

18

Identification of Proteins Targeted Into the Endoplasmic Reticulum by cDNA Library Screening

Takeaki Ozawa and Yoshio Umezawa

Summary

Protein targeting from cytosol into the endoplasmic reticulum (ER) in mammalian cells is an initial step in the biogenesis of most secretory and membrane proteins as well as the proteins localized in the ER, Golgi, and lysosomes. Identification of these proteins is crucial for understanding this biological process, which varies in different mammalian cell types. To identify ER-targeted proteins, we have developed a method for high-throughput screening of genes that encode proteins transported into the ER in living mammalian cells. The principle is based on the reconstitution of two fragments of split green fluorescent protein (GFP) by protein splicing and retrovirus-mediated expression cloning. The method is able to collect ER-targeted proteins systematically and to accurately identify many novel proteins. This method will facilitate understanding of the secretory pathway and intercellular communication in living mammalian cells.

Key Words: Green fluorescent protein; intein; protein splicing; endoplasmic reticulum; cDNA library; fluorescence-activated cell sorting.

1. Introduction

Secreted and cell-surface proteins play important roles in cellular inter-actions and are potential therapeutic targets for agonistic and antagonistic reagents *(1,2)*. These proteins are synthesized as precursors that are charac-terized by a short N-terminal polypeptide known as the ER signal sequence *(3–5)*. The signal sequence itself contains several elements required for optimal function: a hydrophobic core of 4–15 amino acids in length, basic amino acid residues preceding the hydrophobic core, and a pair of uncharged amino acids

From: *Methods in Molecular Biology, Vol. 390: Protein Targeting Protocols Second Edition*
Edited by: M. van der Giezen © Humana Press Inc., Totowa, NJ

at the C-terminal end for cleaving the signal sequence. The sequence serves to direct a nascent polypeptide chain to the cellular secretory pathway and to mediate its translocation across lipid bilayers. Although there are many programs to predict the ER signal sequences *(6–8)*, it is hard to accurately identify ER-targeted proteins by searching in DNA or protein databases for ER signal sequences. Therefore, experimental identification of the signal sequences is crucial for accurate determination of distinct protein compositions in the secretory pathway. Up to now, a majority of localization studies have been undertaken in yeast, primarily as a result of the ease of generating proteins fused to green fluorescent protein (GFP) or an epitope for antibody *(9,10)*. A list of protein localizations in yeast has thus been compiled and used for proteome analysis. For mammalian cells, however, the GFP- or epitope-tagged approach is time consuming for systematic analysis.

Expression cloning is another valuable method for isolating proteins targeted to the ER *(11,12)*. This consists of transfection of cDNA library fragments fused to a reporter and expression of the corresponding proteins inside cells. The presence of the reporter in the extracellular space is a marker for the cDNA fragment that encodes an ER signal sequence. The need to select cells greatly limits the usefulness and throughput of the method, because selection is achieved either by limiting dilution or colony picking.

Here we describe a genetic method for high-throughput screening of genes that encode proteins targeted into the ER in living mammalian cells *(13)*. The principle is based on the reconstitution of split enhanced GFP (EGFP) fragments by protein splicing with a DnaE intein derived from *Synechocystis* sp. PCC6803 *(14)*. The DnaE intein has a characteristic feature of natural splicing ability to ligate accompanying protein fragments *in trans (15)*. The basic scheme is shown in **Fig. 1**. A tandem fusion protein containing an ER-targeting signal (ERTS) and C-terminal fragments of DnaE and EGFP localizes in the lumen of the ER in living mammalian cells. cDNA libraries generated from mRNAs are genetically fused to the sequences encoding the N-terminal halves of EGFP and DnaE. The cDNAs are converted into retrovirus libraries and used to infect the mammalian cells. If test proteins expressed from the cDNA libraries contain a functional ERTS, the fusion products translocate into the ER and bring the N- and C-terminal halves of DnaEs close enough to fold correctly, thereby initiating protein splicing to link the concomitant EGFP halves with a peptide bond. The cells harboring reconstituted EGFP are screened and collected by fluorescence-activated cell sorting (FACS) (**Fig. 2**). From each clone thus collected, cDNA is retrieved by polymerase chain reaction (PCR) and its sequence is analyzed. Thousands of cDNAs can be identified by this single experimental procedure.

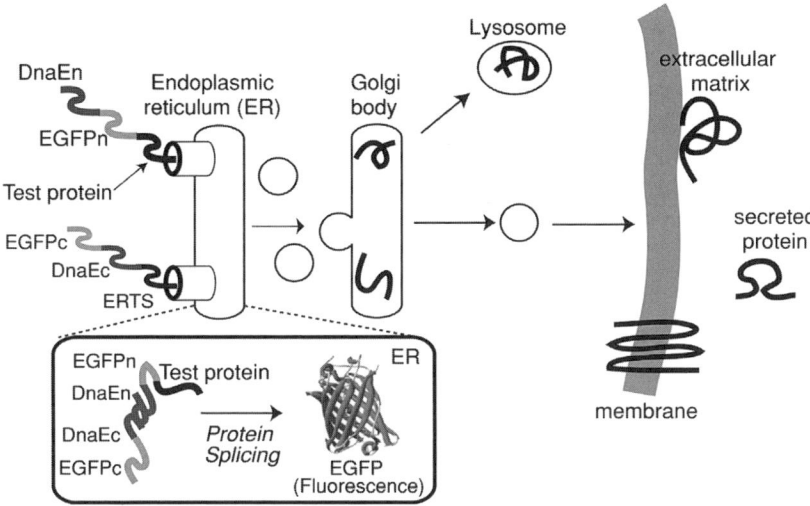

Fig. 1. Scheme of the basic principle for detecting translocation of a test protein into the endoplasmic reticulum (ER) using protein splicing of split EGFP. C-terminal EGFP is connected with C-terminal DnaE and ERTS, which is predominantly localized in the ER. A test protein is connected with N-terminal halves of EGFP and DnaE, which is expressed in the cytosol. When the test protein translocates into the ER, the N-terminal DnaE interacts with the C-terminal one, resulting in protein splicing. The N- and C-terminal EGFPs are linked together by a peptide bond, and the reconstituted EGFP recovers its fluorescence. With this system, secreted proteins and proteins localized in the ER, Golgi body, lysosome, membrane, and extracellular matrix can be identified. ERTS, an ER-targeting signal derived from mouse preprolactin; EGFPn, N-terminal half (1–157 aa) of EGFP; EGFPc, C-terminal half (158–238 aa) of EGFP; DnaEn and DnaEc, N- and C-terminal DnaEs.

2. Materials

2.1. Plasmids and Cells

1. Plamids: pMX-ERTS/DEc(Neo), pMX-ER/LIB-ERTS, pMX-ER/LIB-CaM, pMX-ER/LIB (*see* **Note 1**) (**Fig. 3**). Retrovirus packaging cell line, PlatE cells (*see* **Note 2**) *(16)*.

2.2. Cell Culture, Infection, and Transfection

1. Dulbecco's modified Eagle's medium (DMEM) (Gibco/BRL, Gaithersburg, MD) supplemented with 10% fetal bovine serum (FBS; Gibco/BRL).

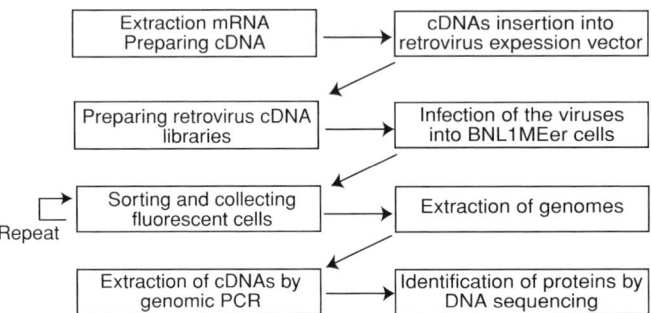

Fig. 2. Strategy for identifying endoplasmic reticulum (ER)-targeting proteins. cDNA libraries, each of which is connected with EGFPn and DnaEn, are prepared from mRNAs and converted into retrovirus libraries. Cultured mammalian cells including EGFPc-DnaEc (BNL1MEer cells) are infected with the retrovirus libraries with 20% infection efficiency. Fluorescent cells are sorted by FACS and collected on 48-well plates. cDNAs integrated in the genome are extracted by PCR and their sequences are identified by DNA sequencing.

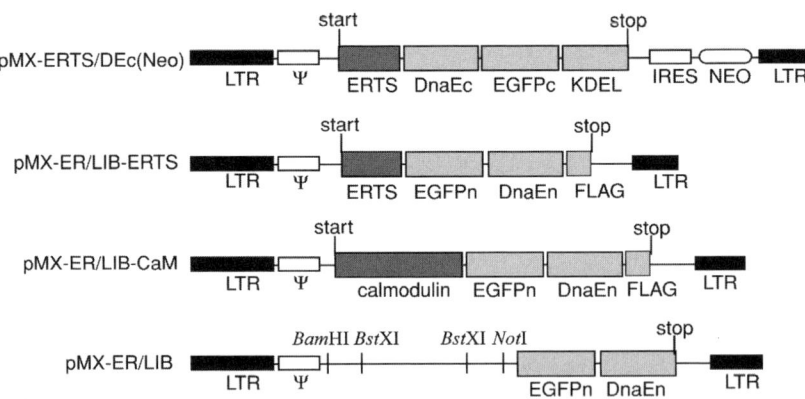

Fig. 3. Schematic structures of the plasmids. ERTS, an ER-targeting signal derived from mouse preprolactin; EGFPn, N-terminal half (1–157 aa) of EGFP; EGFPc, C-terminal half (158–238 aa) of EGFP; DnaEn and DnaEc, N- and C-terminal DnaEs; IRES, internal ribosome entry site; NEO, neomycin resistance; LTR, long terminal repeat; Ψ, retrovirus-packaging signal; FLAG, FLAG epitope (DYKDDDDK).

2. Solution of trypsin (0.25%) (Gibco/BRL).
3. Geneticin liquid (50 mg/mL) (Invitrogen, Carlsbad, CA).
4. OPTI-MEM medium (Gibco/BRL).
5. Lipofectamine 2000 (Invitrogen).
6. Phosphate-buffered saline (PBS; Sigma, St. Louis, MO).
7. Polybrene (hexadimethrine bromide; Sigma) dissolved at 10 μg/mL in sterilized water. Store in aliquot at −30°C.

2.3. Construction of the cDNA Library

1. FastTrack kit (Invitrogen).
2. SuperScript Choice System (Invitrogen).
3. SuperScript III RT (Invitrogen).
4. *Bst*XI adaptor (Invitrogen) dissolved in DEPC-treated water (1 mg/mL).
5. cDNA Size fractionation columns (Invitrogen) equilibrated with TEN buffer (*see* **step 9**) at room temperature before use.
6. Phenol:chloroform:isoamyl alcohol (25:24:1) (Sigma).
7. 47.5 M ammonium actate (NH_4OAc).
8. 70% (v/v) Ethanol stored at −30°C.
9. TE buffer: 10 mM Tris-HCl (pH7.5), 1 mM ethylenediamine tetraacetic acid (EDTA).
10. TEN buffer: 10 mM Tris-HCl (pH7.5), 0.1 mM EDTA, 25 mM NaCl.
11. MAX Efficiency DH10B Competent Cells (Invitrogen).
12. Qiagen Maxi Prep Kit (Qiagen, Hilden, Germany).

2.4. Sorting and Identification of cDNA

1. Cell strainer, 40 μm Nylon (BD Falcon, Bedford, MA).
2. Wizard Genomic DNA Purification Kit (Promega, Madison, WI).
3. LA Taq polymerase (Takara, Shiga, Japan).
4. BigDye Terminator Cycle Sequencing Kits (Applied Biosystems, Foster City, CA).
5. PCR primers for cDNA extraction; primer AGGACCTTACACAGTCCT-GCTGACC (forward) and GCCCTCGCCGGACACGCTGAACTTG (reverse).
6. 310 or 3100 Genetic analyzer (Applied Biosystems).

2.5. Antibodies for Western Blot

1. Mouse monoclonal anti-GFP antibody (Roche Applied Science, Mannheim, Germany). Alkaline phosphatase-conjugated secondary antibody (Jackson, West Grove, PA).
2. Wizard Genomic DNA Purification Kit (Promega).
3. LA Taq polymerase (Takara).

3. Methods

3.1. Preparation of the Cells that Include C-Terminal Half of EGFP in the ER

We describe here a method for preparing a cell line of mouse liver cells, BNL1ME, as an example. The method can be applied to any kind of cell of your choice, provided that the cells can be sorted with a FACS.

1. Seed BNL1ME cells (1×10^6 cells) onto a 6-cm dish 1 d before the transfection, and incubate the cells at 37°C in a CO_2 incubator.
2. Prepare the following solutions in 1.5-mL tubes: solution A, 2 μg of pMX-ERTS/DEc(Neo) in 100 μL of OPTI-MEM medium; and solution B, 5 μL of lipofectamine 2000 in 100 μL of OPTI-MEM medium. Incubate the mixtures at room temperature for 5 min.
3. Add solution B to solution A, vortex gently, and incubate at room temperature for 20 min.
4. During the incubation, replace the medium in the 6-cm dish into a fresh DMEM medium containing 10% of FCS.
5. Add the mixture of solution A and solution B in the 6-cm dish.
6. Incubate the cells on the dishes for 2 d at 37°C in a CO_2 incubator.
7. Remove the medium including the transfection mixture, and add a DMEM medium containing 0.5 mg/mL geneticin and 10% of FCS.
8. Incubate the cells for 5 d at 37°C in a CO_2 incubator.
9. Strip the cells with 500 μL of the trypsin solution, dilute the suspension with a DMEM medium containing 0.5 mg/mL geneticin and 10% of FCS, and then seed the cells onto 10-cm dishes. Adjust 20–50 cells per 10-cm dish.
10. Incubate the cells until the colony of the cells can be observed.
11. Pick up 20 colonies with a pipet tip, and again proliferate each cloned cell up to confluent in 6-cm dishes.
12. Examine the expression of the C-terminal EGFP connected with DnaE by Western blot with the GFP antibody (*see* **Note 3**), and select a cell line that expressed the protein in the ER (*see* **Note 4**). Hereafter this cell line is called BNL1MEer cells.

3.2. cDNA Library Construction

1. Prepare poly(A)+RNA using a FastTrack kit according to the manufacturer's protocol.
2. Add 0.1 μg of random hexamers to a sterile 1.5-mL microcentrifuge tube.
3. Add 5 μg of mRNA to the tube and adjust the total volume to 8 μL using DEPC-treated water.
4. Heat the mixture at 70°C for 3 min, and quickly chill on ice.
5. Add 4 μL of 5X first strand buffer, 2 μL of 0.1 *M* dithiothreitol (DTT), and 1 μL of 10 m*M* dNTP mix.

6. Mix the contents by gentle vortex, and incubate the tube at 37°C for 2 min to equilibrate the temperature.
7. Add 5 μL of Superscript III RT (total volume of the reaction mixture is 20 μL) (*see* **Note 5**), mix gently, and incubate the reaction mixture at 50°C for 1 h.
8. Place the tube on ice to terminate the reaction.
9. Keep the tube on ice, and add 93 μL of DEPC water, 30 μL of 5X second strand buffer, 3 μL of 10 m*M* dNTPmix, 1 μL of *E. coli* DNA ligase (10 units/μL), 4 μL of *E. coli* DNA polymerase I (10 units/μL), and 1 μL of RNase H (2 units/μL) in that order.
10. Vortex the tube, and incubate to complete the reaction for 2 h at 16°C.
11. Add 2 μL of T4 DNA polymerase (5 units/μL), and continue the incubation at 16°C for 5 min.
12. Place the tube on ice, and add 10 μL of 0.5 *M* EDTA.
13. Add 150 μL of phenol:chloroform:isoamyl alcohol (25:24:1), vortex thoroughly, and centrifuge at room temperature for 5 min at 14,000*g*.
14. Vortex the mixture thoroughly, and centrifuge at room temperature for 30 min at 14,000*g*.
15. Remove the supernatant, and wash the pellet with 0.5 mL of 70% ethanol (−20°C).
16. Centrifuge for 5 min at 14,000*g*, and remove the supernatant carefully.
17. Dry the cDNA to completely evaporate residual ethanol.

3.3. cDNA Insertion into pMX-ER/LIB Vector

1. Dissolve the cDNA pellet in 12 μL of DEPC water, and add 2 μL of *Bst*XI adaptor, 4 μL of 5X T4 DNA ligase buffer, 2 μL of T4 DNA ligase (1 unit/μL).
2. Mix gently, and incubate the reaction mixture at 16°C for more than 24 h.
3. Heat the reaction at 70°C for 10 min to inactivate the ligase, and chill on ice.
4. Add 30 μL of DEPC water and 50 μL of phenol:chloroform:isoamyl alcohol (25:24:1), vortex thoroughly, and centrifuge at room temperature for 5 min at 14,000*g*.
5. Carefully remove 45 μL of the upper aqueous layer and transfer it to a fresh 1.5-mL tube.
6. Add 5 μL of 7.5 *M* NH₄OAc, 1 μg of yeast tRNA, and 150 μL of ethanol (−30°C).
7. Vortex the mixture, and centrifuge for 20 min at 14, 000*g* (4°C).
8. Remove the supernatant carefully, and wash the pellet with 0.25 mL of 70% ethanol.
9. Centrifuge for 2 min at 14,000*g*, and remove the supernatant.
10. Dry the cDNA for 10 min to evaporate residual ethanol completely.
11. Dissolve the cDNA in 100 μL of a TEN buffer, apply it to the top of the equilibrated cDNA size fractionation column, discard the effluent.
12. Add 100 μL of a TEN buffer to the column top, discard the effluent.
13. Repeat **step 12** one more time.

14. Add 100 μL of a TEN buffer and collect the effluent into a 1.5-mL microcentrifuge tube.
15. Repeat **step 14** and then add 150 μL of a TEN buffer and collect the effluent into the 1.5-mL microcentrifuge tube that now contains 350 μL of the adaptor-ligated cDNA solution.
16. Add 35 μL of 7.5 *M* NH$_4$OAc, 1 μg of yeast, and 700 μL of ethanol (−30°C) into the tube.
17. Repeat **steps 7–10** for cDNA precipitation.
18. Add 40 μL of a TE buffer to the cDNA pellet.
19. Separate the cDNAs through a 0.8% SeaPlaque gel. Cut out the gel fragment between 0.6 and 10 kbp, and extract cDNA fragments using QiaexII according to the manufacturer's protocol.
20. Resuspend the size-fractionated cDNAs in 10 μL TE buffer.
21. To ligate the cDNAs with the pMX-Mito/LIB vector, prepare the following solution: 2 μL of the cDNAs, 4 μL of a 5X T4 DNA ligase buffer, 1 μL of the pMX-Mito/LIB vector (100 ng), 1 μL of T4 DNA ligase, and 12 μL of water.
22. Incubate the solution at 16°C for more than 16 h.
23. Heat the reaction mixture at 70°C for 10 min to inactivate the ligase.
24. Add 30 μL of distilled water, 5 μL of 3 *M* NaOAc, 1 μg yeast tRNA, and 60 μL of ethanol (−30°C).
25. Repeat **steps 7–10** for cDNA precipitation.
26. Resuspended the pellet in 12 μL TE buffer (*see* **Note 6**).

3.4. Production of Retrovirus Stock

1. PlatE cells (1 × 10^6 cells) are seeded onto a 10-cm dish 1 d before the transfection and incubated at 37°C in a CO$_2$ incubator.
2. Prepare the following solutions in 1.5-mL tubes: solution A, 5 μg of library DNA in 200 μL OPTI-MEM medium; and solution B, 18 μL of lipofectamine 2000 in 200 μL OPTI-MEM medium. Incubate the mixtures at room temperature for 5 min.
3. Add solution B to solution A, vortex gently, and incubate at room temperature for 20 min.
4. During the incubation, replace the medium in the 10-cm dish with a DMEM medium containing 10% of FCS.
5. Add the mixture of solution A and solution B in a 10-cm dish.
6. Incubate the cells on the dishes for 12 h at 37°C in a CO$_2$ incubator.
7. Remove the medium including the transfection mixture, and add 5 mL of a fresh DMEM containing 10% of FCS.
8. Incubate the cells for 48 h at 37°C in a CO$_2$ incubator, and then the supernatant is collected in 1.5-mL tube (*see* **Note 7**), which is used for infection of target cells.

3.5. Infection of Recombinant Retrovirus and Collection of Fluorescent Cells

1. BNL1MEer cells (1×10^6 cells) are seeded onto a 6-cm dish 24 h before the infection and incubated at 37°C in a CO_2 incubator.
2. Add 5 µL of the polybrene solution, and incubate the cells at 37°C for 10 min in a CO_2 incubator.
3. Add 20 µL of the virus stock prepared in **Subheading 3.4.**, and incubate the cells at 37°C for 36 h in a CO_2 incubator.
4. Move the 6-cm dish in another CO_2 incubator set at 30°C, and incubate for 12 h.
5. Strip the cells with 500 µL of a trypsin solution, collect the cells by centrifugation for 2 min at 800*g*, and resuspend in 1 mL PBS (*see* **Note 8**). To remove aggregated cells, pass the cell suspension through a cell strainer (*see* **Note 9**) and collect the cells in a round-bottom tube.
6. The cells are subjected to cell analysis with a FACS (*see* **Note 10**). A region that includes fluorescent cells is determined (see region L in **Fig. 4**).
7. Collect the fluorescent cells in a round-bottom tube filled with the DMEM medium including 10% FCS (*see* **Note 11**). Immediately, spread the cells on 6-cm dishes and incubate the cells at 37°C for 48 h in a CO_2 incubator.
8. Repeat **steps 4–6**. Collect the fluorescent cells onto a 48-well microtiter plate (*see* **Note 12**), and incubate the cells at 37°C in a CO_2 incubator until confluent.

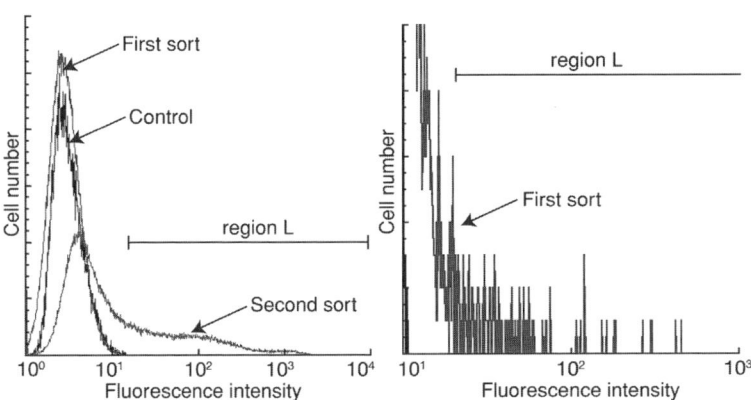

Fig. 4. FACS profiles of BNL1MEer cells infected with retrovirus cDNA libraries. BNL1MEer cells were infected with the cDNA retrovirus libraries with an infection efficiency of 20%. Five days after incubation, the cells were stripped and sorted by FACS (First sort). The enlarged FACS profiles in region L is shown on the right. The fluorescent cells within region L were collected and again sorted by FACS (Second sort). Uninfected cells were inserted to show the background fluorescence (Control).

3.6. Sequencing of the Genome-Integrated cDNA that Encode ER-Targeting Proteins

1. Genomic DNA was extracted with a Wizard Genomic DNA Purification Kit according to the manufacturer's (Promega) protocol.
2. One hundred ng of each genomic DNA is subjected to PCR. The PCR run for 30 cycles (30 s at 98°C for denaturation, 30 s at 58°C for annealing, and 4 min at 72°C for extension) using LA Taq polymerase (*see* **Note 13**).
3. The resulting fragments are sequenced using a BigDye Terminator Cycle Sequencing Kit and analyzed by a genetic analyzer.
4. Compare the obtained cDNA sequence to the databases at the National Center for Biotechnology Information (NCBI) using BLASTn (http://www.Ncbi.nlm.nih.gov/BLAST/).

Notes

1. All the plasmids can be obtained from our laboratory (e-mail address: umezawa@chem.s.u-tokyo.ac.jp).
2. The PlatE cell line can be obtained from Dr. Toshio Kitamura (The Institute of Medical Science, The University of Tokyo).
3. We have found that the GFP antibody (Roche) is excellent for blotting the C-terminal half of EGFP. The other monoclonal-GFP antibodies (Clonetech or TAKARA BIO Inc., Tokyo) are also available for the blotting.
4. The cell line must meet the following criteria: GFP reconstitution in the ER is positive for ER-targeting proteins and negative for cytosolic ones. To check this criteria, two plasmids, pMX-ER/LIB-ERTS (positive) and pMX-ER/LIB-CaM, are used. According to **Subheading 3.4.**, the plasmids are converted into retroviruses, and then the retroviruses are infected into a selected cell line that includes C-terminal EGFP. The fluorescence intensities of the cells are examined by FACS (*see* **Subheading 3.5,**).
5. It is also possible to use SuperScript II Reverse Transcriptase. However, Super-Script III Reverse Transcriptase has several advantages in the synthesis of first-strand cDNAs at 50°C, providing increased specificity, higher yields of cDNA, and longer cDNA products than SuperScript II Reverse Transcriptase.
6. In this step it is very important to completely dry the pellet. Inclusion of residual ethanol will greatly decrease the transformation efficiency. To avoid this, do not use more than $10\,\mu L$ of QiaexII beads in one tube in DNA extraction.
7. The retrovirus library can be stocked for 6 month at $-80°C$. Because only $20–50\,\mu L$ of the retrovirus is used for the subsequent infection procedure, divide the retrovirus-library solution into small portions and stock them at $-80°C$.
8. Because residual trypsin solution affects the viability of the cells, it is important to remove the trypsin solution completely. If necessary, repeat the washing of the cells with the PBS buffer.

9. The size of mesh of a cell strainer depends on the types of the cells. To avoid the cells to stick in a flow cell of the FACS, it is much better to use smaller mesh size.

10. The FACS is generally equipped with collection modes—a mode of higher collection efficiency or for higher purity. In this first sorting procedure, the FACS is adjusted to set a mode of higher collection efficiency in order to collect as many fluorescent cells as possible. In contrast, a mode of high purity is used in the second sorting procedure for the purpose of accurate identification of the ER-targeting proteins.

11. Prevent the collection tube from contamination with antibiotics. Antibiotics affect the cells upon sorting by FACS and kill the collected cells.

12. If the FACS you use is not equipped with a system for collecting cells onto microtiter plates, it is otherwise possible to collect the cells in a round-bottom tube supplemented with a DMEM medium including 10% FCS. The cells are cloned by limiting dilution, and their genomic DNA is collected.

13. Any Taq polymerase can be used.

Acknowledgments

This work was supported by grants from Core Research for Evolutional Science and Technology (CREST) of Japan Science and Technology (JST) and the Ministry of Education, Science, and Culture, Japan.

References

1. Sitia, R., and Braakman, I. (2003) Quality control in the endoplasmic reticulum protein factory. *Nature* **426**, 891–894.
2. Bankaitis, V. A., and Morris, A. J. (2003) Lipids and the exocytotic machinery of eukaryotic cells. *Curr. Opin. Cell Biol.* **15**, 389–395.
3. von Heijne, G. (1985) Signal sequences. The limits of variation. *J. Mol. Biol.* **184**, 99–105.
4. Zheng, N. and Gierasch, L. M. (1996) Signal sequences: the same yet different. *Cell* **86**, 849–852.
5. Martoglio, B., and Dobberstein, B. (1998) Signal sequences: more than just greasy peptides. *Trends Cell Biol.* **8**, 410–415.
6. Bendtsen, J. D., Nielsen, H., von Heijne, G., and Brunak, S. (2004) Improved prediction of signal peptides: SignalP 3.0. *J. Mol. Biol.* **340**, 783–795.
7. Nielsen, H., Engelbrecht, J., Brunak, S., and von Heijne, G. (1997) Identification of prokaryotic and eukaryotic signal peptides and prediction of their cleavage sites. *Protein Eng.* **10**, 1–6.
8. Horton, P. and Nakai, K., (1997) Better prediction of protein cellular localization sites with the nearest neighbors classifier. *Proc. Int. Conf. Intell. Syst. Mol. Biol.* **5**, 147–152.

9. Kumar, A., Agarwal, S., Heyman, J. A., et al. (2002) Subcellular localization of the yeast proteome. *Genes Dev.* **16**, 707–719.

10. Huh, W. K., Falvo, J. V., Gerke, L. C., et al. (2003) Global analysis of protein localization in budding yeast. *Nature* **425**, 686–691.

11. Chen, H. and Leder, P. (1999) A new signal sequence trap using alkaline phosphatase as a reporter. *Nucleic Acids Res.* **27**, 1219–1222.

12. Kojima, T. and Kitamura, T. (1999) A signal sequence trap based on a constitutively active cytokine receptor. *Nat. Biotechnol.* **17**, 487–490.

13. Ozawa, T., Nishitani, K., Sako, Y., and Umezawa, Y. (2005) A high-throughput screening of genes that encode proteins transported into the endoplasmic reticulum in mammalian cells. *Nucleic Acids Res.* **33**, e34.

14. Ozawa, T., Sako, Y., Sato, M., Kitamura, T., and Umezawa, Y. (2003) A genetic approach to identifying mitochondrial proteins. *Nat. Biotechnol.* **21**, 287–293.

15. Wu, H., Hu, Z., and Liu, X. Q. (1998) Protein trans-splicing by a split intein encoded in a split DnaE gene of Synechocystis sp. PCC6803. *Proc. Natl. Acad. Sci. USA* **95**, 9226–9231.

16. Morita, S., Kojima, T., and Kitamura, T. (2000) Plat-E: an efficient and stable system for transient packaging of retroviruses. *Gene Ther.* **7**, 1063–1066.

19

Trafficking Through the Early Secretory Pathway of Mammalian Cells

Theresa H. Ward

Summary

The use of green fluorescent protein (GFP) chimeras to illuminate the secretory pathway in living cells has provided a wealth of information on the mechanisms of protein retention, sorting, and recycling. A wide variety of microscopic techniques, including time-lapse imaging, double-labeling, quantitation, photobleaching, and energy transfer approaches, have been utilized to explore the organization of the early secretory pathway. In this chapter we focus on the application of GFP technology to gain insight into the dynamics of ERGIC-53, a putative cargo receptor localized to the early secretory pathway, and the way in which photobleaching approaches have provided insight into its transport.

Key Words: Confocal microscopy; green fluorescent protein; secretory pathway; endoplasmic reticulum; ER exit; Golgi; microtubules, ERGIC-53.

1. Introduction

The secretory membrane system enables regulated delivery of newly synthesized proteins and lipids to the cell surface. The pathway is made up of distinct organelles—the endoplasmic reticulum (ER), the Golgi complex, and the plasma membrane—between which transport is mediated by tubulovesicular transport intermediates. Proteins resident to the early secretory pathway must be correctly targeted and sorted from anterograde-directed secretory cargo to avoid incorrect localization or transport, for example, to the cell surface. This sorting is accomplished by the sequential action of two coat complexes, COPII and COPI (1,2).

From: *Methods in Molecular Biology, Vol. 390: Protein Targeting Protocols: Second Edition*
Edited by: M. van der Giezen © Humana Press Inc., Totowa, NJ

The use of green fluorescent protein (GFP) chimeras is now ubiquitous in all fields of mammalian cell biology for the study of dynamic processes *in vivo* *(3)*. Because proteins can be tagged with GFP, mostly without altering their targeting and function, the spatial and temporal dynamics of a protein of interest can be resolved. In this chapter, we cover the application of GFP technology to the study of p58, the rat homolog of ERGIC-53 *(4,5)*. This type I unglycosylated transmembrane protein is a putative cargo receptor responsible for aiding the export of some secretory cargo from the ER *(6)*, and cycles through the early secretory pathway *(7)*. Cycling is attained through the concerted action of the terminal amino acids KKFF on the cytosolic tail: the terminal double-phenylalanine motif mediates binding to COPII and facilitates ER exit *(8)*, while the KKXX double-lysine motif binds to COPI *in vitro* and is required for recycling to the ER *(8,9)*.

With the construction of a p58-GFP fusion chimera *(10)*, the behavior of the protein in living cells can be analyzed by live-cell imaging techniques, focusing on photobleaching approaches. The physicochemical properties of GFP are such that upon irradiation with a strong laser pulse, its fluorescence is irreversibly eliminated, and recovery of fluorescence is only viewed by exchange of photobleached molecules for unbleached chimeras moving from outside the region of bleaching *(11)*. Altering the localization motifs of p58 and comparing the wild-type and mutant proteins' behavior can lead to insights into the control mechanisms acting at ER exit sites. Manipulating the transport pathways and comparing p58 to other ER-to-Golgi associated factors can lead to conclusions on the role and behavior of different types of proteins in the early secretory pathway *(1,10)*.

2. Materials

2.1. Construction of GFP Chimeric Proteins

1. Cloning vector with fluorophore-encoding cDNA incorporated on one side of multiple cloning site. Many are available from Clontech (BD Biosciences Clontech, Oxford, UK), Perkin-Elmer (Boston, MA), and Gateway® Technology (Invitrogen, Paisley, Scotland).
2. cDNA of interest.
3. Molecular biology toolkit (e.g., restriction enzymes, PCR ingredients, competent bacteria, etc.)

2.2. Cell Culture and Transfection

1. Dulbecco's modified Eagle's medium (DMEM) (Sigma-Aldrich, Poole, UK) supplemented with 10% fetal bovine serum (FBS) (Sigma-Aldrich), 2 m*M*

glutamine, and penicillin/streptomycin (supplied as 100x; Sigma-Aldrich). With supplements this is referred to as complete DMEM.

2. Solution of trypsin (0.5 g/L) and ethylenediamine tetraacetic acid (EDTA) (0.5 mM) (Sigma-Aldrich).

3. G418 sulfate (geneticin) (PAA, Yeovil, Somerset, UK) dissolved at 500 mg active G418/mL in sterile distilled water (*see* **Note 1**). Store in aliquots at −20°C.

4. FuGENE 6 Transfection Reagent (Roche Diagnostics Ltd., Lewes, East Sussex, UK) used according to manufacturer's instructions.

5. Autoclaved toothpicks.

6. Freezing media: complete DMEM + 10% dimethylsulfoxide (DMSO; Sigma-Aldrich). This can also be supplemented with an additional 10% FBS (i.e., total 20%).

7. Nocodazole (Sigma-Aldrich, Poole, UK) dissolved at 10 mg/mL in dimethyl sulfoxide (DMSO) and stored in aliquots at −20°C.

8. Brefeldin A (BFA) (Epicentre Biotechnologies, from Cambridge Bioscience, Cambridge, UK) dissolved at 10 mg/mL in ethanol and stored in aliquots at −20°C.

9. Phosphate-buffered saline (PBS, pH 7.5). Prepare 10x PBS stock from 115 g disodium hydrogen orthophosphate, anhydrous, 29.6 g sodium dihydrogen orthophosphate, 58.4 g sodium chloride, make up to 1000 mL with distilled water. Adjust pH to 7.5 with HCl.

10. 2x Nonreducing sample buffer for sodium dodecyl sulfate–polyacrylamide gel electrophoresis (SDS-PAGE): 3.55 mL dH$_2$O, 1.25 mL 0.5 M Tris-HCl, pH 6.8, 2.5 mL glycerol, 2.0 mL 10% (w/v) SDS, 0.2 mL 0.5% (w/v) bromophenol blue. For reducing sample buffer, add 50 μL β-mercaptoethanol to 950 μL nonreducing sample buffer.

11. Cell scrapers (Nalge Nunc, available through Scientific Laboratory Supplies Ltd., Nottingham, UK).

2.3. SDS-PAGE

1. Resolving buffer: 1.5 M Tris-HCl pH 8.8.

2. Stacking buffer: 0.5 M Tris-HCl pH 6.8.

3. 30% Acrylamide/bis solution (37.5:1 mixture) (highly toxic when unpolymerized, handle with care) (Helena Biosciences, Sunderland, UK) and N,N,N',N'-tetramethylethylene diamine (TEMED, Sigma-Aldrich, Poole, UK).

4. Ammonium persulfate (APS): prepare 10% (w/v) solution in water and use immediately.

5. 10x Running buffer: 30.3 g Tris-HCl, 144.0 g glycine, 10.0 g SDS dissolved in dH$_2$O to a total volume of 1000 mL.

6. Prestained Rainbow molecular weight markers (Amersham Biosciences, Little Chalfont, UK).

2.4. Western Blotting for p58 Oligomerization

1. Transfer buffer: 48 mM Tris-HCl, 39 mM glycine, 1.3 mM SDS, 20% methanol pH 9.2. Dissolve 5.82 g Tris-HCl, 2.93 g glycine, 3.75 mL of 10% SDS in dd H_2O and make up to 800 mL. Add 200 mL methanol.
2. Hybond-P PVDF membrane (Amersham Biosciences, Little Chalfont, UK) and filter paper (Mini Trans-Blot, Bio-Rad Laboratories, Hemel Hempstead, UK).
3. Ponceau S solution: 0.1% Ponceau S (w/v) in 5% acetic acid (v/v) (Sigma-Aldrich).
4. PBS-Tween (PBS-T): 1x PBS pH 7.5 solution made up with sterile distilled water from the 10x PBS stock adding 0.1% (v/v) Tween 20.
5. Blocking buffer: 5% (w/v) nonfat dry milk in PBS-T.
6. Primary antibody: rabbit anti-green fluorescent protein (anti-GFP) (Molecular Probes, available from Invitrogen, Paisley, Scotland).
7. Secondary antibody: anti-rabbit IgG conjugated to horseradish peroxidase (HRP-anti-rabbit) (Amersham Biosciences).
8. Enhanced chemiluminescent (ECL) Western blotting substrate (Pierce, through Perbio Science) and X-Omat AR film (Kodak, available through Sigma-Aldrich, Poole, UK).

2.5. Confocal Microscopy

1. Microscope glass cover slips (No. 1 13-mm diameter) (VWR), or LabTek 4-well glass cover slip chambers (Nalge Nunc available through Scientific Laboratory Supplies Ltd., Nottingham, UK). Cover slips are cleaned before use (*see* **Note 2**).
2. Imaging medium: DMEM with no phenol red (Sigma-Aldrich) supplemented with 10% FBS, 2 mM glutamine, 25 mM HEPES (pH 7.4).
3. For short-term experiments cover slips can be mounted on a low-tech imaging chamber as described in **Fig. 1**.

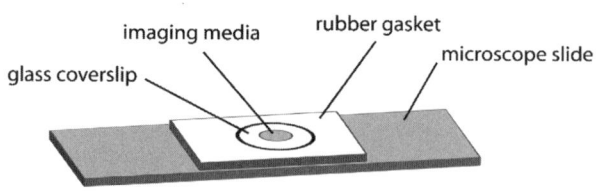

Fig. 1. Imaging chamber for short-duration experiments. A piece of silicone rubber sheet is cut to fit a microscope slide. A hole punch creates the correct sized chamber. The rubber gasket is attached to the slide with petroleum jelly. A drop of imaging media is placed in the hole, and a cover slip to which cells have previously been adhered is inverted onto the liquid (cells facing media). Excess liquid on the top of the cover slip is dried off with tissue. The cover slip adheres through capillary action.

3. Methods

To visualize the behavior of a protein in living cells, it is first necessary to create a GFP chimeric construct. This must be introduced into a relevant cell type, and then tested for functionality and/or localization effectiveness. In the case of p58-GFP, functionality was tested by the ability of the fusion protein to oligomerize, which has been shown to be a requirement of its ability to exit the ER *(12)*. Localization can be compared to the known distribution of the endogenous protein *in vivo* (not shown here, but published in **refs. 7,13**, and *14*).

The dynamics of the p58-GFP fusion protein can be visualized using time-lapse imaging, and the kinetics of its trafficking route *(15)* can be analyzed by photobleaching techniques *(11)* and by disruption of ER-to-Golgi trafficking *(10,16,17)*).

3.1. Construction of Targeted p58-GFP Fusion Protein

1. As GFP-fusion proteins become omnipresent in cell biology, the routes to making them become ever simpler (e.g., Gateway® Technology, Invitrogen). The molecular biology protocols are beyond the scope of this chapter and are assumed to be background knowledge.
2. GFP can be incorporated as a fusion protein at either the N-terminus or the C-terminus of the protein of interest. It can also be placed in a potential linker region between functional domains. Any and all of these fusion constructs can disrupt the function and/or localization of a protein, compared to the endogenous protein, and this can only be assessed on a trial basis. p58 has a motif at the C-terminus of KKFF. The double lysine motif (KKXX) is known to bind the coat complex COPI and may be required for retrieval from the Golgi to the ER *(6,9)*, while the double phenylalanine motif (XXFF) is required for interaction with COPII to enable ER exit *(8)*. This means that tagging at the C-terminus of p58 is not a possibility. One recent publication has now used intradomain GFP insertion to create a p58-GFP chimera *(18)*.
3. p58 is a Type I transmembrane protein in the secretory pathway. To achieve this orientation, a translated signal sequence is required to enable the nascent protein entry through the translocon into the lumen of the ER. This is found at the N-terminus of the protein. Attachment of GFP directly upstream of the N-terminus of p58 would therefore mask the ER-targeting signal sequence and thus disrupt the chimeric protein localization. ss-GFP *(19)* incorporated the hen egg lysozyme signal sequence upstream of EGFP to target GFP into the secretory pathway. ss-GFP was recloned in place of EGFP in the Clontech pEGFP-C3 vector. p58 cDNA was then amplified by PCR without its signal sequence and cloned into the multiple cloning site of the modified C3 vector *(10)*. An alternative approach *(20)* used

the signal sequence of the p58 homolog ERGIC-53 upstream of GFP to target an ERGIC-53-GFP chimera correctly.

4. Other aspects to take into account when making fusion constructs include:

   a. Spectral Choice: Cells are more damaged by lower wavelength light, so for a protein expressed at low levels, which would need higher laser power for imaging, high levels of the blue laser at 405 nm to illuminate cyan fluorescent protein (CFP), the blue-shifted variant of GFP, would be detrimental over relatively short time periods. Certain cell types may have autofluorescent qualities that may create a high background with certain fluorophores, and matching spectra should therefore be avoided.

   b. Fluorophore Choice: GFP and its original spectral variants have dimerizing characteristics and have now been engineered as monomeric variants *(21)*. The red fluorescent proteins from corals also have oligomerising tendencies. Even though mRFP1 was engineered as a monomeric fluorophore *(22)*, certain N-terminal constructs were found to behave oddly *(23)*.

   c. Concatamerized GFP can increase levels of fluorescence *(23–25)*.

   d. Photoactivatable variants are an additional tool for tracking a protein in the cell, although this approach is unlikely to be very tractable to a cytosolic protein.

   e. Choice of Vector: Controllable promoters may be useful for certain fusion proteins. In dual transfected cells (i.e., introducing two different fluorescent constructs), different promoters create less competition for synthesis of the two chimeras.

3.2. Transfection and Construction of Stable Cell Lines

1. Normal rat kidney (NRK) fibroblasts were grown at 37°C with 5% CO_2 in complete DMEM (**Subheading 2.2.**). At near confluence they are passaged with trypsin-EDTA to provide new maintenance cultures in 75-cm^2 flasks (Nunclon™; Scientific Laboratory Supplies Ltd., Nottingham, UK) and experimental cultures as required. A 1:15 split provides sufficient cells to approach confluence after 48 h. For stable transfections, cells are plated onto 100-mm tissue dishes containing a single cover slip to check for transfection efficiency, while for transient transfections cells are plated onto either cleaned cover slips (*see* **Note 2**) in 6-well dishes or onto LabTek chamber slides.

2. Plasmid constructs are transfected into NRK cultures, which have approximately 40% confluency (i.e., approx 24 h after passaging). FuGENE is used according to the manufacturer's protocol. In brief, for p58-GFP transfection into a 100-mm dish of NRK, 9 μL FuGENE is added to 191 μL serum-free DMEM (total volume 200 μL) and incubated for 5 min at room temperature. The FuGENE-DMEM mixture is then pipetted dropwise into a fresh Eppendorf containing 3 μg p58-GFP plasmid DNA. The tube is mixed gently by tapping and incubated at

room temperature for 15 min. This mixture is then pipetted over the cells in the 100-mm dish, and the dish is transferred to the incubator.

3. After 16–24 h, transfection efficiency is checked by mounting the cover slip on the imaging chamber shown in **Fig. 1** to ensure that fluorescent cells are present.

4. G418 is added to the NRK culture to a final concentration of 500 μg/mL. This is sufficient to completely kill all untransfected NRK cells within a week, but must be titrated according to a specific cell line (*see* **Note 3**).

5. If necessary, media is refreshed (incorporating G418), and after approx 7 d colonies should be visible. These can be seen by eye, although care must be taken not to spill media depending upon what light source is used.

6. Individual colonies are identified with marker pen on the dish. A 96-well flat-bottomed plate should be prepared; 150 μL fresh DMEM + G418 should be added to sufficient wells for the number of colonies to be screened. Media is then suctioned off the 100-mm dish. Colonies are gently scraped with sterile toothpicks and cells transferred to a 96-well plate by twizzling the toothpick in a single well.

7. Once colonies are apparent in the 96-well plate, they can be transferred by trypsinization (approx 50 μL trypsin-EDTA needed) to a 12-well dish, previously prepared with a single cover slip, and 1.5 mL DMEM + G418 in each well.

8. Individual colonies can then be screened for fluorescence and correct localization, as described below (**Subheading 3.5.**).

9. Selected colonies should be grown as quickly as possible and aliquots frozen down to prevent overpassaging, as some GFP chimeras are prone to losing expression levels over time. Once a maintenance culture in a 75-cm^2 dish is obtained, G418 levels may be reduced to 250 μg/mL, but certain constructs require maintained higher levels of G418.

10. To freeze down cells from a 75-cm^2 dish, cells are grown to approx 80% confluency, then trypsinized. Trypsinized cells are centrifuged at 700 g for 5 min to pellet, the supernatant removed, and the cells resuspended in freezing medium (5–10 mL depending on confluency). Cryotubes should be placed in a Cryo 1°C Freezing Container (an isopropanol-containing canister that cools more slowly than tubes placed directly into the freezer—but not essential) (Nalgene™), and chilled in a −80°C freezer. After 24–48 h, the cells are transferred to liquid nitrogen.

11. To regrow an aliquot of cells, a 75-cm^2 flask containing 15 mL complete DMEM + 250 μg/mL G418 is prewarmed to 37°C. Cells removed from liquid nitrogen (following safety procedures) should be warmed quickly for 1–2 min in a 37°C waterbath, then added to the flask of prewarmed media.

3.3. Preparation and Separation of Reduced and Nonreduced Samples

1. Stable NRK cell lines expressing p58-GFP are passaged as described in **Subheading 3.2.** Experimental samples are plated in duplicate into 6-well plates. One well is required for each sample.

2. When cells reach 80% confluency, medium is removed from each well with aspiration, cells are rinsed with 1x PBS. Fifty μL of the appropriate sample buffer (reducing or nonreducing) is added. Cells are scraped off the dish and into a labeled microcentrifuge tube.

3. Samples are boiled at 95°C 5 min and are then spun for 2 min at 13,000 rpm in a microfuge to pellet debris. Samples can be allowed to cool to room temperature and are then ready for separation by SDS-PAGE. Alternatively, the samples can then be stored at −20°C until required.

4. To perform SDS-PAGE, the Bio-Rad Mini-PROTEAN® 3 Cell is described here, assembled according to the manufacturer's instructions.

5. A 10% gel is prepared by mixing 4.1 mL ddH$_2$O, 3.3 mL acrylamide mix, 2.5 mL resolving gel buffer, and 0.1 mL 10% (w/v) SDS. Immediately prior to pouring the gel, 50 μL APS and 5 μL TEMED are added. This is gently poured between the gel plates to approx 1 cm below the depth of the wells. A small amount of ddH$_2$O is then pipetted gently onto the upper surface to exclude air bubbles. The gel will take about 30 min to polymerize, at which point the water should be drained off the gel by tilting the casting apparatus and drawing off with filter paper.

6. Quantities for the stacking gel are the same, except stacking gel buffer is used. To polymerize, 50 μL APS and 10 μL TEMED are added. The stacking mix is then poured to the top of the glass plates, and the wells inserted. Polymerization takes approx 30 min.

7. Five hundred mL running buffer is prepared by diluting the 10x stock solution with dH$_2$O.

8. The gel unit is assembled, and running buffer is added to the upper and lower chambers. Twenty μL each sample is loaded into the wells. Reduced samples should be run together, nonreduced together, with two to three wells of separation between them to ensure that the reducing agent does not affect the nonreduced samples. Prestained markers are loaded alongside samples.

9. The gel is run at a constant voltage of 200 V for approx 35 min until the blue dye front is close to the end of the gel.

3.4. Western Blotting to Check Correct Oligomerization

1. Once samples have been separated by SDS-PAGE, they are transferred to PVDF membrane. We use the Bio-Rad Trans-Blot® SD Semi-Dry transfer cell, following the manufacturer's instructions.

2. A sheet of PVDF membrane is cut to slightly larger than the size of the resolving gel. The membrane is first wetted in 100% methanol for 10 s, then washed in distilled water for 5 min. Finally it is equilibrated in transfer buffer for 10 min.

3. The gel unit is disassembled. The stacking gel is removed. The resolving gel is orientated by removing a corner and briefly rinsed in transfer buffer.

4. Two pieces of extra thick filter paper are completely saturated with transfer buffer. To assemble the blot, the layers must be placed on the base of the transfer unit

in the following order, taking care to roll out the air bubbles using a pipet after each layer. One piece of filter paper is placed onto the base of the transfer unit. Next the prewetted blotting membrane is placed on top of filter paper, then the equilibrated gel, then finally the second piece of filter paper.

5. For a single gel, transfer conditions are 12 V for 30 min. Once the unit is disconnected, discard the filter paper and gel.
6. The membrane can be checked by covering with Ponceau S for 1 min to reveal protein bands. This is then rinsed off with 2–3 changes of distilled water.
7. The membrane is incubated with blocking buffer for 1 h with gentle agitation, then washed with PBS-T for 5 min with agitation.
8. The blot is incubated with anti-GFP diluted 1:1000 in blocking buffer for 1 hr, then washed with PBS-T three times for 5 min with agitation.
9. The blot is incubated with HRP-anti-rabbit diluted 1:1000 in blocking buffer for 1 h, then washed with PBS-T three times for 10 min with agitation.
10. The Pierce ECL substrate requires mixing two reagents in equal parts to make the substrate working solution. This is prepared immediately before use. In a dark room, the blot is incubated with working solution for 1 min at room temperature.
11. The blot is removed from solution, removing excess liquid with tissue. It is placed into a film cassette lined with clingfilm. The blot is covered with a layer of plastic wrap, avoiding bubbles. It is exposed to X-ray film for 1–30 min. An example of the results is shown in **Fig. 2**.

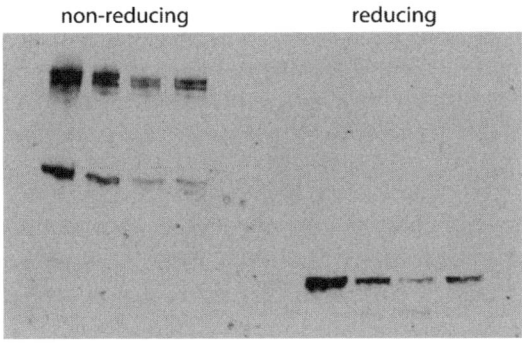

Fig. 2. Western blot of p58-GFP in four separate stable cell lines detected using rabbit anti-green fluorescent protein (anti-GFP). Samples run under reducing conditions show a single band of monomeric p58-GFP. Samples run under nonreducing conditions show multiple bands, larger than the monomer. This shows that the GFP chimeric protein is able to oligomerize correctly in the endoplasmic reticulum, and the GFP moiety is not impeding correct protein folding. This also demonstrates the variable levels of chimeric protein in different stable cell lines.

3.5. Time-Lapse Imaging of Living Cells

1. NRK cells (either stable cell lines or transiently transfected cells) are prepared on LabTek chambers or cover slips as described in **Subheading 3.2.**
2. For live cell imaging, the microscope must be equipped with a means to heat the stage to 37°C. This should be preheated for an hour or so before use to allow the system to stabilize. This avoids focus problems resulting from expanding metal components of the stage and also chilling the cells coming from the incubator.
3. The medium is changed for imaging medium. If the stage is equipped with CO_2 control, HEPES need not be added to the medium. For long periods of imaging in the absence of CO_2, the LabTek chamber should be sealed with petroleum jelly in the lid.
4. An objective is selected. For intracellular organelle resolution, objectives of $40\times$ and above are recommended, with high numerical aperture (NA). In **Fig. 3** a $63\times$ Plan-Apochromat oil immersion lens with 1.4 NA was used. Cells are placed on microscope stage.
5. Single images of cells are acquired using microscope software. These should be optimized to use the full pixel depth, i.e., laser and gain levels should be optimized such that the acquired image is neither too saturated nor too dark. The software should provide a palette to identify black and white levels. Line averaging allows acquisition of high-quality images, typically 4 is used. Zoom is applied to focus on a cell/group of cells of interest.
6. To acquire a time series, the number of images and the interval between them must be selected. To avoid focus drift during the course of a long experiment, some software may have an autofocus function incorporated. Alternatively, for intervals of more than 1 min, it may be necessary to acquire individual images manually, although these must be carefully numbered sequentially to enable later combining to a series.

3.6. Effect of Protein Export Signals on ER Exit Analyzed With FLIP

Fluorescence loss in photobleaching (FLIP) enables a region of the cell to be bleached at regular time intervals *(11,26)*. The bleach region is often referred to as the region of interest (ROI). FLIP can be applied to test if a population of fluorescent proteins is confined to one region of the cell and is primarily a qualitative method.

1. NRK cells expressing p58-GFP are prepared and mounted on the microscope as described in **Subheading 3.5.** Conditions for imaging are optimized with a single scan image (*see* **Note 4**).
2. A field of view with two cells is chosen. Thus, the cell that is not photobleached can be used to compensate focal drift and to quantitate for total fluorophore bleaching over the course of an experiment.

3. The FLIP parameters are set up. In the Zeiss LSM software, this requires the MultiTime Series package. Iterations for the bleach scan are defined by the number of scans necessary to completely bleach out the GFP in the region of interest. The area to be bleached is selected manually. Laser power for the bleach is set to 100%. The number of images and time interval is selected.
4. Take a single image to check focus.
5. Press start, and the computer should automate the image acquisition process, incorporating a bleaching step prior to each single image scan.
6. If not automated, the series is saved before taking any more images. A sample result is shown in **Fig. 3**.

3.7. Organelle Photobleaching to Determine Kinetics of Protein Transport

Selective photobleaching of organelles can be used to visualize and quantify trafficking or flux through an organelle. Whether a protein is associating stably

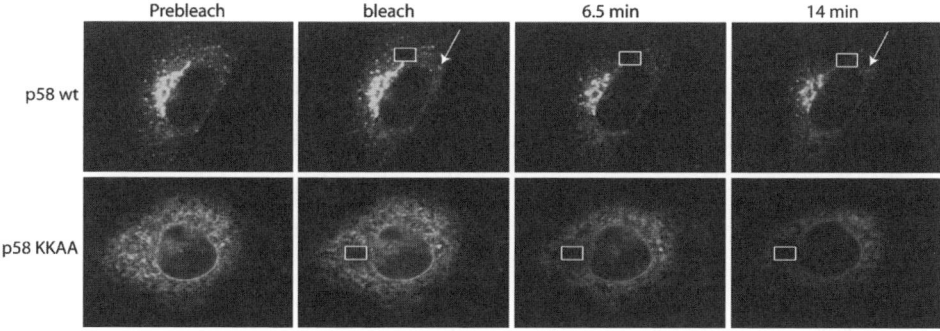

Fig. 3. Repetitive photobleaching (FLIP) of wild-type and mutant p58-GFP in normal rat kidney (NRK) cells. The mutant p58 KKAA is missing the double-phenylalanine motif (KK*FF*) required for endoplasmic reticulum (ER) exit. NRK cells were transiently transfected with the GFP-tagged constructs. The experiment was performed using the photobleaching program on the Zeiss LSM 510 confocal microscope with a 40× 1.2 NA oil-immersion objective. Prebleach images for the two chimeras are shown. The wild-type p58-GFP localizes to the Golgi and punctate structures. This protein cycles through the early secretory pathway, and its localization reflects this. Because the mutant p58 KKAA is unable to exit the ER, it shows a reticular localization. Repeated high intensity laser is applied to the region designated by the white box, and images acquired after each bleach point (every 15 s). This shows that p58 KKAA is confined to the ER and does not emerge into punctate ER exit sites (arrow in p58 wt). Another example where this technique has been applied to look at ER exit is found in **ref. 19**.

with an organelle or exchanging between the organelle and another part of the cell (cytoplasm, or other organelle) can be established in living cells using fluorescence recovery after photobleaching (FRAP) *(11)*. This requires a single photobleaching step (unlike FLIP), after which a time series follows the recovery (or not) of fluorescence to the area.

1. NRK cells expressing p58-GFP are prepared and mounted on the microscope as described in **Subheading 3.5.** Conditions for imaging are optimized with a single scan image (**Subheading 3.6.;** *see* **Note 4**). For photobleaching larger organelles, such as the Golgi, the 25× or 40× lens with a fully open pinhole is used to give a thick confocal slice that encompasses the full depth.
2. The area to be bleached is selected manually, and the bleach parameters are set such that the fluorescent signal of the photobleached ROI will be at background intensity levels. To ensure that a prebleach image is taken, in the Zeiss software the box "Acquire bleach after scan" must be selected. The bleach is incorporated into a time series by selecting the number of images in the time series window and pressing StartB**.
3. The time series is saved. A sample result is shown in **Fig. 4A**.

3.8. Quantitation of Exchange Between the Golgi and Rest of Cell

1. Once an image sequence has been acquired, the data can be analyzed *(15)*. A basic recovery curve can be plotted by calculating the mean ROI pixel value over time, for example, in the Zeiss software or similar image analysis software (*see* **Note 5**).
2. For comparative valuations, quantitate the following ROI over the time series: (1) the photobleached organelle (o), (2) a background region in the cell close to the object of interest (b), (3) the total cell (c), and (4) an extracellular background region (e). The fluorescence (f) and area (a) of each ROI should be recorded.
3. Calculate Golgi fluorescence as follows:

$$(f_o - f_b)a_o / (f_c - f_e)a_c$$

This gives a value based on pixel intensity.
4. To make the data comparable to other proteins/experiments, this can additionally be refined as follows:

$$\text{Relative Golgi fluorescence} = \frac{\text{Golgi fluorescence at time } x}{\text{Golgi fluorescence at time } 0} \times 100$$

This gives the prebleach Golgi fluorescence a value of 100%. An example is given in **Fig. 4B**. The recovery curve gives estimates of the amount of protein that is mobile in the organelle by the plateau value.

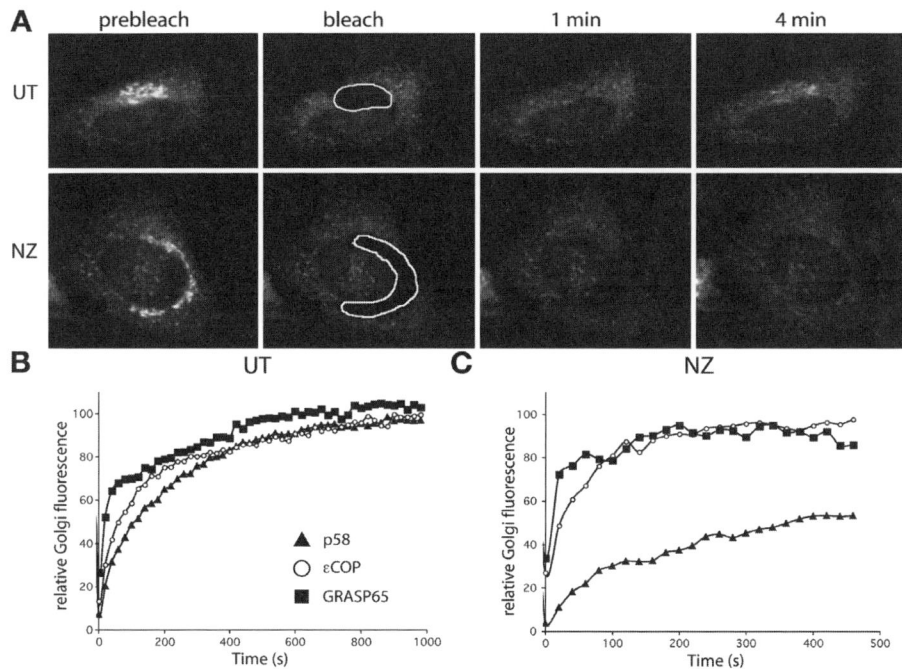

Fig. 4. Membrane trafficking in the early secretory pathway is dependent on intact microtubules (MT). (**A**) FRAP experiment of a cell expressing p58-GFP (untreated, UT) shows that exchange of Golgi fluorescence with other intracellular pools is clearly apparent after 4 min. In contrast, in cells expressing p58-GFP treated with nocodazole (NZ), recovery to the Golgi is not apparent within the same time frame, because p58 in membrane-bound carriers requires intact microtubules to translocate into the Golgi region. Cells were incubated on ice for 15 min to depolymerize MT, then media is exchanged for prewarmed imaging media including NZ (2 μg/mL) to prevent MT repolymerizing. FRAP experiments were undertaken immediately upon rewarming, i.e., before chronic NZ effects such as Golgi fragmentation take place. (**B**) Quantitation of fluorescence recovery in the Golgi region following a FRAP experiment such as that shown in (**A**) (UT). Recovery curves shown are an average of at least five experiments. p58 (closed triangle) is compared to εCOP-GFP (open circle), a coat protein that exchanges between the Golgi and cytosol, and GRASP65-GFP (closed square), a Golgi protein with a myristoyl group thought to tether it to the membrane. (**C**) Quantitation of FRAP recovery in cells treated with NZ. This shows that recovery of fluorescence into the Golgi region remains very fast for εCOP and GRASP65, while p58 does not recover in the same way because of the lack of microtubules. This indicates that GRASP65, like εCOP, exchanges between Golgi membranes and the cytosol and does not depend on membrane-bound carriers from the ER, like p58. Figure key as for (**B**). [(**A**) (UT) from **ref. *10*** with permission from Dr. J. Lippincott-Schwartz.]

3.9. Disruption of the Early Secretory Pathway to Investigate Trafficking Pathways

Cargo sorting, and the formation of membrane-bound transport intermediates, in the early secretory pathway require the sequential action of two coat complexes, COPII and COPI, and the presence of intact microtubules to facilitate traffic from the ER to the Golgi. Inhibition of these events can give information about the roles of different protein–protein interactions, or localization changes, in the activity of a protein of interest.

1. To disrupt microtubules, cells are incubated on ice for 15 min. Cells are then warmed to 37°C by the addition of imaging medium containing nocodazole (at 2–5 µg/mL). This acute treatment disrupts the microtubules, but secondary effects to membranes are not seen immediately. For example, redistribution of the Golgi apparatus to ER exit sites occurs slowly over approx 2 h (chronic treatment *[17]*).
2. Acute nocodazole treatment can be used in conjunction with FRAP (**Subheading 3.7.**) to visualize the effect of microtubule disruption on the association of a protein with the Golgi apparatus, for example. The effect of nocodazole on the trafficking of p58-GFP is shown in **Fig. 4A**, with quantitative comparisons to proteins that follow alternative exchange routes shown in **Fig. 4B**.
3. An alternative method to alter ER–Golgi trafficking pathways is by the addition of Brefeldin A (BFA), a fungal metabolite that causes the Golgi to disassemble and redistribute into the ER by blocking coatomer (COPI) activity *(16)*. This process can be visualized by incubating living cells in imaging medium containing BFA at 5 µg/mL over a time period of 15–30 min (not shown here; but see **refs.** *10* and *27*).
4. Both drug treatments are reversible by changing the imaging medium three times *(16,17)*.

Notes

1. G418 powder comes with an activity quantity stated on the bottle. This is the amount used, rather than total weight, e.g., if the bottle states an activity of 650 mg/g, then 0.769 g G418 should be made up to 1 mL with double-distilled water to get an end concentration of 500 mg/mL active G418.
2. Cover slips are first cleaned by washing in 70% ethanol and then either air-drying or holding the cover slip with tweezers and burning off excess ethanol in a bunsen flame before placing into a culture dish.
3. Other cell types may need less or more G418. This should be optimized by treating untransfected cells with G418 at different concentrations. Optimally all cells should be killed within 7 d.
4. In general, for photobleaching techniques, the 25× or 40× lens with a fully open pinhole is used to give a thick confocal slice that would encompass the cell depth. This enables total cell fluorescence to be acquired for quantitation, and it also

avoids some focal drift problems. However, if better resolution is required, then the 63× lens can be used, with the caveat that fluorescence throughout the cell will not be acquired with a single image, potentially necessitating z-sectioning.

5. If focus drifts, data should be discarded and not used for FRAP quantitation (although it may be usable for qualitative images).

Acknowledgments

This work was supported by a Royal Society Dorothy Hodgkin Fellowship.

References

1. Altan-Bonnet, N., Sougrat, R., and Lippincott-Schwartz, J. (2004) Molecular basis for Golgi maintenance and biogenesis. *Curr. Opin. Cell Biol.* **16**, 364–372.
2. Lee, M. C. S., Miller, E. A., Goldberg, J., Orci, L., and Schekman, R. (2004) Bi-directional protein transport between the ER and Golgi. *Annu. Rev. Cell Dev. Biol.* **20**, 87–123.
3. Ward, T. H. and Lippincott-Schwartz, J. (2006) The uses of green fluorescent protein in mammalian cells, in *Green Fluorescent Protein: Properties, Applications, and Protocols*, 2nd ed. (Chalfie, M., and Kain, S. R., eds.), John Wiley, Hoboken, NJ, pp. 305–337.
4. Schweizer, A., Fransen, J. A., Bächi, T., Ginsel, L., and Hauri, H. P. (1988) Identification, by a monoclonal antibody, of a 53-kDa protein associated with a tubulo-vesicular compartment at the cis-side of the Golgi apparatus. *J. Cell Biol.* **107**, 1643–1653.
5. Lahtinen, U., Dahllöf, B., and Saraste, J. (1992) Characterization of a 58 kDa *cis*-Golgi protein in pancreatic exocrine cells. *J. Cell Sci.* **103**, 321–333.
6. Hauri, H. P., Kappeler, F., Andersson, H., and Appenzeller, C. (2000) ERGIC-53 and traffic in the secretory pathway. *J. Cell Sci.* **113**, 587–596.
7. Klumperman, J., Schweizer, A., Clausen, H., et al. (1998) The recycling pathway of protein ERGIC-53 and dynamics of the ER-Golgi intermediate compartment. *J. Cell Sci.* **111**, 3411–3425.
8. Kappeler, F., Klopfenstein, D. R. C., Foguet, M., Paccaud, J. P., and Hauri, H.-P. (1997) The recycling of ERGIC-53 in the early secretory pathway: ERGIC-53 carries a cytosolic endoplasmic reticulum-exit determinant interacting with COPII. *J. Biol. Chem.* **272**, 31801–31808.
9. Itin, C., Schindler, R., and Hauri, H.-P. (1995) Targeting of protein ERGIC-53 to the ER/ERGIC/*cis*-Golgi recycling pathway. *J. Cell Biol.* **131**, 57–67.
10. Ward, T. H., Polishchuk, R. S., Caplan, S., Hirschberg, K., and Lippincott-Schwartz, J. (2001) Maintenance of Golgi structure and function depends on the integrity of ER export. *J. Cell Biol.* **155**, 557–570.
11. Lippincott-Schwartz, J., Altan-Bonnet, N., and Patterson, G. H. (2003) Photobleaching and photoactivation: following protein dynamics in living cells. *Nat. Cell Biol.* **5**, S7–S14.

12. Nufer, O., Kappeler, F., Guldbrandsen, S., and Hauri, H.-P. (2003) ER export of ERGIC-53 is controlled by cooperation of targeting determinants in all three of its domains. *J. Cell Sci.* **116**, 4429–4440.

13. Lippincott-Schwartz, J., Donaldson, J. G., Schweizer, A., et al. (1990) Microtubule-dependent retrograde transport of proteins into the ER in the presence of brefeldin A suggests an ER recycling pathway. *Cell* **60**, 821–836.

14. Saraste, J. and Svensson, K. (1991) Distribution of the intermediate elements operating in ER to Golgi transport. *J. Cell Sci.* **100**, 415–430.

15. Phair, R. D. and Misteli, T. (2001) Kinetic modelling approaches to *in vivo* imaging. *Nat. Rev. Mol. Cell Biol.* **2**, 898–907.

16. Klausner, R. D., Donaldson, J. G., and Lippincott-Schwartz, J. (1992) Brefeldin A: insights into the control of membrane traffic and organelle structure. *J. Cell Biol.* **116**, 1071–1080.

17. Cole, N. B., Sciaky, N., Marotta, A., Song, J., and Lippincott-Schwartz, J. (1996) Golgi dispersal during microtubule disruption: regeneration of Golgi stacks at peripheral endoplasmic reticulum exit sites. *Mol. Biol. Cell* **7**, 631–650.

18. Sannerud, R., Marie, M., Nizak, C., et al. (2006) Rab1 defines a novel pathway connecting the pre-Golgi intermediate compartment with the cell periphery. *Mol. Biol. Cell* **17**, 1514–1526.

19. Nehls, S., Snapp, E. L., Cole, N. B., et al. (2000) Dynamics and retention of misfolded proteins in native ER membranes. *Nat. Cell Biol.* **2**, 288–295.

20. Ben-Tekaya, H., Miura, K., Pepperkok, R., and Hauri, H.-P. (2005) Live imaging of bidirectional traffic from the ERGIC. *J. Cell Sci.* **118,** 357–367.

21. Zacharias, D. A., Violin, J. D., Newton, A. C., and Tsien, R. Y. (2002) Partitioning of lipid-modified monomeric GFPs into membrane microdomains of live cells. *Science* **296**, 913–916.

22. Campbell, R. E., Tour, O., Palmer, A. E., et al. (2002) A monomeric red fluorescent protein. *Proc. Natl. Acad. Sci. USA* **99**, 7877–7882.

23. Shaner, N. C., Campbell, R. E., Steinbach, P. A., Giepmans, B. N. G, Palmer, A. E., and Tsien, R. Y. (2004) Improved monomeric red, orange and yellow fluorescent proteins derived from *Discosoma* sp. red fluorescent protein. *Nat. Biotechnol.* **22**, 1567–1572.

24. Zaal, K. J. M., Smith, C. L., Polishchuk, R. S., et al. (1999) Golgi membranes are absorbed into and reemerge from the ER during mitosis. *Cell* **99**, 589–601.

25. Gerlich, D., Beaudouin, J., Kalbfuss, B., Daigle, N., Eils, R., and Ellenberg, J. (2003) Global chromosome positions are transmitted through mitosis in mammalian cells. *Cell* **112**, 751–764.

26. Cole, N. B., Smith, C. L., Sciaky, N., Terasaki, M., Edidin, M., and Lippincott-Schwartz, J. (1996) Diffusional mobility of Golgi proteins in membranes of living cells. *Science* **273**, 797–801.

27. Sciaky, N., Presley, J., Smith, C., et al. (1997) Golgi tubule traffic and the effects of brefeldin A visualized in living cells. *J. Cell Biol.* **139**, 1137–1155.

20

Studying Protein Export From the Endoplasmic Reticulum in Plants

Sally L. Hanton, Loren A. Matheson, and Federica Brandizzi

Summary

Understanding the mechanisms of protein sorting and targeting through the plant secretory pathway has become the focus of many research laboratories. The development of a model system whereby recombinant genes can be transiently expressed in protoplasts has facilitated the study of protein transport signals. Experimental strategies combining a protoplast expression system with endoglycosidase H, vacuole purification, and pulse-chase analyses are used to investigate aspects of specific proteins as they pass through the secretory system. This chapter provides details of protoplast preparation and electroporation as well as techniques to study protein trafficking from the endoplasmic reticulum to the Golgi apparatus or vacuolar compartments. Recommendations as to how to troubleshoot problems that can arise while following these protocols are also discussed in this chapter.

Key Words: Plant; secretory pathway; endoplasmic reticulum; Golgi apparatus; protoplasts; pulse chase; immunoprecipitation; vacuole; endoglycosidase H.

1. Introduction

The secretory pathway of plants is a complex system dedicated to the synthesis, modification, and transport of proteins to various locations within or outside the cell. Because this system is essential, cell biologists have taken a keen interest in studying the structure and function of the components. These include proteins that maintain the integrity and mobility of the secretory organelles, enzymes that play a role in posttranslational modification, and receptors that direct vesicle traffic. Regulated trafficking of proteins is fundamental to the correct functioning of the secretory pathway and requires specific sorting signals, which have been the focus of much recent attention *(1–3)*.

From: *Methods in Molecular Biology, Vol. 390: Protein Targeting Protocols: Second Edition*
Edited by: M. van der Giezen © Humana Press Inc., Totowa, NJ

The secretory pathway of all eukaryotes is made up of functionally distinct organelles. For example, the endoplasmic reticulum (ER) is involved in the synthesis, folding, processing, assembly, and storage of secretory proteins (reviewed in **ref. 4**). Export of correctly folded proteins from the ER is the first step in the transfer of cargo through compartments of the secretory pathway. Most of the protein cargo is shipped to the Golgi apparatus from specialized domains of the ER, called ER export sites. The Golgi apparatus, which is the key organelle in sorting proteins, sends the cargo forward, in the direction of the cell surface or vacuolar compartments, or backwards to the ER. In plants, vacuoles have a central role for the storage or degradation of a variety of macromolecules.

In plants, how proteins are exported from the site of synthesis to their final destination is an active field of investigation. In particular, the characterization of the regulation mechanisms for protein exchange between the ER and Golgi is exceptionally challenging because the passage through the Golgi apparatus is not a requirement for all secretory proteins to reach their final destination. Certain proteins leave the ER and are degraded in the cytosol; others reach the vacuole directly from the ER. Studying protein trafficking in plant cells is particularly fascinating because of the complexity of the secretory pathway.

This chapter describes optimized protocols for studying protein trafficking in plant cells, including plant protoplast preparation and transient gene expression. We also describe detailed methods for determining whether specific proteins have been exported via the Golgi apparatus as well as techniques for utilizing metabolic labeling to investigate protein turnover and vacuolar degradation.

2. Materials

Reagents used are generally available from a variety of suppliers; where specified, the reagent is the one that we have found the most effective, but other sources may provide similar or improved results. Safety data and MSDS sheets for each reagent should be consulted for proper usage instructions before beginning any of the protocols outlined in this chapter.

2.1. Protoplast Preparation and Electroporation (see Note 1)

1. Murashige and Skoog (MS) growth medium (5): 0.43% (w/v) MS salts (Sigma), supplemented with 2% (w/v) sucrose and pH adjusted to 5.7 with 1 M KOH. 0.8% (w/v) technical agar is added and the medium autoclaved for 20 min at 121°C to sterilize.
2. Transient expression (TEX) buffer: 0.31% (w/v) Gamborg's B5 salts (6) (Sigma), 2.4 mM morpholino ethanesulfonic acid (MES), 5 mM CaCl$_2$, 3 mM NH$_4$NO$_3$,

400 m*M* sucrose; pH adjusted to 5.7 with 1 *M* KOH, and filter-sterilized through a 0.2-μm-pore filter.

3. Leaf digestion mix (10X): 2% (w/v) macerozyme R-10, 4% (w/v) cellulase R-10 (both from Yakult Pharmaceutical Ind. Co. Ltd, Tokyo, Japan) are dissolved in TEX buffer by mixing at room temperature for 1 h. The mixture is centrifuged at 3000*g* for 15 min to remove insoluble particles. The supernatant is then filter-sterilized through a 0.2-μm-pore filter and stored as 5-mL aliquots at −80°C. Working solutions are prepared by 10-fold dilution in TEX buffer.

4. 100-μm-Pore nylon mesh.

5. Electroporation buffer: 10 m*M* 4-(2-hydroxyethyl)-1-piperazine ethane sulfonic acid (HEPES), 80 m*M* KCl, 4 m*M* CaCl$_2$, 400 m*M* sucrose; pH adjusted to 7.2 with 1 *M* KOH, and filter-sterilized through a 0.2-μm-pore filter.

6. Plasmid DNA encoding the gene to be expressed and appropriate promoter and 3′ sequences; dissolved in TE buffer (10 m*M* Tris-HCl, 0.1 m*M* EDTA, pH 8.0), at a minimum concentration of 2 μg/μL.

2.2. Endoglycosidase H Treatment

1. 250 m*M* NaCl.

2. Extraction buffer: 50 m*M* Tris-HCl, pH 7.5, 2 m*M*EDTA, pH 8.0, supplemented with 1 m*M* β-mercaptoethanol.

3. Denaturation buffer: 100 m*M* sodium citrate, 1% (w/v) sodium dodecyl sulfate (SDS), supplemented with 200 m*M* β-mercaptoethanol.

4. Brefeldin A (BFA, Sigma) is dissolved at 10 mg/mL in dimethylsulfoxide (DMSO) and stored in 10-μL aliquots at −20°C. Working solutions are prepared by appropriate dilution in TEX buffer.

5. Endoglycosidase H (EndoH, New England Biolabs P0702S), diluted 1/125 in extraction buffer immediately prior to use to give an activity of 0.4 IUB units/mL (*see* **Note 2**).

6. Phenylmethylsulfonyl fluoride (PMSF) is dissolved at 100 mg/mL in 95% ethanol and stored in 1-mL aliquots at -20°C.

2.3. Purification of Vacuoles

1. Lysis buffer: 0.2 *M* mannitol, 10% (w/v) Ficoll-400® (Sigma), 20 m*M* EDTA, 2 m*M* dithiothreitol (DTT), 5 m*M* HEPES; pH adjusted to 8.0 with 5 *M* NaOH and supplemented with 10 μg/mL neutral red and 150 μg/mL bovine serum albumin (BSA). Stored in single-use aliquots at -80°C, supplemented with 0.1 μg/mL pepstatin before use.

2. Vacuole buffer: 0.6 *M* betaine, 10 m*M* HEPES; pH adjusted to 7.5 with 5 *M* NaOH, then supplemented with 150 μg/mL BSA. Stored in single-use aliquots at −80°C, supplemented with 0.1 μg/mL pepstatin before use.

3. α-Mannosidase extraction buffer: 250 m*M* Na-acetate buffer, pH 4.6.

4. α-Mannosidase substrate: 6 m*Mp*-nitrophenol-α-D-mannopyranoside, dissolved in extraction buffer.
5. α-Mannosidase stopping buffer: 1 *M* Na$_2$CO$_3$.

2.4. In Vivo Labeling and Immunoprecipitation

1. ^{35}S-Cysteine/methionine (Promix, Amersham Biosciences/GE Healthcare) diluted to a stock activity of 500 μCi/mL in TEX buffer.
2. 250 m*M* NaCl solution.
3. Homogenization buffer: 200 m*M* Tris-HCl, pH 8.0, 300 m*M* NaCl, 1% (v/v) Triton X-100, 1 m*M* EDTA, pH 8.0, supplemented with 2 m*M* PMSF.
4. Antiserum to the protein of interest.
5. Protein A- or G-conjugated agarose (*see* **Note 3**), Protein A–Sepharose® (Sigma) was used in this example, hydrated prior to use.
6. NET buffer: 50 m*M* Tris, pH 7.5, 150 m*M* NaCl, 1 m*M* EDTA, 0.1% (v/v) Igepal (Sigma), 0.02% (w/v) NaN$_3$.
7. NET gel: as for NET buffer but supplemented with 0.25% (w/v) gelatin (*see* **Note 4**).
8. Sample buffer: 0.1% (w/v) bromophenol blue, 3.75 m*M* EDTA, 15 m*M* Tris-HCl, pH 8.8, 750 m*M* sucrose, 2.5% SDS, 17 m*M*DTT.

3. Methods

The methodology presented here for transient expression of recombinant genes in plant protoplasts (**Subheading 3.1.**) offers the advantage of studying processes quantitatively in vivo while maintaining the ability to control events through the extracellular environment and mutational analyses of specific proteins. The gene product of interest can be studied to determine its level of modification and ultimate targeting. If the protein is transported to the Golgi apparatus, its glycans will be modified, making them resistant to digestion by EndoH (**Subheading 3.2.**). If the protein has not come into contact with Golgi-processing enzymes, the endoglycosidase will remove unprocessed Asn-linked glycans. This creates a smaller molecular weight protein product, which can be detected through SDS-PAGE and Western blotting. Resistance to digestion by EndoH is used as a tool to investigate whether a protein has been transported beyond the *cis*-Golgi *(7)*.

To determine whether a protein of interest is targeted to a vacuolar compartment, a purified vacuole preparation is necessary. **Subheading 3.3.** describes the protocol for preparing purified vacuoles for analysis with SDS-PAGE. Whether the protein of interest has been transported to the vacuole or other areas of the endomembrane system, it is of use to monitor both the arrival of the protein as well as the half-life of the protein and its various modified forms. **Subheading 3.4.** describes a protocol for metabolically labeling protoplasts followed by instructions on how to prepare the samples for immunoprecipitation to investigate a single protein of interest.

3.1. Protoplast Preparation and Electroporation

1. Removal of the rigid cell wall in plant cells results in spherical membrane-bound protoplasts, which are dependent on an isotonic or hypertonic medium to maintain integrity. Plasmid-borne genes can be introduced to protoplasts by exposure to a transient electric field (*see* **Note 5**).
2. Tobacco plants (*Nicotiana tabacum* cv. Petit Havana) are grown in sterile culture in solid MS medium with a 16-h day length at 22°C. Propagation is carried out through transfer of apical or internodal subheadings to fresh medium or through sowing seeds (*see* **Note 6**).
3. Cuts are made with a sterile blade in the lower surface of a leaf (*see* **Note 7**), with care being taken not to cut all the way through the leaf. The central vein is removed to allow the leaf to lie flat, and the two halves are placed with the cut side downward in an 11- × 60-mm Petri dish containing 7 mL of 1X leaf digestion mix. The Petri dishes are incubated in the dark for 12–16 h.
4. After incubation the plates are gently shaken to release the protoplasts from the cuticle (*see* **Note 8**). The resulting heterogeneous mixture is then filtered through 100-μm nylon mesh and washed with electroporation buffer. The protoplast suspension obtained is centrifuged in 50-mL conical tubes in a swing-out rotor at 100 rcf for 20 min (acceleration 6, deceleration 0), resulting in a floating band of living protoplasts and a pellet of dead cells. The floating band is carefully penetrated with a Pasteur pipet connected to a peristaltic pump to remove the underlying medium and the dead cells (*see* **Note 9**). The living cells are washed in 25 mL of electroporation buffer and centrifuged at 100 rcf for 5 min. The washing process is repeated twice more, after which the protoplasts are resuspended in a final volume of electroporation buffer appropriate to the number of electroporations required.
5. Five hundred μL of protoplast suspension is gently pipetted into a 1.5-mL semi-micro cuvette (Sarstedt) using a 1000-μL tip with the extreme end removed. Plasmid DNA is diluted in electroporation buffer to a final volume of 100 μL, added to the cuvette, and mixed by gentle shaking (*see* **Note 10**). This mixture was incubated at room temperature for 5 min. The electroporation procedure is carried out at 120 V for 0.5 s (*see* **Note 11**), after which the protoplasts are incubated at room temperature for a further 20–30 min. The cell suspension is transferred to a 60- × 15-mm Petri dish and diluted with 2 mL of TEX buffer, then placed in the dark for approximately 24 h. It is then possible to observe fluorescent fusions directly at a microscope (*see* **Fig. 1**) or to proceed with an alternative protocol such as those outlined below.

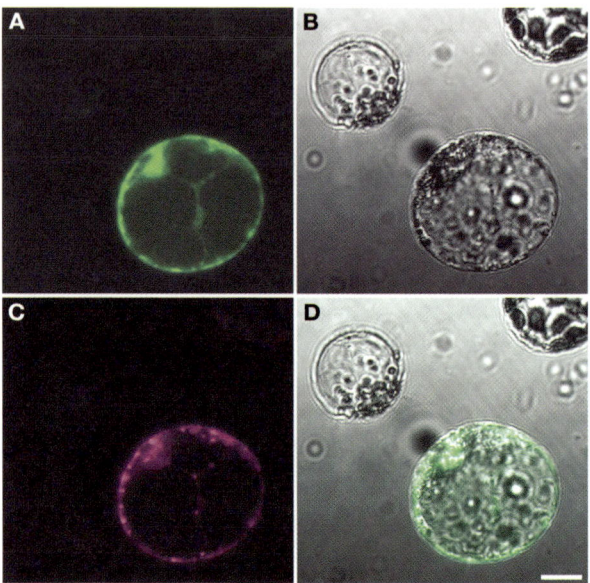

Fig. 1. Photomicrographs of protoplasts after transformation via electroporation demonstrating a mixed population of cells. (**A**) Fluorescence detection of a protoplast expressing a secretory form of green fluorescent protein (GFP) retained in the endoplasmic reticulum *(13)*. (**B**) A transmitted light image of the same field of view reveals the presence of three protoplasts indicating that the transformation was not 100% efficient. (**C**) However, the transformed protoplast is also expressing a yellow fluorescent protein (YFP) tagged to a marker for the Golgi apparatus *(14)*. This demonstrates that double transformation of a single protoplast is possible, confirmed by the overlay in (**D**). Bar = 10 μm.

3.2. Endoglycosidase H Treatment

1. EndoH is an enzyme that cleaves the internal *N*-acetylglucosamine residues of high-mannose carbohydrates, resulting in a shift in apparent molecular weight on SDS-polyacrylamide gels to a position very close to that of the nonglycosylated species (*see* **Fig. 2**, lane 4). Those N-linked glycan moieties that have passed to or beyond the medial Golgi are modified to a complex form by Golgi-localized enzymes and become insensitive to the enzyme *(8)*. It can therefore be used as a diagnostic tool to determine whether a protein has been transported from the ER to the Golgi. To be sure that the assay is working correctly, it is essential to include a control experiment that has been treated with BFA, a drug that causes the redistribution of Golgi proteins to the ER *(9)*.This demonstrates that the glycan under investigation is capable of being modified to an EndoH-insensitive form (*see*

Fig. 2, lane 6). These instructions assume that protoplasts have been prepared, transformed, and incubated in TEX buffer for at least 16 h prior to beginning this protocol.

2. Each protoplast sample is divided into two aliquots using a 1000-μL tip with the extreme end removed. One half of the sample is incubated with and the other without 10 μg/mL BFA for 5 h (*see* **Note 12**).

3. After incubation, the protoplasts are diluted 10-fold with 250 m*M* NaCl and centrifuged in 15-mL conical tubes for 5 min at 100 rcf. The supernatant is discarded and the pellet resuspended in an appropriate volume of extraction buffer; this depends on factors such as the number of protoplasts and the anticipated level of protein expression. One hundred μL is suggested as a starting volume. The mixture is transferred to 1.5-mL microcentrifuge tubes and sonicated briefly (1–2 s, amplitude 5 μm) using an ultrasonic homogenizer (*see* **Note 13**).

4. A volume of denaturation buffer equal to that of extraction buffer is added to each tube and the samples are boiled for 10 min. The samples are allowed to cool to room temperature; this should not be done on ice.

5. Twenty-nine μL of extract are placed in a fresh microcentrifuge tube, to which is added 1 μL of PMSF (*see* **Note 14**) and 10 μL of EndoH solution, giving a final activity of 0.1 IUB U/mL. After mixing, the samples are incubated at 37°C for 16 h.

6. An equal volume of protein sample buffer is added and the samples are analyzed by SDS-PAGE and Western blotting. If working with integral membrane proteins, 1 μL of 10% (v/v) Triton X-100 should be added to each sample before adding protein sample buffer in order to release the proteins from the membrane for easier analysis.

3.3. Purification of Vacuoles

1. In order to determine if a particular protein is targeted to the lytic vacuole, a vacuolar purification must be carried out prior to immunoblotting for the protein of interest. This method also allows an investigation of vacuolar degradation products.

2. These instructions assume that protoplasts have been prepared, transformed, and incubated in TEX buffer for at least 16 h prior to beginning this protocol. It is usually necessary to electroporate 5–10 samples for each purification, dependent on the yield of protein desired (*see* **Note 15**).

3. The protoplast suspension is transferred to a 50-mL conical tube and centrifuged for 5 min at 100 rcf (*see* **Note 16**). Dead cells and debris are removed from below the floating band of live cells using a Pasteur pipet. The cells can then be collected as a pellet by adding 250 m*M* NaCl to a total volume of 50 mL, mixing by inversion of the tube, and centrifugation for 3 min at 200 rcf. The supernatant is discarded and the pellets are placed on ice. Lysis buffer (5 mL, preheated to 42°C) is added and mixed gently for 1 min by swirling the contents of the tube, after which the cells are replaced on ice.

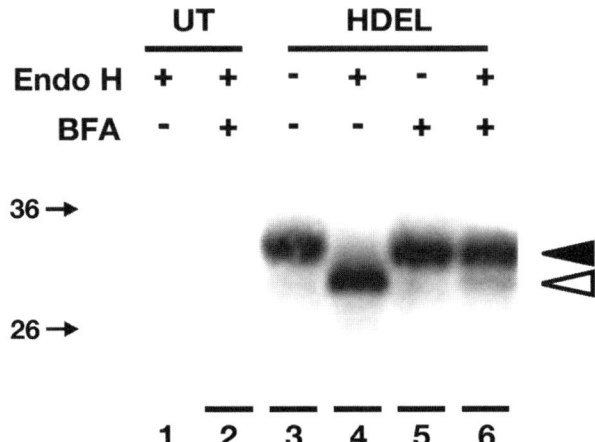

Fig. 2. EndoH treatment of protoplasts expressing spNGFP-HDEL. Protoplasts transformed with DNA encoding spNGFP-HDEL were treated with EndoH as described in **Subheading 3.2.** Equal quantities of extracts were loaded onto 10% sodium dodecyl sulfate gels and run until the gel front reached a point 6 cm below the stacking gel, then analysed by Western blotting with anti-GFP serum (AbCam). The glycan on spNGFP-HDEL is sensitive to digestion by EndoH under normal conditions, resulting in a significant shift in molecular weight (lane 4, open arrowhead) from the glycosylated form (lane 3, black arrowhead). In the presence of Brefeldin A (BFA), Golgi enzymes are redistributed to the endoplasmic reticulum *(9)* and are able to modify the glycan moiety, making it resistant to the action of the endoglycosidase and therefore maintaining the original molecular weight of the protein (lane 6, black arrowhead). UT, untransformed protoplasts. Approximate molecular weights are shown at left in kDa.

4. The sample is transferred into a 15-mL conical centrifuge tube and 3 mL of a 50:50 mix of lysis buffer:vacuole buffer is layered carefully on top. Vacuole buffer (1 mL) is subsequently layered on top of this and the sample is centrifuged at 3000 rcf for 20 min at 4°C.

5. The purified vacuoles are visible as a pinkish band at the interphase between the top two layers. These are removed with a micropipet, diluted 50:50 with vacuole buffer in a 1.5-mL microcentrifuge tube, and centrifuged at maximum speed for 5 min at 4°C, after which the supernatant is removed and discarded.

6. In order to compare samples, similar amounts of vacuole extracts must be present. Approximate quantification can be performed by measuring the activity of the vacuolar enzyme α-mannosidase. The pellet is resuspended in 200 µL of α-mannosidase extraction buffer and a portion is diluted 50:50 in protein sample buffer for analysis by SDS-PAGE.

7. Ninety μL of the remaining sample is transferred to a 1.5-mL microcentrifuge tube for each assay; depending on the yield of vacuoles, it may be necessary to dilute the sample first. 10 μL of substrate is added to start the reaction and the mixture is incubated at 30°C for a suitable length of time; again, this depends on the yield of vacuoles. Thirty minutes is a good starting point but may need to be altered. The reaction is stopped by the addition of 160 μL of stop buffer, and 200 μL of the mixture is transferred into a 96-well microtiter plate. The absorbance of the sample at λ = 405 nm is measured in a plate reader, and the activity per minute and per mL of purified vacuoles is calculated using the following formula:

$$\text{Activity} = \frac{\text{absorbance} \times 1000}{\text{volume of undiluted extract (μL)} \times \text{length of assay (min)}}$$

8. The concentrations of the samples should be altered with α-mannosidase extraction buffer until each contains a similar level of α-mannosidase activity before analysis via Western blotting.

3.4. In Vivo Labeling and Immunoprecipitation

1. Metabolically labeling the pool of *de novo* protein and immunoprecipitating the protein of interest over a time course facilitates the study of precursor–protein relationships in a time-dependent fashion *(10)*. This allows an investigation of the various molecular weight forms, either glycosylation or degradation products, of a specific protein.

2. These instructions assume that protoplasts have been prepared, transformed by electroporation, and diluted in TEX buffer immediately before beginning this protocol. Care should be taken to follow local rules regarding the safe use and disposal of radioactive substances.

3. The protoplast suspension is transferred to a 15-mL conical tube and centrifuged for 5 min at 100 rcf. Dead cells and excess TEX buffer are removed from below the floating band of live cells using a Pasteur pipet, concentrating the suspension of living cells to a final volume of 100–200 μL.

4. ^{35}S-Cysteine/methionine (500 μCi/mL in TEX buffer) is added and mixed by gentle shaking to give a total activity of 100–200 μCi/mL. The suspension is incubated at 25°C for the required labeling period (usually a minimum of 2 h).

5. 250 m*M* NaCl is added to a final volume of 10 mL, followed by centrifugation in a swing-out rotor at 100 rcf for 5 min, after which the supernatant is discarded *(see* **Note 17**). The pellet is resuspended in 500 μL of homogenization buffer, transferred to 1.5-mL microcentrifuge tubes and then centrifuged at maximum speed for 10 min at 4°C. Three hundred μL of the resulting supernatant is added to 600 μL of NET gel and a suitable volume (usually 1 μL; *see* **Note 18**) of the appropriate antiserum, mixed well and incubated on ice for a minimum of 1 h.

6. During this incubation, the protein A–Sepharose is hydrated by making a 0.35% (w/v) suspension of protein A–Sepharose in sterile distilled water (2.5 mg of

protein A–Sepharose is required for each sample). The suspension is allowed to settle on ice for 10 min, then centrifuged at 500*g* for 2 min, and the supernatant is removed carefully and replaced with fresh distilled water. The washing step is repeated twice and the pellet resuspended in NET buffer to a final volume of 100 μL per sample. Transfer of the hydrated resin to individual microcentrifuge tubes is achieved using a 200-μL pipet tip with the extreme end removed. Care should be taken to maintain a homogeneous mixture while pipetting in order that identical amounts of resin are placed into each tube.

7. After incubation the samples are centrifuged at maximum speed in a microcentrifuge for 10 min at 4°C. Eight hundred μL of the supernatant is removed, added to 100 μL of protein A–sepharose suspension, and mixed well. The samples are then incubated at 4°C with slow rotation for a minimum of 1 h.

8. The samples are centrifuged briefly, the supernatant is discarded, and the pellet resuspended in 900 μL of NET buffer. This washing step is repeated a further three times. After the final wash, the pellet is dried and resuspended in 15 μL of sample buffer for loading on SDS-polyacrylamide gels.

9. Proteins can then be transferred to nitrocellulose membrane by electroblotting, or the gel can be dried and used directly for analysis.

10. Immunoprecipitated proteins are then visualized with a PhosphorImager, according to the manufacturer's instructions.

4. Notes

1. Sterile technique is essential because of the amount of sucrose present in the medium.

2. Note that 10,000 NEB units = 1 IUB unit.

3. The choice of agarose conjugate depends on the species origin and isotype of the primary antibody. In general, protein A has the best affinity for rabbit capture antibodies *(11)*.

4. Gelatin is dissolved in warm water with gentle stirring to prevent clumps from forming. Gelatin stocks of 2.5% (w/v) can be stored at -20°C and thawed prior to use by microwaving briefly.

5. This protocol has been optimized for tobacco leaf protoplasts, but has also been used successfully with potato and *Arabidopsis* protoplasts *(4)*.

6. Ensure that sterile technique is used throughout; seed sterilization can be done by treatment with 100% ethanol or 30% bleach for 5 min, followed by thorough washing and plating on MS medium.

7. Optimal protoplast yield can be achieved by using leaves from plants that are between 4 and 6 wk old.

8. Digestion efficiency can be improved by incubating the leaves at 25°C overnight and by leaving the plates for 30 min after shaking.

9. Care should be taken that the floating band of cells is not disturbed and that those cells adhering to the sides of the tube are quickly washed with fresh buffer, because they are extremely sensitive to dehydration.

10. Increasing the amount of DNA usually results in increased expression of the protein of interest, but no more than 40 μg should be used because levels above this cause significant cell mortality.

11. These conditions are used in conjunction with a BTX Electro Square Porator® ECM830 electroporator. If using a different electroporation device, optimization of the electroporation conditions for the best expression levels is recommended. Note that increasing the voltage may result in better transformation efficiency, but will also increase the mortality rate of the protoplasts.

12. In some cases it is necessary to use 20 μg/mL BFA for 6 h in order to achieve significant redistribution of Golgi enzymes *(12)*.

13. The sonication conditions used depend on the model of homogenizer available; the conditions given are appropriate to a VirTis VirSonic® 600 Ultrasonic Cell Disrupter with a 1/8-in. microtip.

14. PMSF is unstable in aqueous solution and should be added directly to the sample.

15. The yield required usually depends on the antibody available to the protein of interest; if the antibody is poor, more protein is required in order for it to be detectable on a Western blot.

16. It is advisable to set aside a small amount of the protoplast suspension for comparison with the purified vacuoles so that any enrichment of the vacuolar protein can be estimated.

17. The pellet can be frozen at −80°C at this stage for extraction at a later date.

18. The binding conditions will vary greatly from protein to protein; the appropriate binding conditions for each application should be determined empirically.

References

1. Yuasa, K., Toyooka, K., Fukuda, H., and Matsuoka, K. (2005) Membrane-anchored prolyl hydroxylase with an export signal from the endoplasmic reticulum. *Plant J.* **41**, 81–94.

2. Hanton, S. L., Bortolotti, L. E., Renna, L., Stefano, G., and Brandizzi, F. (2005) Crossing the divide—transport between the endoplasmic reticulum and Golgi apparatus in plants. *Traffic* **6**, 267–277.

3. Contreras, I., Yang, Y., Robinson, D. G., and Aniento, F. (2004) Sorting signals in the cytosolic tail of plant p24 proteins involved in the interaction with the COPII coat. *Plant Cell Physiol.* **45**, 1779–1786.

4. Vitale, A. and Denecke, J. (1999) The endoplasmic reticulum-gateway of the secretory pathway. *Plant Cell* **11**, 615–628.

5. Murashige, T. and Skoog, F. (1962) A revised medium for rapid growth and bioassays with tobacco tissue cultures. *Physiol. Plantar.* **15**, 473–497.

6. Gamborg, O. L. and Eveleigh, D. E. (1968) Culture methods and detection of glucanases in suspension cultures of wheat and barley. *Can. J. Biochem.* **46,** 417–421.

7. Crofts, A. J., Leborgne-Castel, N., Hillmer, S., et al. (1999) Saturation of the endoplasmic reticulum retention machinery reveals anterograde bulk flow. *Plant Cell* **11,** 2233–2248.

8. Maley, F., Trimble, R. B., Tarentino, A. L., and Plummer, T. H., Jr. (1989) Characterization of glycoproteins and their associated oligosaccharides through the use of endoglycosidases. *Anal. Biochem.* **180,** 195–204.

9. Lippincott-Schwartz, J., Yuan, L. C., Bonifacino, J. S., and Klausner, R. D. (1989) Rapid redistribution of Golgi proteins into the ER in cells treated with brefeldin A: evidence for membrane cycling from Golgi to ER. *Cell* **56,** 801–813.

10. Crofts, A. J., Leborgne-Castel, N., Pesca, M., Vitale, A., and Denecke, J. (1998) BiP and calreticulin form an abundant complex that is independent of endoplasmic reticulum stress. *Plant Cell* **10,** 813–824.

11. Masters, S. C. (2004) Co-immunoprecipitation from transfected cells. *Methods Mol. Biol.* **261,** 337–350.

12. Hanton, S. L., Renna, L., Bortolotti, L. E., Chatre, L., Stefano, G., and Brandizzi, F. (2005) Diacidic motifs influence the export of transmembrane proteins from the endoplasmic reticulum in plant cells. *Plant Cell* **17,** 3081–3093.

13. Brandizzi, F., Hanton, S., DaSilva, L. L., et al. (2003) ER quality control can lead to retrograde transport from the ER lumen to the cytosol and the nucleoplasm in plants. *Plant J.* **34,** 269–281.

14. Brandizzi, F., Frangne, N., Marc-Martin, S., Hawes, C., Neuhaus, J. M., and Paris, N. (2002) The destination for single-pass membrane proteins is influenced markedly by the length of the hydrophobic domain. *Plant Cell* **14,** 1077–1092.

21

Imaging the Golgi Apparatus in Living Mitotic Cells

Nihal Altan-Bonnet

Summary

Live-cell imaging is a powerful tool which allows the observation of dynamic cellular processes while maintaining the native organization of the cell. Its advantages over other methods that disrupt cell integrity are abundantly evident in the study of cell division, where multiple subcellular organelles and molecules are involved in dynamic, spatio-temporally regulated processes such as Golgi and nuclear envelope disassembly/reassembly, spindle apparatus formation, chromosome condensation and segregation, and cytoplasmic division. This chapter will describe practical methods for cell synchronization, selection of fluorescent markers for transfection, and setting up imaging conditions and microscope parameters for acquiring time-lapse images of the Golgi apparatus in mitotic cells. These are general methods that can be applied to the study of many different types of organelles and molecules in dividing cells.

Key Words: Golgi apparatus; mitosis; confocal microscopy; live-cell imaging; cell synchronization; photobleaching; fluorescence recovery after photobleaching.

1. Introduction

Mitosis is a critical event in the life of a cell during which it divides its nuclear and cytoplasmic contents. Nuclear division is comprised of spindle assembly, concerted chromosome movement, alignment, and segregation *(1)*. Cytoplasmic division relies on partitioning and breakdown/reassembly mechanisms for segregating membrane-bound organelles and protein assemblies.

The specific stages of nuclear and cytoplasmic division and the cellular machinery that regulates these events have been studied extensively using a variety of experimental techniques, including biochemical fractionation to purify and assay protein activities, immunofluorescence to visualize the distribution of compartments and molecules at the light level, and electron

From: *Methods in Molecular Biology, Vol. 390: Protein Targeting Protocols: Second Edition*
Edited by: M. van der Giezen © Humana Press Inc., Totowa, NJ

microscopy to visualize at the ultrastructural level. These procedures have provided insight into mitosis, but one common drawback of such aforementioned techniques has been that they all significantly perturb native cellular physiology and subcellular organization. In fractionation-based approaches, the cells are broken apart, resulting in the loss of native organization of organelles and molecules. In immunofluorescence and electron microscopy-based techniques, the cells are fixed and processed, preventing the experimenter from obtaining any information on the spatio-temporal dynamics of organelles and molecules. Moreover, for biochemical assays, where large populations of cells are needed, an assorted mitotic population of cells from different mitotic stages is collected because of the difficulty harvesting large numbers of cells arrested at a single mitotic stage. Hence the activities assayed are products of a mixture of molecules and membranes collected from different mitotic stages. Thus, these experiments need to be interpreted cautiously and cannot be solely relied upon to conclude about events and activities taking place at specific stages of mitosis.

Time-lapse imaging, photobleaching, and photoactivation *(2–5)* are powerful tools for investigating living mitotic cells because they do not disrupt the native spatio-temporal organization and dynamics of subcellular components. Live-cell imaging has been successfully applied to mitotic cells, including investigations of chromosome dynamics and cytoplasmic division *(6)*, and methods have been developed to successfully maintain dividing cells during microscopic observation *(7)*.

The Golgi apparatus is a central organelle of the secretory pathway that is involved in protein processing and sorting *(8)*. It is also one of the few organelles to undergo a dramatic stepwise disassembly and reassembly in mitosis (**Fig. 1**) *(9–12)*. Currently the mitotic fate of Golgi proteins and lipids are intensely debated *(13)*. Two models of breakdown and reassembly for the Golgi apparatus have emerged that are essentially outgrowths of whether the Golgi apparatus can be considered as an autonomous organelle in the cell. In the first model the Golgi apparatus is considered as an autonomous organelle responsible for its biogenesis *(14)*, which becomes gradually vesiculated in mitosis into small (approx 50–100 nm) Golgi vesicles, which are then stochastically segregated between the daughter cells and eventually find each other and coalesce to form a new Golgi apparatus *(10,11,15)*. In the second model, Golgi inheritance occurs through the intermediary of the ER. Golgi fragmentation/dispersal results from mitotic changes in the membrane-trafficking pathways controlling Golgi

GalT-YFP/ Histone 2B-CFP

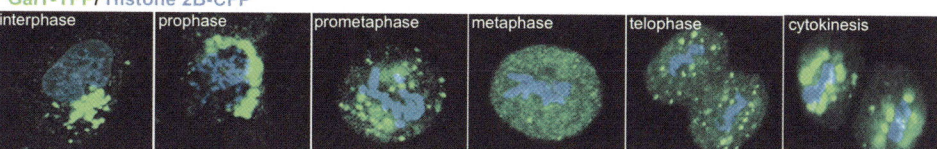

Fig. 1. A panel of images taken of a normal rat kidney (NRK) cell co-expressing GalT-YFP and histone 2B-CFP as markers for the Golgi apparatus and chromatin, respectively. Note that both the Golgi and chromosomes undergo dramatic changes in morphology during mitosis.

outgrowth and absorption into and from the ER *(9,12,16)*. This leads to Golgi proteins redistributing into the ER or to ER export domains at different stages of mitosis.

Understanding the fate of the Golgi apparatus in mitosis is an important cell biological problem because Golgi division is necessary for regulating entry into mitosis, normal chromosome segregation, and cytokinesis *(17,18)*. Moreover, finding out the fate of the Golgi apparatus in mitosis and the molecular machinery regulating its breakdown and reassembly has bearing on the fundamental issues of autonomy in organelle maintenance and organelle biogenesis. In this chapter we describe practical methods to be able to prepare, handle, and image the Golgi apparatus in living mitotic cells. These methods can also be applied to investigations of other organelles and molecules in mitotic cells.

2. Materials

1. Mammalian cell lines (e.g., HeLa, normal rat kidney [NRK], CHO, COS, etc.) maintained and passaged in the appropriate tissue culture media (e.g., DMEM, RPMI).
2. Tissue culture media, appropriate for cell type, supplemented with fetal bovine serum (FBS).
3. Tissue culture media, appropriate for cell type, without serum, to be used in the Fugene transfection protocol.
4. DNA plasmids encoding green fluorescent protein (GFP), cyan fluorescent protein (CFP), yellow fluorescent protein (YFP) or other fluorescent protein-tagged Golgi resident enzymes (e.g., galactosyltransferase-GFP, mannosidase II-GFP).
5. Fluorescently labeled lipid probes NBD-ceramide and BODIPY-sphingomyelin (Invitrogen).
6. Lab-Tek (Nalgene Nunc International) or Mat-Tek (Mat-Tek Corporation) brand coverglass bottom, covered imaging chambers allow for growing cells and visualizing them using high-numerical-aperture objectives.

7. HEPES buffer, 1 *M* pH7.3 (Invitrogen) to maintain pH during imaging.
8. Fugene 6 DNA transfection reagent (Roche Pharmaceuticals) for mammalian cells.
9. Aphidicolin and nocodazole (Sigma). Both Nocodazole and aphidicolin were made up in dimethyl sulfoxide (DMSO) as 5 mg/mL stock solutions and kept frozen at −20°C.
10. Brefeldin A (Sigma), to test for the Golgi localization of a marker, was made up in Ethanol as a 5 mg/mL stock solution and kept frozen at −20°C.
11. Hoechst 33342 (Molecular Probes, Invitrogen) to stain DNA was made up in water as a 5 mg/mL stock solution and kept at 4°C.
12. 2% Formaldehyde solution. Dissolve 2 g of paraformaldehyde powder (Sigma) in 100 mL of 1X phosphate-buffered saline by heating to 70°C while stirring in a fume hood. Cool solution to room temperature, then adjust pH to 7.4 using HCL or NaOH.
13. Vaseline (Cheeseborough Inc.) to seal imaging chambers.
14. Nevtek Airstream incubator ASI 400 to heat up microscope stage (Nevtek Corp).
15. Digital thermometer and probe to measure stage temperature (Omega Instruments).
16. Laser scanning confocal microscope system such as the LSM510 series by Zeiss Instruments (Carl Zeiss) (*see* **Note 1**) equipped with Argon and HeNe lasers providing 458-, 488-, 514-, 543-, and 633-nm excitation lines. For excitation of photoactivatable GFP, the system can be equipped with a Krypton laser providing 413-nm excitation or with a Ti:sapphire laser such as the Chameleon laser (Coherent Instruments) for multiphoton excitation.

3. Methods

3.1. Choosing Cell Types

In all mammalian cells the Golgi apparatus is disassembled and reassembled during mitosis. Thus, in principle any mammalian cell type can be chosen for investigation. However, in practice the choice of cell type(s) used for any study will be influenced by the ease of transfection with plasmids for that cell type, cell synchronization efficiency, cell size, Golgi apparatus size, and the morphology of the cell during mitosis (which will affect the quality of the images acquired, i.e., does it stay relatively flat and attached to the dish, or is it nonadherent in mitosis). Furthermore, in some cell types, such as PtK cells, there is less extensive dispersal of Golgi resident proteins in metaphase then in other cell types (*e.g.,* HeLa and CHO) *(9,19,20)*. Therefore, it is advisable to investigate the breakdown and reassembly of the Golgi apparatus in a few different cell types.

3.2. Labeling the Golgi Apparatus with Fluorescent Reporters in Living Cells

The Golgi apparatus is a complex organelle whose components include transmembrane proteins, peripheral membrane proteins, and lipids. Any one or more of these components can in principle be labeled with a fluorescent reporter, and imaging representatives from each component group will provide a more complete representation of the fate of the whole Golgi apparatus in mitosis.

3.2.1. Labeling the Golgi with Fluorescent Proteins

The surface of the Golgi is host to thousands of different cytoplasmic proteins, including signalling, cytoskeletal, and nuclear regulatory proteins. These proteins included MAP kinases, tankyrase, dynein, myosin isoforms, protein kinase A, calmodulin kinase, casein kinase, etc. These proteins all bind to the cytoplasmic leaflet of the Golgi membrane bilayer (16,18,21). Many of these proteins are released from the Golgi apparatus into the cytoplasm as early as prophase and reassociate with Golgi membranes at the end of mitosis in telophase (16). Thus, they cannot be relied upon solely as a reporter for the Golgi apparatus during mitosis.

To specifically follow Golgi membranes, one can choose a transmembrane protein or a lipid. Almost all Golgi resident enzymes are transmembrane proteins. Common markers chosen as reporters for the Golgi membranes have been galactosyltransferase (GalT), *N*-acetylgalactosaminyltransferase (GalNac), and mannosidase (9–12,18,22). These proteins can be tagged with a spectrum of fluorescent protein tags including GFP (23), its spectral variants, cyan fluorescent protein CFP and yellow fluorescent protein YFP, and the respective brighter variants Cerulean (24) and mCitrine (25); monomeric red fluorescent protein (mRFP1) (26) and its modifications with different hues such as mCherry, mbanana, anmOrange (27); and a photoactivatable green fluorescent protein (PA-GFP) (28).

One rapid way to ensure that a GFP-tagged protein is targeted properly to the Golgi apparatus is by treating cells with Brefeldin A (BFA). BFA is a fungal toxin that rapidly and reversibly blocks anterograde transport of membrane out of the ER and accelerates retrograde transport of membranes from the Golgi back to the ER (29). The end result, Golgi apparatus redistributing to the ER, can be visualized by microscopy.

An additional way to determine if the tagged protein is targeted properly to the Golgi apparatus is by colocalization with a native Golgi protein. Cells expressing the GFP-tagged protein can be fixed with a 2% formaldehyde solution (the relatively low percentage of formaldehyde in this solution is intended to minimize dimming of GFP fluorescence) and immunostained with an antibody against the resident nontagged Golgi protein *(30)*.

3.2.2. Transfection

There is a variety of transfection reagents on the market. We have frequently used Fugene 6 transfection reagent, a lipid-based reagent, to transfect cells that we will image for mitosis primarily because one can add this reagent directly to the media containing the synchronization agent during the synchronization time and there is minimal cell death associated with keeping Fugene on the cells for long periods (even up to 3–4 d) of time. The protocol for transfection should be followed as directed by the manufacturer. For a typical transfection, utilize excess Fugene reagent volume to DNA in micrograms (i.e., 3–4 μL Fugene: 1 μg DNA). Mix Fugene with serum-free media that is buffered to pH 7–7.2 at room temperature and make sure to expel Fugene reagent directly into the media because it loses activity by binding to the sides of polypropylene tubes. Incubate for 5 minutes at room temperature, then:

1. Pipet mixture into separate tube containing DNA.
2. Mix again by inverting tube several times.
3. Let stand 15–30 min at room temperature.
4. Pipet the mixture into chambers containing cells that are in the media with synchronization agent. For a chamber with 1.5 mL of media, pipet approx 100 μL of Fugene/serum-free media/DNA mixture. A chamber with approx 500 μL of media in each well will need at least 50 μL of the transfection mixture per well.

3.2.3. Labeling the Golgi Apparatus with Fluorescent Lipids

The Golgi apparatus is rich in sphingolipids and ceramide *(31)*, and thus these lipids provide another source of Golgi reporters for the investigator. Fluorescent lipids, such as NBD-ceramide and BODIPY sphingomyelin, are rapidly taken up by cells from the culture media and concentrated at the Golgi apparatus *(32)*. Imaging the Golgi apparatus via fluorescent lipids has the advantage of allowing the investigator to not only be able to monitor the dynamics and metabolism of Golgi lipids but also of labeling the Golgi membranes acutely in the cell cycle. Potential caveats to working with fluorescent lipids include fluorescent tags (e.g., NBD or BODIPY) having relatively broad emission spectrums that can limit the use of secondary fluorescent labels such as GFP/YFP/mRFP-tagged proteins within the same cells and that over a number of h the lipids can be transported into other cellular membranes (e.g., plasma membrane, ER) because of membrane trafficking.

To label the Golgi apparatus with NBD-ceramide:

1. Cells are incubated 5 μ*M* NBD-ceramide in serum-free media containing 20 m*M* HEPES pH 7.3 at 4°C for 10 min. This lipid will become incorporated into the outer leaflet of the plasma membrane bilayer at 4°C. This 4°C incubation should be done prior to entry into mitosis such that the cold temperature does not interfere with microtubule dynamics and the kinetics of mitotic progression.
2. Wash cells twice with DMEM/HEPES/10% FBS.
3. Incubate cells at 37°C for at least 30 min. During this time the NBD-ceramide will be taken up via endocytosis, and the lipid will move to the Golgi apparatus. At the Golgi it will become metabolized into sphingomyelin and temporarily trapped.
4. The fluorescent lipid can be imaged using 488-nm excitation light and 515-nm emission light filters.

3.3. Synchronizing Mammalian Cells to Enrich the Mitotic Population

Numerous methods for synchronizing mammalian cells have been reported in the literature *(33–35)*. Common to all these methods is being able to *reversibly* stall or stop the cells at a specific stage of the cell cycle and then release. In theory, after the cells have been released from the treatment, they should all progress in synchrony towards mitosis. In practice this is not the case and the number of cells undergoing mitosis in synchrony will be based on a variety of factors, including the efficacy of the synchronizing agent and the health of the cells after treatment with the agent. The strength of live-cell imaging over other methods of assaying mitotic cells is that it allows the experimenter to locate and focus on only the groups of cells undergoing mitosis on the dish. Therefore, even if the synchronization is not highly efficient, there will be usually sufficient number of mitotic cells for the experimenter.

There are several methods for synchronization. Serum starvation for 12–48 h will force cells into the quiescent state G0 *(36)*. Cells with longer doubling times will need longer serum-free media incubation times. The major disadvantages of this method are that it takes long periods of time to achieve quiescence and the lack of serum during this time may be detrimental to the physiology of some cell types.

Cells can be also synchronized with low levels (typically 25–50 ng/mL) of the microtubule-depolymerizing agent nocodazole. The cells will be blocked after they have gone into mitosis at the prometaphase stage. The advantage of this method is that very large numbers of cells will be in mitosis when the nocodazole is washed off the cells and hence can be followed under the microscope. But washing off the nocodazole from mitotic cells is tricky because the cells are rounded and not firmly attached to the dish, which usually results in a loss of great number of cells from the dish. On the other hand, insufficient washing will prevent the normal progression through mitosis. Moreover, synchrony in practice is not maintained well for the next round of mitosis, and this could potentially be a problem for the experimenter, who needs to follow the cells from the very beginning of mitosis and to treat the cells with reagents acutely prior to entry into mitosis *(37)*. An additional disadvantage of synchronization by depolymerizing microtubules is that ER-to Golgi-membrane trafficking pathways are dependent on microtubules, and nocodazole treatment will back up Golgi proteins at ER exit sites *(38)* and potentially create a phenotype unique to nocodazole arrest.

An alternative method is to block cells with a chemical inhibitor earlier in the cell cycle. Aphidicolin can block cells at the G1/S boundary by reversibly inhibiting DNA replication. The advantage of a block early in the cell cycle is that the inhibitor can be washed out long before the cells enter mitosis, lessening the chance there may be remaining effects of the inhibitor on mitosis. One potential disadvantage is that the time the cells enter mitosis will be relatively broad (e.g., 30–60 min) compared to nocodazole, where the cells are already arrested in mitosis, when the block is removed.

Typical aphidicolin concentrations reported in the literature range from 2 to 15 μg/mL and incubation times from 12 to 24 h. The experimenter should empirically determine the concentration and time for aphidicolin treatment for their specific cell type. The optimal time will be the time after which there is minimal cell death but maximum cell synchronization and entry into mitosis. Below is a protocol that we use for NRK cells.

3.3.1. Sample Aphidicolin Synchronization Protocol for NRK Cells

1. Cells are plated on Lab-Tek or MatTek chambers 24–36 h prior to the commencement of synchronization at 50–70% density (*see* **Note 2**). These chambers are deep-welled chambers with coverglass bottoms in order to be able to do high-magnification and high-resolution microscopy with oil or water immersion objectives. Most cells will directly adhere to the glass, but if cells do not adhere, the coverglass can be coated with polylysine or another extracellular matrix material.
2. If Fugene is used as a transfection reagent, then the reagent/DNA mixture can be added directly into the chambers with the aphidicolin and left on for the duration of the aphidicolin treatment.
3. The duration of aphidicolin treatment depends on the sensitivity of the cell type to aphidicolin treatment and the average time it takes for that cell type to complete one cell cycle. For NRK cells we incubate with 10 μg/mL of aphidicolin for approx 16–18 h, which provides sufficient time for the majority of the cells to come to a

block at G1/S boundary from whatever stage of the cell cycle they were at when aphidicolin was added, with minimal cell toxicity.

4. After 16–18 h the aphidicolin-containing media is washed away from the NRK cells with at least six changes of fresh cell culture medium to ensure that all the aphidicolin has been removed. Leaving aphidicolin behind will result in cell death and/or low numbers of cell entry into mitosis.

5. We follow the cells for the next 6–12 h to determine the time at which cells begin undergoing mitosis. Once this time is determined it should vary little from experiment to experiment as long as similar numbers of cells were plated. Typically, for NRK cells, more than 80% of the cells undergo mitosis between 8 and 10 h post-aphidicolin washout.

3.4. Preparing Cells for Imaging and Temperature Control of the Stage

1. A few hours before entry into mitosis, change the media of the cells into an imaging medium, which is typically the tissue culture media in which the cells are normally maintained (containing serum) and buffered. The imaging media can also be placed on the cells in the final wash after aphidicolin incubation. The imaging medium should be buffered with 20–50 mM HEPES pH 7.3. HEPES is a buffer well tolerated by many cell types.

2. In most applications in light-level microscopy where the excitation range is approx 400–500 nm, the background fluorescence from phenol red, a pH indicator commonly found in tissue culture media, is not strong enough to interfere with the acquisition of the fluorescence emission from the sample, and indeed having phenol red can be a convenient pH indicator of the culture media during the course of the experiment. However, in some samples to be imaged there may be low levels of fluorescently tagged reporters and/or weakly emitting fluorescent tags, and therefore it is up to the investigator, by acquiring some preliminary scans on the microscope, to determine if the background fluorescence from phenol red is too high for their particular application. If so, then media can be replaced with one missing phenol red.

3. The imaging chambers should be filled as much as possible with the media. This ensures a large volume above the cells that will not evaporate over the course of 6–12 h of imaging. Evaporation of media will cause changes in the salt concentrations and pH of the media.

4. To minimize evaporation of the media, the inside of the chamber lid should be smeared with a coat of Vaseline, and the lid should then be fitted on the chamber. Alternatively, the lid can be placed on the chamber, then the outside edges glued to the chamber.

5. There are several options for temperature control of the microscope stage. A Nevtek Airstream incubator in combination with a digital thermometer with a thermocouple whose temperature sensor is attached to the head of the condenser or to a site on

the stage adjacent to the cell chamber is a convenient setup for holding steady temp ($+/-0.5°C$) in most applications requiring imaging up to 12 h *(9,12,39,40)*. It also allows easy access to the cell chamber for adding reagents into the cells in the course of the experiment while they are on the stage without shaking the cell chamber or stage, both of which could result in the loss of the position of the cell being imaged. The drawback is that because of exposure to ambient CO_2 and humidity, over time the media will gradually evaporate and become basic in pH. Thus, if one is planning on imaging for time periods greater than 10–12 h, it is advisable to invest in an enclosed chamber setup, one that either can be mounted on the stage (e.g., Bioptechs perfusion chamber, Bioptechs Inc., Butler, PA) or one that encloses the whole body of the microscope where humidified, heated air and CO_2 can be piped in to the chamber. These types of chambers can usually be obtained from the manufacturer of the microscope or even custom-manufactured to the desired specifications (Precision Plastics Inc., Columbia City, IN).

3.5. Setting Up the Microscope for Imaging: Parameters for Choosing Objectives, Pinhole Size, Laser Intensity, and Frequency of Acquisition

1. About 30 min to 1 h before cells begin to enter mitosis, the cell will be brought to the microscope stage, which should already have been heated to the desired temperature. Warming the temperature of the stage and objectives prior to imaging will minimize focus shifts as a result of the warm-up of the objective and stage during acquisition and also minimize the effects of changes in temperature on the physiology of the cells.
2. For acquiring high-resolution images of the Golgi in one or few cells it is best to use high magnification and high-numerical-aperture objectives such as: water immersion 40X/1.2 NA or 63X/1.2 NA; oil immersion 40X/1.3NA or 63X/1.4NA objectives.
3. Because cells or tissues to be imaged are often in an aqueous solution, water immersion objectives will provide the images with the least optical aberrations. Oil immersion objectives provide images that are comparable in resolution and brightness to water immersion objectives only when samples are close to the glass bottom of the chamber. When samples are thick or removed,tens of micrometers from the surface of the coverglass, optical aberrations, resulting from refractive index mismatch, become more obvious *(41)*. Especially when acquiring high-resolution z-stack series from mitotic cells, which often are significantly rounded up off the cover slip surface, it is advisable to use water immersion objectives to minimize optical aberrations.
4. When appropriate mitotic cells have been identified (*see* **Subheading 3.6.**) for imaging, quick scans should be acquired while adjusting the focus, the pinhole, laser intensity, and detector gain to determine the optimal image settings.
5. The pinhole setting should be reflective of the type of imaging to be done. For experiments where the goal is to acquire the highest resolution images possible, it

is better to choose a high-numerical-aperture objective and a small pinhole setting, typically liss than 2 Airy units. Objectives with large numerical aperture are more efficient in light collection relative to low-numerical-aperture objectives. Setting small pinhole diameter will minimize the collection of fluorescence emission out of the focal plane. However, going below 1 Airy unit is usually not recommended because there is very little resolution gained to compensate for the significant loss in light collection from the sample *(41)*. In **Fig. 2** we illustrate the dramatic difference in the visual information that can be obtained from a sample simply when the pinhole is changed. We have taken images of the Golgi resident enzyme marker galactosyltransferase tagged with YFP (GalT-YFP) in a metaphase cell taken with two different pinhole settings; all other settings were the same. In the top panel, the pinhole is wide open (8 Airy units, {GT}10 μm thickness, 63X/1.4N.A. objective), whereas in the bottom panel the pinhole is closed down to 1.2 Airy units (*see* **Note 3**).

6. Greater resolution can be discernible if the scan area is comprised of as many pixels as available. For example, more detail will be observed on a scan area of 1024×1024 pixels compared to a scan area of 256×256 pixels. However, greater scan area leads to slower acquisition times, which may not be optimal in experiments where the sample being imaged is moving relatively fast.

7. For quantitative imaging, in which the fluorescence per pixel in the images will be measured after the image is obtained, it is advisable to use a wide-open pinhole setting to collect fluorescence from the entire depth of field (*see* **Note 4**). In addition high magnification and high-numerical-aperture objectives are limited in their depth of light collection; therefore, lower-numerical and lower-magnification objectives are recommended for accurate fluorescence quantification.

8. The dynamic range of the detectors, which are typically photomultiplier tubes, should be as great as possible (e.g., 12-bit detectors will give you a dynamic range of 4096 gray levels) to capture both very dim objects and very bright objects. The gain on these detectors should be optimized on sample scans such that dim objects can be visualized without the saturation of pixels for the bright objects. If saturation occurs, quantitation of the fluorescence within the saturated pixels will not be accurate.

9. If a time series is to be acquired with frequent image acquisition, it is important to minimize the laser excitation as much as possible in order to decrease photo-bleaching the fluorophore and to not block entry or progression in mitosis. However, a low laser excitation intensity with high detector gain can give noisier images. Thus, it is a trade-off, and the optimal parameters should be determined for every sample to be imaged (*see* **Note 5**).

3.6. Identifying Mitotic Cells

Once cells are placed on the microscope, using either brightfield or epifluo-rescence, the field should be browsed for cells about to undergo mitosis. Mitotic

Metaphase

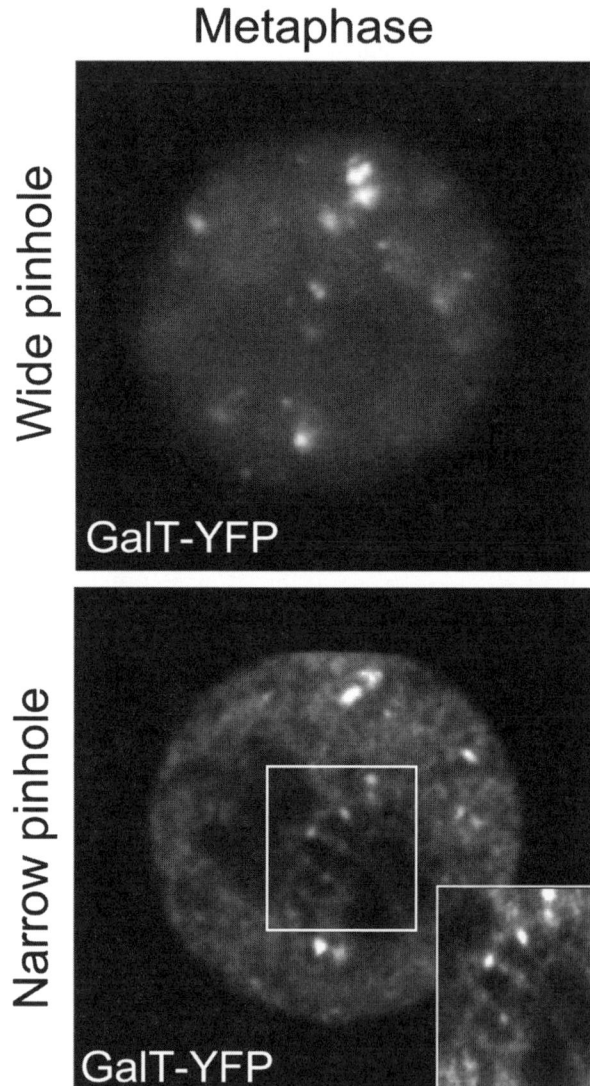

Fig. 2. A mitotic normal rat kidney (NRK) cell expressing GalT-YFP was imaged at metaphase using a 63X/1.4 NA oil immersion obejective and two different pinhole settings. (**Top panel**) The pinhole is fully open, 8 Airy units; (**bottom panel**) the pinhole is narrow, 1.2 Airy units.

cells and the mitotic stages can be identified by a number of independent criteria, including the degree of chromosome condensation, the extent of cell rounding-up, centriole and spindle organization, and the morphology of the Golgi apparatus.

The extent of chromosome condensation in cells can be assessed by phase contrast light microscopy, by co-expression of histone 2B-YFP/CFP (**Fig. 1**), and/or by staining with an intercalating DNA dye such as Hoechst 33342 *(18)*, but note that the latter will block cells in mitosis once it is added. Cells typically round up as they undergo mitosis, with rounding becoming most apparent at metaphase (**Fig. 1**, metaphase) (*see also* **Note 6**).

Another method for staging mitotic cells is to look at the organization of the microtubules. This can be done in living cells without hindering mitotic progression by microinjecting in to cells prior to mitosis a fluorescent tag for tubulin (e.g., tubulin GFP or tubulin FITC). The two centrioles will move apart from each other in early prophase, around the nuclear envelope, to opposite ends of the nucleus. In prometaphase there is a rapid depolymerization of cytoplasmic microtubules and the growth of the spindle is apparent *(42)*. In anaphase/telophase the spindle will be elongated *(43)*.

Finally, the distinct morphological changes undergone by the Golgi membranes are an excellent identifier of the mitotic stage of the cell (**Fig. 1**). In prophase the Golgi membranes are extended around the nuclear envelope. In prometaphase Golgi membrane proteins are found in scattered fragments across the cell. In metaphase these proteins are dispersed, and few fragments remain in the cell. In telophase Golgi membrane proteins reemerge in fragments, which coalesce to form a Golgi in cytokinesis.

3.7. Highlighting the Golgi Pool of Fluorescence in the Cell

Because almost all Golgi resident membrane proteins are cycling between the Golgi and ER compartments, at any point in time there will be Golgi proteins in the ER. There will also be Golgi proteins in the ER because of new synthesis. One way to exclude the newly synthesized pool is by pretreating cells with cycloheximide to block protein synthesis. However, this treatment is not always applicable because it can interfere with the progress of mitosis in many types of cells. There are, however, two alternative methods that will not hinder the progression of mitosis but still allow the investigator to follow exclusively the Golgi pool of enzymes.

The first method consists of simply photobleaching all fluorescent labels outside the Golgi apparatus. It is termed performing an inverse fluorescence recovery after photobleaching, or iFRAP *(5)*. This can be done easily on laser scanning confocal microscopes such as the Zeiss LSM510 (Carl Zeiss) by "drawing" a region of interest (ROI) inside the cell using the software provided by the microscope manufacturer. This ROI will include everything in the cell except the Golgi apparatus. Then, with a short laser pulse, typically approx15 s, depending on the size of the region to be photobleached, of high-intensity laser beam of the correct wavelength for excitation of the fluorophore-tagged enzyme, all non-Golgi pools of fluorescence can be photobleached (**Fig. 3A**). It is also important to perform a whole-cell photobleach experiment, where the region of interest includes the entire cell that is about to undergo mitosis, and then monitor the recovery of fluorescence to determine the time it takes for fluorescence to revert, as a result of new protein synthesis, folding of newly synthesized pools of protein or even to some reversible fluorescence recovery. In mitotic cells new protein synthesis is for the most part inhibited, so very little, if any, fluorescence should come back to the photobleached area.

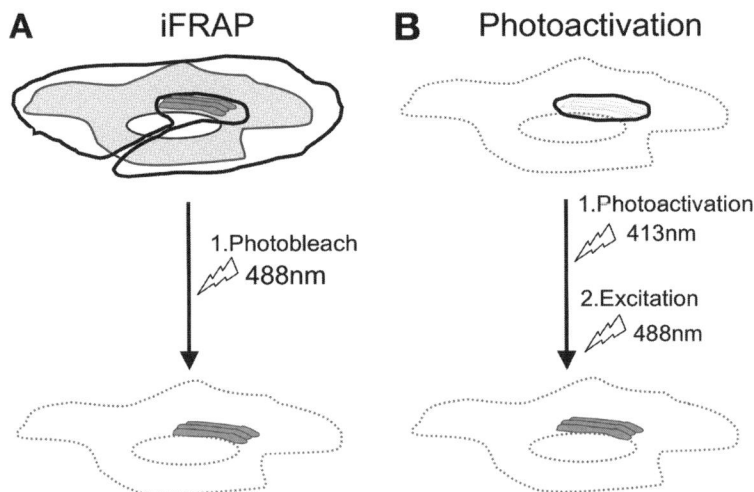

Fig. 3. Two ways of highlighting the Golgi pool of fluorescence. (**A**) Cell is expressing a Golgi resident protein tagged with green fluorescent protein (GFP) that normally cycles between the Golgi and endoplasmic reticulum compartments. To monitor only the Golgi pool of GFP-tagged protein, we photobleach the non-Golgi pool of fluorescence with a high-intensity laser beam of 488 nm. (**B**) We can also express the same Golgi resident protein with a photoactivatable GFP tag. Using a 413-nm high-intensity laser light, we photoactivate only the Golgi pool of protein in the cell, followed by excitation of this Golgi pool with 488-nm light to obtain GFP fluorescence emission.

The second method involves labeling the cells with a photoactivatable GFP-tagged Golgi resident protein *(28)*. This tag, unless photoconverted with 413-nm light, will not emit much green fluorescence when excited with 488 nm. Once it is converted, it can be excited with 488-nm light and emit green fluorescence ({GT}505 nm) (*see* **Note 7**). The increase in fluorescence is greater than 60-fold after photoactivation *(28)*. Once the Golgi region of interest is photoactivated, one can follow exclusively the fate of this pool of proteins (**Figs. 3B** and **4B**) without worrying about interference from newly synthesized or folded pools of Golgi resident proteins at the ER (*see* **Note 8**).

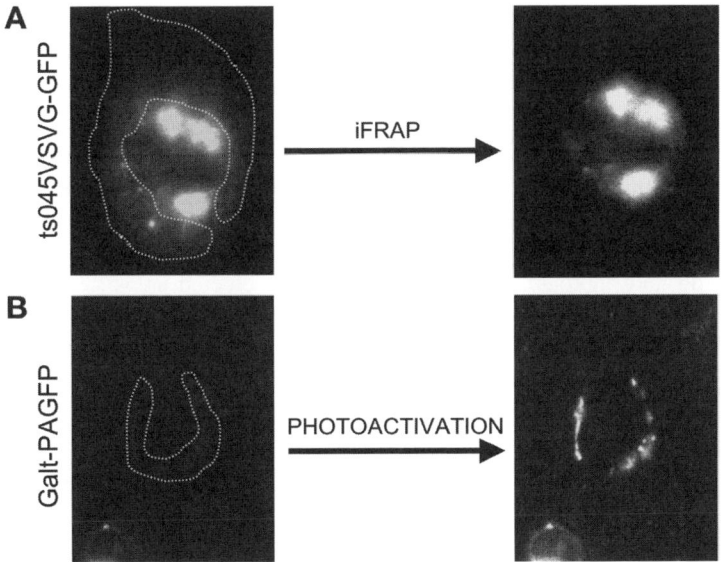

Fig. 4. Examples of the use of iFRAP and photoactivatable green fluorescent protein (GFP) to highlight Golgi pools of proteins in cells entering mitosis. (**A**) Cell is expressing the temperature-sensitive ts045VSVG-GFP that has accumulated at the Golgi after a 15-min temperature shift from 40 to 32°C. Note that there is still ts045VSVG-GFP present in the non-Golgi pool consisting of the endoplasmic reticulum. The non-Golgi pool of ts045VSVG-GFP (dashed line) is photobleached with a high-intensity 488-nm light, leaving behind only the Golgi pool of ts045VSVG-GFP fluorescence. (**B**) Cell is expressing a photoactivatable GFP-tagged galactosyltransferase enzyme, Galt-PAGFP. The Golgi region (dashed line) is photoactivated by a short pulse ({LT}2 s) of 413-nm light. Immediately following the photoactivation, the cell is excited with 488-nm light, revealing the Golgi pool of GalT-PAGFP fluorescence.

The Golgi pool should be highlighted prior to entry into mitosis (in late G2 phase) or in early prophase before the Golgi membranes have disassembled. However, if the highlighting is done too early ({GT}30 min prior to mitosis), trafficking between the Golgi and ER compartments will again have dispersed the Golgi pools of fluorescence into the ER compartments. Once the Golgi pool of fluorescent molecules is highlighted either by iFRAP or by photoactivation, the dynamics of this pool of fluorescent molecules can be exclusively followed throughout the course of mitosis (**Fig. 4**) by using time-lapse microscopy where images are acquired at defined intervals of time.

Notes

1. One of the most important decisions investigators will make will be choosing the right imaging system to fit their experimental needs. A number of laser scanning–confocal microscope systems are available in addition to deconvolution and spinning disc microscope systems. All have their advantages and disadvantages, and ultimately the choice will depend on the types of experiments being performed. For imaging the dynamics of the Golgi apparatus as well as other organelles and molecular machinery in living cells, we have successfully used the Zeiss LSM510 confocal laser scanning system, which has allowed us to acquire fast time-lapse images in $x - y - z$ planes, easily implement a variety of photobleaching and photoactivation experiments on selected areas within cells and the capability to write macros that can combine a sequence of time-lapse imaging and photobleaching functions on many cells by marking the position of cells on the stage and thereby acquiring data on multiple cells under the same experimental conditions.

2. The density of cells plated is an important parameter for aphidicolin-based synchronization protocols. If too few cells are plated ({LT}50% of surface area covered with cells), aphidicolin treatment is not successful and even toxic. It is not well understood, but the presence of a sufficient number of cells may provides a conditioned medium for survival and continuation of the cell cycle during and after aphidicolin treatment. However, if too many cells are plated, then the efficacy of synchronization is diminished. The optimal density is between 50 and 70% confluence on the dish.

3. When the pinhole setting is reduced, greater detail becomes discernible, such as the reticular staining pattern of the Golgi enzymes (*see* **Fig. 2**, inset), whereas when the pinhole is wide open, the image is of a haze.

4. The thickness of the sample to be imaged is important to determine in order to set the correct pinhole size in experiments where ultimately the fluorescence in volume of interest will be quantitated. For example, if the volume to be quantitated is 10 μm in thickness, the pinhole will need to be adjusted such that the focal plane thickness will be at least 10 μm. In addition, high-magnification and high-

numerical-aperture objectives are limited in their depth of light collection, and therefore lower-numerical and lower-magnification objectives are recommended for accurate fluorescence quantification.

5. GFP and CFP tags are more resistant to photobleaching than YFP tags. Hence it is recommended to minimize the excitation light as much as possible when imaging YFP-tagged proteins. In addition, exciting cells with 413-nm light to image a CFP-tagged marker can at times slow down the mitotic progression or even halt it altogether. Therefore, when imaging CFP-tagged proteins, especially when they are attached to nuclear markers such as histone 2B (**Fig. 1**), it is recommended that the 413-nm excitation be kept to a minimum or to excite with a longer wavelength of light, i.e., 458-nm laser light (this wavelength of light is emitted by Argon lasers), which will be less damaging to mitotic progression *(44)*.

6. However, the extent of this behavior depends on cell type and should not be solely relied upon as the basis of identification of mitotic cells because apoptotic dying cells will also frequently round up.

7. Prior to photoactivation, the Golgi region will appear as dim fluorescence with the GFP filters. This may be sufficient for drawing a region of interest around the Golgi apparatus for photoactivation. Alternatively, one can co-express a different fluorescently tagged Golgi marker (perhaps a CFP- or RFP-tagged marker) to be able to independently locate the Golgi apparatus.

8. One potential drawback of photoactivation at the onset of mitosis is that the high-energy 400-nm light used for photoactivation can arrest the cell cycle. To avoid this problem one should assay and record the settings for different intensities of 400-nm light utilized in order to obtain maximum photoactivation and minimum cell-cycle arrest. Another option to minimize cell damage and mitotic progression is to photoactivate using a multiphoton laser line of approx 800–850 nm.

Acknowledgments

The author would like to thank Jennifer Lippincott-Schwartz for initiating and guiding the development of many of the techniques presented here and Gregoire Altan-Bonnet for critically reading this manuscript.

References

1. Nasmyth, K., Peters, J. M., and Uhlmann, F. (2000) Splitting the chromosome: cutting the ties that bind sister chromatids. *Science* **288**, 1379–1384.
2. Bulina, M. E., Chudakov, D. M., and Britanova, O. V., et al. (2006) A genetically encoded photosensitizer. *Nat. Biotechnol.* **24**, 95–99.
3. Chudakov, D. M., Verkhusha, V. V., Staroverov, D. B., Souslova, E. A., Lukyanov, S., and Lukyanov, K.A. (2004) Photoswitchable cyan fluorescent protein for protein tracking. *Nat. Biotechnol.* **22**, 1435–1439.
4. Lippincott-Schwartz, J. (2004) Dynamics of secretory membrane trafficking. *Ann. N.Y. Acad. Sci.* **1038**, 115–124.

5. Lippincott-Schwartz, J., Altan-Bonnet, N., and Patterson, G. H. (2003) Photo-bleaching and photoactivation: following protein dynamics in living cells. *Nat. Cell Biol.* Suppl:S7–14.

6. Mitchison, T. J. and Salmon, E. D. (2001) Mitosis: a history of division. *Nat. Cell Biol.* **3**, E17–21.

7. Khodjakov, A. and Rieder, C. L. (2006) Imaging the division process in living tissue culture cells. *Methods* **38**, 2–16.

8. Marsh, B. J. and Howell, K. E. (2002) The mammalian Golgi-complex debates. *Nat. Rev. Mol. Cell Biol.* **3**, 789–795.

9. Zaal, K. J., Smith, C. L., Polishchuk, R. S., et al. (1999) Golgi membranes are absorbed into and reemerge from the ER during mitosis. *Cell* **99**, 589–601.

10. Pecot, M. Y. and Malhotra, V. (2004) Golgi membranes remain segregated from the endoplasmic reticulum during mitosis in mammalian cells. *Cell* **116**, 99–107.

11. Axelsson, M. A. and Warren, G. (2004) Rapid, endoplasmic reticulum-independent diffusion of the mitotic Golgi haze. *Mol. Biol. Cell* **15**, 1843–1852.

12. Altan-Bonnet, N., Sougrat, R., Liu, W., Snapp, E. L., Ward, T., Lippincott-Schwartz, J. (2006) Golgi inheritance in mammalian cells is mediated through endoplasmic reticulum export activities. *Mol Biol Cell* **17**, 990–1005.

13. Barr, F. A. (2004) Golgi inheritance: shaken but not stirred. *J. Cell Biol.* 164, 955-958.

14. Pelletier, L., Stern, C. A., Pypaert, M., et al. (2002) Golgi biogenesis in *Toxoplasma gondii*. *Nature* **418**, 548–552.

15. Shorter, J. and Warren, G. (2002) Golgi architecture and inheritance. *Annu. Rev. Cell. Dev. Biol.* **18**, 379–420.

16. Altan-Bonnet, N., Sougrat, R., and Lippincott-Schwartz, J. (2004) Molecular basis for Golgi maintenance and biogenesis. *Curr. Opin. Cell. Biol.* **16**, 364–372.

17. Sutterlin, C., Hsu, P., Mallabiabarrena, A., and Malhotra, V. (2002) Fragmentation and dispersal of the pericentriolar Golgi complex is required for entry into mitosis in mammalian cells. *Cell* **109**, 359–369.

18. Altan-Bonnet, N., Polishchuk, R., Phair, R. D., Weigert, R., and Lippincott-Schwartz, J. (2003) A role for Arf1 in mitotic Golgi disassembly, chromosome segregation and cytokinesis. *Proc. Natl. Acad. Sci. USA* **100**, 13314–13319.

19. Schroeter, D., Ehemann, V., and Paweletz, N. (1985) Cellular compartments in mitotic cells: ultrahistochemical identification of Golgi elements in PtK-1 cells. *Biol. Cell.* **53**, 155–163.

20. Shima, D. T., Cabrera-Poch, N., Pepperkok, R., and Warren, G. (1998) An ordered inheritance strategy for the Golgi apparatus: visualization of mitotic disassembly reveals a role for the mitotic spindle. *J. Cell Biol.* **141**, 955–966.

21. De Matteis, M. A. and Morrow, J. S. (2000) Spectrin tethers and mesh in the biosynthetic pathway. *J. Cell. Sci.* **113**, 2331–2343.

22. Ward, T. H., Polishchuk, R. S., Caplan, S., Hirschberg, K., and Lippincott-Schwartz, J. (2001) Maintenance of Golgi structure and function depends on the integrity of ER export. *J. Cell. Biol.* **155**, 557–570.

23. Chalfie, M. (1995) Green fluorescent protein. *Photochem. Photobiol.* **62**, 651–656.

24. Rizzo, M. A., Springer, G. H., Granada, B., and Piston, D. W. (2004) An improved cyan fluorescent protein variant useful for FRET. *Nat. Biotechnol.* **22**, 445–449.

25. Griesbeck, O., Baird, G. S., Campbell, R. E., Zacharias, D. A., and Tsien, R. Y. (2001) Reducing the environmental sensitivity of yellow fluorescent protein. Mechanism and applications. *J. Biol. Chem.* **276**, 29188–29194.

26. Bevis, B. J. and Glick, B. S. (2002) Rapidly maturing variants of the Discosoma red fluorescent protein (DsRed). *Nat. Biotechnol.* **20**, 83–87.

27. Shaner, N. C., Campbell, R. E., Steinbach, P. A., Giepmans, B. N., Palmer, A. E., and Tsienm, R.Y. (2004) Improved monomeric red, orange and yellow fluorescent proteins derived from *Discosoma* sp. red fluorescent protein. *Nat. Biotechnol.* **22**, 1567–1572.

28. Patterson, G. H. and Lippincott-Schwartz, J. (2002) A photoactivatable GFP for selective photolabeling of proteins and cells. *Science* **297**, 1873–1877.

29. Lippincott-Schwartz, J., Yuan, L. C., Bonifacino, J. S., and Klausner, R. D. (1989) Rapid redistribution of Golgi proteins into the ER in cells treated with brefeldin A: evidence for membrane cycling from Golgi to ER. *Cell* **56**, 801–813.

30. Sciaky, N., Presley, J., Smith, C., et al. (1997) Golgi tubule traffic and the effects of brefeldin A visualized in living cells. *J. Cell. Biol.* **139**, 1137–1155.

31. van Meer, G. and Sprong, H. (2004) Membrane lipids and vesicular traffic. *Curr. Opin. Cell. Biol.* **16**, 373–378.

32. Lipsky, N. G. and Pagano, R. E. (1985) Intracellular translocation of fluorescent sphingolipids in cultured fibroblasts: endogenously synthesized sphingomyelin and glucocerebroside analogues pass through the Golgi apparatus en route to the plasma membrane. *J. Cell. Biol.* **100**, 27–34.

33. Cooper, S. (2003) Rethinking synchronization of mammalian cells for cell cycle analysis. *Cell. Mol. Life. Sci.* **60**, 1099–1106.

34. Amon, A. (2002) Synchronization procedures. *Methods Enzymol.* **351**, 457–467.

35. Krek, W. and DeCaprio, J. A. (1995) Cell synchronization. *Methods Enzymol.* **254**, 114–124.

36. Prather, R. S., Boquest, A. C., and Day, B. N. (1999) Cell cycle analysis of cultured porcine mammary cells. *Cloning* **1**, 17–24.

37. Cooper, S., Iyer, G., Tarquini, M., and Bissett, P. (2006) Nocodazole does not synchronize cells: implications for cell-cycle control and whole-culture synchronization. *Cell Tissue Res.* **324**, 237–242.

38. Cole, N. B., Sciaky, N., Marotta, A., Song, J., and Lippincott-Schwartz, J. (1996) Golgi dispersal during microtubule disruption: regeneration of Golgi stacks at peripheral endoplasmic reticulum exit sites. *Mol. Biol. Cell.* **7**, 631–650.

39. Rabut, G. and Ellenberg, J. (2004) Automatic real-time three-dimensional cell tracking by fluorescence microscopy. *J. Microsc.* **216**, 131–137.

40. Dabiri, G. A., Turnacioglu, K. K., Sanger, J. M., and Sanger, J. W. (1997) Myofibrillogenesis visualized in living embryonic cardiomyocytes. *Proc. Natl. Acad. Sci. USA* **94**, 9493–9498.

41. Pawley, J. B. (1995) *Handbook of Bological Confocal Microscopy*, 2nd ed. Plenum Press, New York.

42. Zhai, Y., Kronebusch, P. J., Simon, P. M., and Borisy, G. G. (1996) Microtubule dynamics at the G2/M transition: abrupt breakdown of cytoplasmic microtubules at nuclear envelope breakdown and implications for spindle morphogenesis. *J. Cell. Biol.* **135**, 201–214.

43. Waterman-Storer, C., Desai, A., and Salmon, E. D. (1999) Fluorescent speckle microscopy of spindle microtubule assembly and motility in living cells. *Methods Cell Biol.* **61**, 155–173.

44. Altan-Bonnet, N., Phair, R. D., Polishchuk, R. S., Weigert, R., and Lippincott-Schwartz, J. (2003) A role for Arf1 in mitotic Golgi disassembly, chromosome segregation, and cytokinesis. *Proc. Natl. Acad. Sci. USA* **100**, 13314–13319.

22

Evaluating Yeast Biosynthetic Vacuolar Transport

Brian A. Davies, Darren S. Carney, and Bruce F. Horazdovsky

Summary

Study of the lysosomal protein transport system has been facilitated through dissection of the analogous vacuolar protein sorting (VPS) pathway in *Saccharomyces cerevisiae*. Resident enzymes of the yeast vacuole are synthesized as inactive precursors and are cleaved to their mature forms upon delivery to this compartment. Quantitative assessment of this delivery can be achieved through the use of pulse-chase experiments monitoring the cleavage of zymogens to their mature forms. The experimental procedures for analysis of carboxypeptidase Y (CPY) and carboxypeptidase S (CPS) maturation are described.

Key Words: Carboxypeptidase Y; CPY; carboxypeptidase S; CPS; vacuolar protein sorting; VPS; yeast vacuole; ESCRT; MVB; endosome.

1. Introduction

The mammalian lysosome serves a vital function in maintaining cellular homeostasis (reviewed in **ref. *1***). In addition to functioning in the metabolism of various proteins and lipids delivered to the compartment via autophagy or phagocytosis, the lysosome serves to downregulate signal transduction pathways through the degradation of activated cell surface receptors delivered to the lumen of the lysosome via the endocytic pathway (reviewed in **ref. 2**). To accomplish these degradative functions, the lysosome is populated with a repertoire of hydrolases, including proteases, lipases, and phosphatases. The proper delivery of these enzymes to the lysosome is important for both maintaining the activity of the compartment itself as well as restricting the inappropriate localization and activation of these potentially harmful activities.

From: *Methods in Molecular Biology, Vol. 390: Protein Targeting Protocols: Second Edition*
Edited by: M. van der Giezen © Humana Press Inc., Totowa, NJ

The yeast *Saccharomyces cerevisiae* has provided a model system in which to dissect the machinery mediating transport to the mammalian lysosome through the study of protein targeting to the analogous yeast organelle, the vacuole (reviewed in **ref. 3**). The yeast vacuole is a hydrolytic compartment involved in macromolecular degradation as well as pH and ion homeostasis. Similar to the lysosome, the yeast vacuole contains a variety of hydrolases that are transported to the compartment as inactive precursors. The delivery of these proenzymes to the proteolytic environment of the vacuole results in cleavage of the propeptide and activation. Thus, function of the vacuolar protein sorting (VPS) system can be assessed by pulse-chase experiments monitoring the cleavage of precursors to their active forms over time.

Carboxypeptidase Y (CPY) and carboxypeptidase S (CPS) are two particularly useful cargoes of the VPS biosynthetic pathway. CPY is a soluble zymogen that transits the early stages of the secretory pathway through the endoplasmic reticulum (ER) and Golgi complex (*4*). During this process, the core-glycosylated CPY precursor (p1CPY) is modified further by Golgi-resident enzymes, resulting in the p2 precursor form (p2CPY). The vacuolar protein sorting receptor Vps10 then binds p2CPY in a trans-Golgi compartment and diverts the zymogen from the secretory pathway (*5*). Vps10 and p2CPY are transported to an endosomal compartment, after which p2CPY continues on to the vacuole to be activated by removal of the pro-peptide.

Delivery of CPS to the lumen of the vacuole is accomplished utilizing much of the same pathway with an additional sorting process required. CPS is synthesized as a transmembrane protein with a short amino-terminal cytoplasmic tail (*6*). Following transit in the early secretory pathway through the ER and Golgi complex, CPS is transported to an endosomal compartment. Ubiquitylation of the CPS cytoplasmic tail and recognition of this ubiquitin-modified zymogen by the endosome-associated complexes required for protein sorting (ESCRTs) facilitates the inclusion of CPS into membrane invaginations that give rise to the intralumenal vesicles of the multivesiclar body (MVB) (reviewed in **ref. 7**). This process in conjunction with heterotypic fusion of the MVB with the vacuole allows delivery of the CPS precursor into the lumen of the vacuole, where CPS is cleaved to its active form.

The kinetic analysis of CPY and CPS maturation by pulse-chase analysis allows a reproducible, quantitative means to assess function of the VPS pathway. Biosynthetic labeling of a small pool of zymogen is accomplished by adding ^{35}S-methionine and -cysteine to rapidly growing yeast. Following a 10-min pulse with the ^{35}S-methionine/cysteine, excess unlabeled amino acids are added and the ^{35}S-labeled zymogens are chased through the VPS system.

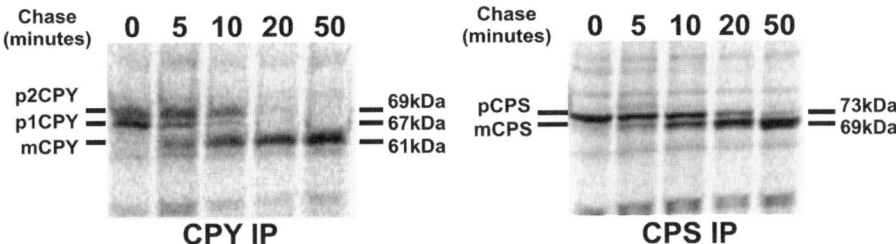

Fig. 1. Maturation of carboxypeptidase Y and carboxypeptidase S. A representative experiment analyzing CPY and CPS processing is shown, although different from the time course recommended in the experimental procedures. Wild-type yeast labeled with ^{35}S-Pro-mix was harvested 0, 5, 10, 20, and 50 min after addition of Chase, and CPY and CPS were immunoprecipitated from extracts generated at each time point. Samples were resolved by SDS-PAGE and detected using a PhosphorImager system. Bands corresponding to the endoplasmic reticulum (ER)-modified (p1CPY; 67 kDa), Golgi-modified (p2CPY; 69 kDa) ,and mature CPY (mCPY; 61 kDa) are indicated (**left panel**). CPS immunoprecipitation samples were treated with EndoH$_f$ prior to SDS-PAGE, and the deglycosylated precursor (pCPS; 73 kDa) and mature CPS (mCPS; 69 kDa) species are indicated (**right panel**). (Courtesy of Andrea J. Oestreich.)

Maturation of CPY is generally accomplished within 30 min in wild-type yeast, while maturation of CPS may take slightly longer (**Fig. 1**). A suitable time course usually involves harvesting samples at 0, 10, 20, 30, 60, and 90 min after the addition of chase. Samples are then lysed under denaturing conditions and subjected to immunoprecipitation with polyclonal antisera raised against CPY or CPS. The immunoprecipitated samples are then resolved by sodium dodecyl sulfate–polyacrylamide gel electrophoresis (SDS-PAGE), and gels are visualized and quantitated using a phosphoimaging system. The percentage of precursor remaining is then plotted to determine the maturation kinetics.

2. Materials

2.1. Cell Growth and Labeling

1. Minimal medium: 0.67% Difco yeast nitrogen base without amino acids (BD Biosciences, San Jose, CA), 2% glucose, 25 mM potassium phosphate, pH 5.4, and appropriate amino acids for the strains to be analyzed. Dissolve the yeast nitrogen base and glucose in water and autoclave. Make 100x amino acid stock solutions (0.2% histidine, 0.3% leucine, 0.3% lysine, 0.2% tryptophan, 0.2% adenine, and 0.2% uracil as complete), leaving out particular amino acids as required, filter-sterilize, and store aliquots at −20°C. Add amino acids and 1M potassium phosphate buffer (pH 5.4) to appropriate final concentrations (*see* **Note 1**).

2. Redivue Pro-mix L-(^{35}S)-methionine, cysteine (GE Healthcare Life Sciences, Piscataway, NJ).

3. 25x Chase: 5% yeast extract, 125 mM methionine, 25 mM cysteine. Dissolve in water, filter-sterilize, and store 1-mL aliquots at $-20°$C.

2.2. Sample Processing and Immunoprecipitation

1. Trichloroacetic acid (TCA; Fischer Scientific, Fair Lawn, NJ).
2. Acetone should be cold.
3. Urea cracking buffer: 6 M urea, 50 mM Tris-HCl, pH 7.5, 1% sodium dodecyl sulfate (SDS).
4. 0.5-mm Glass beads (BioSpec Products, Bartlesville, OK).
5. Tween-20 IP buffer: 50 mM Tris-HCl, pH 7.5, 150 mM NaCl, 0.5% Tween-20, 0.1 mM EDTA.
6. 100 mg/mL Bovine serum albumin fraction V, protease-free (BSA; Roche Diagnostics, Indianapolis, IN) in water. Store 1-mL aliquots at $-20°$C.
7. Protein A slurry: swell 0.4 g Protein A–Sepharose 4B CL (GE Healthcare Life Sciences, Piscataway, NJ) in 11.2 mL 10 mM Tris-HCl, pH 7.5, 1 mg/mL BSA, 1 mM NaN$_3$ in a 15-mL conical overnight at 4°C. Spin 1 min at 300g. Aspirate supernatant and replace with fresh buffer. Store at 4°C.
8. Tween-20 urea buffer: 100 mM Tris-HCl, pH 7.5, 200 mM NaCl, 2 M urea, 0.5% Tween-20.
9. TBS: 50 mM Tris-HCl, pH 7.5, 150 mM NaCl.
10. Glycosidase buffer: 1% β-mercaptoethanol, 0.5% SDS.
11. EndoH Cocktail: 250 mM sodium citrate, 20 U/μL EndoH$_f$ (New England Biolabs, Ipswich, MA). Make just before use by diluting 0.2 μL EndoH$_f$ in 8 μL 250 mM sodium citrate per sample.
12. 5x Laemmli sample buffer: 0.312 M Tris-HCl, pH 6.8, 10% SDS, 25% β-mercaptoethanol, 0.05% bromophenol blue.

2.3. Sample Separation and Quantitation

1. Benchmark Prestained Protein Ladder (Invitrogen, Carlsbad, CA).
2. Gel fix solution: 10% methanol, 10% acetic acid.
3. Chromatography paper (Fischer Scientific).
4. Storage Phosphor Screen (GE Healthcare Life Sciences)

3. Methods
3.1. Cell Growth and Labeling

1. Grow 10 mL culture overnight in minimal media at 30°C with 200–250 rpm agitation in 25-mm glass culture tubes.
2. Determine the optical density at 600 nm (OD$_{600}$) and dilute culture to 0.2 OD$_{600}$/mL in 70 mL minimal media in a 250-mL flask (*see* **Note 2**).

3. Incubate at 30°C until $OD_{600} = 0.5$–0.8 (log phase growth; generally 3–5 h after dilution).

4. Harvest 35 OD_{600} in a 50-mL Falcon conical tube by spinning 3 min at 1700g in the Allegra X-22 benchtop centrifuge (Beckmann Coulter Inc., Fullerton, CA). If the OD_{600} is less than 0.7, this will require two successive spins.

5. Resuspend the sample in 7 mL minimal media (at 5 OD_{600}/mL).

6. Incubate 10 min in a 30°C shaking waterbath at 200–250 rpm.

7. Initiate the labeling by adding 35 μL Pro-mix (1 μL/OD_{600}), staggering samples by 20 s.

8. Continue culturing in 30°C shaking waterbath.

9. Ten minutes after the addition of label, add 280 μL 25x Chase maintaining the 20-s stagger between samples. Vortex briefly and transfer 1 mL to a 1.5-mL Eppendorf tube containing 100 μL TCA and place on ice to precipitate the sample. Return remaining culture to the 30°C waterbath.

10. Continue incubating the culture at 30°C, removing 1-mL samples at 10, 20, 30, 60, and 90 min after addition of Chase. At each time point, briefly vortex the sample before transferring 1 mL to an Eppendorf tube containing 100 μL TCA. Place the TCA precipitation tubes on ice.

11. Allow the samples to sit on ice for at least 10 min, and then process for immuno-precipitation.

3.2. Sample Processing and Immunoprecipitation

1. Spin the samples 3 min at 13,000g in a Microcentrifuge 5415C (Eppendorf North America, Westbury, NY) to pellet the yeast.

2. Aspirate the supernatants and add 1 mL cold acetone.

3. Sonicate the samples with a Branson Ultrasonic Cleaner 2510 (Branson Ultrasonic Corporation, Danbury, CT) to resuspend the TCA pellet (*see* **Note 3**).

4. Spin the samples 3 min at 13,000g in a Microcentrifuge, and repeat acetone wash with sonication.

5. Spin the samples 3 min at 13,000g in a Microcentrifuge and aspirate the super-natants.

6. Dry the TCA pellets in a Speedvac Centrifuge (Thermo Electron Corporation, Waltham, MA) for more than 5 min without heating.

7. Resuspend the pellets in 100 μL urea cracking buffer by sonicating.

8. Carefully add 100 μL 0.5-mm glass beads (*see* **Note 4**).

9. Vortex the samples 10 min in a MT-360 Microtube Mixer (Tomy Tech USA, Fremont, CA). Alternatively, vortex individually for 2 min.

10. Heat the samples at 65°C for 4 min.

11. Add 1 mL Tween-20 IP buffer and 10 μL 100 mg/mL BSA. Vortex the samples briefly.

12. Spin the samples 10 min at 13,000g in a Microcentrifuge to pellet insoluble material.

13. Transfer 1 mL of the supernatants to new 1.5-mL Eppendorf tubes and repeat the spin (*see* **Note 5**).
14. Transfer 950 μL of the supernatants to new 1.5-mL Eppendorf tubes, taking care not to disturb the pellets.
15. Transfer 3 μL of the samples to Beckmann Ready Caps with Xtalscint (Beckmann Coulter Inc., Fullerton, CA) and determine label incorporation using a scintillation counter (*see* **Note 6**).
16. Add 2 μL anti-CPY or anti-CPS polyclonal antisera and incubate samples overnight at 4°C with gentle mixing on a rocking platform or end-over-end rotator.
17. In the morning, add 100 μL Protein A–Sepharose slurry to the samples (*see* **Note 7**).
18. Incubate the samples 1.5–2 h at 4°C with gentle mixing.
19. Spin the samples 5 s at 13,000g in a Microcentrifuge to pellet beads.
20. Remove the supernatants, transferring to new 1.5-mL tubes for additional immunoprecipitations (*see* **Note 8**).
21. Continuing with washes, add 1 mL Tween-20 urea buffer to each tube and invert the samples to mix.
22. Spin the samples 5 s to pellet the beads and aspirate the supernatants (*see* **Note 9**).
23. Repeat 1-mL Tween-20 urea buffer wash, aspirating the supernatants after pelleting.
24. Add 1 mL Tween-20 IP Buffer and invert the samples to mix.
25. Pellet the beads, aspirate the supernatants, and repeat the 1-mL Tween-20 IP Buffer wash.
26. Wash the beads with 1 mL TBS. Invert the samples to mix and briefly spin. Aspirate the supernatants and briefly spin again. Carefully aspirate most of the remaining supernatants.
27. Dry the samples in the Speedvac Centrifuge at 50°C for 30 min to remove remaining buffer.
28. For CPS immunoprecipitation, proceed to **step 31**. For CPY immunoprecipitations, add 50 μL 5x Laemmli sample buffer to the beads and briefly vortex.
29. Incubate the samples at 95°C for 4 min.
30. Spin the samples briefly and store at −20°C until ready to load onto SDS-polyacrylamide gel.
31. CPS immunoprecipitation samples require glycosidase treatment to better resolve the mature and precursor forms. To deglycosylate CPS, add 32 μL glycosidase buffer to the beads and briefly vortex.
32. Incubate the samples at 95°C for 4 min and spin for 5 s.
33. Place the samples on ice to cool for 30 s.
34. Add 8.2 μL EndoH Cocktail to the samples.
35. Incubate the samples at 37°C for 1hr.
36. Add 10 μL 5x Laemmli sample buffer and briefly vortex the samples.
37. Incubate the samples at 65°C for 4 min.

38. Spin the samples briefly and store at −20°C until ready to load onto SDS-polyacrylamide gel.

3.3. Sample Separation and Quantitation

1. Remove the samples from −20°C and incubate at 65°C for 4 min.
2. Spin the samples 2 min at 13,000g in a Microcentrifuge to pellet the beads.
3. During the spin, flush the wells of an 8% polyacrylamide gel.
4. Load 10 μL of the samples onto gel, taking care not to transfer the beads themselves. Also combine 3 μL Benchmark Prestained Protein Ladder with 7 μL 5x Laemmli sample buffer and load onto gel (*see* **Note 10**).
5. Run the gel at 200 V until the 30-kDa band has run off the gel.
6. Place the gel in gel fix solution and incubate for 10 min with gentle agitation on an orbital shaker.
7. Pour off the gel fix solution and use chromatography paper to transfer the gel to a slab gel dryer.
8. Dry the gel for 1 h at 80°C.
9. Load the dried gel in a cassette and expose to Storage Phosphor Screen at room temperature overnight or for a few days, depending on the level of label incorporation (*see* **Note 11**).
10. Scan the Storage Phosphor Screen with a PhosphorImager system, such as the Storm or Typhoon (GE Healthcare Life Sciences).
11. Utilize the ImageQuant software package (GE Healthcare Life Sciences) to determine the density of the precursor and mature forms at each time point, using a reference object to subtract the background signal.
12. Calculate the percent precursor by dividing the amount of precursor (above background) by the sum of the precursor and mature forms (above background) and plot vs time to determine the rate of maturation.

Notes

1. The addition of potassium phosphate to buffer the media is optional. In some mutants defective for function of the VPS pathway, the secreted CPY precursor is subject to an aberrant cleavage event in the absence of the buffering agent. This aberrant form complicates quantitation of CPY sorting, and thus buffering the media is preferred. The amino acid mixture described here is sufficient to support growth of the SEY6210 yeast strain. Different yeast stains may require the inclusion of additional amino acids depending on their genotypes.
2. Although it is an inexact measure of cell quantity, 1 OD_{600} tends to correspond to 2–3×10^7 yeast cells. To provide good aeration for the yeast, the culture volume should not exceed 40% of the vessel volume.
3. To facilitate good sonication, fill the ultrasonic cleaner with an inch of water and then tilt to achieve a gradient of sonication. This tends to generate 2 "hot spots" that can be used to resuspend the pellet. Individually position the sample

such that decavitation within the tube is observed. Usually, once you find a good position for resuspending one sample, that same position will work well for all samples, with the pellet being resuspended within 30 s. Thorough resuspension during the first acetone wash makes later resuspensions much easier and improves sample recovery. Use of Sherlock tube closures (USA Scientific, Ocala , FL) is recommended to prevent inadvertent tube opening.

4. To dispense the glass beads, generate a 100-μL scoop by trimming the lower portion from a 1.5-mL Eppendorf tube and affixing it to a wooden applicator stick (VWR, West Chester, PA) with epoxy resin. This facilitates reproducible addition of beads. To prevent spilling glass beads, fashion a micro-funnel by trimming the end off of a P-1000 pipet tip. Use this funnel to prevent glass beads from compromising the Eppendorf tube closure, as this could cause workspace contamination.

5. If any of the glass beads or precipitated material are transferred into the final immunoprecipitation tube, this labeled material will also be present following the washing of the Protein A–Sepharose beads. The resulting enhanced background signal will prevent quantitation of the specific immunoprecipitated bands. Draw the supernatant from the top of the liquid, and leave at least 50 μL remaining in the tube untransferred, even if this results in less than 950 μL of supernatant in the final immunoprecipitation tube.

6. The determination of label incorporation serves two purposes: to ensure that equal recovery of the all samples has been achieved during the TCA precipitation and acetone washing steps, and to approximate the time required for adequate exposure of the Storage Phosphor Screen. Label incorporation of greater than 1×10^7 cpm per OD_{600} is usually achieved and allows an exposure time of 1 d to be sufficient for reliable quantitation.

7. To enable equal transfer of Protein A–Sepharose, invert the 15-mL conical to resuspend the slurry often as the beads settle rapidly. In addition, trimming off the end of a P-200 pipet tip facilitates equal transfer.

8. Before continuing with additional immunoprecipitations, any residual antibodies should be cleared. Add 100 μL of Protein A–Sepharose slurry to these supernatants and incubate 1.5–2 h with gentle mixing. Spin the samples 5 s to pellet beads and transfer the supernatants to new tubes to which the next antisera is then added. The procedure can then be continued from **Subheading 3.2., step 16**. Other VPS pathway cargoes examined include Protein A (PrA), vacuolar alkaline phosphatase (ALP), and Sna3.

9. During the washing of the Protein A–Sepharose beads, remove all but approx 50 μL of the liquid, taking care not to remove any of the beads.

10. We usually use the Benchmark Prestained Ladder because the single pink band at approx 60 kDa allows easy orientation of the other blue surrounding bands and migrates in a similar size range as CPY.

11. Use of the Storage Phosphor Screen and PhosphoImager system allows good quantitation of immunoprecipitated protein over a wide window of labeled-protein concentrations. It is also possible to use X-ray film to detect isolated protein and then use densitometry of the scanned films to determine protein amounts; however, care must be taken to ensure that the linear sensitivity of the X-ray film has not been exceeded or the analysis will be compromised. In general, use of the PhosphoImager system is preferred for this quantitative analysis.

References

1. Futerman, A. H. and van Meer, G. (2004) The cell biology of lysosomal storage disorders. *Nat. Rev. Mol. Cell Biol.* **5**, 554–565.
2. Katzmann, D. J., Odorizzi, G., and Emr, S. D. (2002) Receptor downregulation and multivesicular-body sorting. *Nat. Rev. Mol. Cell Biol.* **3**, 893–905.
3. Graham, T. R. and Nothwehr, S. F. (2002) Protein transport to the yeast vacuole., in *Protein Targeting, Transport and Translocation* (Dalbey, R. E. and von Heijne, G., eds.), Academic Press, London, pp. 322–357.
4. Stevens, T., Esmon, B., and Schekman, R. (1982) Early stages in the yeast secretory pathway are required for transport of carboxypeptidase Y to the vacuole. *Cell* **30**, 439–448.
5. Marcusson, E. G., Horazdovsky, B. F., Cereghino, J. L., Gharakhanian, E., and Emr, S. D. (1994) The sorting receptor for yeast vacuolar carboxypeptidase Y is encoded by the VPS10 gene. *Cell* **77**, 579–586.
6. Spormann, D. O., Heim, J., and Wolf, D. H. (1992) Biogenesis of the yeast vacuole (lysosome). The precursor forms of the soluble hydrolase carboxypeptidase yscS are associated with the vacuolar membrane. *J. Biol. Chem.* **267**, 8021–8029.
7. Babst, M. (2005) A protein's final ESCRT. *Traffic* **6**, 2–9.

23

Targeting to Lysosomes in Mammalian Cells

The Biosynthetic and Endocytic Pathways

Ann H. Erickson and Jeffrey P. Bocock

Summary

Endogenous or ectopically expressed lysosomal proteins can be detected in their biosynthetic or endocytic pathways by Western blotting of biosynthetic forms in cells, cell fractions, or their culture medium, by pulse-chase radiolabeling accompanied by immunoprecipitation, or by electron or immunofluorescence microscopy. Western blotting and microscopy reveal the steady-state distribution of a protein, whereas pulse-chase studies are required both to identify transient forms and to define the relationship of the biosynthetic forms detected. Targeting to lysosomes can be dramatically affected by synthesis levels and carbohydrate modification, whether the synthesis is upregulated naturally, for example, by cell transformation, or whether it results from ectopic expression. This occurs because a lysosomal protein, unlike a protein expressed in the cytoplasm, must interact with receptors and be packaged into vesicles that mediate its transport though the secretory pathway. Use of microscopy to establish localization is, therefore, a key aspect of characterization of the cellular pathways utilized by lysosomal proteins.

Key Words: Lysosomes; endosomes; secretory pathway; Western blotting; pulse-chase; immunoprecipitation; cell fractionation; confocal immunofluorescence microscopy.

1. Introduction

Lysosomal proteins are synthesized in the endoplasmic reticulum (ER) on membrane-bound ribosomes. Translocation across the endoplasmic reticulum is mediated by transient signal peptides via a signal receptor protein (SRP)-mediated mechanism. Because import into the ER lumen occurs co-translationally, the enzymes are not detected in the cytoplasm except under rare circumstances when initiation occurs at methionines located C-terminal

From: *Methods in Molecular Biology, Vol. 390: Protein Targeting Protocols: Second Edition*
Edited by: M. van der Giezen © Humana Press Inc., Totowa, NJ

to the signal peptide *(1,2)*. After cleavage of the signal peptide and acquisition of high-mannose carbohydrate, N-acetyl glucosamine-phosphate is added to the sugar chains and the proteins are transported to the Golgi apparatus, where complex sugars are frequently added. From the *trans*-Golgi network (TGN), lysosomal proteins can be either secreted constitutively or targeted to endosomes. In certain transformed cells, protease precursors can additionally be targeted to and stored in multivesicular endosomes *(3)*. Transport from endosomes to enzymatically active low-pH lysosomes results in activation of those enzymes transported as precursors, primarily the proteases. Lysosomal proteins also reach lysosomes via uptake from the cell exterior by endocytosis and transport through endosomes to terminal lysosomes.

Various techniques can be utilized to characterize lysosomal proteins in these biosynthetic and endocytic pathways. Lysosomal proteases, more than the sugar- and lipid-processing enzymes, are susceptible to proteolytic processing steps that characteristically occur at specific sites along the pathway. The inactive proform of a protease can usually be distinguished from an active protease lacking the activation peptide on the basis of size. Arrival at the enzyme-rich mature lysosome is frequently revealed by cleavage into a light and a heavy chain, often accompanied by exopeptidase removal of a few amino acids at the cleavage sites. These biosynthetic forms of lysosomal enzymes can usually be readily resolved by polyacrylamide gel electrophoresis and detected by Western blotting or isolated by immunoprecipitation and detected by fluorography.

Pulse-chase radiolabeling followed by immunoprecipitation is the classic way to determine the relationship of protein forms visualized at steady state by Western blotting or by continuous radiolabeling. It is ultimately more sensitive than Western blotting because immunoprecipitation allows detection of all the protein in a large amount of cellular protein. Western blotting without immunoprecipitation is eventually limited by the amount of cellular protein that can be loaded in a lane of a polyacrylamide gel. Immunoprecipitates can be detected by Western blotting, but this is often complicated by the detection of the immunoglobulin, which, being unlabeled, is not visualized when radiolabeled proteins are detected. Finally, gels resolving radiolabeled proteins can be exposed to film for several wk, whereas lability of the peroxidase reagent limits the detection time during Western blotting.

Confocal immunofluorescence microscopy reveals the steady-state distribution of a protein. High-expression levels can result in artificial accumulation in the ER because of a shortage of protein chaperones to aid folding, in increased secretion because of a shortage of mannose-phosphate receptors in the TGN, and, at least in certain transformed cells, in storage of proenzyme in multivesicular endosomes *(3)*. While characterization of endogenous enzyme is therefore preferable, ectopic expression enables analysis of targeting of mutant proteins and is useful when endogenous protein levels are low or when specific antibodies are not available. When a lysosomal protein is expressed from a plasmid, and especially if it is designed with an epitope tag, microscopy should be utilized to establish that the protein being studied correctly traverses the secretory pathway to lysosomes. Lysosomal proteases, which characteristically undergo multiple proteolytic processing steps, can generally be assumed to have reached lysosomes if the cleavage products normally generated in lysosomes, for example, cleavage of the single chain into a light and a heavy chain, can be detected by Western blotting. In the absence of diagnostic cleavage steps, localization by microscopy is mandatory.

Mature lysosomes are dense perinuclear vesicles that contain active proteases and lack mannose-phosphate receptors. Lysosomal proteins can, however, be present in a variety of other cellular vesicles. These are often mistakenly called lysosomes simply because some form of a lysosomal enzyme is present. Too often the activity of the enzymes in that compartment is not established. Many of the lysosomal enzymes that digest sugars and lipids are synthesized and targeted as active enzymes. The proteases, however, are transported through the cell as inactive precursors that are generally only activated in late endosomes or lysosomes. Protease activity is, therefore, a good marker for mature lysosomes. Protease activation can be established by an enzyme assay of a cell fraction, by demonstration of the lack of the proenzyme in the protein pattern obtained on polyacrylamide gel electrophoresis of a cell lysate, by use on Western blots or in microscopy of an antiserum specific for the activation peptide, or by use of active site-labeling reagents coupled with analysis by polyacrylamide gel electrophoresis (PAGE) *(4)*. The absence of mannose-phosphate receptors, which travel to late endosomes and are returned from there to the Golgi, can be established by Western blotting a vesicle preparation, but microscopy is obviously required to establish that individual vesicles lack the receptor. There is some evidence that lysosomal proteases can be active in early endosomes *(5)*, but because these enzymes need low pH for optimal activity and could be highly damaging if active in a common cellular pathway, proteases are probably only activated in specialized endosomes or domains of multivesicular endosomes in specialized cells, for example, in cells that process antigens.

Multiple additional techniques can be employed to further characterize the various biosynthetic forms of the lysosomal proteins. Detection of complex carbohydrate acquisition is characteristically utilized to demonstrate that a protein has reached or traversed the Golgi apparatus. Presence of a mannose-phosphate recognition marker can be revealed by labeling with [^{32}P]orthophosphate. Brief endoglycosidase treatment can reveal how many asparagines acquire carbohydrate chains. Secreted protein can be concentrated from cell culture medium by acid precipitation. Proteins anchored in membranes can be distinguished from those soluble within membrane-bound vesicles through the preparation and lysis of microsomes. Proteins peripherally associated with the membrane can be distinguished from integral membrane proteins by washing the microsomes with buffers of different pH. Finally, vesicles can be fractionated by gradient centrifugation and identified though co-localization with marker proteins. The techniques detailed below are classic approaches to the characterization of lysosomal enzymes in intact cells and in cell fractions.

2. Materials

2.1. Cell Culture and Lysis

1. Cell culture: culture human HeLa, monkey COS, and mouse NIH 3T3 fibroblasts in Dulbecco's modified Eagle's medium (DMEM) containing high glucose (4.5 g/L)-containing 10% fetal bovine serum (FBS) and 0.0005% gentamycin and 0.0025% kanamycin.
2. Cell culture medium for harvest of secreted proteins: DMEM containing glutamine and antibiotics and supplemented with insulin, selenium, and transferrin (ITS; Sigma). The 1000x ITS stock is stored frozen in 50-μL aliquots that, once defrosted, can be held at 4°C for about 2 wk. Serum is omitted and bovine serum albumin-containing media should be avoided to reduce the protein load on the gel lane.
3. Cycloheximide chase: 1 mM cycloheximide in ethanol; store frozen in aliquots.
4. PBS+ wash buffer: 0.01 M phosphate-buffered saline (PBS) containing 0.01% Ca^{2+} and 0.01% Mg^{2+}. Use this form of PBS when it is important to maintain an intact cell monolayer.
5. Cell harvest buffer for PAGE: 0.1 M Tris base, 2% (w/v) sodium dodecyl sulfate (SDS), 24% sucrose or 10% glycerol, and 0.02 M EDTA. This buffer can be stored indefinitely at room temperature, but to increase its rate of denaturation of cellular proteins, it can be incubated in a boiling water bath prior to use. Obviously it is important to denature lysosomal proteases released during vesicle lysis before they mediate cleavage.

2.2. BCA Protein Assay

1. Reagent A: Pierce (cat. no. 23231).
2. Reagent B: 4% (w/v) bicinchoninic acid, anhydrous sodium salt (BCA) Pierce (cat. no. 23230) in dH$_2$O. Rock solution for 30 min at room temperature to dissolve BCA. The solution is stable about 2 wk at room temperature and turns yellow when old.

3. Reagent C: 4% (w/v) cupric sulfate, pentahydrate, in dH$_2$O is stable at room temperature indefinitely.
4. Protein standards: 1 mg/mL dH$_2$O stock solution of ovalbumin or bovine serum albumin, frozen in aliquots.

2.3. SDS-PAGE

1. Sample buffer: 0.8 *M* Tris-HCl, 0.17 *M* EDTA, 20% sucrose, 0.008% bromophenol blue (BPB), 3.3% SDS, 17 m*M* dithiothreitol (DTT). Add DTT from a 0.5 *M* stock frozen in aliquots at −80°C. Omit the dye if the protein concentration of the sample is to be determined prior to PAGE. Instead add 2 μL 1% BPB/100 μL sample immediately before loading on a gel lane.
2. Acrylamide stock: 30% acrylamide, 0.8% bis-acrylamide. Decolorize by adding a small spatula of charcoal (Norit-A, Fisher C-176), stirring 30 min at room temperature, and filtering though a 1.2 μm (Millipore Type RA) filter. Store at 4°C in a brown bottle or in a bottle covered with foil to reduce exposure to light.
3. Separating gel buffer: 2 *M* Tris-HCl, pH 8.8. Store at 4°C.
4. Stacking gel buffer: 0.5 *M* Tris-HCl, pH 6.8. Store at 4°C.
5. 10% (w/v) Ammonium persulfate (APS) in dH$_2$O. Prepare 10 mL and store at 4°C.
6. Additional reagents: 10% SDS. Store stock solution at room temperature indefinitely. TEMED (Bio-Rad). Store stock at 4°C.
7. 10x Gel run buffer: 1.92 *M* glycine, 0.25 *M* Tris base, 1% SDS. Store 10x and 1x buffer at room temperature.
8. Destain: 35% methanol, 10% acetic acid in water.

2.4. Western Blotting

1. Transfer buffer: 0.01 *M* 2-cyclohexylamino-1-propanesulfonic acid, pH 11.0 (CAPS); 10% methanol. Store 10x CAPS stock at room temperature.
2. Tween-saline wash buffer: 0.01 *M* Tris, pH 7.5, 0.05% Tween-20, 0.15 *M* NaCl. Store in a large vat at room temperature.
3. Blocking buffer: 0.5 to 1% Carnation nonfat dry milk dissolved in Tween saline. Rock the solution during the blotting period to ensure that the milk dissolves completely. Make this fresh because it is prone to bacterial contamination even if stored at 4°C.
4. Strip buffer: 62.5 m*M* Tris, pH 6.7, 2% (w/v) SDS, 0.1 M β-mercaptoethanol. The Tris-SDS can be stored at room temperature, but the reducing reagent should be added fresh for each use.

2.5. Chemiluminescence Detection of Blotted Protein

1. 6.8 m*M* P-Coumaric acid (4-hydroxycinnamic acid). Dissolve the P-coumaric acid in 0.1 *M* Tris-HCl, pH 9.0. Store frozen.
2. 50 mg/mL Luminol (5-amino-2,3-dihydro-1,4-phthalazinedione). Dissolve the luminol in dimethylsulfoxide (DMSO), wrapped in foil. Store frozen.

3. Reagent A: mix 40 mL 0.1 *M* Tris-HCl, pH 9.0, 0.8 mL 6.8 m*M* P-coumaric acid and 0.6 mL 50 mg/mL luminol. This solution is stable at least 1 mo when wrapped in foil and stored at 4°C.
4. Reagent B: 40 mL 0.1 *M* Tris-HCl, pH 9.0, mixed with 0.8 mL stabilized peroxide Pierce (cat. no. 34062). The solution is stable at least 1 mo when wrapped in foil and stored at 4°C. The peroxide stock is stable about 1 year at 4°C.

2.6. Pulse-Chase of Radiolabeled Proteins

1. Starvation medium (1 mL): 910 µL DMEM lacking methionine and cystine Gibco/Invitrogen (cat. no. 21013-024), 80 µL dialyzed FBS, 10 µL 200 m*M* glutamine, 10 µL 100x gentamycin/kanamycin. To remove unlabeled methionine from serum, dialyze 100 mL FBS against 4 L 0.01 *M* Tris-HCl, pH 7.4, or PBS with stirring overnight at 4°C with two buffer changes. Store the serum frozen in aliquots.
2. Isotope: [^{35}S]methionine (e.g., ICN/MP Tran^{35}S-Label) at 500 µCi/mL in starve medium. Store methionine in aliquots at −80°C.
3. Chase medium (1 mL): 980 µL DMEM containing 10% FBS, 20 µL 240 m*M* methionine. Store methionine in aliquots at −80°C.
4. ^{32}P-Labeling of mannose phosphate recognition marker: [^{32}P]Orthophosphate at 500 µCi/mL.

2.7. Immunoprecipitation (IP)

1. SDS cell harvest buffer: 0.5% SDS, 50 m*M* Tris-HCl, pH 9.0, 100 m*M* NaCl, 2 m*M* EDTA, pH 8.0. The high pH is to reduce the activity of lysosomal enzymes, which are optimally active at acidic pH. The buffer can be stored at room temperature indefinitely.
2. Nonionic detergent cell harvest buffer: 1% Triton X-100, 50 m*M* Tris-HCl, pH 8.0, 5 m*M* EDTA, aprotinin to 20 µg/mL.
3. IP SDS wash buffer: 0.1% (w/v) SDS, 0.15 *M* Tris-HCl, pH 8.0, 0.15 *M* NaCl, 0.005 *M* EDTA, pH 8.0.
4. IP nonionic detergent wash buffer: 2% Triton X-100, 150 m*M* triethanolamine or 0.01 *M* Tris-acetate, pH 8.5, 150 m*M* NaCl, 5 m*M* EDTA.

2.8. Concentration of Secreted Protein

1. Carrier yeast tRNA Boehringer (cat. no. 109–517). Store frozen in 1 mg/mL aliquots. Add to a final concentration of 25 µg/mL.
2. Trichloroacetic acid (TCA): prepare 100% TCA by adding the appropriate amount of dH$_2$0 directly to the brown reagent bottle. Store at 4°C. Use at a final concentration of 20%.

2.9. Carbohydrate Analysis

1. Endoglycosidase H source: New England Biolabs, Boehinger or Genzyme.
2. Protein A Sepharose elution buffer: 1% SDS, 50 mM Tris-HCl, pH 7.5, 50 mM DTT.
3. Endo H cell harvest buffer: 1% SDS in dH$_2$O.
4. Endo H dilution buffer: 10–50 mM NaH$_2$PO$_4$, pH 5.5, containing 0.5% NP-40.
5. Protease inhibitor cocktail: 20 µg/mL aprotinin (1 mg/mL stock in dH$_2$O); 10 mM PMSF (0.2 M stock in isopropanol); 5 mM TPCK (50 mM stock in ethanol). Store stock solutions in aliquots at −80°C.
6. Peptide N-glycosidase F source: New England Biolabs, Boehinger or Genzyme
7. Endo F cell harvest buffer: 1% SDS.
8. Endo F dilution buffer: 20 mM NaH$_2$PO$_4$, pH 7.2, 50 mM EDTA, 1% MEGA-8 (octanoyl-N-methylglucamide), 1% β-mercaptoethanol.
9. Tunicamycin: 1 mg/mL in DMSO; store frozen in aliquots.

2.10. Microsome Preparation and Membrane Association Assays

1. Homogenization buffer: 3 mM imidazole, pH 7.4, 250 mM sucrose, 1 mM EDTA
2. Protease inhibitors: 400x E64 in distilled water (dH$_2$O) and 400x DCIC (di-chloro-iso-coumarin) in DMSO, both stored in aliquots at −80°C.
3. To strip membranes of peripheral proteins: 100 mM sodium carbonate pentahydrate, pH 11.5.

2.11. Cell Fractionation

1. Percoll gradient: 27–30% Percoll in 0.25 M sucrose.
2. Sucrose gradient: 0.7 and 1.6 M sucrose, both in 20 mM imidazole, pH 7.4.

2.12. Confocal Immunofluorescence (IF) Microscopy

1. Antibodies (*see* **Note 1**) to detect proteins characteristic of various cellular compartments where lysosomal proteins can be localized include:

   a. ER: Protein disulfide isomerase (Affinity Bioreagents) at 0.4 µg/mL or calnexin (Santa Cruz Biotechnology).
   b. Golgi: Golgin97 (Molecular Probes/Invitrogen) at 0.2 µg/mL is used to mark the *trans*-Golgi. To label the *cis*-Golgi, use antibodies to the KDEL receptor.
   c. Lysosomes: LAMP1 or LAMP2 (Developmental Studies Hybridoma Bank) at 1:500 to 1:1000 or antibodies specific for a cathepsin.
   d. Late endosomes: the cation-independent mannose-6-phosphate receptor (Affinity Bioreagents) at 1–2 µg/mL or Rab9 and Rab7.
   e. Early endosome: early endosomal markers include the early endosomal antigen 1 (EEA1) and Rab5.

f. Multivesicular endosomes: multivesicular endosomes have been identified by staining with the tetraspanin CD63 (Developmental Studies Hybridoma Bank) at 1:1000 *(6)*.

g. Recycling endosomes: recycling endosomes are marked by Rab11 (Zymed) at 0.1 µg/mL.

2. Fluorescent secondary antibodies: many fluorescent secondary antibodies are available, including classic dyes such as Texas Red (568 nm) and FITC (488 nm), but newer antibodies incorporating AlexaFluor dyes (available from Molecular Probes/Invitrogen) are recommended.

3. IF Protocol:

a. 4% (w/v) Paraformaldehyde in PBS (*see* **Note 2**).
b. Quench buffer: 1% nonfat dry milk and 150 mM sodium acetate, pH 7, in PBS.
c. Wash buffer: 1% nonfat dry milk in PBS.

3. Methods
3.1. Harvest of Lysosomal Enzymes Within Cells

1. Cell type: because lysosomes are ubiquitous, most lysosomal enzymes can be readily detected in traditional fibroblast or epithelial cell lines such as mouse NIH 3T3 fibroblasts or human HeLa cells. Transformed cells often synthesize and secrete higher levels of lysosomal proteases than do untransformed cells *(3,7)*. It is easier to visualize vesicles (endosomes and lysosomes) in large flat epithelial cells, like HeLa or monkey COS cells, than in elliptical fibroblasts or neurons. Also, epithelial cells usually remain attached to glass cover slips though the wash steps of the staining protocol more efficiently than do elliptical cells, and they are also generally transfected with higher efficiency. The cell choice often depends on the species specificity of the antiserum available.

2. Cell number: the number of cells utilized will vary with the expression level of the enzyme to be detected and the sensitivity of the detection method to be utilized. If the protein is to be visualized by Western blotting, the amount of cellular protein that can be effectively resolved in a single gel lane could be the limiting factor, and that, in turn, is controlled by the size of the gel utilized. Approximately 500 µg of protein (a 35-mm dish or 1 well of a 6-well plate of confluent cells) can be resolved on a standard 1.5-mm 10-well gel without band distortion, although 250 µg or less will yield sharper bands. If the protein under study is not detectable in this amount of cells, then the protein can be immunoprecipitated from a 100-mm dish of cells.

3. Cell harvest method: place culture dishes on ice and wash twice with cold PBS+, using approximately 2 mL/wash/35-mm plate in order to remove serum and secreted proteins. Scrape the cell monolayer with a rubber policeman or a Teflon scrapper Nalge Nunc (cat. no. 179693) into 100-µL hot cell harvest buffer (*see* **Note 3**), which lacks dye (e.g., bromophenol blue) that would interfere with a BCA protein

assay. This viscous lysate is pipetted or dragged into a 1.5-mL microfuge tube, boiled for 5 min, and then sonicated for 30 s (Vibra Cell/Sonics & Materials) to shear the DNA. Samples can be frozen at this point for future use.

4. Cycloheximide chase: to terminate protein synthesis, cycloheximide can be added to the cell culture medium to 10 μM for 1–8 h. This "chases" a protein though the biosynthetic pathway, similar to the chase obtained on removal of a radiolabeled amino acid. It is useful to assess the stability of a protein or to reduce protein accumulation in the ER in order to increase targeting to vesicles.

3.2. Harvest of Secreted Lysosomal Enzymes

1. Most soluble lysosomal enzymes are secreted to some extent, but the amount of the protein in the cell culture medium varies greatly, depending primarily on the synthesis level. This secretion occurs despite the presence of mannose-phosphate on the proteins. Proteases are primarily secreted as inactive precursors, except under certain physiological conditions. Active proteases are secreted, for example, by activated macrophages *(3)* or when an increase in intracellular free calcium and cAMP induces fusion of lysosomes with the plasma membrane *(8)*. Secreted proteins can be concentrated for analysis by acid precipitation, spin concentration, or immunoprecipitation.

2. Incubate a cell monolayer overnight in serum-free culture medium containing ITS. Keep the medium volume small; for example, use 4 mL on a 100-mm plate. Generally the amount of protein in the cell medium is low so that all of the proteins secreted overnight by cells on a 100-mm dish can be resolved in one gel lane as long as that medium lacks serum proteins.

3. Pipet the cell culture medium into microfuge tubes and spin for 2 min at full speed to pellet floating dead cells. Transfer the supernatant to clean tubes. A 9-in. Pasteur pipet pulled out in a Bunsen burner flame is useful for this step. Wash the remaining cell monolayers, harvest by scraping, and determine the protein concentration as described below.

4. Concentrate the secreted proteins by acid precipitation. Add yeast tRNA to 25 μg/mL as a carrier. This aids precipitation but is invisible on a gel. Add cold TCA to a final concentration of 20%. Incubate on ice for 30 min to 1 h.

5. Spin in a microfuge at full speed, in the cold room, for 15–30 min. Pellets will be tiny or invisible, so orient the tubes so that you know where to expect the pellet. For example, place the lid attachment point of each tube on the outside circumference of the rotor.

6. Suck off the TCA, using a pulled-out Pasteur pipet, being careful to avoid the tiny pellets. Do a 30-s spin to wash TCA off the tube walls and remove this acid. Add gel sample buffer (about 75 μL) containing bromophenol blue dye, a pH indicator. Vortex. If the sample is yellow, add 1–2 μL untitrated 1 *M* Tris. Add Tris until the sample remains blue, indicating the pH is neutral. The protein will not resuspend efficiently if the buffer remains acidic.

7. TCA pellets can be difficult to resuspend. Sonicate the samples. Check that the pellet is actually resuspended by centrifuging the tubes for 2 min. If a pellet is observed indicating insoluble protein is present, incubate the tubes at 37°C for 2–4 h with occasional vortexing. When the pellets are resuspended, add DTT to 20 mM, boil 5 min, and immediately load on a gel.

8. Acid precipitation is not efficient for concentrating protein from dilute solutions or large volumes. An alternative approach, useful when there is more than 4 mL of serum-free medium, is to utilize spin filters, such as Vivaspin/Sartorius, to concentrate the medium prior to acid precipitation.

9. The amount of a protein secreted to the medium is a function of the number of cells on the plate. If multiple samples are to be compared, determine the amount of medium to be treated with TCA based on a protein assay of the cells. For example, if one cell monolayer has twice as much protein as another, precipitate protein from all of the medium from one plate and from half the medium from the other.

3.3. Pulse-Chase of Lysosomal Proteins

1. Suck off the cell culture medium and wash cell monolayers twice with PBS+. Add 1 mL starve medium/60-mm dish. Incubate at 37°C for 30 min to 1 h to reduce the amount of free methionine present in the cells.

2. Add radiolabeled methionine to the starve medium at 500 µCi/mL. Because the culture will only be maintained for a short time, this can be done on a lab bench to prevent contamination of the tissue culture hood. To conserve label, the medium volume can be reduced to 600 µL, but the plates should be rocked every 5 min or continuously during the labeling period to ensure that the monolayer does not dry out. The pulse should be carried out in a sealed chamber to reduce contamination of the incubator by volatile components in the isotope preparation. A Modular Incubator Chamber (Billups-Rothenberg) can be flooded with 5% CO_2, sealed, and then placed in an incubator. A disposable charcoal filter is utilized on the exhaust line and within the chamber to capture volatile radioactivity. Place a dish of water in the chamber to maintain humidity. To detect the protein in the ER, pulse the cells for 5–15 min (*see* **Note 4**). By the end of a 30-min pulse, protein molecules have reached the Golgi and are undergoing secretion.

3. If the cells are to be harvested at the end of the pulse period, the plate should be immediately placed on ice and washed twice with cold PBS+. If the isotope is to be chased into intermediate and mature biosynthetic forms of the protein under study, the PBS+ should be 37°C. Add 1 mL 37°C chase medium/60-mm dish and incubate 2–24 h.

4. Continuous label: To label all biosynthetic forms equally, cells can be labeled continuously for 4–5 h (*see* **Note 5**). Overnight continuous labeling often results in some chasing.

5. Labeling of mannose-6-phosphate: labeling a biosynthetic form of a lysosomal protein with [^{32}]P is suggestive of the presence of the mannose-phosphate

recognition marker, especially if that label is lost on removal of N-linked sugar chains. The protocol is similar to that for protein labeling, expect the cells are washed in phosphate-free buffers, starved for phosphate, and incubated 4 h in phosphate-free medium containing [^{32}P]orthophosphate. Relatively long exposures to film may be required, especially if the protein has only one or two carbohydrate chains.

3.4. Immunoprecipitation

1. Wash the cell monolayer twice with 37°C PBS containing 0.01% Ca^{2+} and Mg^{2+} (PBS+). Add 0.5 mL hot SDS cell harvest buffer/60-mm dish and scrape the cells off the dish (*see* **Note 3**). Transfer to a microfuge tube, vortex, and boil 5 min. Sonicate about 30 s to shear the DNA. Microfuge 3–5 min. The tiny white pellet obtained is aggregated protein and should be discarded. Immediately transfer the supernatant to a clean tube with a pulled-out Pasteur pipet. If the protein concentration is to be assayed, take out an aliquot now, before addition of protease inhibitors that are proteins, for example, aprotinin. Adjust the samples to 8 mM iodoacetamide (IAA), 4% (w/v) Triton X-100 (to reduce the SDS concentration prior to antibody addition) and 26 µg/mL aprotinin, by adding 6.5 µL 0.5 M IAA, 65 µL 20% Triton X-100, and 6.5 µL 2 mg/mL aprotinin and mix well. Antiserum may be added or the samples may be frozen for later probing.
2. If harvest under nondenaturing conditions is desired, wash the monolayers twice with cold PBS+, scrape into 0.5 mL cold Triton harvest buffer, and sonicate briefly on ice. Heating, either by boiling or during sonication, may cause a pellet to form.
3. To immunoprecipitate secreted proteins from medium on cells during a pulse or chase, remove the medium to a microfuge tube, and spin 2 min to pellet dead and floating cells. To use SDS denaturation, add to 1 mL of this supernatant 25 µL 25% SDS, 75 µL 1 M Tris-HCl, pH 9.0, 13 µL 0.2 M EDTA and methionine to 10 mM, either as powder or from a frozen stock. Boil, spin, transfer, and add 13 µL 0.5 M IAA, 166 µL 20% Triton X-100, and 13 µL 2 mg/mL aprotinin.
4. To immunoprecipitate secreted proteins from medium on cells during a pulse or chase using a nonionic detergent, add to 1 mL of medium 60 µL 20% Triton X-100, 25 µL 200 mM EDTA, 50 µL 1 M Tris-HCl, pH 8.0, and 10 µL 2 mg/mL aprotinin. Spin, transfer, and add antiserum. Do not boil.
5. The amount of antiserum to efficiently precipitate the desired protein from a given amount of cellular protein needs to be titrated for each antiserum. Polyclonal antisera often give better yields than affinity-purified sera. Combine the cell lysate from three dishes and then divide into three equal parts, saving a sample to protein assay. Initially try 5, 10, and 25 µL of polyclonal serum/60-mm dish or 1–3 µL affinity-purified serum. After addition of a specific antiserum, incubate at 37°C for 1 h and then at 4°C overnight, with rotating (*see* **Note 6**).
6. Microfuge the tubes for 5 min and transfer to clean tubes with a pulled-out Pasteur pipet. Discard the pellets comprised of precipitated protein.

7. Add Protein A- or Protein G-Sepharose (PAS/PGS), depending on the subclass of the antiserum utilized. Protein A/G beads (Pierce) bind all human IgG subclasses. Hydrate the resin and store at 4°C in dH_2O containing 0.02% sodium azide. Before use, remove broken beads by washing twice with dH_2O. Add dH_2O, rock resin, and pellet by spinning 1 min at 1000g in a cold centrifuge. After the final spin, prepare a 1:1 slurry by adding dH_2O equal to the volume of the resin. Rock to resuspend the resin evenly and pipet immediately. If the Pipetman tips are highly tapered, cut off the end with a razor blade. Add resin equal to 4 times the volume of serum. If using affinity-purified serum, add at least 10 µL of resin. Confirm that an equal amount of resin was added to each tube by spinning the tubes briefly in a microfuge.
8. Rotate the tubes for 3 h at room temperature.
9. Spin briefly in a microfuge to pellet the resin. Remove the supernatant with a pulled-out Pasteur pipet. This supernatant can now be reprobed by the addition of a second antiserum.
10. Wash the resin four times by vortexing and pelleting by a brief microfuge spin. Suction off most of the the supernatant each wash using a pulled-out Pasteur pipet or a Pasteur pipet having a 200-µL Pipetman tip held on the end by the suction. After the last wash, remove the wash buffer completely, using a Hamilton syringe.
11. Add gel sample buffer containing dye and DTT, vortex, boil 5 min, pellet resin for 2 min in a microfuge, and remove the supernant with a Hamilton syringe (*see* **Note 7**).
12. Immunoprecipitates from unlabeled cells contain relatively little protein and thus can be visualized with sensitive chemiluminescence reagents, such as Pierce Super-Signal West Pico (cat. no. 34079), without obtaining a high background.
13. Radioactive immunoprecipitates should be visualized on gels impregnated with PPO. Soak gel for 30 min in destain, 2 times for 30 min in DMSO, and 3 h in 20% PPO dissolved in DMSO. Rinse gel with water, which precipitates the PPO, and soak in 3% glycerine in water for 3–5 min (gel will swell). Dry on a gel-drying plate connected to a trapped vacuum pump. The PPO-DMSO can be reused until a precipitate develops.

3.5. Microwave BCA Protein Assay

1. To ensure reproducibility and enable comparison between gel lanes, load equal amounts of cellular protein in each gel lane.
2. Prepare protein standards diluted from a frozen stock of ovalbumin or bovine serum albumin, using 0, 5, 10, 25, and 50 µg each in 500 µL.
3. Preparation of working reagent: mix 1 volume of reagent A with 1 volume of reagent B. Add 1/50 volume of reagent C. This solution is green. A precipitate forms initially but will go into solution with vortexing or rocking. Prepare 0.5 mL for each sample to be assayed, including the standards.

4. Dilute 2 μL of sonicated cell extract from a 35-mm dish with 498 μL dH$_2$O. The sample size can be varied depending on the protein concentration of the extract. Run duplicates of each sample and average the results. Mix 500 μL sample with 500 μL of working reagent (final sample dilution is 1:500). Float the standards and samples in 200 mL water in a 1-L plastic beaker. Microwave on high for 20 s. Remove from water and cool to room temperature.

5. Immediately read the absorbance at 562 nm in plastic disposable 1-mL cuvettes.

6. Graph the A562 values of the standards on the *y*-axis vs concentration (μg/mL) on the *x*-axis. Draw a standard curve, the intercept of which is set to zero. Measure the slope of this line. Calculate the concentration of the starting sample by dividing the sample absorbance by the slope of the standard line and then multiplying by the dilution factor.

7. To demonstrate that equal amounts of cellular protein were loaded in each gel lane, blot for a constitutively synthesized protein, for example, calnexin or actin, after blotting with the primary serum specific for the protein of interest.

3.6. SDS-PAGE

1. These instructions assume use of a Hoefer SE 400 or SE 600 gel system but can be easily adapted to other formats. Wash the glass plates with a mild detergent and brush, rinse in dH$_2$O, followed by ethanol, and allow to air-dry (*see* **Note 8**).

2. Prepare a 1.5-mm-thick 12% gel by mixing 12 mL acrylamide stock, 5.6 mL separating gel buffer, 0.3 mL 10% SDS, and 12 μL TEMED. Bring volume to 30 mL with dH$_2$O. Just before pouring the gel add 85 μL 10% APS. Cover cylinder top with Parafilm, invert to mix, and pour between the gel plates. Leave space for a stacking gel (the gel shrinks slightly as it polymerizes) and overlay with distilled dH$_2$O dripped from a 10-mL syringe with needle. Allow the gel to polymerize at room temperature.

3. Prepare a 5% stacking gel: mix 1.7 mL acrylamide stock, 2.5 mL stacking gel buffer, 100 μL 10% SDS, and 5 μL TEMED. Bring volume to 10 mL with dH$_2$O and add 83 μL 10% APS. Cover cylinder top with Parafilm, invert to mix, and pour between the gel plates. Insert a 10-well comb and allow the gel to polymerize at room temperature (about 30 min).

4. Dilute the run buffer to 1x. Assemble gel and add run buffer to top and bottom chambers. Use a syringe with a bent needle or a large kitchen basting bulb to remove bubbles from the bottom edge of the glass plates.

5. Load the samples using a 500-μL Hamilton syringe. Load prestained molecular weight markers in the first lane.

6. Assemble the gel unit, connect to a power supply and run overnight at 70 V constant voltage. Judge the length of the run by the migration of the prestained markers (*see* **Note 9**). The voltage can be increased to at least 120 V without cracking the glass plates due to uneven heating.

7. Polymerized gels can be stored at 4°C for at least a week by placing a wet paper towel over the comb and sealing the glass/gel sandwich in a large plastic bag.

3.7. Western Blotting

1. This protocol assumes use of a semi-dry blotting apparatus (e.g., ISS-Enprotech, W.E.P. Company). Run dH_2O over the top and bottom electrode plates of the blotting apparatus for about 5 min. Cut a piece of Immobilon-P PVDF membrane (Millipore) exactly equal to the size of the gel, typically 11.5×14 cm. Thoroughly wet this membrane with 5 mL of methanol, then add 5 mL CAPS transfer buffer and 40 mL dH_2O. Cut four similarly-sized pieces of Whatman 1 or 3MM paper.

2. Disassemble the gel and cut off the stacking gel with a pizza cutter.

3. To assemble the blot, wet two pieces of Whatman paper in the CAPS-methanol buffer and, wearing gloves, lay them in the center of the bottom electrode, rolling each down to avoid trapping air bubbles. Add the membrane, the gel, and then two more sheets of wet filter paper. Roll a pencil over the sandwich to force out air bubbles. Pour the remaining buffer over the stack.

4. Add the blotter top, connect to a power supply and transfer (for a 1.5-mm-thick gel) with constant current for 2–2.5 h at 150 W (about $1 \, mA/cm^2$). During this time, make up the blocking buffer, 50 mL of 1% nonfat dry milk in Tween-saline, and rock the solution to ensure that the milk dissolves completely.

5. Remove the blotter top and determine if the prestained molecular weight markers have transferred by carefully peeling only lane 1 of the gel off the membrane, without disturbing the rest of the stack. If the marker proteins have transferred, disassemble, mark each prestained marker with a pencil line, and block the membrane by soaking 1 h at room temperature in the milk solution. Longer transfer times may be needed for efficient transfer of proteins greater than 60 kDa.

6. Briefly rinse the blot in Tween-saline buffer. If the antiserum to be used is expensive or limited in supply, put the blot into a heat-sealed bag using, for example, a Rival Seal-A-Meal Vacuum heat sealer (*see* **Note 10**). If multiple antisera are to be utilized, lay the membrane on a glass plate and cut into strips using a pizza cutter run along a ruler. Trim the lower right corner off each piece to aid orientation and place in a bag. Seal the bag tightly around the membrane, removing bubbles before sealing the last edge by dragging the bag over the edge of the bench and pinching air bubbles out the top.

7. The antiserum dilution will vary with the serum and antigen concentration and needs to be determined by incubating identical gel strips in different amounts of primary serum. A 1:20,000 dilution works well for a self-prepared high-titer serum, whereas 1:6000 is appropriate when using a commercial anti-FLAG peroxidase serum (Sigma) and 1:1000 is a good starting point for commercial affinity-purified antisera. Incubate the blot for 1–2 h at room temperature with rocking or overnight at 4°C.

8. Rinse twice in Tween-saline buffer and then wash twice for 5 min, rocking in a plastic box close to the size of the blot (e.g., Tupperware sandwich box).

9. The secondary antiserum also needs to be titered. Commonly 1.25 μL in 25 mL Tween-saline buffer for 1 h at room temperature, with shaking, is appropriate. Repeat the wash steps as for the primary serum. Visualize the protein bands using chemiluminescence.

10. As for immunoprecipitation, controls are mandatory when utilizing an antiserum. Staining with preimmune or normal serum of the same species can be used to detect nonspecific protein bands. Alternatively, if available, run cell homogenate lacking the antigen, for example, untransfected cells if a tagged protein is being expressed.

11. If the blot is to be reprobed with a second antiserum, store the blot soaking in Tween-saline buffer or wrapped in Saran wrap at 4°C. Add sodium azide to 0.02% to inhibit bacterial growth. Do not allow the PVDF membrane to dry out. If the second primary antiserum is from the same species as the first primary antiserum, strip off the antibodies by rocking the blot submerged in strip buffer in a 70°C oven for 1 h. Wash five times for 5 min in Tween-saline.

3.8. Chemiluminescence

1. Rinse the blot briefly in dH$_2$O. Mix 2 mL solution A with 2 mL solution B in the corner of a tilted sandwich-sized pan. Rock the blot in this solution for 1 min at room temperature. Remove excess liquid by dragging the blot over the edge of the pan four times, turning and using a clean side of the pan each time.

2. Place the blot face-down on a piece of Saran wrap and wrap the blot, leaving the overlap on the back side so that the front is smooth.

3. Place the blot in a film cassette face-up, snug in the upper left corner. Bend the lower right corner of the film when adding it to the cassette to aid orientation. Expose to film for 1–10 min. If desired, add another piece of film and leave it overnight at room temperature. If the signal is too strong in 0.5–1 min, the exposure process can be carried out after 1–2 h.

4. After developing the film and before moving the blot from the cassette, lay the film over the blot and mark the location of the prestained molecular weight markers.

3.9. Carbohydrate Analysis

1. Sugar chains added to lysosomal enzymes help keep the proteins soluble, provide a site for generation of the mannose-phosphate targeting motif, and help protect the proteins from proteolysis within lysosomes. Proteins still in the ER possess high-mannose carbohydrate chains that can be cleaved by treatment with endoglycosidase H. Once a protein reaches the Golgi, mannose residues are removed and complex sugars such as *N*-acetylglucosamine and sialic acid are added. High-mannose and sugar chains containing complex sugars can be distinguished by their sensitivity to particular endoglycosidases. Endoglycosidase H (endo H) removes

high-mannose but not complex sugar chains. Peptide N-glycosidase F (also known as PNGase F or N-glycosidase F) cleaves both types of sugar chains. Thus, biosynthetic forms of a protein that are resistant to endo H but sensitive to endo F have probably reached or passed through the Golgi. Carbohydrate chains modified with mannose-phosphate remain sensitive to endoglycosidase H (*9*). Thus, a lysosomal enzyme that possess only a single carbohydrate addition site will remain totally endo H sensitive, even after it has traversed the Golgi (*see* **Note 11**).

2. Whole cell extracts or immunoprecipitates can be treated with endoglycosidases. The former approach requires more enzyme, because many cellular proteins possess N-linked sugar chains; thus, treatment of an immunoprecipitate is often more efficient, uses less enzyme, and produces a cleaner result.

3. Endo H treatment of immunoprecipitates: Elute protein from PAS/PGS in 50 μL of 1% SDS, 50 mM Tris-HCl, pH 7.5, and 50 mM DTT by incubating in a boiling water bath for 5 min. Spin in a microfuge to pellet the beads and then transfer the supernatant to a clean tube with a Hamilton syringe. Add 120 μL 0.3 M citrate, pH 5.5, to the PAS pellet, vortex, spin, and transfer 100 μL to the tube containing the original supernatant. This wash of the PAS improves the yield and adjusts the citrate concentration to 0.2 M. Add 17 μL of 10x protease inhibitors and 2 μL endo H (stock at 1 mU/μL) and incubate overnight at 37°C. Terminate the reaction and remove citrate by TCA precipitation of the protein (*see* **Note 12**).

4. Endoglycosidase treatment of whole cell extracts: scrape cells on a 35-mm dish into 100 μL 1% SDS. Boil 5 min, then sonicate to shear the DNA. Dilute 1:10 in the appropriate buffer to reduce the SDS concentration to less than 0.2%. Determine the protein concentration. For endo H, dilute 1:10 with 10 50 mM NaH$_2$PO$_4$, pH 5.5, containing 0.5% NP-40. Add endo H to 1 mU/μg cellular protein and incubate at 37°C overnight. For PNGF, add 900 μL dilution buffer and 1.2 mU PNGF/μg cellular protein and incubate overnight at 37°C. Depending of the volume and the gel well size, add concentrated gel loading buffer (e.g., 4x) or TCA precipitate the samples (*see* **Note 13**).

5. Determination of the number of N-linked carbohydrate chains added to a protein: Add 2.5 μL endo H and incubate at 37°C for 2.5 h. Terminate the reaction by TCA-precipitating the proteins. This brief digest yields proteins that have lost varying numbers of their carbohydrate chains. A protein ladder is usually obtained, comprised of protein that has lost no sugar, one chain, two chains etc. The number of asparagines modified with sugar chains can be determined by counting the bands in the ladder. This analysis is best undertaken on a pulse-labeled sample so that the newly synthesized protein remains in the ER and has not acquired complex sugars.

6. Tunicamycin treatment: cell monolayers can be treated with the antibiotic tunicamycin, which inhibits the dolichol phosphate-mediated sugar addition machinery in the ER. Lysosomal enzymes totally lacking N-linked sugar chains both aggregate, detected by PAGE as high-mass protein bands, and undergo increased secretion due to their inability to bind mannose-phosphate targeting receptors. Add the drug to 1–1.5 μg/mL culture medium for 12–24 h.

3.10. Microsome Preparation and Membrane Association Assays

1. Because lysosomal proteins are sequestered within the secretory pathway, they can be isolated from cellular extracts by the preparation of microsomes. Lysosomal proteins can either be within the lumen of microsomes and vesicles or be integral membrane proteins of microsomes and vesicles.
2. Place culture dishes on ice, wash cells twice with PBS+, and twice with cell homogenization buffer. Add protease inhibitors to an aliquot of homogenization buffer and scrape cells using a rubber policeman or a Teflon cell scraper Nalge Nunc (cat. no. 179693) into 500 μL of this homogenization buffer. Incubate 10 min on ice. Transfer to a 7-mL Dounce homogenizer (Wheaton) and homogenize in an ice bucket using 20 strokes with a tight-fitting A pestle (*see* **Note 14**).
3. Transfer to a microfuge tube and spin 2 min. Remove supernatant and save on ice. Add more fresh buffer to the pellet and repeat the homogenization. Avoid repeated homogenization of the supernatant as this may lyse vesicles. Pool all supernatants and repeat the spin. Transfer this postnuclear supernatant (PNS) to a Beckman TL100 microfuge tube (cat. no. 357448). Save the pellet, which contains unbroken cells and nuclei, as a whole cell control sample.
4. Place microfuge tubes in a TLA100.3 fixed-angle rotor using plastic Beckman adaptors. Spin at 120,000*g* (60,000 rpm) at 4°C for 15–30 min. Orient tubes so you will know where to expect the pellet.
5. Remove the supernatant with a pulled-out Pasteur pipet. This sample contains cytoplasm. The pellet contains microsomes and vesicles with their fluid content.
6. To isolate membranes, these vesicle must be lysed or rendered leaky. Add pH 11.5 carbonate buffer to the pellet and vortex well. Incubate on ice for 30 min. Buffers of a lower pH can be utilized if it is desirable to leave peripheral membrane proteins associated with the membranes.
7. Repeat the TL100 spin. The supernatant contains the soluble contents of vesicles and microsomes as well as peripheral membrane proteins released in the high-pH buffer that collapses the membranes into sheets *(10,11)*. The carbonate wash can be repeated to ensure efficient wash of peripheral proteins from the membranes.
8. The supernatant fractions can be concentrated by acid precipitation. The pellet can be solubilized directly in gel loading buffer. Microsome pellets can also be quick-frozen in liquid nitrogen or a dry ice–ethanol and stored at −80°C.

3.11. Cell Fractionation by Sucrose Gradient Centrifugation

1. Mature lysosomes, dense vesicles containing active lysosomal proteases, can be isolated from other cellular vesicles by virtue of their density using relatively simple sucrose or Percoll gradients. Terminal lysosomes can also be purified based on their ability to concentrate endocytosed magnetic dextran beads *(12)*. Lysosomal enzymes are usually also detected in a lighter fraction, often called "light lysosomes," that is a mixture of ER, Golgi, endosomes, and plasma membrane-derived vesicles. This

gradient band primarily possesses the precursor form of lysosomal proteases, which establishes that these vesicles are not mature lysosomes.

2. Prepare a PNS from six to eight confluent 100-mm dishes of cells by homogenization, as described above, using 4×0.6 mL homogenization buffer containing protease inhibitors. Save about 150 μL of the cell homogenate as a control for enzyme assays performed to localize organelles.

3. Pour a 10-mL continuous sucrose gradient in a clear SW41 tube. Add 5 mL 0.7 M sucrose to the right chamber of a gradient maker. Open the connection between the two chambers long enough to allow sucrose to fill the connection and force out any air bubbles. With a Pasteur pipet, return any sucrose in the 1.6 M sucrose chamber to the 0.7 M chamber. Add 5 mL 1.6 M sucrose to the left chamber. Open the connection between the two chambers and immediately begin to stir the left chamber with a short stir bar. Sucrose should run down the SW41 tube in a steady stream, not by drops. This is easier to achieve with reused SW41 tubes. Alternatively, scratch the inside of the tube with a Pasteur pipet and let the sucrose run down that part of the tube wall.

4. Carefully overlay the sucrose with the PNS, using a Pasteur pipet or a syringe having a piece of thin tubing attached to a needle. Centrifuge at 96,000g, 4°C for 3 h. Collect 0.5-mL fractions from the top using an Auto Densi-Flow IIC fractionator (Haake Buchler).

5. To read the A_{280} of the fractions, dilute 50 μL with 950 μL 1% SDS to open vesicles.

6. Run about 100 μL of each fraction/gel lane. Vesicle populations can be identified by blotting for the same marker proteins used to identify organelles by immunofluorescence or assays for enzymes characteristic of organelles can be performed *(13)*. Enzymes unique to the various endosome populations have not been identified, however.

3.12. Cell Fractionation by Percoll Gradient Centrifugation

1. Add 27–30% Percoll to a Beckman 1×3.5 inch Quick-Seal centrifuge tube using a 10-mL syringe and an 18g needle. Underlay with a 2 mL 2.5 M sucrose cushion using a long Pasteur pipet. Overlay the PNS (up to 8 mL) using a syringe and a needle fitted with a piece of tubing. Add 0.125 M sucrose to fill the tube (2–4 mL). Heat-seal.

2. Centrifuge in a Beckman VTi50 rotor for 30 min at 33.000g at room temperature.

3. With a sharp razor blade, cut off the top of the tube without squeezing the tube. Collect 1-mL fractions from the top with an Auto Densi-Flow IIC apparatus.

4. Small (100-μL) samples can be diluted with 4x PAGE loading buffer, boiled, and loaded on a gel, even if they are cloudy. For larger samples or IP proteins, remove the Percoll by centrifugation. Percoll does not interfere with enzyme assays.

3.13. Immunofluorescence Staining Protocol (14)

1. Plate cells on 22-mm × 22-mm cover slips at the density desired (*see* **Note 15**).

2. Wash cells once with PBS (*see* **Note 16**). Incubate cells on ice 5 min with 4% paraformaldehyde. Wash once with PBS.

3. To permeabilize the cells to enable antibody uptake, incubate cover slips on ice 30 s with ice-cold ($-20°$C) methanol (*see* **Note 17**). Wash cells once with PBS.
4. Wash cells three times for 5 min with quench buffer.
5. Wash cells three times for 5 min with wash buffer.
6. Move the cover slips to a flat surface. Incubate with primary antiserum diluted in wash buffer for 1 h at room temperature (*see* **Note 18**). When using a new antibody for the first time, titration from 0.1 to 10 µg/mL is recommended.
7. Wash cells three times with wash buffer for 5 min each.
8. Dilute secondary antibody to desired concentration in wash buffer. Incubate cover slips with diluted secondary for 1 h at room temperature in the dark (e.g., cover with an ice bucket cover).
9. Wash coverslips five times for 5 min by incubating with PBS in the dark.
10. To fix cells to a slide for microscope viewing, place a drop of FluorSave Reagent (Calbiochem) on a microscope slide. Place cover slip with the cell monolayer facing toward the slide on the FluorSave drop. Suction off excess liquid around the coverslip.
11. Incubate in the dark for at least 1 h before storing in the dark at 4°C until viewing.

3.14. Endocytosis of Fluorescent Transferrin, a Marker for Endocytic Compartments

1. Incubate cells in culture medium containing 20 µg/mL fluorescently labeled transferrin in the dark (e.g., cover with foil) at 4°C to synchronize uptake.
2. Place the cells in 37°C to allow uptake for the length of time desired (*see* **Note 19**).
3. Fix and counterstain cells (if desired). Poor results have been obtained following permeabilization with methanol. For best results, permeabilize with saponin.

Notes

1. While LAMP1 and LAMP2 are frequently used as marker proteins for lysosomes, they actually have a broader distribution that includes multivesicular endosomes and late endosomes. Lysosomes are better described as compartments that are positive for LAMP1 and LAMP2 and negative for mannose-6-phosphate receptors. Similarly, while CD63 is considered to mark multivesicular endosomes, it can also be found in late endosomes and on the plasma membrane. Fluorescently labeled transferrin (Molecular Probes/Invitrogen) can be used to label multiple endosomal compartments depending on the time of internalization.
2. Because paraformaldehyde is not very soluble in water, both heating and addition of small amounts of concentrated NaOH is necessary. Autofluorescence of paraformaldehyde increases with the age of the solution. When solutions that are initially colorless begin to take on a yellowish cast, they should be replaced.
3. Harvest in hot, high-pH buffers containing detergent is employed to inactivate lysosomal proteases released as soon as the cells are lysed. If the samples are to

be directly loaded on a gel, harvest in an SDS-containing buffer, because nonionic detergents can generate artifacts on gels. The volume of harvest buffer utilized depends on the size of the gel available and the amount of cellular protein to be solubilized by the detergent.

If the protein under study is to be immunoprecipitated, the cells can be harvested in an SDS or a nonionic detergent-containing buffer. SDS is more effective at denaturing lysosomal proteases but will also denature added immunoglobulin unless the SDS concentration is reduced, after the initial denaturation, by the addition of a nonionic detergent that traps SDS in detergent micelles. SDS harvest and wash in an SDS-containing buffer generally yields cleaner immunoprecipitates than those obtained with nonionic detergents. The yield of specific bands may be slightly lower, but the nonspecific background is also reduced. If, however, the antiserum used was raised against a protein in its native conformation, it will probably be necessary to use a nonionic buffer to retain conformational determinants.

4. Protein possessing a signal peptide cannot be visualized by Western blotting or by radiolabeling, because the signal peptide is cleaved cotranslationally. This initial form of the protein can only be isolated if mRNA is translated in vitro in the absence of microsomal membranes.

5. Medium utilized for pulses can be removed and centrifuged to pellet floating dead cells, and the supernatant can be frozen for 1–2 wk. There is sufficient radiolabeled Met left in this medium for continuous labeling.

6. Controls are mandatory for immunoprecipitates as no IP is perfectly clean. The signal-to-background ratio simply varies. Always run a mock sample with preimmune or a normal serum at the same IgG protein concentration as the specific serum. Samples can be precleared by immunoprecipitating with preimmune serum, if available, but often background results from protein precipitation during the IP process rather than crossreaction of the antiserum with another protein. If whole serum is used rather than an IgG fraction or affinity-purified serum, immunoglobulin heavy chain may migrate on PAGE as a broad band that pushes down on minor proteins migrating just below it, producing an artifactual band about 43 kDa. This is especially noticeable in radiolabeled samples, where the 43-kDa band just below the cold IgG band characteristically appears to "smile."

7. Freeze the tubes containing the used resin. To regenerate the resin, combine pellets and wash the used resin in 5% SDS, 0.1 M Tris-HCl, pH 7.4, 25 mM DTT, and 20 mM EDTA. Boil twice for 5 min in a volume at least 1:1 with the resin. Repeat using the same buffer with 8% Triton X-100 substituted for the SDS. Wash four times. Wash twice with dH$_2$O. Store the resin at 4°C as a slurry in dH$_2$O containing 0.02% Na azide.

8. Before adding the acrylamide solution to the assembled glass plates, test the seals by squirting dH$_2$O between the plates and observing for 10–15 min. If persistent

leaking occurs, the bottom edge of the glass sandwich can be sealed, while vertical in the gel apparatus, with a solution of hot 1% agarose.

9. If the protein of interest is 50 kDa or greater, resolution can be improved by running one or more prestained markers completely off the gel, altering the gel percentage, pouring a gradient gel, or by using longer (10-cm) gel plates (increase the gel solution volume by 1.7 times) in the same apparatus.

10. Avoid plastic bags that have a surface pattern, such as those sold for freezing food, as these produce artifact patterns on the blot. Smooth plastic is available in a roll from Fisher (Kopak Roll Stock Pouches 01-812-26E).

11. The molecular mass of a protein is reduced by approx 2 kDa for each N-linked sugar chain removed. Overexpression can swamp the sugar addition machinery, resulting in the appearance of some relatively stable protein, presumably aggregated in the ER protein, which lacks all sugars.

12. Removal of carbohydrate may cause some proteins to precipitate out of solution. If a pellet is obtained at this point after a microfuge spin, it may contain the desired deglycosylated protein. Resuspend in gel sample buffer and omit the TCA step.

13. Mega-8 will not distort the protein bands on a gel as does Triton X-100 or NP-40.

14. Dounce pestles wear down. If there is not slight resistance, replace the pestle. The efficiency of cell lysis can be monitored with a phase microscope.

15. Depending on the scarcity of both cells and reagents, cells can be plated on the $0.7 \, cm^2$ of culture slides (BD Biosciences) or on cover slips that fit, for example, a 6-well culture cluster (Corning). Cover slips can be conveniently autoclaved in a glass Petri dish. The cell density utilized will depend on the cell type and whether the cells are to be transfected. Confluency percentages below 100% produce better images because isolated cells are easier to find.

16. Pipetting solutions directly onto the cells can shear the cells off the cover slip. Pipetting onto the cover slip at the same place each time can minimize this.

17. Alternative protocols utilize detergents, such as 0.5% Triton X-100 or saponin to induce permeabilization *(2)*. If the staining is weak or the background high with methanol permeabilization, try detergent permeabilization.

18. Cover slips can be stained on a piece of Parafilm. If antibody is precious, the volume of antibody utilized can be decreased by placing the coverslip face down on a bead of diluted antibody. This technique also helps if the cells being manipulated are not very adherent. The optimal amount of primary and secondary antiserum must usually be determined by titration. As with immunoprecipitation and Western blotting, it is important to stain cells in parallel with preimmune or normal serum of the same species at the same protein concentration as the primary serum in order to assay for nonspecific staining.

19. Approximate times are 5–10 min to detect transferrin in early endosomes and 30 min for late endosomes/lysosomes, although signal in these terminal compartments will decrease as transferrin is degraded.

Acknowledgments

Many of the protocols described originated in Gunter Blobel's laboratory, The Rockefeller University. The authors would also like to thank UNC colleagues Jean Cook for the chemiluminescence protocol, and JoAnn Trejo for sharing immunofluorescence microscopy methodology. This work was supported by research grant MCB-0235680 from the National Science Foundation.

References

1. Mehtani, S., Gong, Q., Panella, J., Subbiah, S., Peffley, D. M., and Frankfater, A. (1998) *In vivo* expression of an alternatively spliced human tumor message that encodes a truncated form of cathepsin B. Subcellular distribution of the truncated enzyme in COS cells. *J. Biol. Chem.* **273**, 13236–13244.
2. Goulet, B., Baruch, A., Moon, N.-S., Goulet, B., Baruch, A., Moon, N.-S., Poirier, M., Erickson, A., Bogyo, M., and Nepveu, A. (2004) A cathepsin L isoform that is devoid of a signal peptide localizes to the nucleus in S phase and processes the CDP/Cux transcription factor. *Mol. Biol. Cell* **14**, 207–219.
3. Collette, J., Bocock, J. P., Ahn, K., Collette, J., Bocock, J. P., Ahn, K., Chapman, R. L., Godbold, G., Yeyeodu, S., and Erickson, A. (2004) Biosynthesis and alternate targeting of the lysosomal cysteine protease cathepsin L. *Int. Rev. Cytol.* **241**, 1–51.
4. Bogyo, M., Verhelst, S., Bellingard-Dubouchaud, V., Toba, S., and Greenbaum, D. (2000) Selective targeting of lysosomal cysteine proteases with radiolabeled electrophilic substrate analogs. *Chem. Biol.* **7**, 27–38.
5. Berg, T., Gjoen, T., and Bakke, O. (1995) Physiological functions of endosomal proteolysis. *Biochem. J.* **307**, 313–326.
6. Ahn, K., Yeyeodu, S., Collette, J., Madden, V., Arthur, J., Li, L., and Erickson, A. H. (2002) An alternate targeting pathway for procathepsin L in mouse fibroblasts. *Traffic* **3**, 147–159.
7. Collette, J., Ulku, A. S., Der, C. J., Jones, A. S., and Erickson, A. H. (2004) Enhanced cathepsin L expression is mediated by different Ras effector pathways in fibroblasts and epithelial cells. *Int. J. Cancer* **112**, 190–199.
8. Stinchcombe, J., Bossi, G., and Griffiths, G. M. (2004) Linking albinism and immunity: the secrets of secretory lysosomes. *Science* **305**, 55–59.
9. Varki, A. and Kornfeld, S. (1980) Structural studies of phosphorylated high mannose-type oligosaccharides. *J. Biol. Chem.* **255**, 10847–10858.
10. Fujiki, Y., Hubbard, A. L., Fowler, S., and Lazarow, P. B. (1982) Isolation of intracellular membranes by means of sodium carbonate treatment: application to endoplasmic reticulum. *J. Cell Biol.* **93**, 97–102.
11. Howell, K. E. and Palade, G. E. (1982) Hepatic Golgi fractions resolved into membrane and content subfractions. *J. Cell Biol.* **92**, 822–832.

12. Duvvuri, M. and Krise, J. P. (2005) A novel assay reveals that weakly basic model compounds concentrate in lysosomes to an extent greater than pH-partitioning theory would predict. *Mol. Pharm.* **2**, 440–448.

13. Yeyeodu, S., Ahn, K., Madden, V., Chapman, R., Song, L., and Erickson, A. H. (2000) Procathepsin L self-association as a mechanism for selective secretion. *Traffic* **1**, 724–737.

14. Paing, M. M., Stutts, A. B., Kohout, T. A., Lefkowitz, R. J., and Trejo, J. (2002) beta-Arrestins regulate protease-activated receptor-1 desensitization but not internalization or down-regulation. *J. Biol. Chem.* **277**, 1292–1300.

24

Monitoring Autophagy in Yeast

The Pho8Δ60 Assay

Daniel J. Klionsky

Summary

Autophagy is an ubiquitous degradative process in eukaryotic cells *(1,2)*. It is involved in various developmental programs and is also implicated in human pathophysiology *(3,4)*. The basic process involves the formation of a cytosolic double-membrane vesicle, termed an autophagosome *(5,6)*. The autophagosome sequesters bulk cytosol, and after completion its outer membrane fuses with the limiting membrane of the lysosome/vacuole. The fusion event releases the inner single-membrane vesicle into the lysosome/vacuole lumen, where the vesicle is now termed an autophagic body. The autophagic body and its cargo are typically degraded, and the resulting macromolecules are recycled for subsequent use in the cytosol. Autophagy is the only pathway with the capacity to degrade entire organelles. Accordingly, it may play a critical role in preventing pathologies that result from damaged organelles including the mitochondria (7), or from the accumulation of large protein aggregates, such as occur in certain types of neurodegenerative diseases *(8)*. This article describes an assay to monitor bulk autophagy in yeast. The marker protein is a cytosolic derivative of the vacuolar enzyme alkaline phosphatase, Pho8. Following uptake into the vacuole, the precursor enzyme is cleaved at the C terminus to generate the active form. Cells expressing the cytosolic form of Pho8, Pho8Δ60, are assayed for alkaline phosphatase activity before and after shifting to conditions that induce autophagy.

Key Words: Alkaline phosphatase; lysosome; protein degradation; vacuole.

1. Introduction

Autophagy is generally considered to be a nonspecific degradative process *(9)*; however, there are specific types of autophagy in which particular proteins *(10,11)* or organelles *(12)* are either targeted for degradation or delivered to

From: *Methods in Molecular Biology, Vol. 390: Protein Targeting Protocols: Second Edition*
Edited by: M. van der Giezen © Humana Press Inc., Totowa, NJ

the lysosome/vacuole as resident hydrolases. In higher eukaryotes there are relatively few methods for specifically measuring autophagy *(13)*. In contrast, several methods have been described in yeasts, and one of the most useful assays for monitoring nonspecific autophagy is the Pho8Δ60 assay.

The *PHO8* gene encodes the vacuolar enzyme alkaline phosphatase *(14)*. Pho8 is a type II integral membrane protein containing an N-terminal transmembrane domain that acts as an internal uncleaved signal sequence allowing translocation into the endoplasmic reticulum; like most resident vacuolar hydrolases, Pho8 is delivered to the vacuole through a portion of the secretory pathway, and the transmembrane domain retains it in the limiting membrane *(15)*. The type II topology of Pho8 results in a small cytosolic domain corresponding to the N terminus. Following vacuolar delivery, the C-terminal propeptide is removed within the lumen, resulting in the mature, active form of the enzyme.

Deletion of the N-terminal transmembrane domain from Pho8 generates a species referred to as Pho8Δ60 *(16)*. This protein is no longer able to enter the endoplasmic reticulum and remains within the cytosol *(17)*. The only way for Pho8Δ60 to reach the vacuole is through autophagy, and if that occurs the C-terminal propeptide is removed. Accordingly, vacuolar delivery of Pho8Δ60 via the autophagic pathway can be measured enzymatically, which forms the basis of the assay *(16)*. In brief, cells are assayed for Pho8Δ60-dependent alkaline phosphatase activity after shifting cells to conditions that induce autophagy, such as nitrogen starvation. Wild-type cells display background levels of activity immediately after the shift, but show an increase in activity after 1 to several hours (**Fig. 1**). A strain that is defective in autophagy will maintain the basal level of activity after switching to the inducing condition.

2. Materials

2.1. Cells, Culture, and Lysis

1. This assay can be used to determine whether a particular gene product is involved in autophagy. The gene of interest is deleted from the chromosome in the Pho8Δ60 assay strain. Positive and negative control strains should also be used (*see* **Note 1**).
2. Growth medium (YPD): 1% yeast extract, 2% peptone, and 2% glucose or synthetic minimal medium (SMD): 0.67% yeast nitrogen base, 2% glucose and auxotrophic amino acids, nucleosides and/or vitamins as needed. All specialized media components are available from ForMedium™ (Norwich, UK). These media are stable for months at room temperature.
3. Starvation medium (SD-N): 0.17% yeast nitrogen base without ammonium sulfate or amino acids, containing 2% glucose (*see* **Note 2**). This medium is stable for months at room temperature.

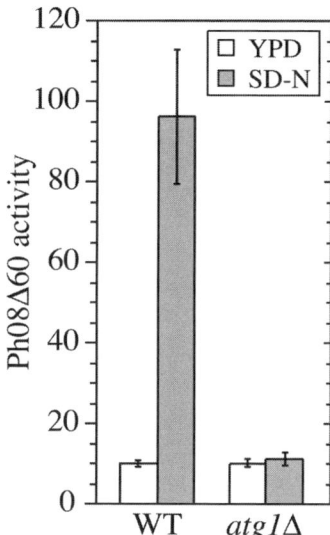

Fig. 1. Pho8Δ60 assay for monitoring nonspecific autophagy. Wild-type and *atg1*Δ strains were grown in growth medium (YPD) and shifted to starvation medium (SD-N) for 4 h. Samples were collected and protein extracts assayed for alkaline phosphatase activity. The wild-type strain displays background levels of activity before shifting to autophagy-inducing conditions, and then shows a substantial increase in activity after 4 h of starvation. The *atg1*Δ mutant is defective for autophagy and retains a background level of activity even after nitrogen starvation.

4. Lysis buffer: for 50 mL of lysis buffer, mix from the indicated stock solutions: 1 mL of 1 *M* PIPES (Research Organics, Cleveland, OH) (20 m*M* final conc.), 2.5 mL of 10% Triton X-100 (Sigma-Aldrich, St. Louis, MO) (0.5% final conc.), 2.5 mL of 1 *M* KCl (50 m*M* final conc.), 5 mL of 1 *M* potassium acetate (100 m*M* final conc.), 0.5 mL of 1 *M* MgSO₄ (10 m*M* final conc.), 50 μL of 10 m*M* ZnSO₄ (10 μ*M* final conc.), 0.5 mL of 100 m*M* PMSF in 95% ethanol (1 m*M* final conc.) (Roche Applied Science, Indianapolis, IN); the PMSF should be added just before the lysis buffer is added to the cells in **Subheading 3.1., step 9**. The stock solution of PMSF is stable for several months at room temperature (*see* **Note 3**).

5. Glass beads for lysis (0.4–0.6 mm soda lime; Thomas Scientific, Swedesboro, NJ) are prepared as follows:

   a. Wash beads in 1 L of 1 *N* HCl by soaking overnight with occasional mixing.
   b. Rinse with water 10–15 times, with distilled water 5 times, and with Millipore filtered water (or equivalent) 5 times.

c. Dry overnight in an oven. It is best to dry the glass beads in the final storage container to avoid the need to transfer them later.

2.2. Enzyme Assay

1. Alkaline phosphatase substrate solution (18): prepare a 100 mM stock of p-nitrophenyl phosphate (pNPP) dissolved in Millipore filtered water. This solution can be kept frozen at −20°C for a few months or at 4°C for 1 or 2 d. For the final substrate solution, dilute pNPP to 1.25 mM in reaction buffer. For 50 mL of reaction buffer, mix from the indicated stock solutions: 12.5 mL of 1 M Tris-HCl, pH 8.5 (250 mM final conc.), 2 mL 10% Triton X-100 (0.4% final conc.), 0.5 mL of 1 M $MgSO_4$ (10 mM final conc.), 50 μL of 10 mM $ZnSO_4$(10 μM final conc.) (see Note 4).
2. Stop buffer: 1 M glycine/KOH, pH 11.0.
3. Standard curve: p-nitrophenol 10 mM solution (Sigma N-7660). The product of the enzymatic cleavage of p-nitrophenyl phosphate is p-nitrophenol, which is yellow at alkaline pH.
4. Protein concentration is measured by the BCA assay (Pierce Chemical Co., Rockford, IL).

3. Methods
3.1. Cells, Culture, and Lysis

1. Strains are grown in 5–10 mL of YPD or SMD medium at 30°C in flasks. Cells are typically grown to log phase, then diluted and grown again to log phase (A_{600} = 1.0) (see Note 5).
2. Centrifuge at 1000g (3000 rpm) in a SA-600 rotor or equivalent for 5 min.
3. The supernatant is discarded and the cell pellet is resuspended (washed) in 5 mL of SD-N medium. Centrifuge at 1000g for 5 min.
4. The supernatant is discarded and the cell pellet is resuspended in 5 mL of SD-N.
5. The cell culture is incubated in SD-N at 30°C for 4 h (see Note 6).
6. An aliquot of cells equivalent to 2–5 A_{600} units (1 mL of culture at A_{600} = 1.0 is equivalent to 1.0 A_{600} unit) is removed from the culture and pelleted by centrifugation at 1000g for 5 min.
7. The supernatant is decanted and discarded and the cell pellet is resuspended (washed) in 1 mL of water. Centrifuge at 1000g for 5 min.
8. The supernatant is decanted and discarded and the cell pellet is resuspended (washed) in 2 mL of ice cold 0.85% NaCl containing 1 mM PMSF. Centrifuge at 1000g for 5 min at 4°C.
9. The supernatant is removed by aspiration and the cell pellet is resuspended in 200–500 μL of ice-cold lysis buffer (final cell suspension is at 1.0 A_{600} unit/100 μL, or A_{600} = 10/mL) and transferred to a 1.7-mL microcentrifuge tube (see Note 7).

10. A half-volume of glass beads is added and the cells are lysed by mixing on a vortex for 1–10 min at 4°C (*see* **Note 8**).
11. Centrifuge the lysed cells at 15, 7000*g* (13,000 rpm) in a microcentrifuge for 5 min at 4°C. Remove supernatant for analysis of alkaline phosphatase activity.

3.2. Enzyme Assay

1. Prewarm alkaline phosphatase substrate solution at 37°C, allowing 400 µL per reaction and per blank.
2. Prepare standard curve samples in 100 µL of Reaction buffer at 0–100 nmol *p*-nitrophenol in 1.7-mL microcentrifuge tubes. One µL of the 10 m*M* stock is equivalent to 10 nmol of *p*-nitrophenol. These concentrations should generate A_{400} values between 0 and 2.0.
3. Prepare enzyme and substrate blanks. For enzyme blank, place 100 µL of reaction buffer in tube. For substrate blank, place 100 µL of cell extract sample in a tube and add 400 µL of reaction buffer without *p*-nitrophenyl phosphate; do not add alkaline phosphatase substrate solution in **step 5**.
4. Place 50–100 µL of lysed samples into microcentrifuge tubes. Bring final volume to 100 µL with lysis buffer. Keep tubes on ice until ready to begin assay, then place all tubes at 37°C (*see* **Note 9**).
5. Add 400 µL of prewarmed alkaline phosphatase substrate solution at 15-s intervals to each tube and incubate at 37°C for 5–20 min; start timing after the first addition. Add 400 µL of reaction buffer without *p*-nitrophenyl phosphate to the substrate blank tube. The length of time for the incubation may need to be determined empirically based on the results with the positive control strain extract. The 15-s intervals between additions of substrate solution ensure that each sample is incubated the same length of time, and allows for the time needed to add the substrate solution, and later the stop buffer.
6. Add substrate solution to the blanks. Add stop buffer to the blanks.
7. Stop the reaction by adding 500 µL of stop buffer at 15-s intervals (*see* **Note 10**).
8. Centrifuge the tubes at maximal speed for 2 min to remove any precipitate or debris.
9. Measure the absorbance of 1.0 mL of the blanks and the samples at A_{400} (*see* **Note 11**).

3.3. Calculate Specific Activity

1. Determine protein concentration: set up a standard curve using a 2 mg/mL stock solution of bovine serum albumin (BSA); set up 8–10 tubes with a range between 0 and 1.0 mg/mL of BSA. Calculate the protein concentration in the samples using 50 µL of the lysate.
2. Subtract the enzyme and substrate blanks from the absorbance readings of the samples.
3. Calculate the concentration in nmol of *p*-nitrophenol in the samples by graphing the adjusted A_{400} values relative to the standard curve.
4. Calculate specific activity as nmol *p*-nitrophenol/min/mg protein.

Notes

1. Yeasts have genes for two alkaline phosphatases, *PHO8* and *PHO13* *(14,19)*. Pho8 is a repressible vacuolar enzyme, whereas Pho13 is a cytosolic enzyme. The *PHO8* gene should be replaced by integrative recombination with *pho8Δ60* *(16)*. Optimally, the *PHO13* gene should be deleted in the assay strain to eliminate background activity; that is, the only alkaline phosphatase activity in the cell should correspond to Pho8Δ60. The positive control should be a wild-type strain (*pho8::pho8Δ60 pho13Δ*) and the negative control should be the same strain with a deletion of one of the *ATG* genes that is required for nonspecific autophagy (e.g., *atg1Δ*). All strains that are going to be compared should have the same auxotrophies/prototrophies; some differences may affect the absolute Pho8Δ60 activities, presumably as a result of variations in cell growth rates and/or responses to starvation media.

2. Shifting cells to medium lacking carbon instead of nitrogen can also be used to induce autophagy; however, media lacking nitrogen may give the best response in terms of Pho8Δ60 activity.

3. PMSF is highly unstable in aqueous solutions. An alternative is to use water-soluble Pefabloc SC (100 mM stock in water), which can be stored for at least 2 mo at $-20°C$; however, PMSF inhibits both serine and cysteine proteases, whereas Pefabloc SC only inhibits the former.

4. It is possible to purchase *p*-nitrophenyl phosphate tablets (Sigma N-9389), which can be dissolved in reaction buffer. The instructions on the box suggest dissolving the 5 mg tablets in 5 mL; however, we find that dissolving in 8 mL works well in the assay.

5. It is important that cells be harvested at the same density for reproducibility within and between assays. If a time course is going to be analyzed rather than a single time point (*see* **Note 6**), a larger volume of culture should be prepared, allowing for a minimum of $1.0 A_{600}$ unit of cells for each time point.

6. The assay can also be performed as a time course, starting with a larger volume of cells, and removing samples every hour. Alkaline phosphatase activity typically starts to increase after approx 1 h of starvation, but does not reach an appreciable level until 3–4 h following the induction of autophagy.

7. Approximately 100 µL of lysate will be used in the assay, and some of the volume will be lost within the volume of the glass beads used for cell lysis. Therefore, a minimum of 200 µL cell suspension should be prepared. The cell pellet or the cells resuspended in lysis buffer can be frozen at $-20°C$ at this point, and the procedure continued at a later time.

8. To measure out the glass beads, create a "scoop" by cutting a microcentrifuge tube at the appropriate mark (half the volume of the resuspended cells, 100–250 µL), leaving a strip of plastic along one side to use as a handle. The beads should not be more than one-half of the cell suspension volume;

lysis is inefficient if the beads go above the liquid level in the tube, and the assay requires the removal of the supernatant from the bead mixture. Also, mixing on a vortex with glass beads can generate a substantial amount of heat; care must be taken to avoid denaturing the enzyme. It may be necessary to empirically determine the length of time needed for efficient lysis. Heating can also be reduced by using the vortex for 1-min intervals, followed by cooling on ice.

9. To ensure that the reaction is in the linear range, use at least two concentrations of sample and/or two time points for incubation with substrate.

10. The specific activity calculation will adjust for the length of the incubation; however, the reaction needs to be in the linear range (*see* **Note 9**), and the absorbance reading for *p*-nitrophenol also needs to be in the linear range of the spectrophotometer. The time to stop the incubation can be judged visually based on the color of the sample tubes, but the sample tubes should not be vibrant yellow or they may be beyond the linear range.

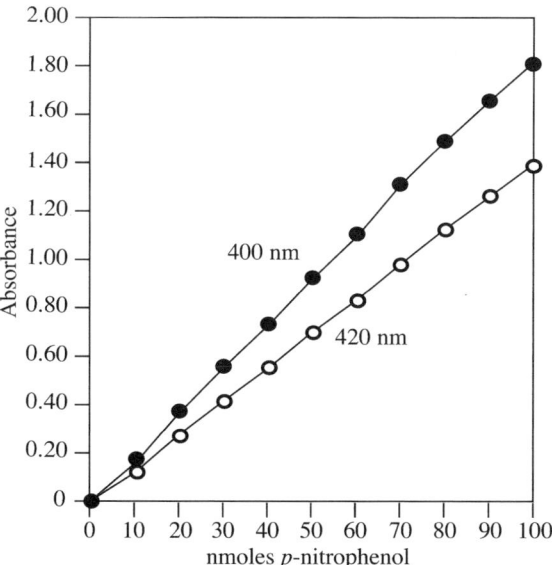

Fig. 2. Standard curve. Eleven 1.7-mL microcentrifuge tubes were used to generate the standard curve in duplicate. Five hundred μL of reaction buffer (minus the volume of the *p*-nitrophenol) and 500 μL of stop buffer were added to each tube. Zero to 10 μL of 10 m*M* *p*-nitrophenol were added to each tube, and the contents were mixed by vortex. The absorbance was determined at 400 nm (filled circles) and 420 nm (open circles).

11. Different wavelengths are used in this assay in various publications, including 400, 405, 410, and 420 nm. The optimal wavelength appears to be 400 nm, although there is relatively little difference at 405 and 410 nm; however, 420 nm results in a substantially lower absorbance reading (**Fig. 2**).

Acknowledgments

The author would like to thank Su Chen, Heesun Cheong, Zhiping Xie and Drs. Julie Legakis and Usha Nair for helpful comments. This work was supported by Public Health Service grant GM53396 from the National Institutes of Health.

References

1. Klionsky, D. J. and Emr, S. D. (2000) Autophagy as a regulated pathway of cellular degradation. *Science* **290**, 1717–1721.
2. Yorimitsu, T. and Klionsky, D. J. (2005) Autophagy: molecular machinery for self-eating. *Cell Death Diff.* **12**, 1542–1552.
3. Levine, B. and Klionsky, D. J. (2004) Development by self-digestion: molecular mechanisms and biological functions of autophagy. *Dev. Cell* **6**, 463–477.
4. Shintani, T. and Klionsky, D. J. (2004) Autophagy in health and disease: a double-edged sword. *Science* **306**, 990–995.
5. Kim, J. and Klionsky, D. J. (2000) Autophagy, cytoplasm-to-vacuole targeting pathway, and pexophagy in yeast and mammalian cells. *Annu.. Rev. Biochem.* **69**, 303–342.
6. Klionsky, D. J. and Ohsumi, Y. (1999) Vacuolar import of proteins and organelles from the cytoplasm. *Annu. Rev. Cell Dev. Biol.* **15**, 1–32.
7. Jin, S. (2006) Autophagy, mitochondrial quality control, and oncogenesis. *Autophagy* **2**, 80–84.
8. Rubinsztein, D. C., DiFiglia, M., Heintz, N., et al. (2005) Autophagy and its possible roles in nervous system diseases, damage and repair. *Autophagy* **1**, 11–22.
9. Klionsky, D. J. (2005) The molecular machinery of autophagy: unanswered questions. *J. Cell Sci.* **118**, 7–18.
10. Onodera, J. and Ohsumi, Y. (2004) Ald6p is a preferred target for autophagy in yeast, *Saccharomyces cerevisiae. J. Biol. Chem.* **279**, 16071–16076.
11. Nair, U. and Klionsky, D. J. (2005) Molecular mechanisms and regulation of specific and nonspecific autophagy pathways in yeast. *J. Biol. Chem.* **280**, 41785–41788.
12. Dunn, W. A., Jr., Cregg, J. M., Kiel, J. A. K. W., et al. (2005) Pexophagy: The selective autophagy of peroxisomes. *Autophagy* **1**, 75–83.
13. Mizushima, N. (2004) Methods for monitoring autophagy. *Int. J. Biochem. Cell Biol.* **36**, 2491–2502.

14. Kaneko, Y., Hayashi, N., Toh-e, A., Banno, I., and Oshima, Y. (1987) Structural characteristics of the *PHO8* gene encoding repressible alkaline phosphatase in *Saccharomyces cerevisiae*. *Gene* **58**, 137–148.
15. Klionsky, D. J. and Emr, S. D. (1989) Membrane protein sorting: biosynthesis, transport and processing of yeast vacuolar alkaline phosphatase. *EMBO J.* **8**, 2241–2250.
16. Noda, T., Matsuura, A., Wada, Y., and Ohsumi, Y. (1995) Novel system for monitoring autophagy in the yeast *Saccharomyces cerevisiae*. *Biochem. Biophys. Res. Commun.* **210**, 126–132.
17. Klionsky, D. J. and Emr, S. D. (1990) A new class of lysosomal/vacuolar protein sorting signals. *J. Biol. Chem.* **265**, 5349–5352.
18. Mitchell, J. K., Fonzi, W. A., Wilkerson, J., and Opheim, D. J. (1981) A particulate form of alkaline phosphatase in the yeast, *Saccharomyces cerevisiae*. *Biochim. Biophys. Acta* **657**, 482–494.
19. Kaneko, Y., Toh-e, A., Banno, I., and Oshima, Y. (1989) Molecular characterization of a specific *p*-nitrophenylphosphatase gene, *PHO13*, and its mapping by chromosome fragmentation in *Saccharomyces cerevisiae*. *Mol. Gen. Genet.* **220**, 133–139.

25

Protein Targeting to Yeast Peroxisomes

Ida van der Klei and Marten Veenhuis

Summary

Peroxisomes are important organelles of eukaryote cells. Although these structures are of relatively small size, they display an unprecedented functional versatility. The principles of their biogenesis and function are strongly conserved from very simple eukaryotes to humans. Peroxisome-borne proteins are synthesized in the cytosol and posttranslationally incorporated into the organelle. The protein-sorting signal for matrix proteins, peroxisomal targeting signal (PTS), and for membrane proteins (mPTS), are also conserved. Several genes involved in peroxisomal matrix protein import have been identified (*PEX* genes), but the details of the molecular mechanisms of this translocation process are still unclear. Here we describe procedures to study the subcellular location of peroxisomal matrix and membrane proteins in yeast and fungi. Emphasis is placed on protocols developed for the methylotrophic yeast *Hansenula polymorpha*, but very similar protocols can be applied for other yeast species and filamentous fungi. The described methods include cell fractionation procedures and subcellular localization studies using fluorescence microscopy and immunolabeling techniques.

Key Words: Peroxisome; yeast; *Hansenula polymorpha*; cell fractionation; fluorescence microscopy; immunolabeling.

1. Introduction

Peroxisomes belong to the class of microbodies, which are cell organelles that are present in virtually all eukaryotic cells. These organelles are composed of a proteinaceous matrix that is enclosed by a single membrane. The matrix is composed of enzymes that are involved in a great variety of metabolic pathways. Peroxisomes typically contain at least one hydrogen peroxide producing oxidases together with catalase. A second typical function of peroxisomes is the β-oxidation of fatty acids (for a review, *see* **ref. 1**).

From: *Methods in Molecular Biology, Vol. 390: Protein Targeting Protocols: Second Edition*
Edited by: M. van der Giezen © Humana Press Inc., Totowa, NJ

All peroxisomal proteins are encoded by nuclear genes and synthesized on free ribosomes in the cytosol. It is generally assumed that both peroxisomal membrane and matrix proteins are posttranslationally imported into peroxisomes. However, recent findings suggest that at least some peroxisomal membrane proteins (PMPs) may traffic to peroxisomes via the endoplasmic reticulum (reviewed in **refs.** *2,3*).

Most peroxisomal matrix proteins are routed to their target organelle via the PTS1 protein import pathway. The PTS1 signal is located at the extreme C terminus of peroxisomal matrix proteins and consists of three amino acids: -SKL or conserved variants thereof. The PTS1 signal is recognized by the C-terminal tetratricopeptide (TPR) repeat domain of the receptor protein, Pex5p (reviewed in **refs.** *4,5*).

Relatively few peroxisomal matrix proteins are sorted to peroxisomes by a PTS2 sequence. The PTS2 consists of a nonapeptide with the consensus (R/K)-(L/V/I)-X_5-(H/Q)-(L/A), which is located at the N terminus of peroxisomal matrix proteins. PTS2 proteins are recognized by the receptor protein Pex7p. However, this receptor requires auxiliary proteins, which contain a conserved Pex7p binding box and most likely form the genuine PTS2 receptor together with Pex7p (*4,5*). In most yeast and filamentous fungi, the auxiliary protein is Pex20p (*2,4*). In *Saccharomyces cerevisiae*, however, two homologous proteins, Pex18p and Pex21p, perform the same function as Pex20p, whereas in human cells the long isoform of Pex5p, Pex5pL, is required for PTS2 recognition together with Pex7p.

Some peroxisomal matrix proteins are imported in a Pex5p-dependent manner, but lack a typical PTS1 sequence or contain a PTS1 sequence that is redundant for sorting. These proteins are recognized by a still unknown domain in the N-terminal half of Pex5p, instead of the C-terminal TPR domain. Examples are alcohol oxidase of *Hansenula polymorpha* (*6*) and acyl-CoA oxidase of *S. cerevisiae* (*7,8*). The exact sorting signal of these matrix proteins that mediates binding of the protein to the N-terminus of Pex5p is not known yet.

Peroxisomal proteins that lack a PTS can also be imported via formation of a hetero-oligomeric complex with another PTS-containing protein. An example of this so-called piggyback mechanism is the import pathway of *S. cerevisiae* Dci1p and Eci1p. These two PTS1 proteins can be imported as a hetero-oligomeric complex. When the PTS1 of Eci1p is removed, this protein is co-transported into peroxisomes with Dci1p (*9*).

So far, no general conserved consensus sequence is known for PMPs. However, a typical feature of the target information of PMPs (the so-called mPTSs) is the presence of a stretch of positively charged residues in conjunction with at least one transmembrane region *(1,2,10)*. Two peroxins, Pex3p and Pex19p, have been suggested to play a role in recognition and insertion of PMPs (for a review, *see* **ref. *11***). Pex3p is a peroxisomal membrane protein, which is capable to recruit Pex19p to the peroxisome. Pex19p is a soluble peroxin that has the capacity to specifically bind PMPs. At present two models exist for the function of Pex19p. The first model predicts that Pex19p represents the mPTS receptor, because of its specific interactions with PMPs. Upon binding to a newly synthesized PMP in the cytosol, Pex19p is predicted to associate to Pex3p at the peroxisomal membrane, followed by insertion of the PMP into the membrane. Using synthetic peptide scans and yeast two-hybrid analyses, Pex19p-binding sites have been identified in two *S. cerevisiae* PMPs *(12)*. These sites turned out to be composed of a short helical motif with a minimal length of 11 amino acids and most likely represent (part of) the mPTSs.

The second model of Pex19p function predicts that Pex19p is not the mPTS receptor, but assists in formation of functional PMPs or PMP-complexes, suggesting a chaperone-like function for Pex19p *(11)*. The latter option is in line with the observation that in some yeast species (*H. polymorpha, Yarrowia lipolytica*) PMPs are normally sorted to peroxisomal membranes in the absence of Pex19p. However, these peroxisomal membranes are not fully functional *(13,14)*.

In contrast to most other protein targeting and translocation processes, attempts to completely reconstitute peroxisomal protein sorting *in vitro* using isolated yeast peroxisomes have been without success. However, a few examples exist of *in vitro* assays in which (part of) the peroxisomal protein sorting pathway is probably reconstituted. These include import of peroxisomal matrix proteins in plant peroxisomes *(15)*, *in vitro* binding and release assays for Pex5p *(16–18)* and *in vitro* insertion of PMPs into membranes of mammalian peroxisomes *(19)*.

In this chapter we present methods that are used to localize peroxisomal membrane and matrix proteins in yeast. We focus on techniques used in the methylotrophic yeast *H. polymorpha*, but very similar procedures can be applied for other yeast (*Pichia pastoris, S. cerevisiae, Y. lipolytica*) and filamentous fungi.

First, we describe the isolation of peroxisomes by sucrose density centrifugation. This allows one to analyze the protein composition of the organelles and also to separate the peroxisomal proteins into matrix- and membrane-bound ones. Second, we describe fluorescence microscopy procedures to analyze sorting of PTS1-, PTS2-, and peroxisomal membrane proteins. Finally, we present methods to determine the subcellular location of peroxisomal proteins by electron microscopy using immunogold labeling procedures.

2. Materials

2.1. Cultivation of Yeast Cells

2.1.1. Media Components

1. Minimal medium: per liter add 0.2 g $MgSO_4$, 0.7 g K_2HPO_4, 3.0 g NaH_2PO_4, and 0.5 g yeast extract, add 1 mL of a 1000x concentrated trace element solution. After autoclaving the medium, add 1 mL of the filter-sterilized vitamin stock solution (1000x). When ammonium has to be used as nitrogen source: add 2.5 g $(NH_4)_2SO_4$ per liter medium *(20)*. When methylamine is used as nitrogen source: add 2.5 g methylamine and 1 g K_2SO_4 per liter. When required, uracil or amino acids are added to a final concentration of 30 μg/mL. For growth on agar plates the media are supplemented with 1.5% agar.
2. A thousand fold concentrated trace elements stock solution: dissolve 10 g EDTA in 950 mL distilled water. Add 4.4 g $ZnSO_4 \cdot 7H_2O$ under continuous stirring. When the solution is clear, adjust the pH to 6 with 5 M NaOH. Subsequently add the other trace elements one by one under continuous stirring and adjustment of the pH to 6. The trace elements to be added are per liter stock solution: 1.0 g $MnCl_2 \cdot 4H_2O$, 0.32 g $CoCl_2 \cdot 6H_2O$, 0.32 g $CuSO_4 \cdot 5H_2O$, 0.22 g $(NH_4)_6Mo_7O_{24} \cdot 4H_2O$, 1.47 g $CaCl_2 \cdot 2H_2O$ and 1.0 g $FeSO_4 \cdot 7H_2O$. When all compounds are dissolved, water is added to a final volume of 1 L. The solution can be stored in the dark at 4°C.
3. A thousand fold concentrated vitamin stock solution: dissolve 20 mg biotin in 20 mL 0.1 M NaOH. Add 980 mL 20 mM potassium phosphate buffer pH 7.5 and subsequently 600 mg thiamine. When both vitamins are dissolved, sterilize the solution by filtration through a 0.2-μm filter. The solution can be stored at 4°C.

2.2. Cell Fractionation

1. Lytic enzyme for protoplast preparation: Zymolyase®-20T.
2. Preincubation buffer: 100 mM Tris-HCl buffer pH 8.0 supplemented with 50 mM EDTA, 140 mM β-mercaptoethanol, and 1.2 M sorbitol (prepare fresh).
3. Protoplasting buffer: 50 mM potassium phosphate buffer pH 7.2 containing 1.2 M sorbitol.
4. Protoplast disruption buffer: 5 mM MES buffer pH 5.5 containing 1.2 M sorbitol, 1 mM PMSF, 5 mM NaF, and 2.5 μg/mL leupeptin (freshly prepared).
5. Buffer A: 5 mM MES pH 5.5, 0.1 mM EDTA and 1 mM KCl.
6. Stock solution of 65% (w/w) sucrose (855.7 g sucrose/liter) in buffer A.
7. Peroxisome lysis buffer: 0.1 M Tris-HCl, pH 8.0.
8. High-salt buffer: 0.1 M Tris-HCl, pH 8.0, supplemented with 1 M NaCl.
9. Carbonate extraction buffer: 0.1 M carbonate pH 11.0 (freshly prepared).
10. Trichloroacetic acid (TCA) solution: 50% TCA in water.
11. 1% Socium dodecyl sulfate (SDS) in a solution of 0.1 M sodium hydroxide.
12. Sucrose solutions for sucrose density centrifugation: mix buffer A and calculated amounts of the 65% sucrose stock solution in buffer A to obtain solutions with

Table 1
Preparation of Sucrose Stock Solutions

Final sucrose concentration (% w/w)	Sucrose stock (mL)	Buffer A (mL)	Total volume (mL)
52	22.6	7.4	30
50	21.6	8.4	30
48	47.8	22.2	70
46	26.0	14.0	40
44	30.8	19.2	50
40	16.5	13.5	30
35	14.1	15.9	30

Table 2
Discontinuous Sucrose Gradients

Gradient for PNS		Gradient for P3	
mL	%	mL	%
6.5	65	6.5	65
2.5	52	2.5	52
2.5L	50	2.5	50
5.0	48	7.5	48
3.0	46	4.0	46
4.0	44	5.0	44
	5–8 mg PNS protein		5–8 mg P3 protein resuspended in 40% sucrose
Overlay	35%	Overlay	35%

varying sucrose concentrations as indicated in **Table 1**. Pour sucrose gradients in centrifuge tubes (36-mL polypropylene tubes to be used with a Sorvall SV288 rotor and a Sorvall RC 5C$^+$ centrifuge) for fractionation of a postnuclear supernatant (PNS) or $30,000g$ organellar pellet (P3) fractions (*see* **Table 2**).

2.3. Fluorescence Microscopy

2.3.1. Plasmids

1. pHIPZ4-DsRed-T1.SKL (*21*).
2. pHIPX5-Thio$_{N50}$-GFP (*22*).

3. pHIPX4-PEX3$_{1-50}$GFP (pFEM75) *(10)*.
4. pHIPX10-PEX14GFP *(23)*.

2.3.2. Transformation

1. YPD medium (1% yeast extract, 1% peptone, 1% glucose).
2. YND medium (0.67% yeast nitrogen base without amino acids [DIFCO] supplemented with 1% glucose and 0.25% ammonium sulfate as nitrogen source).
3. TED solution (100 m*M* Tris-HCl, 50 m*M* EDTA, 25 m*M* DTT, pH 8.0).
4. STM solution (270 m*M* sucrose, 10 m*M* Tris-HCl, 1 m*M* MgCl$_2$, pH 8.0). Autoclave solution and cool it on ice before use.
5. Agar plates with appropriate selective YND medium.
6. Electroporation apparatus (BTX ECM600).
7. Electroporation cuvettes (2-mm gap; BTX Disposable Cuvettes Plus No. 620 Blue).

2.4. Electron Microscopy

2.4.1. Fixation and Embedding of Fungal Cells in Resins

1. 1.5% (w/v) KMnO$_4$ in water.
2. 0.5% (w/v) Uranylacetate in water.
3. Ethanol solutions: 50% (v/v); 70% (v/v), 96% (v/v), and 100% ethanol in water.
4. Epon embedding resin: mix 100 g of Epon 812 (glycid ether) with 92 g of methylnadicanhydride (MNA), then add 2.3 g of 2,4,6-tri(dimethylaminomethyl) phenol (DMP-30). Prepare fresh.
5. 0.1 *M* Na-Cacodylate buffer, pH 7.2.
6. 50% (v/v) Glutaraldehyde, electron microscopy grade.
7. 40% (w/v) Formaldehyde in water (freshly prepared). Add 20 g of paraformaldehyde to 40 mL of water. Warm slowly to 60°C under continuous stirring. Then carefully add 10 *M* NaOH until the solution clarifies and add demineralized water to a final volume of 50 mL.
8. 0.4% (w/v) Na-Periodate in water.
9. 1% (w/v) NH$_4$Cl in water.
10. Unicryl embedding resin (Aurion; Wageningen, The Netherlands).

(*See* **Note 1**.)

2.4.2. Immunolabeling

1. Phosphate-buffered saline (PBS)–glycine buffer (10x concentrated stock solution): Per 500 mL distilled water: 40 g NaCl, 1 g KCl, 7.2 g Na$_2$HPO$_4$ · 2H$_2$O, 1.14 g NaH$_2$PO$_4$ H$_2$O, 7.5 g glycine and 0.5 g NaN$_3$.
2. Bovine serum albumin (BSA).
3. Poststaining solution: 1% (w/v) uranylacetate and 0.2% (w/v) methylcellulose in distilled water.

3. Methods

3.1. Cultivation of Yeast Cells

For cell fractionation and microscopy experiments (*see* **Note 2**), cells are extensively precultivated in shake flasks in mineral medium containing glucose as carbon source (0.5%).

1. Inoculate glucose/ammonium sulfate medium with colonies from a fresh glucose-containing agar plate. Place the flask in a shaker incubator (200 rpm) at 37°C.
2. Once the culture is in the late exponential growth phase, use it as inoculum to start a new culture (starting optical density at 660 nm [OD_{660}] of 0.1) in mineral medium containing glucose/ammonium sulfate.
3. Incubate this culture until it reaches the late exponential growth phase.
4. Dilution of a late exponential glucose culture into fresh glucose medium at OD_{660} 0.1 is repeated three times.
5. The last glucose culture (at the late exponential growth phase) is used to inoculate the final culture of mineral medium containing methanol (0.5%) as the sole carbon source in the presence of ammonium sulfate or methylamine as nitrogen source.
6. Cells are grown until an optical density of 1.5–2.0 (approx 16 h, which results in proper induction of peroxisomes/peroxisomal proteins).

3.2. Cell Fractionation

1. Harvested cells by centrifugation (5-min 5000*g* at room temperature) (*see* **Note 2**).
2. Resuspend in preincubation medium (approx 0.06 g wet weight/mL) and incubate for 15 min at 37°C.
3. Harvested cells again by centrifugation (5 min 5000*g* at room temperature).
4. Wash pellet once in protoplasting buffer and resuspend in the same buffer (0.06 g wet weight/mL buffer).
5. Prepare protoplasts by adding the lytic enzyme Zymolyase 20T, which degrades the yeast cell walls, to a final concentration of 1 mg/mL.
6. Incubate the cell suspension at 37°C for 30–120 min (gently shaking in a water bath at max 50 rpm).
7. Monitor the progress of protoplast formation by inspection of samples of the cell suspension by light microscopy. To facilitate this, dilute small aliquots 1:1 with water, which causes osmotic swelling and disruption of protoplasts.
8. Stop the incubation when approx 75% of the cells are subject to disruption in water and hence have been converted into protoplasts, cool the suspension on ice.

All subsequent steps are performed at 4°C.

9. Collect protoplasts by centrifugation (8 min, 4500*g*).
10. Wash the intact pellet once in protoplast disruption buffer to remove excess lytic enzyme. The pellet is not resuspended, but detached from the wall of the centrifuge tube using a spatula.

11. After a second centrifugation step (8 min at 4500g), carefully resuspend the pellet in protoplast disruption buffer and homogenize using a Potter Elvehjem homogenizer (generally 5–10 strokes).

12. Centrifuge the homogenate for 10 min at 3000g.

13. Collect the supernatant (designated S1) and pour into a separate centrifuge tube. The pellet (P1) contains whole cells, unbroken protoplasts and nuclei and can be discarded.

14. Centrifuge the supernatant S1 again for 10 min at 3000g. Pour the S2 into a fresh centrifuge tube. S2 is designated PNS and contains cytosol and cell organelles. The pellet (P2) can be discarded.

15. The PNS is either centrifuged at high speed in order to obtain an organellar pellet (P3; *see* **Subheading 3.3.**) or loaded onto a discontinuous sucrose density gradient (*see* **Subheading 3.4.**).

3.3. Differential Centrifugation

1. Centrifuge the PNS for 30 min at 30,000g, 4°C, yielding P3 and S3. The PNS, P3, and S3 can be analyzed for the presence of peroxisomal proteins by Western blotting. When equal portions are loaded per lane, the organellar proteins present in the PNS are recovered in the P3 fraction, whereas soluble cytosolic proteins are recovered in the S3 fraction. Like the PNS, the P3 (organellar pellet) can be subjected to sucrose gradient centrifugation for further separation of organelles (*see* **Subheading 3.4.**).

3.4. Sucrose Density Centrifugation

1. Load the PNS or P3 fraction onto the appropriate sucrose gradient (load 5–8 mg protein per gradient; *see* **Table 2**).

2. Pour an overlay solution on top of the sample to completely fill the tube.

3. Centrifuge the gradients in a vertical rotor (e.g., Sorvall SV288 rotor, for 2.5 h, 30,000g, 4°C).

4. Harvest the gradients from the bottom in approx 20–25 fractions ($\sim$ 1.5 mL per fraction) by pinching a hole in the bottom of the tube using an injection needle.

5. Collect the drops in microcentrifuge tubes (1.5 mL).

3.5. Analysis of Gradient Fractions

1. Determine protein concentrations using the Bio-Rad Protein Assay (Biorad GmbH, Munich, Germany) and BSA as a standard.

2. For TCA precipitations, add an equal volume of 50% TCA to aliquots of the fractions. Mix extensively.

3. Incubate the samples for at least 30 min at −20°C to promote protein precipitation.

4. After thawing, pellet the protein by centrifugation (10 min at 16,000g, room temperature).

5. Wash twice with ice-cold 80% acetone.

6. Air-dry the sample and dissolve in a small volume of SDS/NaOH solution.
7. Add one-quarter volume of a 5x concentrated SDS–sample buffer *(24)* and boil for 5 min. Use for SDS-polyacrylamide gel electrophoresis (PAGE) and Western blotting. Equal portions of each sample are used per lane. To localize peroxisomes and other cell organelles, Western blots are prepared using various organellar marker proteins (*see* **Fig. 1**). SDS-PAGE *(24)* and Western blotting *(25)* are performed by established procedures.

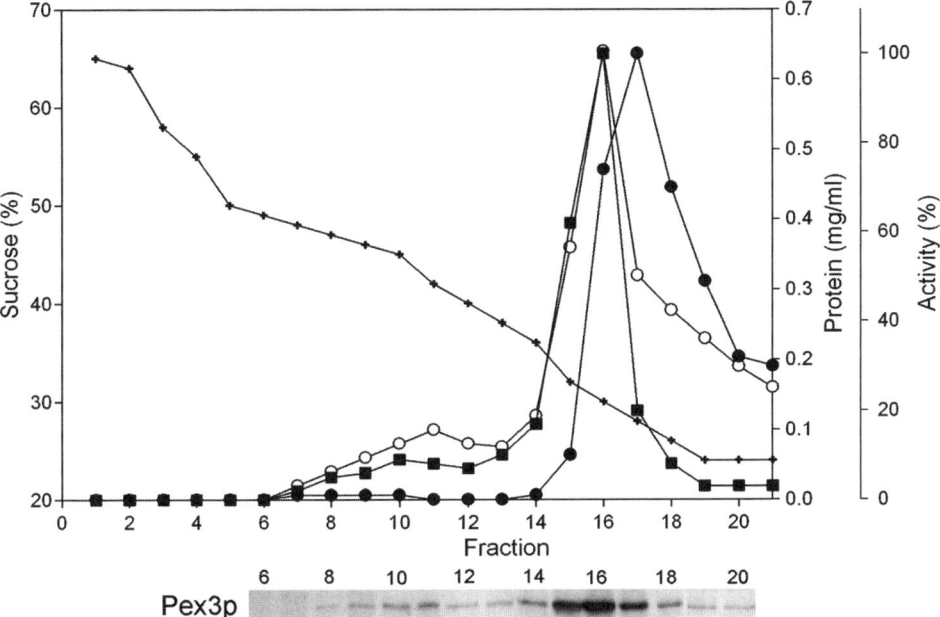

Fig. 1. Sucrose density gradient prepared from a postnuclear supernatant of *H. polymorpha pex2* cells. The top panel shows the distribution of protein (•) in mg/mL, sucrose (+) expressed as percentages (w/w), the enzyme activities of the peroxisomal matrix protein alcohol oxidase (o), and the mitochondrial marker protein cytochrome *c* oxidase (■) expressed as percentages of the activity in the peak fractions, which were arbitrarily set at 100. The bottom panel shows a Western blot of fractions 6–20, which demonstrates the distribution of the PMP Pex3p. Equal volumes of each fraction were loaded per lane. In gradients prepared from wild-type cells, peroxisomes (and hence alcohol oxidase and Pex3p) sediment to high-density fractions (fractions 4–6). However, in cells of the *pex2* mutant, alcohol oxidase is predominantly present in fractions of low density (fractions 16–21), where soluble cytosolic proteins are also present. Pex3p, which is bound to peroxisomal membrane ghosts in these cells, is also present in low-density fractions (fractions 15–17).

3.6. High-Salt and Carbonate Treatment of Peroxisomal Fractions

All steps are performed at 4°C. To determine whether proteins of peroxisomal fractions are soluble, membrane-associated or integral peroxisomal membrane components, successive extractions are performed. Both organellar pellets obtained by differential centrifugation (P3) and peroxisomal peak fractions from sucrose gradients can be used.

1. Mix and pool peroxisomal peak fractions (as determined by the presence of perox-isomal marker proteins).
2. Mix an aliquot of the sample (P3 or pooled peroxisomal fractions) with the same volume of cold 0.1 M Tris-HCl pH 8.0 solution. Peroxisomes swell and lyse due to the sudden decrease in osmotic value. By ultracentrifugation the matrix proteins (supernatant) are separated from the membranes (pellet).
3. Take a sample (T1-total) prior to centrifugation, ultracentrifuge the remaining suspension for 15 min at 200, 000g at 4°C (S1, supernatant; P1, pellet).
4. Use equal portions of T1, S1, and P1 for analysis by Western blotting. The remaining portion of the P1 pellet fraction is subjected to subsequent high-salt treatment, which removes proteins that are loosely associated to the membranes.
5. Resuspend the pellet in 0.1 M Tris-HCl pH 8.0.
6. Add an equal volume of the same Tris buffer containing 1 M NaCl.
7. Take a sample of the mixture prior to centrifugation (T2).
8. Centrifuge the remaining sample as indicated above, and take samples from the pellet (P2) and supernatant (S2) fractions. T2, P2, and S2 can be analyzed by Western blotting, as indicated above using equal portions of the fractions.
9. The P2 pellet can be further fractionated by resuspending it in 0.1 M carbonate solution. By this treatment only integral membrane proteins remain bound to the membrane. Take fractions prior to (T3) and after centrifugation (P3 and S3), and analyze as indicated above.

3.7. Fluorescence Microscopy

1. To analyze import of PTS1 proteins, plasmid pHIPZ4-DsRed-T1.SKL is trans-formed to the *H. polymorpha* strain to be analyzed. This plasmid contains the gene encoding the red fluorescent protein DsRed containing a PTS1 signal at the extreme C-terminus (-SKL). This gene is placed under control of the alcohol oxidase promoter that is repressed in glucose-grown cells, but strongly induced during growth of cells on methanol *(21)* (**Fig. 2**).
2. To visualize PTS2 protein import, cells are transformed with plasmid pHIPX5-Thio$_{N50}$-GFP, which allows expression of green fluorescence protein (GFP) containing the PTS2 signal of *S. cerevisiae* thiolase at the extreme N-terminus. This gene is placed under control of the amine oxidase promoter and hence only expressed in media containing methylamine as the sole nitrogen source *(22)*.

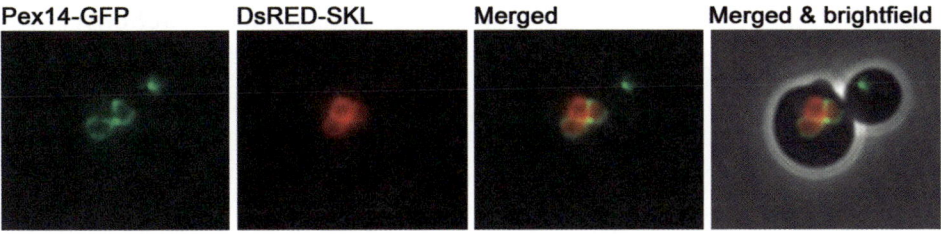

| Pex14-GFP | DsRED-SKL | Merged | Merged & brightfield |

Fig. 2. Fluorescence microscopy images of methanol-grown *Hansenula polymorpha* wild-type cells to show the co-localization of the peroxisomal membrane protein Pex14p, fused to green fluorescent protein (GFP), and DsRed-SKL used as a marker of the peroxisomal matrix.

3. To analyze targeting of the PMP Pex3p, cells are transformed with plasmid pFEM75. This plasmid encodes the first 50 amino acids of Pex3p, which contains an mPTS *(10)* fused to GFP (also designated Pex3$_{N50}$-GFP). The gene is placed under control of the alcohol oxidase promoter. In wild-type cells Pex3$_{N50}$-GFP is localized to peroxisomes. Also in *pex* mutants that contain remnants of peroxisomal membranes, Pex3$_{N50}$-GFP is targeted to these structures. Interestingly, however, Pex3$_{N50}$-GFP also localizes to the nuclear envelope (which is continuous with the endoplasmic reticulum) in *H. polymorpha* mutant cells that lack peroxisomal membranes (i.e., in *H. polymorpha pex3* [26] and *pex19* [27]). At high expression levels, this resulted in the formation of numerous small peroxisomal membrane vesicles in *H. polymorpha pex3* cells *(26)* or peroxisomal structures in *pex19* cells *(14,27)*.

Plasmid pHIPX10-PEX14GFP contains a gene encoding a Pex14-GFP fusion protein under control of its own promoter *(23)*. In *H. polymorpha* WT cells and in cells of *pex* mutants that contain peroxisomal membrane remnants, Pex14-GFP is localized to peroxisomal membranes (**Fig. 2**). However, in *pex3 (28)* or in *pex19* cells *(14)*, Pex14-GFP is mistakenly localized to mitochondria.

After introduction of the required plasmid to the yeast cells by transformation (*see* **Subheading 3.8.**), cells from the exponential growth phase (*see* **Subheading 3.1.** and **Note 2**) are analyzed by fluorescence microscopy using standard microscopy techniques (for examples, *see* **Fig. 2**).

3.8. Transformation of H. polymorpha

Transformation of *H. polymorpha* cells and site-specific integrations of single or multiple copies of plasmid DNA is performed as according to the procedures developed by Faber and colleagues *(29,30)*.

1. Inoculate cells from a fresh agar plate into YPD medium and incubate overnight at 37°C in a shaker incubator (200 rpm).

2. Inoculate 2 mL of the overnight culture into 200 mL prewarmed YPD medium and grow at 37°C to an optical density (at 600 nm) of 1.2–1.5 (generally takes 4–5 h).
3. Harvest the cells by centrifugation (10-min 4,000g at room temperature).
4. Resuspend the cells in 50 mL TED solution and incubate for 15 min at 37°C in a shaker incubator at 200 rpm.
5. Wash the cells by resuspending them in 200 mL ice-cold, sterile STM solution followed by centrifugation (10-min 4,000g at 4°C).
6. Wash the cells by resuspension into 100 mL ice-cold STM solution and subsequent centrifugation (10-min 4,000g at 4°C).
7. Resuspend the cells in 1 mL ice-cold STM solution and prepare 60-μL aliquots of the suspension. Keep the required number of aliquots on ice. Remaining aliquots can be frozen in liquid nitrogen and stored at −80°C for later use.
8. Add plasmid DNA (max. 3–4 μL per aliquot) to the cell suspension, mix gently, and keep the suspension on ice (*see* **Note 3**). Use linearized plasmid DNA for integration of the plasmid into the genome.
9. Transfer the suspension to a sterile electroporation cuvette and apply an electro-pulse. Pulse settings: 50 μF, 129 Ω, 1.5 kV (7.5 kV/cm).
10. Immediately add 940 μL YPD medium and transfer the suspension to a 2.2-mL microcentrifuge tube.
11. Incubate the tube for 1 h at 37°C in a shaker (200 rpm).
12. Collect the cells by centrifugation (2 min 4,000g in a microcentrifuge at room temperature).
13. Wash the cells by resuspension in 1 mL selective YND medium and subsequent centrifugation (2-min 4,000g microcentrifuge).
14. Resuspend the cells in 1 mL selective medium and plate 1, 10, and 89% of the suspension on selective YND agar plates.
15. Incubate the plates at 37°C. Colonies will appear after 2–3 d.
16. Correct integration in the *H. polymorpha* genome can be checked by Southern blotting or PCR using established procedures.

3.9. Electron Microscopy

3.9.1. Morphology

3.9.1.1. KMnO$_4$ Fixation

1. Harvest 20 OD units (e.g., 20 mL of a culture with an OD$_{660}$ = 1) of a fresh batch culture of exponentially growing cells by centrifugation (3 min 5000g at room temperature) (*see* **Note 2**).
2. Wash the cells twice with 5 mL demineralized water by resuspension/ centrifu-gation and then resuspend pellet in 5 mL KMnO$_4$ solution (*see* **Note 1**).
3. Incubate the cell suspension for 20 min at room temperature and shake gently every 5 min.

4. After incubation, collect the cells by centrifugation and wash with demineralized water until the supernatant is colorless (three times with 5 mL water each normally suffices).
5. Resuspend the pellet in 5 mL uranylacetate solution (*see* **Note 1**) and centrifuge for 15 min at 5000*g* to obtain a firm pellet. The supernatant should not be discarded, but left on top of the pellet for at least 4 h or maximally overnight at room temperature.

3.9.1.2. DEHYDRATION/EMBEDDING

The cells are dehydrated in a graded ethanol series at room temperature.

1. Decant the uranyl acetate supernatant from the pellet.
2. Incubate the pellet with solutions of increasing ethanol concentrations according to the scheme below: 15 min in 50% ethanol (the pellet remains intact), 15 min in 70% ethanol (the pellet is broken into small pieces (approx 1–5 mm^3) using a spatula, 15 min in 96% ethanol (mix carefully, do not use a vortex, the small pieces should stay intact), 15 min in 100% ethanol (mix carefully). 30 min in 100% ethanol (mix carefully).

In the subsequent steps, the cells are impregnated with Epon resin (*see* **Note 1**). During these incubations the tubes are continuously mixed using a slowly rotating incubator. The fixed material should be carefully mixed. After each incubation step, the tubes are placed in a rack and when the cell material is settled, the Epon/ethanol mixture is decanted and the next solution is added.

3. Incubate the samples (i.e., pieces of cell material) with Epon/ethanol mixtures according to the following scheme: 4–8 h in a 1:1 mixture of 100% ethanol and Epon, overnight in a 1:3 mixture of 100% ethanol and Epon, 1 h in pure Epon solution, 6 h in pure Epon solution.
4. Fill gelatin capsules three-quarters with pure Epon and load one piece of fixed cell material onto the Epon in the capsule; it will readily sink.
5. Polymerize the Epon by incubating the capsules for 24 h at 80°C.
6. Prepare sections using a diamond knife.
7. Inspect the sections in a transmission electron microscope (*see* **Fig. 3**).

3.9.2. Aldehyde Fixation and Unicryl Embedding for Immunolabeling Experiments

3.9.2.1. ALDEHYDE FIXATION

All steps are performed at 4°C.

1. Harvest at least 20 OD units (*see* **Subheading 3.9.1.1.**) of a fresh culture by centrifugation (3 min 5000*g*).

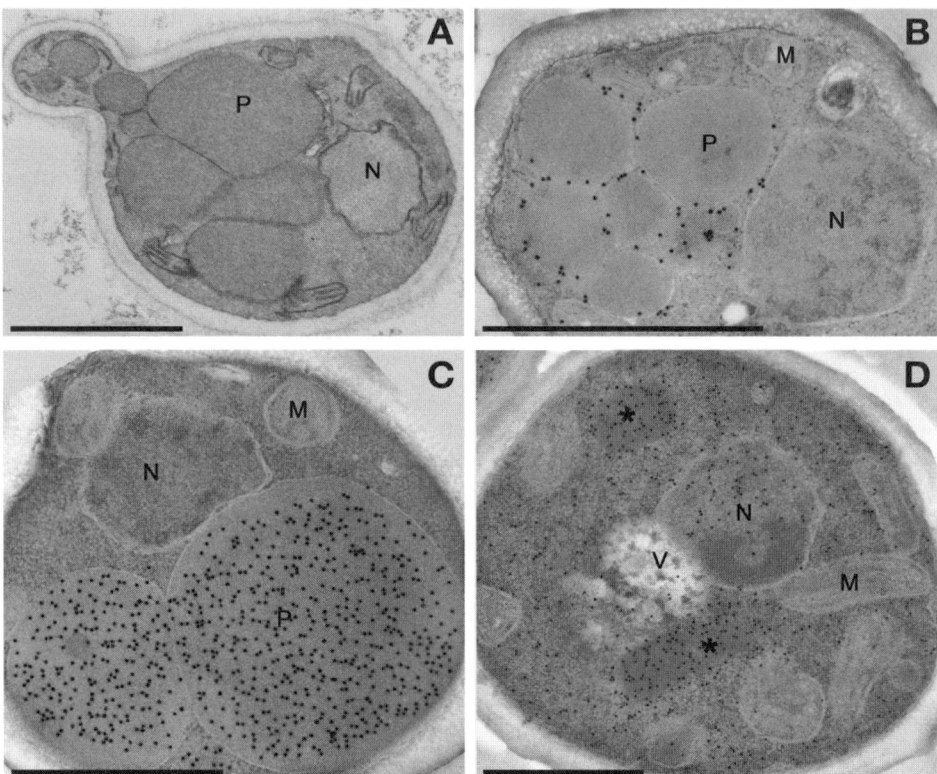

Fig. 3. Electron micrographs to show the overall morphology of *Hansenula polymorpha* cells grown on methanol. (**A**) KMnO$_4$ fixation: note the massive induction of peroxisomes, of which small ones are transferred into the developing bud. (**B–D**) Immunocytochemical experiments to demonstrate the location of Pex3p on the peroxisomal membrane (**B**), alcohol oxidase in the peroxisomal matrix of wild-type cells (**C**), and alcohol oxidase in the cytosol and nucleus of a *H. polymorpha pex6* mutant (**D**). (**B–D**) Glutaraldehyde fixed cells. Bar = 0.5 μm.*, cytosolic alcohol oxidase crystalloid. M, mitochondrion; N, nucleus; P, peroxisomes.

2. Wash the cells twice with water and add 5 mL of one of the fixation solutions indicated below: 3% glutaraldehyde in 0.1 *M* Na-cacodylate, pH 7.2, 3% formaldehyde in 0.1 *M* Na-cacodylate, pH 7.2, 0.5% glutaraldehyde +2.5% formaldehyde in 0.1 *M* Na-cacodylate, pH 7.2 (*see* **Note 1**).

Three percent glutaradehyde is the preferred fixative. However, glutaraldehyde may affect the antigenicity of specific proteins. In that case formaldehyde or a mixture of formaldehyde/glutaraldehyde mixture can be used.

3. Carefully resuspend the cells in the fixative solution and incubated for 2 h at 4°C. Mix the suspension every 15 min.
4. Discard the supernatant upon collecting the cells by centrifugation, add fresh Na-cacodylate buffer. The pellet should be kept intact.

3.9.2.2. DEHYDRATION AND EMBEDDING IN UNICRYL

All steps are performed at room temperature.

1. Wash the intact pellet of fixed yeast cells twice with demineralized water.
2. Add freshly prepared Na-periodate solution to the pellet (keep pellet intact).
3. After incubation for 15 min on a slowly rotating incubator, wash the pellet twice with demineralized water.
4. Incubate the pellet in a solution of 1% NH_4Cl for 15 min at room temperature.
5. Decant the NH_4Cl solution and wash the pellet once with water.
6. Dehydrate the cell material is now in a graded ethanol series as follows: add 50% ethanol solution to the pellet and centrifuge for 15 min at $10,000g$ to obtain a very tight pellet.
7. Replace the 50% ethanol solution by a solution of 70% ethanol. At this stage the pellet is broken into small pieces (pieces of approx 1–5 mm³; discard excess material and save approx 10 pieces of the appropriate size). Incubate the material according to the following scheme (for details, *see* **Subheading 3.9.1.2.**): 15 min in 70% ethanol, 15 min in 96% ethanol, 15 min in 100% ethanol, 30 min in 100% ethanol.
8. For infiltration with Unicryl (*see* **Note 1**), incubate the pieces successively as indicated below (*see also* **Subheading 3.9.1.2.**): 3 h in a 1:1 solution of 100% ethanol and Unicryl, 1 h in pure Unicryl, overnight in pure Unicryl, 6 h in pure Unicryl.
9. Embed the material in gelatin capsules filled three-quarters with Unicryl. Use only carefully dried gelatin capsules!
10. Polymerize the Unicryl for 4 d using ultraviolet light at 4°C.
11. After 2 d of polymerization, fill the capsules completely with Unicryl.
12. Cut sections using a diamond knife and use for immunolabeling.

3.9.3. Immunolabeling

Immunocytochemical staining methods are performed on ultrathin sectons that are collected on Fromvar-coated nickel grids (do not use copper grids). The incubation steps are performed by floating the grids, section-side-down, on top of small droplets of the solution on a sheet of Parafilm. All steps are performed at room temperature unless stated otherwise.

1. Incubate the grids with the following solutions: 0.5% BSA in PBS-glycine buffer for 5 min as a blocking step.

2. Transfer grids to a droplet of appropriately diluted primary antibody (*see* **Note 4**) in PBS-glycine buffer containing 0.5% BSA and incubate for 1 h at room temperature (alternatively, this step can be performed overnight at 4°C). The appropriate dilution of the primary antibody is generally 10 times less than that used for Western blotting (e.g., 1:100 if a 1:1000 dilution is the optimal dilution for Western blotting).
3. Rinse the grids with PBS-glycine buffer, three times 5 min each.
4. Incubate the grids with a solution of secondary antibodies conjugated to gold in PBS-glycine buffer containing 0.5% BSA (use dilution as recommended by the manufacturer). Use the appropriate secondary antibodies (i.e., goat–anti-rabbit (GAR) gold when the primary antibodies were raised in goat or goat–anti-mouse gold when the primary antibodies were raised in mice).
5. Rinse the grids in PBS-glycine buffer, six times 5 min each.
6. Rinse grids in distilled water, four times 5 min each.
7. Remove excess of water by carefully tipping one side of the grid (section-side-up) onto filter paper.
8. Poststain the sections by placing the grid (section-side-down) onto a droplet of 1% uranylacetate and 0.2% methylcellulose for 20 s (*see* **Note 1**).
9. Remove excess staining solution using filter paper and allow the grid to dry.

Notes

1. Fixatives, solvents, resins, and poststaining solutions used in electron microscopy are generally very toxic. Take extreme care (wear gloves) and perform fixation and embedding procedures in a fume hood.
2. In the cell fractionation and microscopy experiments, success is strongly related to the cultivation procedures. The best results are invariably obtained when cells are used that have been grown at maximum growth rates (until the mid-exponential growth phase). Therefore, extensive precultivation is a must. Slowly growing cells or cells from the stationary growth phase often develop thick cell walls and storage materials (e.g., glycogen or fat droplets) that can affect protoplast formation and fixation/embedding for electron microscopy. When strains are used that are not able to grow on methanol (for instance *pex* mutants), cells are grown on a mixture of 0.1% glycerol and 0.5% methanol. In this medium cells grow on glycerol as carbon and energy source, whereas methanol functions as inducer of peroxisomal proteins and additional energy source.
3. Always use negative (no DNA applied) and positive transformation controls (i.e., fixed amount of generally used plasmid). The latter can be used to calculate the transformation frequency (i.e., number of transformants per μg DNA).
4. The antigenicity of proteins is affected by the fixatives and can strongly affect immunolabeling experiments. To analyze this inhibitory effect, Western blots can be incubated with fixation solutions before decoration of the blots with primary and secondary antibodies. This can be done as follows: prepare Western blots of crude cell extracts. Load several lanes of an SDS-PAA gel with the same protein

sample. After running of the gel and transfer of the proteins to nitrocellulose, the blot is stained with Ponceau S solution. This allows visualizing the lanes. Separate the individual lanes of the blot (use scissors or knife) and mark the resulting nitrocellulose strips. After blocking of the strips in blocking buffer, the strips are incubated with the different fixative solutions essentially as indicated for fixation of the cells. After fixation, the blots are extensive washed with buffer and, subsequently, the Western blot staining procedure is completed. Inspection of the strips will reveal whether or not certain fixatives affect the antigenicity of the protein to be localized.

References

1. Purdue, P. E. and Lazarow, P. B. (2001) Peroxisome biogenesis. *Annu. Rev. Cell Dev. Biol.* **17**, 701–752.
2. Heiland, I. and Erdmann, R. (2005) Biogenesis of peroxisomes. Topogenesis of the peroxisomal membrane and matrix proteins. *FEBS J.* **272**, 2362–2372.
3. Kunau, W. H. (2005) Peroxisome biogenesis: end of the debate. *Curr. Biol.* **15**, R774–776.
4. Brown, L. A. and Baker, A. (2003) Peroxisome biogenesis and the role of protein import. *J. Cell Mol. Med.*. **7**, 388–400.
5. Baker, A. and Sparkes, I. A. (2005) Peroxisome protein import: some answers, more questions. *Curr. Opin. Plant Biol.* **8**, 640–647.
6. Ozimek, P., Veenhuis, M., and van der Klei, I. J. (2005) Alcohol oxidase: a complex peroxisomal, oligomeric flavoprotein. *FEMS Yeast Res.* **5**, 975–983.
7. Klein, A. T., van den Berg, M., Bottger, G., Tabak, H. F., and Distel, B. (2002) *Saccharomyces cerevisiae* acyl-CoA oxidase follows a novel, non-PTS1, import pathway into peroxisomes that is dependent on Pex5p. *J. Biol. Chem.* **277**, 25011–25019.
8. Schafer, A., Kerssen, D., Veenhuis, M., Kunau, W. H., and Schliebs, W. (2004) Functional similarity between the peroxisomal PTS2 receptor binding protein Pex18p and the N–terminal half of the PTS1 receptor Pex5p. *Mol. Cell Biol.* **24**, 8895–8906.
9. Yang, X., Purdue, P. E., and Lazarow, P. B. (2001) Eci1p uses a PTS1 to enter peroxisomes: either its own or that of a partner, Dci1p. *Eur. J. Cell Biol.* **80**, 26–38.
10. Baerends, R. J. S., Faber, K. N., Kram, A. M., Kiel, J. A. K. W., van der Klei, I. J., and Veenhuis, M. (2000) A stretch of positively charged amino acids at the N terminus of *Hansenula polymorpha* pex3p is involved in incorporation of the protein into the peroxisomal membrane. *J. Biol. Chem.* **275**, 9986–9995.
11. Schliebs, W. and Kunau, W. H. (2004) Peroxisome membrane biogenesis: the stage is set. *Curr. Biol.* **14**, R397–R399.
12. Rottensteiner, H., Kramer, A., Lorenzen, S., Stein, K., Landgraf, C., Volkmer-Engert, R., and Erdmann, R. (2004) Peroxisomal membrane proteins contain

common Pex19p-binding sites that are an integral part of their targeting signals. *Mol. Biol. Cell* **15**, 3406–3417.

13. Lambkin, G. R. and Rachubinski, R. A. (2001) *Yarrowia lipolytica* cells mutant for the peroxisomal peroxin Pex19p contain structures resembling wild-type peroxisomes. *Mol. Biol. Cell* **12**, 3353–3364.

14. Otzen, M., Perband, U., Wang, D., et al. (2004). *Hansenula polymorpha* Pex19p is essential for the formation of functional peroxisomal membranes. *J. Biol. Chem.* **279**, 19181–19190.

15. Baker, A., Charlton, W., Johnson, B, et al. (2000) Biochemical and molecular approaches to understanding protein import into peroxisomes. *Biochem. Soc. Trans.* **28**, 499–504.

16. Gouveia, A. M., Guimaraes, C. P., Oliveira, M. E., Reguenga, C., Sa-Miranda, C., and Azevedo, J. E. (2003) Characterization of the peroxisomal cycling receptor Pex5p import pathway. *Adv. Exp. Med. Biol.* **544**, 219–220.

17. Miyata, N. and Fujiki, Y. (2005) Shuttling mechanism of peroxisome targeting signal type 1 receptor Pex5: ATP-independent import and ATP-dependent export. *Mol. Cell Biol.* **25**, 10822–10832.

18. Platta, H. W., Grunau, S., Rosenkranz, K., Girzalsky, W., and Erdmann, R. (2005) Functional role of the AAA peroxins in dislocation of the cycling PTS1 receptor back to the cytosol. *Nat. Cell Biol.* **7**, 817–822.

19. Matsuzono, Y. and Fujiki, Y. (2006) *In vitro* transport of membrane proteins to peroxisomes by shuttling receptor Pex19p. *J. Biol. Chem.* **281**, 36–42.

20. van Dijken, J. P., Otto, R., and Harder, W. (1976) Growth of *Hansenula polymorpha* in a methanol-limited chemostat. Physiological responses due to the involvement of methanol oxidase as a key enzyme in methanol metabolism. *Arch. Microbiol.* **111**, 137–144.

21. Monastyrska. I., van der Heide, M., Krikken, A. M., Kiel, J.A., van der Klei, I. J., and Veenhuis, M. (2005) Atg8 is essential for macropexophagy in *Hansenula polymorpha*. *Traffic* **6**, 66–74.

22. Otzen, M., Wang, D., Lunenborg, M. G., and van der Klei, I. J. (2005) *Hansenula polymorpha* Pex20p is an oligomer that binds the peroxisomal targeting signal 2 (PTS2). *J. Cell Sci.* **118**, 3409–3418.

23. de Vries, B., Todde, V., Stevens, P., Salomons, F., van der Klei, I. J., and Veenhuis, M. (2006) Pex14p is not required for N-starvation induced microautophagy and in catalytic amounts for macropexophagy in *Hansenula polymorpha*. *Autophagy* **2**, 183–188.

24. LaemmLi, U. K. (1970). Cleavage of structural proteins during the assembly of the head of bacteriophage T4. *Nature* **227**, 680–685.

25. Kyhse-Andersen, J. (1984). Electroblotting of multiple gels: a simple apparatus without buffer tank for rapid transfer of proteins from polyacrylamide to nitrocellulose. *J. Biochem. Biophys. Methods* **10**, 203–209.

26. Faber, K. N., Haan, G. J., Baerends, R. J., Kram, A. M., and Veenhuis, M. (2002) Normal peroxisome development from vesicles induced by truncated *Hansenula polymorpha* Pex3p. *J. Biol. Chem.* **277**, 11026–11033.

27. Otzen, M., Krikken, A. M., Ozimek, P., et al. (2006) In the yeast *Hansenula polymorpha* peroxisome formation from the ER is independent of Pex19p, but involves the function of p24 proteins. *FEMS Yeast Res.* **6**, 1157–1166.

28. Haan, G. J., Baerends, R. J. S., Krikken, A. M., Otzen, M., Veenhuis, M. and van der Klei, I. J. (2006) Re-assembly of peroxisomes in *Hansenula polymorpha pex3* cells upon re-introduction of Pex3p involves the nuclear envelope. *FEMS Yeast Res.* **6**, 186–194.

29. Faber, K. N., Haima, P., Harder, W., Veenhuis, M., and AB, G. (1994). Highly-efficient electrotransformation of the yeast *Hansenula polymorpha*. *Curr. Genet.* **25**, 305–310.

30. Faber, K. N., Swaving, G. J., Faber, F., AB, G., Harder, W., Veenhuis, M., and Haima, P. (1992). Chromosomal targeting of replicating plasmids in the yeast *Hansenula polymorpha*. *J. Gen. Microbiol.* **138**, 2405–2416.

26

Expression, Localization, and Topology
of Fluorescently Tagged Plasma Membrane-Targeted
Transmembrane Proteins

Jeanne Shepshelovich and Koret Hirschberg

Summary

Transmembrane proteins constitute a significant proportion of the total number of proteins encoded in eukaryotic genomes. These proteins are involved in countless processes required for cellular function and homeostasis. Mutations mapping within their genes form the cellular mechanisms for a variety of pathological conditions. The surface expression of polytopic proteins can be limited by their relatively slow folding in the endoplasmic reticulum (ER). This is especially evident in heterologous overexpression systems *(1)*. On the other hand, expression of these proteins in endogenously expressing primary cell lines is generally a challenging task in itself. Here we present a comprehensive scheme to establish arrival at the plasma membrane (PM) of transiently expressed, fluorescently tagged transmembrane proteins and provide two experimental tools to determine whether a distinct domain is cytosolic or extracellular using fluorescence microscopy.

Key Words: Green fluorescent protein; surface expression; transmembrane proteins; encoplasmic reticulum; Golgi apparatus; plasma membrane; saponin.

1. Introduction

Arrival of a green fluorescent protein (GFP)-tagged transmembrane protein at the plasma membrane (PM) indicates that the protein has been folded correctly, it has passed endoplasmic reticulum (ER) quality control *(2)*, and its PM-targeting signals have been recognized by the secretory transport machinery *(3)*. These targeting signals facilitate concentration at ER exit sites and transport of the protein in membrane-bound carriers, which bud from the ER exit sites and translocate on microtubular tracks to the Golgi apparatus *(4)*. After passing

From: *Methods in Molecular Biology, Vol. 390: Protein Targeting Protocols: Second Edition*
Edited by: M. van der Giezen © Humana Press Inc., Totowa, NJ

through the latter's compartments, the proteins are transported to the PM on post-Golgi membrane carriers *(5,6)*. Transient overexpression of PM-targeted transmembrane proteins could result in a significant increase in protein load on the ER's folding machinery *(7)*. The direct consequence is a wide distribution of the fluorescently tagged overexpressed proteins in organelles of the secretory pathway, namely, the ER, Golgi apparatus, PM, and, in many cases, intracellular aggregates. Many transmembrane proteins are exclusively expressed in specialized cell types, where they play a pivotal role, such as chloride transport by the cystic fibrosis transmembrane regulator (CFTR) in lung alveolar epithelial cells *(8)* and iodide transport by pendrin in thyroid epithelial cells *(9)*. Here we provide an electroporation-based method which is particularly efficient for the transfection of primary cells and other hard-to-transfect cell lines. In addition, basic methods for live-cell microscopy and indirect immunofluorescence, optimized to preserve intracellular organelles in cells expressing fluorescently tagged proteins, are described and applied to an analysis of the topology of distinct segments of fluorescent protein (FP)-tagged transmembrane proteins.

2. Materials

2.1. Electroporation

1. Electroporation is carried out in a 0.4-cm electrode gene pulser cuvette (Bio-Rad, Hercules, CA).
2. 1 M Piperazine-1,4-bis(2-ethanesulfonic acid) (PIPES) (Sigma, St Louis, MO) buffer, pH 7.0.
3. 10 mM Calcium acetate.
4. 200 mM Magnesium acetate (Sigma).
5. Preparation of electroporation buffer, for a volume of 100 mL: 2.6 g potassium glutamate (potassium glutamate monohydrate, Sigma), 2 mL 1 M PIPES buffer, pH 7, 0.1 mL 10 mM calcium acetate, 1 mL 200 mM magnesium acetate, add DMEM to a final volume of 100 mL, sterilize by filtration using a 0.2-μm-pore-diameter vacuum filtration unit (Millipore, Bedford MA) and split into 5- to 15-mL screw-cap tubes. Store at −20°C (*see* **Note 1**).
6. Imaging medium: RPMI without phenol red (Gibco) supplemented with 100 units penicillin, 0.1 mg/mL streptomycin (Gibco), 10% (v/v) fetal bovine serum, 50 mM N-2-hydroxyethylpiperazine-N'-2-ethanesulfonic acid (HEPES), pH 7.4 , store at 4°C and use for up to 2 wk.

2.2. Immunofluorescence Microscopy

1. Prepare a 1% (w/v) sodium azide solution in phosphate-buffered saline (PBS). Use gloves throughout and work in a chemical hood as sodium azide is highly toxic.

2. Prepare a 10% (w/v) stock solution of saponin (Sigma); aliquot and store at $-20°C$.
3. Solution A: add 10% fetal bovine serum and 0.05% sodium azide to PBS. Store at 4°C.
4. Solution B: add saponin from a 10% (w/v) stock to Solution A to a final concentration of 0.2% saponin.
5. Solution C: add saponin from a 10% stock to Solution A to a final concentration of 0.1% saponin.
6. Fluoromount G (Southern Biotechnology Associates, Birmingham, AL); store at 4°C.
7. Fluorophore bleaching solution: dissolve sodium hydrosulfite (sodium dithionite, Sigma) in PBS and make a $0.8\,M$ 10X solution.

3. Methods

3.1. Transfection Protocols

There are numerous commercial transfection protocols based on lipophilic transfection reagents that are highly efficient and straightforward to use. However, these are usually only effective for a limited set of highly transformed cell lines. Here we provide an electroporation-based method which was found to be highly efficient for a large variety of transformed and primary cell lines. Subconfluent cells are electroporated in a specific medium that allows the use of high voltage and capacitance.

1. The day before the transfection, cells are replated in a 175-cm^2 flask or on any dish with an equivalent surface area. The cells should remain subconfluent after 24 h.
2. On the day of transfection, cells are trypsinized using the appropriate Trypsin-EDTA solution depending on the cell type. The medium is decanted, cells are briefly washed once with 2–5 mL Trypsin-EDTA solution and incubated with an additional 2 mL Trypsin-EDTA solution for 5–7 min at 37°C. The trypsinization is arrested by the addition of an 8-mL serum-containing medium, and the cells are centrifuged for 5 min on a tabletop centrifuge (200 RCF, Sigma 2-5 Osterode am Harz, Germany).
3. The supernatant is carefully removed, preferably with a Pasteur pipet connected to a vacuum source, and the cells are gently resuspended in the electroporation medium using a volume that will add up to exactly 0.4 mL together with the plasmid DNA solution.
4. Between 20 and 40 µg of plasmid DNA are required for the electroporation, at a concentration of approximately 1 mg/mL in water, not in Tris-EDTA buffer.
5. The cuvette containing the cells and the plasmid DNA, at a total volume of exactly 0.4 mL, is incubated on ice for 10 min.
6. After gently tapping several times on the cuvette to evenly resuspend the cells, electroporation is carried out using the initial values of 400 V and 900 µF (*see* **Note 2**) (Gene pulser II, Bio-Rad, or Easyject plus, Geneflow, Fradley Staffordshire, England).

7. Cells are immediately resuspended in prewarmed medium (37°C) and plated onto either dishes with cover slips or LabTek chambers (Nunc, Naperville, IL) according to the type of experiment. While plating the cells, it should be taken into account that approximately 50% of them survive the electroporation.
8. After 3–4 h, when the cells have settled and adhered to the substrate, gently replace the medium with fresh prewarmed medium to dispose of the dead cells.
9. Cells can be visualized or processed after 18–24 h (*see* **Note 3**).

3.2. Mounting Living Cells for Fluorescence Microscopy

In the case of an inverted microscope, living cells can be visualized in LabTek (Nunc) chambers or in any other commercially available device, such as 35-mm glass-bottom dishes (MatTek, Ashland, MA). Imaging living cells on an upright microscope requires that the cells be plated and transfected on glass cover slips immersed in a Petri dish. A 2- to 5-mm-thick rectangle of flexible silicon rubber sheet is cut to fit the slide width and a hole, not exceeding the dimensions of the cover slip (**Fig. 1**), is excised in the middle.

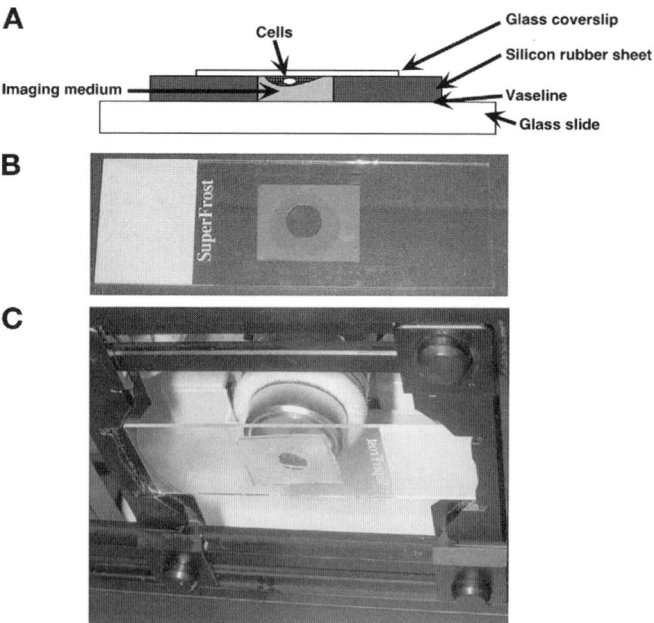

Fig. 1. Imaging living cells in an upright microscope. (**A**) Schematic demonstration of the structure of the slide. (**B**) Top view. (**C**) The slide on the microscope stage.

In the case of a 13-mm-diameter cover slip, the hole can be made with a paper hole-puncher. The sheet is then attached to a glass slide with a minimal amount of Vaseline or any other inert grease. The slide should be pressed firmly against a hard surface to eliminate air trapped between the sheet and the glass slide. The hole is filled with 20–40 µL of imaging medium and the cover slip is applied with the cells facing down. Next, the cover slip is carefully pressed to adhere to the silicon rectangle. The slide is turned and gently pressed onto two or three paper wipes in order to completely dry the outer side of the cover slip from excess imaging medium. It is particularly important to eliminate any residual medium that could end up mixing with the immersion oil and interfere with the microscopy. Cells are viable for at least 2–3 h.

3.3. Membrane-Protein Labeling by Indirect Immunofluorescence

The immunofluorescence protocol presented here is an optimal method for combining immunofluorescence with visualization of FP-tagged proteins while preserving the structure of intracellular membranous organelles. Cells are fixed with freshly made 2–4% formaldehyde solution and membrane permeabilization is performed using the mild steroid detergent saponin.

All washing and addition of solutions should be performed with extra care so as not to disrupt the cell membranes. This protocol is adapted to cells grown on 13-mm cover slips in a 60-mm Petri dish.

1. Wash the cells once with 5 mL PBS.
2. Add 2–4% v/v formaldehyde solution and incubate at room temperature for 20 min. Concentrated formaldehyde (formaldehyde 37%; cat. no. 104002 Merck, Darmstadt, Germany) is diluted in PBS.
3. Wash three times with 5 mL of solution A.
4. Add primary antibody diluted as suggested by the manufacturer in Solution B. Incubate for 1 h at 37°C (*see* **Note 4**).
5. After 1 h incubation, cover slips are replaced in a 60-mm Petri dish and gently washed four times with 5 mL solution C. Each wash should include a 5-min incubation.
6. Secondary antibody conjugated with a fluorophore is prepared in solution B, typically 1:300–500 (v/v). Prior to incubation, centrifuge for 5 min at maximum speed in a mini-centrifuge to pellet aggregates that might nonspecifically label the sample. Incubate for 1 h at 37°C (*see* **Note 4**).
7. Wash as in **step 5** and mount on a slide using fluoromount G. For long-term storage, the cover slips should be sealed with transparent nail polish and left to dry for at least 20 min.

3.4. Determining the Topology

The topology of the fluorescent protein attached to the N- or C-terminus of a protein can be determined by using one of the following two techniques. The first is based on testing whether a membrane-impermeable oxidant can irreversibly bleach the FP (**Fig. 2**). The second is based on indirect immunofluorescence in the presence or absence of a membrane-permeabilization agent (**Fig. 3**). To verify that the externally oriented fluorescent tag is bleached by the oxidant, the protein of interest is co-expressed with a FP-tagged protein such as GPI-anchored FP (**Fig. 3**).

1. Cells are plated directly in a Labtek chamber (Nunc) and co-transfected with the FP-tagged protein of interest and another FP-tagged protein that will serve as a positive control for the activity of the mild oxidant. One such protein is a FP with a glycosylphosphatidylinositol anchor *(10)*.
2. Images are taken before and after the addition of dithionite to a final concentration of 80 m*M* followed by 3-min incubation. The dithionite solution (sodium hydrosulfite, cat. no. 15,795-3, Aldrich Chemical Comp. Milwaukee, WI) is freshly prepared in PBS immediately before use.
3. In the case of an antibody directed against a specific peptide sequence of the FP-tagged protein, indirect immunofluorescence can be carried out as in **Subheading**

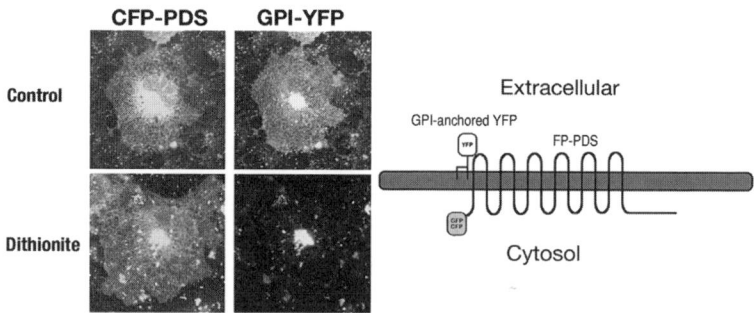

Fig. 2. Determining the topology of the N-terminal fluorescent protein tag of pendrin with a membrane-impermeable oxidant. Cos7 cells co-expressing CFP-PDS *(1)* and GPI-YFP were imaged in Labtek chambers (Nunc) containing imaging medium. Imaging was carried out on a Zeiss LSM PASCAL equipped with an Axiovert 200 inverted microscope. Ar 458- and 514-nm laser lines were used for ECFP and EYFP, respectively. Cells (lower panels) were treated with 80 m*M* dithionite for 3 min prior to image capturing. The scheme on the right-hand side demonstrates the topology of the FP-tagged pendrin and GPI-YFP.

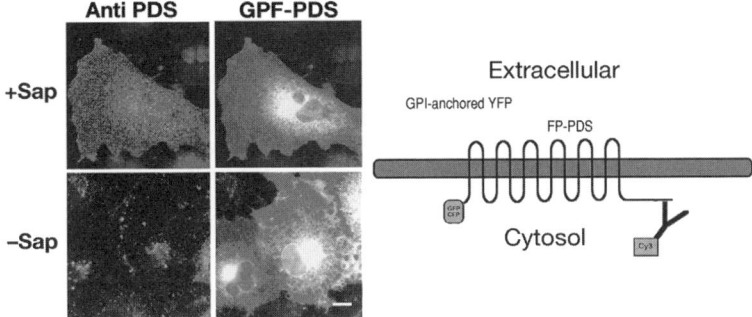

Fig. 3. Determining the topology of the C terminus of GFP-PDS using immuno-fluorescence microscopy. Confocal images of cells expressing GFP-PDS were fixed and probed with anti-pendrin antibody *(17)* raised against amino acids 630–643 of human pendrin in the presence (top panel) or absence (lower panel) of permeabilization agent and with secondary anti-mouse Cy3-tagged monoclonal antibody. GFP and Cy3 were detected with Ar 488-nm and HeNe 543-nm laser lines, respectively. The scheme on the righthand side demonstrates the topology of the FP-tagged pendrin and the segment recognized by the antibody.

3.3., in the presence or absence of saponin-mediated permeabilization. When the peptide sequence recognized by the antibody is cytosolic, it will not be accessible to the antibody in the absence of saponin (**Fig. 3**).

3.5. Dual-Color Live-Cell Confocal Microscopy of the Intracellular Distribution of Transmembrane Proteins in Organelles of the Secretory Pathway

There are several alternatives for choosing pairs of FP variants that can be simultaneously co-expressed in a cell. The most frequently used combination of fluorescent tags is enhanced cyan fluorescent protein (ECFP) and enhanced yellow fluorescent protein (EYFP). However, there are several disadvantages to the use of this pair that need to be considered in the context of the experimental design. The first is that ECFP is a much weaker fluorophore *(11)* than EYFP. Thus, the ability to detect an ECFP-tagged protein will require that its expression be significantly higher than that of the co-expressed, EYFP-tagged protein. Furthermore, it should be noted that the shorter wavelength and consequently higher energy of the light required to excite the ECFP is more harmful to the cell than that required to excite the EYFP. Note that a much brighter version of ECFP is now available under the name Cerulean *(12)*, This version is 2.5 times brighter than ECFP. Another shortcoming of the use of this pair is that partial overlap between the fluorescent signals of these two FPs may occur in most standard confocal systems, where excitation of ECFP and EYFP is carried out with argon laser lines at 458 and 514 nm, respectively. This cross-contamination is minimized in confocal systems that utilize a 405-nm diode laser to excite the ECFP. The use of an ECFP (or Cerulean [*12*]) tag in combination with any of the available red FPs (DsRED or the tandem far-red DHcRED [*13*]) is highly advantageous because the excitation and emission spectra of these fluorophores are considerably separated, allowing simultaneous excitation and detection, even in basic confocal systems, without cross-contamination of the emissions. The use of EGFP in combination with DsRED or the tandem far-red DHcRED is an excellent option as well.

By expressing two fluorescently tagged proteins, one can obtain information on the intracellular distribution and topology of a protein of interest based on a fluorescently tagged marker with well-established localization *(9,14)*. **Figure 4** demonstrates dual-color labeling of an ER and PM-resident transmembrane protein, FP-pendrin, with four different cellular markers. The first pair (**Fig. 4A,B**) consists of soluble markers that label the aqueous intracellular space. The second pair consists of membrane markers of the PM (**Fig. 4C**) and the ER (**Fig. 4D**). **Fig. 4A** shows free DHcRED, a cytosolic marker that indiscriminately labels the entire cytosolic space within the cell, including the inside of the nucleus. **Fig. 4B** shows an ECFP molecule that contains the cleavable signal sequence of hen egg lysosyme (hclss-ECFP), which restricts the distribution of ECFP to the inner, luminal side of the ER.

A useful marker for studying the secretory pathway is the fluorescently tagged tsO45 thermo-reversible mutant vesicular stomatitis virus G protein (VSVG-FP) *(5)*. This membrane glycoprotein has been used to study the dynamics of the constitutive secretory pathway *(3)*. The intracellular localization of VSG-FP can be experimentally restricted to distinct organelles by using specific temperature blocks. Cells transfected with VSVG-FP and incubated overnight at temperatures above 39.5°C will exhibit exclusive ER localization. At temperatures below 37°C, preferably between 32 and 34°C, VSVG-FP will localize to the PM. The advantage of these two temperature blocks is that they are specifically selective for the VSVG-FP and are not expected to affect any other co-expressed protein of interest. VSVG-FP can be localized to the Golgi apparatus by incubating cells for 2–4 h at 19.5°C after overnight incubation at 39.5°C. Similarly, reducing the temperature to 18 or even 15°C can cause the accumulation of VSVG-FP at the ER exit sites *(15)*. However, in this case these low temperatures will nonspecifically accumulate all proteins of the secretory pathway in the same compartments. Localization of

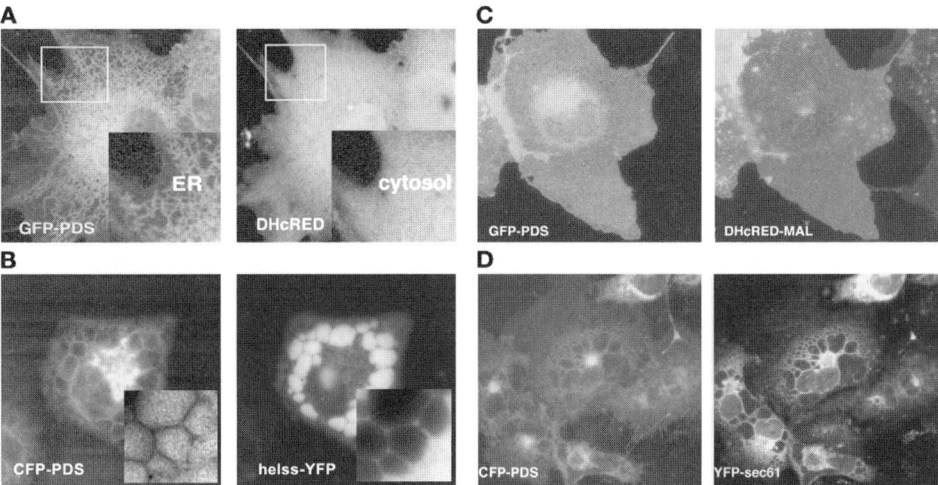

Fig. 4. Dual-color confocal analysis with soluble and transmembrane markers. (**A**) Cos7 cells co-expressing GFP-PDS *(1)* and the cytosol marker DHcRED; twofold magnified inserts demonstrate the ER reticular structure. (**B**) Cos7 cells co-expressing YFP-PDS the ER lumen marker hen egg lysozyme signal sequence (helss-) ECFP. Inverted and magnified inserts exhibit the difference between endoplasmic reticulum (ER) lumen and membrane labeling. (**C**) Cos7 cells co-expressing GFP-PDS *(1)* and the plasma membrane marker DHcRED-MAL *(18)*, treated with 80 mg/mL cycloheximide for 6 h. (**D**) Cos7 cells co-expressing CFP-PDS *(1)* and the ER membrane marker Sec61-YFP. Living cells were imaged in Labtek chambers (Nunc) containing imaging medium. Imaging was carried out on a Zeiss LSM PASCAL equipped with an Axiovert 200 inverted microscope. Ar 458- and 514-nm and HeNe 543-nm laser lines were used for ECFP, EYFP, and DHcRED, respectively.

transmembrane proteins to the Golgi apparatus, such as intracellular aggregates *(9,16)* and late endocytic recycling compartments *(10)*, should be determined with great care as many other cellular structures co-localize to the same perinuclear region near or around the microtubule-organizing center.

Notes

1. The pH of the electroporation buffer tends to rise with time. Use only red-colored buffer: the use of violet-colored buffer will reduce transfection efficiency.
2. Optimization of this method to a specific cell line should focus on three parameters: (1) finding the optimal voltage, (2) capacitance values, and (3) the amount of plasmid DNA.

3. Transient expression of polytopic proteins can perturb various processes of secretory traffic such as folding and quality control in the ER. Such perturbations can derail the protein and cause it to accumulate in organelles different from those it was targeted to. There are several approaches to lower the level of expression:

    a. Reduce the amount of plasmid DNA for transfection. The fluorescently tagged transmembrane protein can thus be co-transfected with a different spectral variant of tagged protein that will serve as an organelle marker.
    b. Perform fluorescent microscopy at shorter periods after transfection.
    c. Preincubate with a protein-synthesis inhibitor such as cycloheximide or pyromycin for 4–5 h to eliminate the ER pool of ER-retained partially folded proteins (1).

4. All steps except **steps 4** and **6** are to be carried out in a 35-mm Petri dish or in a 6-well dish for multiple samples. To avoid using an excess of primary and secondary antibodies, the cover slips are transferred to a humid chamber containing a wet tissue and a Parafilm (American National Can, Menasha, WI) strip attached to the bottom of the box. The diluted antibody (20–40 μL) is carefully applied to the paraffin film, taking care not to touch and puncture it. The cover slips are gently turned over on top of the drop with cells facing down. An alternative is to overlay the cover slip cells facing up, on a glass slide. Prior to application of the diluted antibody (20–40 μL), the edges of the cover slip are marked on the slide with a wax pen (PAP PEN, Zymed, San Francisco, CA) to keep the diluted antibody solution from spreading away from the cover slip.

Acknowledgments

The authors would like to thank Susan Wall (Emory University, Atlanta, GA) for the gift of anti-pendrin antibodies and Edna Zolin and Camille Vainstein for editorial review of the manuscript. This work is supported by the Israeli Science Foundation grant number 679/05.

References

1. Shepshelovich, J., Goldstein-Magal, L., Globerson, A., Yen, P. M., Rotman-Pikielny, P., and Hirschberg, K. (2005) Protein synthesis inhibitors and the chemical chaperone TMAO reverse endoplasmic reticulum perturbation induced by overexpression of the iodide transporter pendrin. *J. Cell Sci.* **118**, 1577–1586.
2. Ellgaard, L. and Helenius, A. (2003) Quality control in the endoplasmic reticulum.*Nat. Rev. Mol. Cell Biol.* **4**, 181–191.
3. Lippincott-Schwartz, J., Roberts, T. H., and Hirschberg, K. (2000) Secretory protein trafficking and organelle dynamics in living cells. *Annu. Rev. Cell Dev. Biol* **16**, 557–589.

4. Presley, J. F., Cole, N. B., Schroer, T. A., Hirschberg, K., Zaal, K. J., and Lippincott-Schwartz, J. (1997) ER-to-Golgi transport visualized in living cells. *Nature* **389,** 81–85.

5. Hirschberg, K., Miller, C. M., Ellenberg, J., et al. (1998) Kinetic analysis of secretory protein traffic and characterization of golgi to plasma membrane transport intermediates in living cells. *J. Cell Biol.* **143,** 1485–1503.

6. Toomre, D., Keller, P., White, J., Olivo, J. C., and Simons, K. (1999) Dual-color visualization of trans-Golgi network to plasma membrane traffic along microtubules in living cells.*J. Cell Sci.* **112 (Pt 1),** 21–33.

7. Harding, H. P., Calfon, M., Urano, F., Novoa, I., and Ron, D. (2002) Transcriptional and translational control in the Mammalian unfolded protein response. *Annu. Rev. Cell Dev. Biol.* **18,** 575–599.

8. Kopito, R. R. (1999) Biosynthesis and degradation of CFTR. *Physiol. Rev.* **79,** S167–173.

9. Rotman-Pikielny, P., Hirschberg, K., Maruvada, P., et al. (2002) Retention of pendrin in the endoplasmic reticulum is a major mechanism for Pendred syndrome. *Hum. Mol. Genet.* **11,** 2625–2633.

10. Nichols, B. J., Kenworthy, A. K., Polishchuk, R. S., et al. (2001) Rapid cycling of lipid raft markers between the cell surface and Golgi complex. *J. Cell Biol.* **153,** 529–541.

11. Patterson, G., Day, R. N., and Piston, D. (2001) Fluorescent protein spectra. *J. Cell Sci.* **114,** 837–838.

12. Rizzo, M. A., Springer, G. H., Granada, B., and Piston, D. W. (2004) An improved cyan fluorescent protein variant useful for FRET. *Nat. Biotechnol.* **22,** 445–449.

13. Gerlich, D., Beaudouin, J., Kalbfuss, B., Daigle, N., Eils, R., and Ellenberg, J. (2003) Global chromosome positions are transmitted through mitosis in mammalian cells. *Cell* **112,** 751–764.

14. Walsh, T., Abu Rayan, A., Abu Sa'ed, J., et al. (2006) Genomic analysis of a heterogeneous Mendelian phenotype: multiple novel alleles for inherited hearing loss in the Palestinian population. *Hum. Genomics* **2,** 203–211.

15. Ward, T. H., Polishchuk, R. S., Caplan, S., Hirschberg, K., and Lippincott-Schwartz, J. (2001) Maintenance of Golgi structure and function depends on the integrity of ER export. *J. Cell Biol.* **155,** 557–570.

16. Kopito, R. R. (2000) Aggresomes, inclusion bodies and protein aggregation. *Trends Cell Biol.* **10,** 524–530.

17. Royaux, I. E., Suzuki, K., Mori, A., et al. (2000) Pendrin, the protein encoded by the Pendred syndrome gene (PDS), is an apical porter of iodide in the thyroid and is regulated by thyroglobulin in FRTL-5 cells. *Endocrinology* **141,** 839–845.

18. Llorente, A., de Marco, M. C., and Alonso, M. A. (2004) Caveolin-1 and MAL are located on prostasomes secreted by the prostate cancer PC-3 cell line. *J. Cell Sci.* **117,** 5343–5351.

III

GENERAL METHODS

27

Intracellular Protein Localization
by Immunoelectron Microscopy

Tetsuaki Osafune and Steven D. Schwartzbach

Summary

Eukaryotic cells are characterized by the presence of a number of membrane-bound organelles which have differing degrees of internal structure. After synthesis in the cytoplasm, proteins must be targeted to the appropriate organelle and localized to the correct suborganellular compartment. We describe a method for immunoelectron microscopy that can be used to localize a protein not only to the correct organelle but to the appropriate suborganellular compartment. Cells are fixed to preserve subcellular structures and ultrathin sections are labeled with a monospecific antibody to the protein of interest. Protein-A gold is used to visualize the antigen–antibody complex by transmission electron microscopy allowing the intracellular location of the antigen to be determined. The methodology described was developed to study protein localization in *Euglena* but it is applicable to most organisms.

Key Words: Complex chloroplasts; *Euglena*; immunoelectron microscopy; intracellular localization.

1. Introduction

Protein trafficking to complex chloroplasts has been studied by fusing a protein to green fluorescent protein (GFP) and determining the intracellular location of the fusion protein by confocal microscopy *(1)*, purification of organelles *(2)*, or immunoelectron microscopy *(3,4)*. Of these methods, immunoelectron microscopy is the most versatile. In contrast to the use of fusion proteins, immunoelectron microscopy can be used on all cells rather than being restricted to cells that can be transfected. Intracellular localization through organelle isolation has restricted utility because organelles can not be isolated from every organism, not every organelle can be isolated free of contamination

From: *Methods in Molecular Biology, Vol. 390: Protein Targeting Protocols: Second Edition*
Edited by: M. van der Giezen © Humana Press Inc., Totowa, NJ

by other organelles, suborganellular compartments often cannot be purified for biochemical characterization and when a protein is recovered in multiple organelles, it is often difficult to distinguish true localization from contamination artifacts. The versatility of immunoelectron microscopy is demonstrated by its use to show for the first time in *Euglena* that the light-harvesting chlorophyll a/b binding protein of photosystem II (LHCPII) is transported to the Golgi apparatus prior to chloroplast localization (*3,4*) and that ribulose 1-5 bisphosphate carboxylase/oxygenase (RUBISCO) relocates in both the *Euglena* (*5*) and dinoflagellate complex plastid (*6*) from the stroma to the pyrenoid at cell cycle phases when enzyme activity is high while the pyrenoid disappears and RUBISCO redistributes back to the stroma at cell cycle phases when enzyme activity is low.

Immunoelectron microscopy methods must resolve the conflict between preserving cellular ultrastructure and retaining protein antigenicity. Cryoultramicrotomy and rapid freeze-replacement fixation methods have the advantage that they retain protein antigenicity by fixing the antigen under in vivo conditions but have the disadvantage of poorly maintaining ultrastructural integrity. Chemically fixed resin-embedded samples maintain structural integrity while sacrificing protein antigenicity. The simplicity of chemical fixation methods and the ability to overcome the loss of antigenicity by using high-titer antibodies or studying abundant antigens make chemical fixation methods the most widely used immunolabeling procedures.

Chemical fixation methods differ according to when immunogold labeling is performed and the type of resin utilized. Pre-embedding methods perform immunogold labeling before ultrathin sections are prepared from resin-embedded samples resulting in greater sensitivity and better microstructure preservation. Postembedding methods perform immunolabeling after ultrathin sections are prepared from resin-embedded samples resulting in decreased antigenicity. Plants and algae contain cell walls, vacuoles, and other structures which present barriers to antibody penetration directly into pre-embedded cells resulting in the phenomenon of highly sensitive antigen detection in a small part of the cell surface layer with little to no ability to detect antigens in the cell interior. The loss of antigenicity in postembedded samples is more than outweighed by the increase in antibody penetration making postembedding the immunolabeling method of choice for plants and algae.

Available embedding resins are the epoxy resins such as epon and the hydrophilic resins such as Lowicryl and LR White. Epoxy resins have the advantage of being easy to use and exhibiting excellent microstructure preservation. Their major disadvantage is that they are polymerized by heating which could compromise antigenicity. The advantages of the hydrophilic resins are their ease of use, excellent membrane visualization, and preservation of antigenicity because heating is not required for polymerization. Disadvantages of the hydrophilic resins are poor preservation of cell microstructure and a tendency for image deformation by the electron beam.

As this brief discussion indicates, the final choice of immunolabeling protocols requires compromises between microstructure and antigen preservation. The postembedding immunolabeling method presented in this chapter using epon was developed to study the most abundant chloroplast proteins, LHCPII *(3,4)* and RUBISCO *(5)*. The loss of antigenicity associated with postembedding immunolabeling of epon ultrathin samples is overcome by the high abundance of these proteins in *Euglena* and the better antibody penetration into postembedded cells.

2. Materials

2.1. Sample Preparation

1. 50% Glutaraldehyde (EM grade, Electron Microscopy Sciences, Hatfield, PA). Store at 0–4°C (*see* **Note 1**).
2. Prepare a 0.1 M potassium phosphate buffer pH 7.2 using 0.1 M KH_2PO_4 to adjust the pH of a 0.1 M solution of K_2HPO_4 to pH 7.2.
3. 2% (w/v) agarose (Bio-Rad, Hercules, CA) prepared in deionized water.
4. 50%, 70%, 90% Ethanol.
5. Acetone.

2.2. Embedding Samples

1. Epon (Embed-812, Electron Microscopy Sciences).
2. Dodecenyl succinic anhydride, (DDSA; Electron Microscopy Sciences).
3. Nadic methyl anhydride, (NMA; Electron Microscopy Sciences).
4. 2,4,6-Tris(dimethylaminomethyl) phenol (DMP-30; Electron Microscopy Sciences).
5. Prepare embedding mix A in a disposable beaker by combining 44 mL Embed-812 with 67 mL DDSA. Prior to combining, warm the resin and anhydride to 60°C to reduce viscosity, combine the two components and mix thoroughly with a wooden spatula (*see* **Note 2**).
6. Prepare embedding mix B in a disposable beaker by combining 67 mL Embed-812 with 56 mL NMA. Prior to combining, warm the resin and anhydride to 60°C to reduce viscosity, combine the two components and mix thoroughly with a wooden spatula (*see* **Note 2**).
7. Prepare fresh resin in a disposable beaker by combining equal volumes of room temperature embedding mix A and embedding mix B and adding DMP-30 for a final concentration of 2%. Mix thoroughly using a wooden spatula and place in

a vacuum desiccator for 15 min to degass. Store tightly sealed at room temperature and use that day (*see* **Note 3**).
8. Gelatin capsules size 00 (Electron Microscopy Sciences).

2.3. Preparation of Ultrathin Sections

1. Formvar/Carbon film-coated 200 mesh nickel grids (Electron Microscopy Sciences).
2. Diamond knife (Electron Microscopy Sciences).
3. Block Trimmer (Electron Microscopy Sciences).
4. Eyelash with handle (Ted Pella Inc., Redding, CA).
5. Whatman #1 filter paper (Fisher Scientific, Pittsburgh, PA).
6. Anti-capillary self closing tweezers (Ted Pella Inc).

2.4. Immunogold Labeling of Sections

1. 0.3% (w/v) Hydrogen peroxide prepared immediately before use by mixing 60 μL 50% (w/v) hydrogen peroxide (Sigma, St. Louis, MO) with 9.94 mL distilled water.
2. Prepare a 0.01 M phosphate buffer pH 7.0 using 0.01 M KH_2PO_4 to adjust the pH of a 0.01 M solution of K_2HPO_4 to pH 7.0.
3. Phosphate-buffered saline (0.01 M PBS pH 7.0) is prepared by adding 0.85 g NaCl to 100 mL 0.01 M phosphate buffer pH 7.0.
4. PBS–bovine serum albumin (BSA) is prepared by adding BSA (Sigma) for a final concentration of 1% to PBS.
5. PBS-Tween is prepared by adding Tween 20 (Sigma) to a final concentration of 0.05% to PBS.
6. Antibody to protein to be localized diluted to appropriate concentration with PBS-BSA immediately before use.
7. Protein-A gold: Protein-A conjugated to 15-nm gold particles (EY laboratories, San Mateo, CA) (*see* **Note 4**).
8. 3% Uranyl acetate (Electron Microscopy Sciences) is prepared by mixing 1.5 g uranyl acetate with 50 mL deionized water. The solution is stirred overnight in a foil-covered container, 10 drops of glacial acetic acid is added and the solution is stored at 4°C for no more than 3 mo.
9. Parafilm (Fisher Scientific).

3. Methods

3.1. Sample Preparation

1. Add 0.9 mL 50% glutaraldehyde to 45 mL *Euglena* cells (*see* **Note 5**) in culture (*see* **Note 6**) for a final concentration of approx 1% glutaraldehyde (*see* **Note 7**) and incubate at 4°C for 60 min. Recover the cells by centrifugation for 2 min in a tabletop centrifuge.
2. Resuspend the cell pellet in 10 mL 0.1 M phosphate buffer pH 7.2, incubate 5 min at room temperature on a specimen rotator, and recover the cells by centrifugation for 2 min in a tabletop centrifuge. Repeat two times (*see* **Note 8**).

3. Resuspend the cell pellet in 1 mL 0.1 *M* potassium phosphate buffer pH 7.2 and transfer to a microfuge tube. Pellet the cells.
4. Embed the cell pellet in 2% (w/v) agarose by resuspending the cell pellet in 60°C 2%(W/V) agarose and immediately centrifuge for 30 s to pellet the cells. Remove the tube and place on ice to solidify the agarose.
5. Remove the agarose plug from the microfuge tube with a needle and cut off the region containing the cells. Cut the agarose into small cubes and transfer to a 15-mL conical centrifuge tube.
6. Dehydrate the sample by incubating on a specimen rotator for 20 min in 10 mL 50% ethanol and recover the sample by gentle centrifugation. Repeat once.
7. Dehydrate the sample by incubating on a specimen rotator for 20 min in 10 mL 70% ethanol and recover the sample by gentle centrifugation. Repeat once.
8. Dehydrate the sample by incubating on a specimen rotator for 20 min in 10 mL 90% ethanol and recover the sample by gentle centrifugation. Repeat once.
9. Resuspend the sample in 10 mL acetone, incubate on a specimen rotator for 20 min and recover the sample by gentle centrifugation. Repeat three times.

3.2. Embedding Samples

1. Resuspend the sample in 3 mL of a 1:2 resin:acetone mixture, place uncapped on a specimen rotator in a hood and incubate 4 h to overnight. Recover the sample by gentle centrifugation and remove the resin using plastic pipets (*see* **Note 9**).
2. Resuspend the sample in 3 mL of a 2:1 resin:acetone mixture, place uncapped on a specimen rotator in the hood and incubate 4 h. Recover the sample by gentle centrifugation and remove the resin using plastic pipets.
3. Resuspend the samples in 3 mL 100% resin and incubate in a vacuum desiccator for 1–2 h. Recover the sample by gentle centrifugation and remove the resin using plastic pipets.
4. Resuspend the sample in a small volume of resin. Fill a gelatin capsule about half full and overlay with the cell sample. Place the capsule in a centrifuge tube positioning it upright in the tube using tissue paper and centrifuge at full speed in a clinical centrifuge for 10 min to pellet the cells to the bottom of the capsule.
5. Remove the capsule from the centrifuge tube, top off the capsule with resin, insert a sample identification label into the resin at the top of the capsule and polymerize by incubation in a 60°C oven for 24 h (*see* **Note 10**).
6. Allow the polymerized block to cool for 24 h. Remove the block from the gelatin capsule by placing in water at 37°C until the capsule dissolves.

3.3. Preparation of Ultrathin Sections

1. Mount the block in a block trimmer and shape the end into a four-sided pyramid with walls at a 45° angle and a 0.5- to 0.75-mm² top surface.

2. Mount the trimmed block on a microtome, making sure that the block face is parallel to the knife edge. Fill the diamond knife trough with distilled water so it is level with the cutting edge.
3. Cut a ribbon of silver ultrathin sections approx 80–90 nm thick (*see* **Note 11**).
4. Use an eyebrow tool to separate the ribbon into five or six sections and align them in the trough.
5. Place a grid held with a pair of tweezers under the sections and raise it positioning the sections in the middle of the grid (*see* **Note 12**).

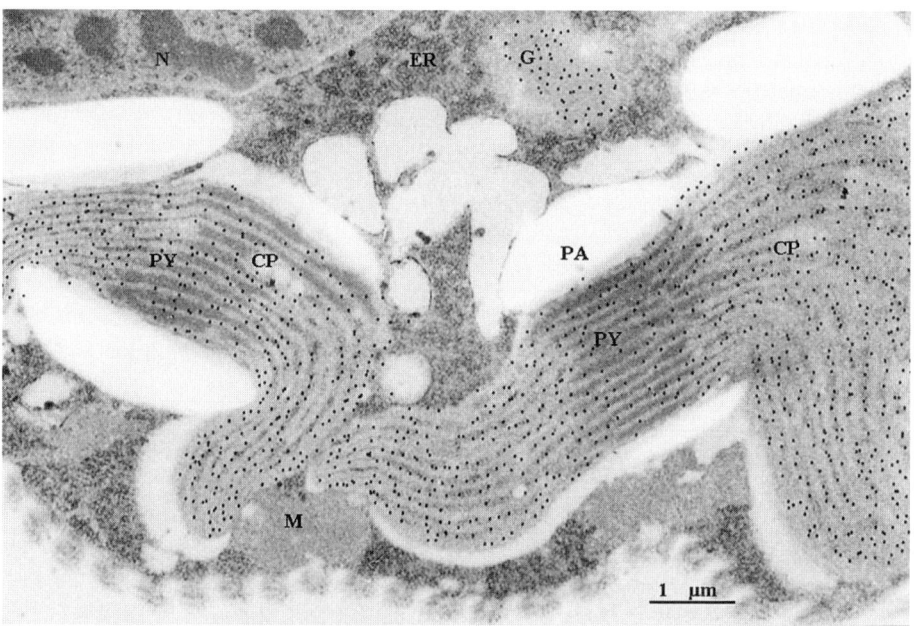

Fig. 1. Immunolocalization of the light harvesting chlorophyll a/b binding protein of photosystem II (LHCPII) to the Golgi apparatus and chloroplasts of synchronously dividing *Euglena*. Cell division was synchronized by growing cells photoautotrophically on an alternating 10:14 light dark cycle and cells were sampled for immunoelectron microscopy 10 h after the start of the light cycle. Immunogold is localized over the chloroplast (CP) and Golgi apparatus (G) with virtually no immunogold over the mitochondria (M), nucleus (N), endoplasmic reticulum (E), or paramylum grains (PA). Within the chloroplast, immunogold can be seen directly over the thylakoid membranes with little immunogold in the space between the individual thylakoid membranes indicative of the subchloroplast localization of LHCPII to thylakoids. Note the reduced amount of immunogold label over the thylakoids spanning the pyrenoid (PY) region of the chloroplast. Bar = 1 μm. (Micrograph courtesy of T. Osafune; unpublished.)

6. Blot the jaws of the tweezers and the bottom of the grid with Whatman #1 filter paper to absorb all of the liquid. Place the grid sample-side-up on a dry piece of filter paper in a Petri dish and allow to dry overnight.

3.4. Immunogold Labeling of Sections

1. Ultrathin sections on grids are floated-section-side-down on a 300-μL drop (*see* **Note 13**) of freshly prepared 0.3%(W/V) hydrogen peroxide solution for 10 min (*see* **Note 14**).
2. Wash ultrathin sections on grids four times with PBS by floating grids section-side-down on a 300-μL drop of PBS for 30 min.

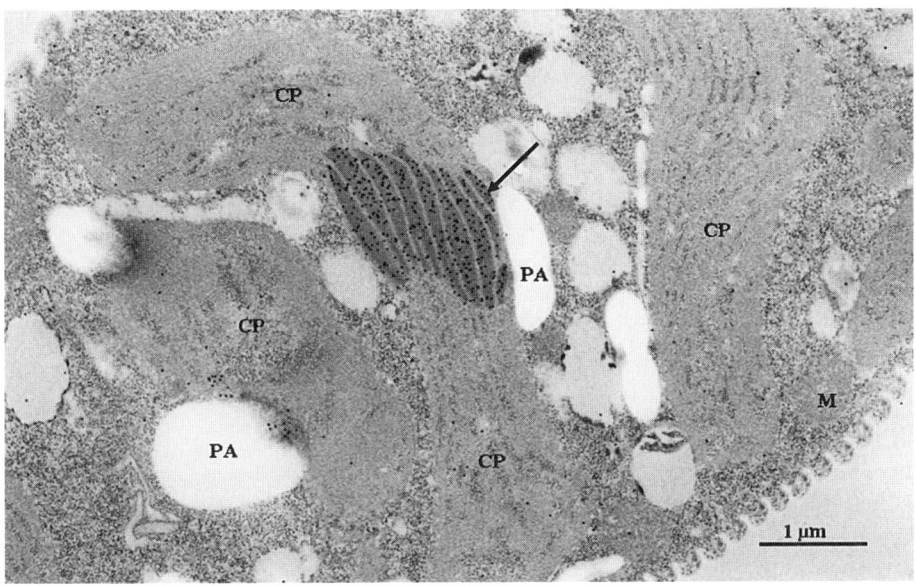

Fig. 2. Immunolocalization of the large subunit of ribulose bisphosphate carboxylase/oxygenase (RUBISCO) to the pyrenoid of synchronously dividing *Euglena*. Cell division was synchronized by growing cells photoautotrophically on an alternating 10:14 light dark cycle and cells were sampled for immunoelectron microscopy 10 h after the start of the light cycle Immunogold is localized over the dense pyrenoid region (arrow) of one of the chloroplasts (CP) with virtually no immunogold over the less dense chloroplast stroma, mitochondria (M) or cytoplasmic regions of the cell. Note the association of a paramylum granule (PA) with the pyrenoid region. Within the pyrenoid, immunogold is found only in the dense region between the thylakoids in contrast to LHCPII (**Fig. 1**) which is found predominately over the thylakoid membranes. Bar = 1 μm. (Micrograph courtesy of T. Osafune; unpublished.)

3. Block ultrathin sections on grids by floating grids section-side-down on a 300-μL drop of PBS-BSA for 30 min.
4. Incubate ultrathin sections on grids with primary antibody by floating grids section-side-down on a 300-μL drop of primary antibody diluted in PBS-BSA and incubating at 37°C for 20 min (*see* **Note 15**).
5. Wash ultrathin sections on grids twice with PBS-Tween by floating grids section-side-down on a 300-μL drop of PBS-Tween for 30 min.
6. Incubate ultrathin sections on grids with protein-A gold by floating grids section-side-down on a 300-μL drop of protein-A gold diluted 1:10 or 1:20 in PBS and incubating for 20 min at room temperature.
7. Wash ultrathin sections on grids twice with PBS-Tween by floating grids section-side-down on a 300-μL drop of PBS-Tween for 10 min.
8. Wash ultrathin sections on grids twice with deionized water by floating grids section-side-down on a 300 μL drop of deionized water for 10 min.
9. Stain ultrathin sections on grids with uranyl acetate by floating grids section-side-down on a 300-μL drop of 3%(W/V) uranyl acetate for 10 min (*see* **Note 16**).
10. Wash ultrathin sections on grids twice with deionized water by floating grids section-side-down on a 300-μL drop of deionized water for 10 min.
11. Blot the bottom of the grid with Whatman #1 filter paper and place sample-side-up on a dry piece of filter paper in a Petri dish and allow to dry overnight.
12. Examine sections in the electron microscope. **Figures 1** and **2** are examples of protein-A gold labeled immunoelectron micrographs of *Euglena* cells stained with antibodies to LHCPII (**Fig. 1**) and the large subunit of RUBISCO (**Fig. 2**).

Notes

1. Glutaraldehyde and the other chemicals used for fixation and staining are extremely hazardous. Solutions should be prepared and used in a hood. Wear gloves and protective eyeware whenever handling chemicals.
2. Unused embedding mix can be stored for up to 6 mo at −20°C but it is preferable to use a freshly prepared mix.
3. Resin waste should be collected and allowed to polymerize before disposal.
4. By using two different sized gold particles for sequential immunolabeling, two antigens can be co-localized on a single section.
5. The minimum number of cells that will provide sufficient material for sectioning is 4×10^7 cells. Using fewer cells produces a small cell pellet which after embedding, is difficult to position in the microtome for sectioning. Starting with a larger number of cells facilitates the sectioning process.
6. Glutaraldehyde fixation can be performed directly in any medium the cells are grown in. Isolated organelles are fixed directly in the final purification buffer.
7. The amount of glutaraldehyde used for fixation represents a compromise between preserving cellular microstructure and retaining antigenicity. Better microstructure

preservation can be obtained at higher glutaraldehyde concentrations but this may result in a loss of antigenicity resulting in a decline in colloidal gold labeling density.

8. Cells can be stored in 0.1 *mM* potassium phosphate buffer pH 7.2 for up to 1 wk at 0–4°C.

9. It is important to add enough acetone resin mixture to the tube so that the specimens containing agarose pieces are not exposed to air when the acetone evaporates from the mixture.

10. Epoxy resins are easy to use and provide excellent preservation of microstructure after sectioning. The necessity to heat the resin for polymerization does however lead to a loss of antigenicity making their use most suitable for localization of high-abundance proteins such as LHCPII and RUBISCO. Hydrophilic resins such as Lowicryl K4M and LR White which can be polymerized by ultraviolet light at low temperature are better at preserving antigenicity. They are more widely used for immunoelectron microscopy although they exhibit a lower level of ultrastucture preservation and the sections are more easily damaged by the electron beam.

11. Thick sections can initially be cut and examined in the light microscope to identify the region of the block containing the sample.

12. The most convenient way to pickup and transfer grids is to use a self closing anti-capillary tweezers. Extra care must be taken to remove all moisture when other types of instruments are used for transfer.

13. A piece of parafilm should be placed in a Petri dish containing a piece of buffer-saturated Whatman #1 filter paper. Place drops of washing solution side by side on the parafilm sheet and move the grid sequentially from drop to drop incubating with the lid closed. Incubation in a water-saturated environment prevents the drops from evaporating. Depending on the scarcity of the reagent and time of incubation, drop size can be varied from 100 to 1000 μL.

14. Incubation in hydrogen peroxide increases immunoreactivity by etching the resin surface making the embedded antigen more accessible to the antibody. Samples embedded in hydrophilic resins such as Lowicryl K4M and LR White do not need to be pretreated with hydrogen peroxide solution and should be incubated directly in PBS.

15. The optimal antibody dilution and incubation conditions must be determined empirically for each antibody. Typical dilutions are 50- to 1000-fold for a 20-min incubation with antibody at 37°C. Lower antibody concentrations can be used with longer incubation times at lower temperatures. Incubation times as long as 24 h at 4°C can be used. The lower the abundance of the antigen, the higher the antibody concentration and/or incubation time needed. When excessive background labeling is seen, increasing the amount of BSA in the blocking solution, lowering the antibody concentration and/or shortening the incubation time may decrease nonspecific labeling. Control experiments with preimmune serum should be routinely performed to verify the specificity of the immunoreaction.

16. Care must be taken to withdraw the solution from the top of the bottle to avoid depositing precipitated crystals on the grid.

Acknowledgments

This work was supported by National Science Foundation Grant MCB-0080345 to S.D.S and Grant in Aid for Scientific Research No. 15570054 from the Minstry of Education, Sciences, Sports and Culture Japan to T.O.

References

1. Apt, K. E., Zaslavkaia, L., Lippmeier, J. C., et al. (2002) *In vivo* characterization of diatom multipartite plastid targeting signals. *J. Cell Sci.* **115,** 4061–4069.
2. Slavikova, S., Vacula, R., Fang, Z., Ehara, T., Osafune, T. and Schwartzbach, S. D. (2005) Homologous and heterologous reconstitution of Golgi to chloroplast transport and protein import into the complex chloroplasts of *Euglena. J. Cell Sci.* **118,** 1651–1661.
3. Osafune, T., Sumida, S., Schiff, J. A., and Hase, E. (1991) Immunolocalization of LHCPII apoprotein in the Golgi during light-induced chloroplast development in non-dividing *Euglena* cells. *J. Electron Microsc.* **40,** 41–47.
4. Osafune, T., Schiff, J. A., and Hase, E. (1991) Stage-dependent localization of LHCP II apoprotein in the Golgi of synchronized cells of *Euglena gracilis* by immunogold electron microscopy. *Exp. Cell Res.* **193,** 320–330.
5. Osafune, T., Yokota, A., Sumida, S., and Hase, E. (1990) Immunogold localization of ribulose-1,5-bisphosphate carboxylase with reference to pyrenoid morphology in chloroplasts of synchronized *Euglena gracilis* cells. *Plant Physiol.* **92,** 802–808.
6. Nassoury, N., Wang, Y., and Morse, D. (2005) Brefeldin a inhibits circadian remodeling of chloroplast structure in the dinoflagellate *Gonyaulax. Traffic* **6,** 548–561.

28

Separation of Proteins by Blue Native Electrophoresis

Olga Randelj, Joachim Rassow, and Christian Motz

Summary

Blue native gel electrophoresis is a native electrophoresis method that can be used for molecular weight determination for most soluble protein complexes as well as for most membrane proteins. Subsequent sodium dodecyl sulfate–polyacrylamide gel electrophoresis (SDS-PAGE) can be used in a second dimension to resolve the complexes into their subunits. The method has been extensively used for the analysis of the respiratory chain complexes, for the determination of intermediates of mitochondrial protein import, and for the identification of the composition of the protein import machinery for mitochondria and chloroplasts. Here we describe the basic method and some applications in the research of mitochondrial protein import.

Key Words: Electrophoresis; polyacrylamide gel electrophoresis; blue native; membrane proteins.

1. Introduction

In the field of protein targeting it is often of interest to analyze the assembly of membrane protein complexes. Blue native polyacrylamide gel electrophoresis (BN-PAGE, **Fig. 1**), a method developed by Schägger and Jagow *(1)*, has proven to be a valuable tool to analyze the assembly of newly imported proteins *(2,3)*, to asses import intermediates, or to investigate the complexes of the import machinery of mitochondria *(4–6)* or chloroplasts *(7)*.

BN-PAGE offers some major advantages:

1. It allows determination of the molecular weight for both positively and negatively charged proteins as well as for membrane proteins.
2. In contrast to gel filtration, BN-PAGE offers the possibility of testing many samples on the same gel, i.e., under identical conditions.

From: *Methods in Molecular Biology, Vol. 390: Protein Targeting Protocols: Second Edition*
Edited by: M. van der Giezen © Humana Press Inc., Totowa, NJ

3. Small samples (electrophoretic amounts down to 5 μg) are possible.
4. The electrophoresis can be performed using standard equipment for SDS-PAGE.

With other native electrophoresis methods, the mobility of the proteins is determined mainly by their intrinsic charge but also to some extent by their size. Therefore, neither the pI nor the molecular weight of the proteins can be exactly determined. BN-PAGE overcomes this difficulty by binding of the anionic dye Coomassie blue G-250 and so inducing a charge shift to the proteins. Thus the electrophoretic mobility depends only on the molecular weight. Schägger et al. *(8)* showed that BN-PAGE allows the analysis of molecular masses of most soluble as well as most membrane proteins. Only a few proteins that do not bind the Coomassie dye and have no or almost no negative charge at pH 7.5 showed significantly different behavior. As a side effect, the negatively charged dye also prevents aggregation of the proteins.

While in SDS-PAGE the SDS:protein ratio and thus the charge:mass ratio is constant for all proteins, the ratio of bound dye to protein shows some variation in BN-PAGE. Integral membrane proteins show a particular increase in their apparent molecular weights compared to soluble proteins because of differences in the affinity for detergent or dye molecules. Heuberger et al. *(9)* showed

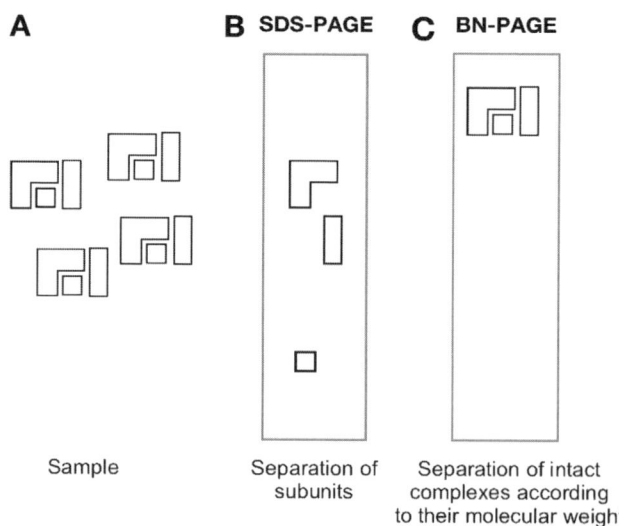

Fig. 1. Principle of the separation with different electrophoretic techniques: (**A**) proteins to be separated; (**B**) SDS-PAGE: the subunits are separated according to their size; (**C**) the intact protein complexes are separated by their size.

for different membrane proteins that bound detergents were replaced by dye molecules during electrophoresis. The apparent molecular weight was increased by a factor of 1.8 compared to the molecular weight of soluble proteins that served as marker proteins.

2. Materials

Prepare all solutions in distilled H_2O unless otherwise stated.

1. Electrophoresis chamber: the gel has to be cooled during electrophoresis. Although running the gel in the cold room in a standard electrophoresis chamber may be sufficient, best cooling is achieved using a system with cooling in a water bath, such as the Hoefer SE600 system (*see* **Note 1**).
2. Cathode buffer A: 50 mM Tricin, 15 mM Bistris, 0.02% Coomassie Brilliant Blue G-250, pH 7.0 at 4°C, do not adjust the pH (*see* **Notes 1** and **2**).
3. Cathode buffer B: 50 mM Tricin, 15 mM Bistris, pH 7.0 at 4°C (*see* **Note 1**).
4. Anode buffer: 50 mM Bistris, adjust pH 7.0 with HCl (*see* **Note 1**).
5. Gel buffer (3x): 1.5 M ε-aminocaproic acid, 150 mM Bistris (*see* **Note 3**).
6. AB-Mix: 49.5% acrylamide, 1,5% bis-acrylamide.
7. APS: 10% (w/v) ammonium persulfate in H_2O.
8. N, N, Nı, Nı-Tetramethylethylene diamine (TEMED).
9. Lysis buffer: 20 mM Tris-HCl, 0.1 mM EDTA, 500 mM ε-aminocaproic acid, 10% glycerol, pH 7.0 (HCl). Complete with the desired detergent (*see* **Notes 3–6**) and add phenylmethylsulfonyl fluoride (PMSF) to 1 mM just before use.
10. PMSF 200 mM in 2-propanol.
11. Sample buffer: 500 mM ε-aminocaproic acid, 100 mM Bistris, 5% Serva Blue G, pH 7.0 (HCl) (*see* **Note 2**).
12. 100% Methanol.
13. Blot buffer: 20 mM Tris, 150 mM glycine, 0.02% sodium dodecyl sulfate (SDS), 20% methanol, do not adjust the pH.

3. Methods

3.1. Blue Native Gel Electrophoresis

3.1.1. Preparing the Gel

1. Prepare the gel assembly according to the manufacturer's instructions.
2. Mix the separating gel solutions (**Table 1**, *see* **Notes 7** and **8**) without APS and fill them into a gradient mixer, filling the higher concentration in the exit chamber. Add the APS and TEMED, then pour the gel. Overlay with 2-propanol or 1-butanol.
3. After the separating gel has polymerized, discard the alcohol, then mix and pour the stacking gel (**Table 1**). Place a comb in the gel to form the wells.

Table 1
Composition of Separating and Stacking Gels

Components	Separating gel:acrylamide concentration					Stacking gel
	4%	6%	13%	16%	20%	
AB-mix	0.75 mL	1.15 mL	2.35 mL	3 mL	3.75 mL	0.3 mL
Gel buffer	3 mL	3 mL	3 mL	3 mL	3 mL	1.25 mL
Glycerol	—	—	1 mL	1 mL	1 mL	—
H$_2$O			Fill to 9 mL			2.2 mL
APS	38 µL	38 µL	30 µL	30 µL	30 µL	30 µL
TEMED	3.8 µL	3.8 µL	3 µL	3 µL	3 µL	3 µL

3.1.2. Sample Preparation

Prior to the electrophoresis the samples are lysed in a buffer-containing detergent. The choice of detergent is crucial because too harsh conditions will dissociate the complexes, whereas too mild conditions may not be sufficient to solubilize the proteins (*see* **Fig. 2B** and **Notes 4–6**).

1. Perform the targeting experiment (*see* **Notes 9** and **10**) as described in Chapter 10 to the stage prior to the sample preparation for the SDS-PAGE.
2. Reisolate the mitochondria (or other organelles or membranes, about 70 µg protein) by centrifugation. Discard the supernatant.
3. Resuspend the samples in 70 µL lysis buffer (i.e., 1 µL/µg protein) containing the detergent and 1 mM PMSF.
4. Incubate 20 min on ice (*see* **Note 6**).
5. Centrifuge 20 min at 20, 000g (minimum) to pellet insoluble aggregates and unlysed mitochondria.
6. Discard the pellet (or keep it for lysis control using SDS-PAGE).
7. Add 7 µL of sample buffer (i.e., 10% of the sample volume) to the supernatant.

3.1.3. Electrophoresis

1. Apply the samples to the gel.
2. Apply the markers solubilized in sample buffer (*see* **Note 11**).
3. Overlay the samples with cathode buffer A, assemble the precooled electrophoresis unit, and fill the cathode buffer A in the upper buffer chamber.
4. Start the electrophoresis at 200 V until the blue front reaches the separating gel, then set the voltage to 500 V with the current limited to 15 mA per gel (*see* **Note 12**).

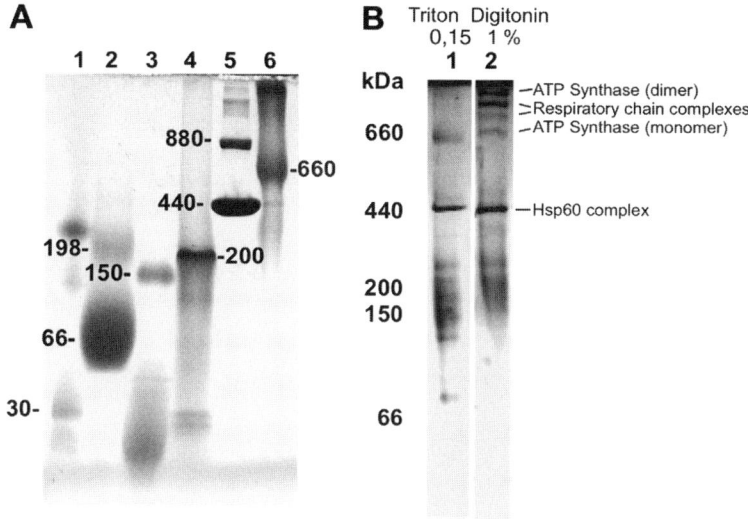

Fig. 2. Separation of protein complexes by BN-PAGE (4–16% acrylamide gradient), staining with Coomassie. (**A**) Marker proteins: lane 1: carbonic anhydrase (30 kDa), 2: BSA (66 kDa), 3: alcohol dehydrogenase (150 kDa), 4: β-amylase (200 kDa), 5: apoferritin (440 kDa), 6: thyroglobulin (660 kDa). Note that some of the markers give more than one band; therefore it is not recommended to mix all the markers in one lane. (**B**) Separation of proteins from isolated yeast mitochondria in BN-PAGE (4–16% gradient): lane 1: lysis of mitochondria with Triton X-100, lane 2: lysis with digitonin. Note that the large complexes of the respiratory chain are visible only when the mitochondria were lysed in digitonin. The ATP synthase dimer is only visible in digitonin.

5. Removing excess dye can be achieved by changing the cathode buffer A to cathode buffer B after about one-third of the run. This is recommended when blotting the gel but can be omitted if the gel will be stained after electrophoresis.

3.1.4. Staining

The Serva Blue G will not stain all proteins, so it will be necessary to stain the gel according to your standard procedures, e.g., staining with Coomassie Blue R-250 or silver staining.

3.1.5. Western Blotting

For Western blotting it is best to use polyvinylidene fluoride (PVDF) membranes, which will allow removing the dye using organic solvents. Note that, unlike nitrocellulose, PVDF needs extra preparation:

1. Prepare the PVDF membrane by immersing for a few seconds in 100% methanol until the membrane is translucent. Wash with water for another 5 min.
2. Incubate the membrane a few minutes in blot buffer. The membrane submerges when it is equilibrated.
3. Equilibrate the gel for a few seconds in transfer buffer.
4. Perform Western blotting according to your standard procedures.
5. After blotting, remove the blue dye by washing the membrane with methanol, then rinse with water.
6. The membrane can be stained or used for antibody detection following your standard procedures. Note that if the membrane becomes dry, you will have to repeat **step 1** before going on.

3.2. Two-Dimensional Gels

1. After running the native gel, disassemble the glass plates and remove the stacking gel.
2. Cut out the lanes of interest.
3. Place the lane on a glass plate at the usual position of the stacking gel. Leave some space at the side for a marker or a control (*see* **Fig. 3A** and **Note 13**).
4. Assemble the glass plates for a standard SDS gel. The excised gel strip will be held in its position by friction.
5. Pour the separating gel. Leave about 0.5–1 cm below the gel strip. Overlay with 2-propanol or 1-butanol (according to your standard procedures).
6. Decant the propanol and pour the stacking gel, avoiding air bubbles under the gel strip. This can be achieved by tilting the gel assembly, and slowly straighten it while pouring the gel. Place a small comb for one or two wells for marker and control (*see* **Note 13**).
7. Place the gel into the electrophoresis apparatus. Apply the marker. If you need reducing conditions, overlay the blue native strip with reducing sample buffer.
8. Perform the electrophoresis according to your standard protocols. The blue dye in the native gel strip will form a thin line. After the line has reached the separating gel, remove the gel strip with a thin spatula. This will avoid an uneven run.
9. The gel can be stained, blotted, etc. according to your standard procedures.

3.3. Using BN-PAGE to Analyze an Import Experiment

Figure 4A shows the analysis of a standard import experiment with blue native electrophoresis. Basically, isolated mitochondria from *Saccharomyces cerevisiae* were either left untreated (left lane) or pretreated with apyrase and oligomycin to deplete the mitochondria of endogenous ATP. Subsequently, both samples were incubated with radiolabeled lysate of the precursor protein of Oxa1 as described in Chapter 10. After the import, the reisolated mitochondria were lysed in lysis buffer with 1% digitonin and subjected to blue native

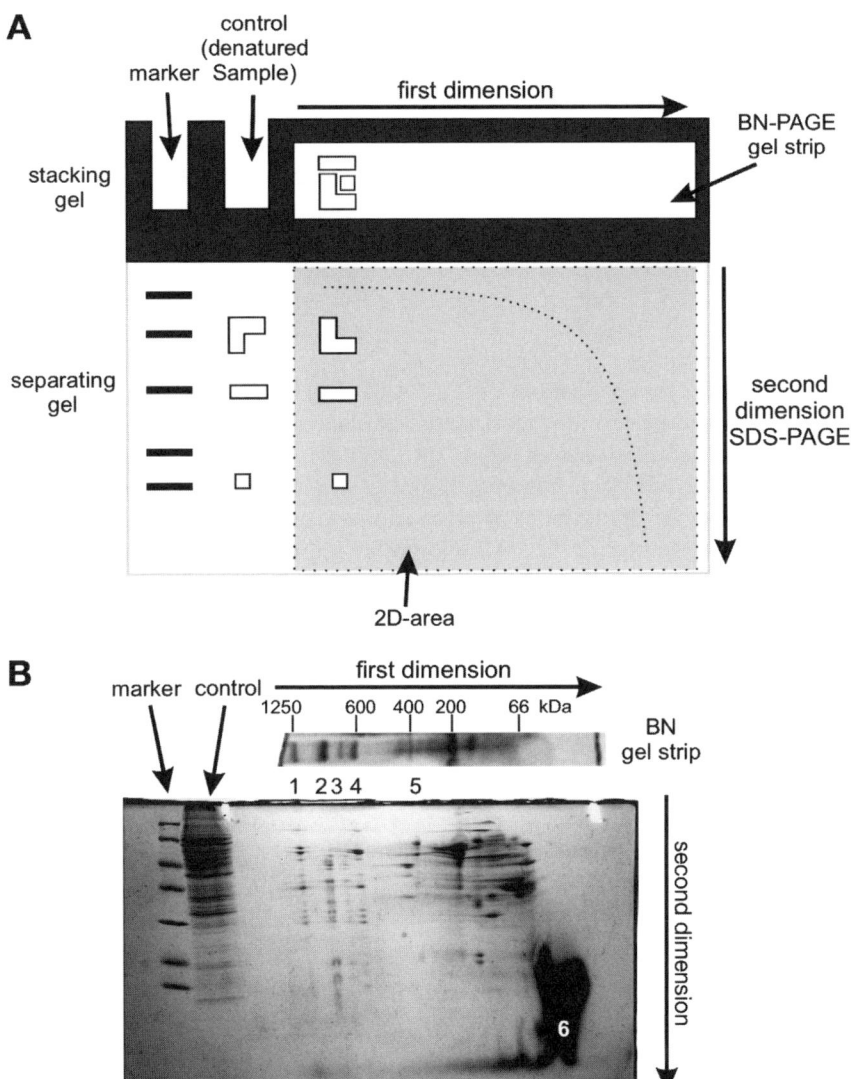

Fig. 3. Two-dimensional gel: (**A**) principle—in the first dimension (BN-PAGE, left to right), the protein complexes are separated, in the second dimension (SDS-PAGE, top-down), the subunits and monomers are separated. The subunits of one complex can be found one below the other. Monomers will be found on the dotted hyperbolic line. (**B**) Example—mitochondria (70 μg protein) lysed in digitonin: lanes: 1, ATP synthase (dimer); 2 and 3, respiratory chain complexes; 4, ATP synthase (monomer); 5, Hsp60 complex; 6, excess Coomassie dye.

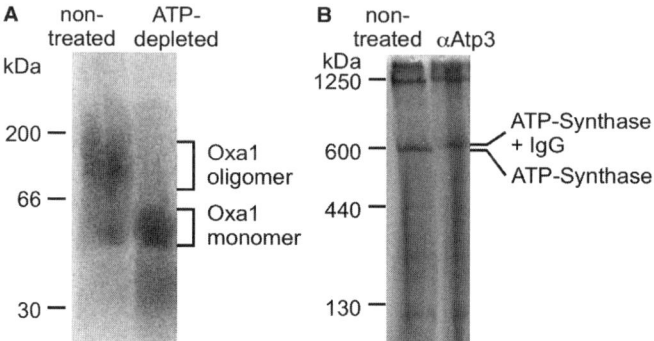

Fig. 4. Examples for applications of BN-PAGE: **(A)** Import of radiolabeled precursor of Oxa1 protein into yeast mitochondria: the left lane shows an import under standard conditions, the right lane shows an import into ATP-depleted mitochondria. **(B)** Autoradiogram of a BN-PAGE with mitochondria after the import of radiolabeled γ-subunit (Atp3) and lysis with digitonin. The right sample is treated with antiserum against Atp3. Because of binding of the IgG molecules to the ATP synthase complex, the corresponding bands are shifted from 600 to 750 kDa.

electrophoresis as described in **Subheading 3.1.** The gels were analyzed using a phosphor imager.

Figure 4A shows the radiolabeled proteins on the blue native gels. The untreated sample show a broad band at approx 150 kDa, representing the functional tetramer of Oxa1 *(10)*, and a smaller band below 66 kDa, corresponding to the unassembled monomers of 42 kDa. With ATP-depleted mitochondria, only the monomeric species can be detected.

3.4. Using BN-PAGE for Assessing Import Intermediates of Carrier Proteins

Carrier proteins are proteins of the inner mitochondrial membrane with six transmembrane helices. The ADP/ATP carrier (AAC) is one of the most abundant proteins in the mitochondrial inner membrane. For *Neurospora crassa*, five different import stages were characterized *(11)*. BN-PAGE was used to monitor the import stages for the AAC of *Saccharomyces cerevisiae* *(12,13)*. Using radiolabeled protein, the AAC can be arrested in three different stages, which can be monitored by BN-PAGE: Stage II is detected as a 400- to 500-kDa complex, consisting of the carrier bound to the receptor of the TOM complex, stage III is the monomer of the carrier (<60 kDa) imported into the intermembrane space. Stage IV, the carrier bound to the TIM22 translocase, is observed as a 400-kDa complex, whereas stage V represents the functional dimer in the inner membrane (>60 kDa).

To monitor the different import stages perform an in vitro import with isolated mitochondria and ^{35}S-labeled lysate as described in Chapter 10 with the following changes: for formation of the stage II intermediate add 25 units/mL apyrase to the lysate and $8\,\mu M$ antimycin A, $20\,\mu M$ oligomycin and $1\,\mu M$ valinomycin to the mitochondria in import buffer *without* externally added ATP and incubate 10 min at 25°C prior to the addition of lysate to the mitochondria. For formation of the stage III intermediate, dissipate the membrane potential by addition of $1\,\mu M$ valinomycin prior to the import. Formation of the stage IV intermediate needs a reduced membrane potential, which can be achieved by addition of 20–$120\,\mu M$ CCCP. For formation of stage V, perform a standard import experiment. After reisolation of the mitochondria by centrifugation, resuspend the mitochondrial pellet in lysis buffer ($1\,\mu L/\mu g$ mitochondrial protein) with 1% digitonin. Run a blue native gel as described above. Use a 6–16% polyacrylamide gradient. After the run, dry the gel and expose it to a PhosphorImager screen.

3.5. Using an Antibody Shift to Determine Interaction Partners

Antibody binding is not affected by BN-PAGE and can thus be detected by a molecular mass shift of about 150 kDa (the molecular weight of a IgG molecule). Wiedemann et al. *(2)* used this method to identify components of the assembly intermediates of the TOM complex. This method can be used as an alternative to coimmune precipitation.

Dissolve the samples in lysis buffer containing 1% digitonin and add 1–5 µL polyclonal antiserum against potential interaction partners. Use a sample with the same amount of preimmune serum as control. Incubate 20 min on ice. Centrifuge, then run the blue native gel as described in **Subheading 3.1.**

Detect your protein of interest by Western blotting or PhosphorImager analysis. Protein bands showing an increase of the molecular weight by 150 or 300 kDa in comparison with the control indicate the binding of one or two IgG molecules, and thus the presence of the protein the antibody was raised against in the protein complex of interest **(Fig. 4B)**.

Notes

1. The Hoefer SE600 system needs 400 mL of cathode buffer and 4 L of anode buffer; both can be reused up to five times.
2. Coomassie Brilliant Blue R-250 cannot be used.
3. Schägger and Jagow *(1)* use a concentration of 500 mM ε-aminocaproic acid in the gel and 750 mM in the lysis buffer. The concentration, however, can be varied in a wide range (100 mM to 1 M).
4. The choice of detergent is a critical step, as shown in **Fig. 2B**. For yeast mitochondria, 1% digitonin will give a good lysis while keeping most protein

complexes together. Also, Triton X-100 in a range from 0.2 to 2% may give good results. Any other detergent may be used; it should be selected to extract the proteins and preserve the protein complexes (*see* **Note 6**).

5. For determination of the protein pattern of the monomers, 1% SDS or 8 *M* urea can be used.

6. The appropriate lysis conditions (lysis time, lysis buffer, detergent) should be determined in a pilot experiment: lyse the samples under different conditions, centrifuge at 20,000*g* (better: 100,000*g*) and analyze the pellet and supernatant on a SDS gel for the protein of interest. Select conditions that keep the protein in the supernatant.

7. A linear gradient range from 4 to 16% will give a good resolution from 1000 Da to 100 kDa; for other molecular weight ranges, gradients between 4 and 20% can be used. For separation of proteins in the range from 50 to 100 kDa, a homogeneous gel with 10% acrylamide has proven suitable *(14)*.

8. The volume of 18 mL for the separating gel is convenient for a gel with the dimensions 18 × 16 cm and 1 mm thickness, as purchased with the Hoefer SE600. For other gels, the appropriate volume has to be determined. Use of spacers with more than 1 mm thickness may result in broadening of the protein bands.

9. Protein amounts of 50–100 μg per sample for a heterogeneous mixture of proteins and 5–30 μg for purified protein are usually satisfactory detectable. For import experiments using mitochondria, use twice as much mitochondria and lysate as you would for a normal experiment analyzed by SDS-PAGE. A high background or smearing on the gel, however, can be reduced by reducing the sample size.

10. When using larger volumes of lysate, free radioactive methionine may be incorporated into mitochondrially translated proteins and thus results in additional bands on the gel. This can be prevented by addition of an excess (4 m*M*) of nonradioactive methionine to the lysate prior to the import.

11. Molecular weight markers for gel filtration calibration (e.g., Sigma MW-GF-1000; *see* **Fig. 2A**) can be used as markers. Some of the markers may give more than one band, so it is best to run each marker in a separate lane first to see which may be mixed without confusing the bands (e.g., BSA and Apoferritin).

12. The gel will run approx 5 h. Set the voltage to 50 V if you want to run the gel overnight.

13. For analyzing protein import with two-dimensional gels, it is good to apply an aliquot from the same experiment in Laemmli buffer (SDS buffer) as control on the side of the gel. A control of a SDS denatured sample that contains the protein of interest will make it easier to find the corresponding spots on the two-dimensional gel. It will also facilitate troubleshooting.

References

1. Schägger, H. and von Jagow, G. (1991) Blue native electrophoresis for isolation of membrane protein complexes in enzymatically active form. *Anal. Biochem.* **199**, 223–231.

2. Wiedemann N., Kozjak V., Chacinska A., Schönfisch B., Rospert S., Ryan M.T., Pfanner N., Meisinger C. (2003) Machinery for protein sorting and assembly in the mitochondrial outer membrane. *Nature* **424**, 565–571.

3. Paschen, S. A., Waizenegger, T., Stan, T., et al. (2003) Evolutionary conservation of biogenesis of β-barrel membrane proteins. *Nature* **426**, 862–866.

4. Dekker, P. J. T., Müller, H., Rassow, J., and Pfanner, N. (1996) Characterization of the preprotein translocase of the outer mitochondrial membrane by blue native electrophoresis. *Biol. Chem.* **377**, 535–538.

5. Dekker, P. J. T., Martin, F., Maarse, A. C., et al. (1997) The Tim core complex defines the number of mitochondrial translocation contact sites and can hold arrested preproteins in the absence of matrix Hsp70/Tim44. *EMBO J.* **16,** 5408–5419.

6. Chacinska, A., Rehling, P., Guiard, B., et al. (2003) Mitochondrial translocation contact sites: separation of dynamic and stabilizing elements in formation of a TOM-TIM-preprotein supercomplex. *EMBO J.* **22,** 5370–5381.

7. Jänsch, L., Kruft, V., Schmitz, U. K., and Braun, H.-P. (1998) Unique composition of the preprotein translocase of the outer mitochondrial membrane from plants. *J. Biol. Chem.* **273**, 17251–17257.

8. Schägger, H., Cramer, W. A., and von Jagow, G. (1994) Analysis of molecular masses and oligomeric states of protein complexes by blue native electrophoresis and isolation of membrane protein complexes by two-dimensional native electrophoresis. *Anal. Biochem.* **217**, 220–230.

9. Heuberger, E. H. M. L., Veenhoff, L. M., Duurkens, R. H., Friesen, R. H. E., and Poolman, B. (2002) Oligomeric state of membrane transport proteins analyzed with blue native electrophoresis and analytical ultracentrifugation. *J. Mol. Biol.* **317**, 591–600.

10. Nargang, F. E., Preuss, M., Neupert, W., and Herrmann, J. M. (2002) The Oxa1 protein forms a homooligomeric complex and is an essential part of the mitochondrial export translocase in Neurospora crassa. *J. Biol. Chem.* **277**, 12846–12853.

11. Pfanner, N. and Neupert, W. (1987) Distinct steps in the import of ADP/ATP carrier into mitochondria. *J. Biol. Chem.* **262**, 7528–7536.

12. Ryan, M. T., Müller, H., Pfanner, N. (1999) Functional staging of ADP/ATP carrier translocation across the outer mitochondrial membrane *J. Biol. Chem.* **274**, 20619–20627.

13. Rehling, P., Model, K., Brandner, K., et al. (2003) Protein insertion into the mitochondrial inner membrane by a twin-pore translocase. *Science* **299**, 1747–1751.

14. Reif, S., Voos, W., and Rassow, J. (2001) Intramitochondrial dimerization of citrate synthase characterized by blue native electrophoresis. *Anal. Biochem.* **288**, 97–99.

29

Computational Prediction of Subcellular Localization

Kenta Nakai and Paul Horton

Summary

It is widely recognized that much of the information for determining the final subcellular localization of proteins is found in their amino acid sequences. Thus the prediction of protein localization sites is of both theoretical and practical interest. In most cases, the prediction has been attempted in two ways: one is based on the knowledge of experimentally characterized targeting signals, while the other utilizes the statistical differences of general sequence characteristics, such as amino acid composition, between localization sites. Both approaches have limitations, and it is recommended to check the results of various prediction methods based on different principles as well as training data. Recently, increased proteomic analyses of localization sites have provided new data to assess the current status of predictive methods. In this chapter we discuss these issues and close with an example illustrating the use of the WoLF PSORT web server for localization prediction.

Key Words: Subcellular localization; signal peptide; sequence analysis.

1. Introduction

To understand living systems through their genome sequences is one of the most fundamental challenges in modern biology. For example, the prediction of protein three-dimensional structure from its amino acid sequences has fascinated many researchers who regard it as the challenge to crack the "second genetic code." The problem, however, has turned out to be rather tough, although there have been significant advances (1). On the other hand, projects determining all the three-dimensional folds of proteins, i.e., structural genomics projects, have been launched (2). Thus, quite possibly we will be handed a complete set of solutions of the prediction problem even before we have discovered how to solve it. The prediction of subcellular localization sites of proteins is in a

From: *Methods in Molecular Biology, Vol. 390: Protein Targeting Protocols: Second Edition*
Edited by: M. van der Giezen © Humana Press Inc., Totowa, NJ

similar situation, more or less; we now know the localization sites of most yeast proteins, for example, and it is likely that the use of sequence homology is the most practical way to predict the localization site for most proteins of other organisms. Nevertheless, there are still some good grounds for believing that the localization prediction problem will continue to be one of the major topics in bioinformatics even with the expected wealth of proteome data: first, even after we know all the final localization sites, it is still important to know their sequence determinants or targeting signals that are expected to be embedded in their amino acid sequence themselves; second, we will want to assess the accuracy of our current models of localization mechanisms; third, proteomic determination of subcellular localization sites is not always precise enough. For example, homology-based prediction will be difficult when the target has isoform(s) localized at different sites *(3)*. Therefore, theoretical methods should be used to complement proteomic data. Finally, localization prediction is one of the best defined problems in sequence analysis, because the answer is well defined compared to more general prediction of protein function (and yet the prediction result is still useful for functional characterization) and because the prediction seems to be independent of the difficulty of precise three-dimensional structure prediction in most cases.

How accurately can we predict the presence/absence of targeting signals of proteins and their final subcellular localization from their amino acid sequences? Even apart from the imperfection of current methodology, there seem to be some fundamental limitations: with the progress of molecular cell biology, it is now evident that the protein sorting pathways are much more complex than expected. Thus, examining the presence of, say, the N-terminal mitochondrial targeting signal is not sufficient to detect all proteins targeted to mitochondria *(4)*. In addition, the assumption that every protein contains its targeting signal as part of its amino acid sequence is not always true: some proteins are transported by binding to another protein that has its own signal (hitchhiking/piggybacking mechanism) while at least some proteins seem to require signals in their mRNA for proper localization *(5,6)*. It is also oversimplifying reality to assume that every protein has a single target site because many proteins have multiple localization sites or dynamically change their localization, some of which are rigorously regulated *(7)*. Nevertheless, these limitations emphasize the importance of predictive approaches that could be useful to assess the generality of each piece of knowledge, which is of crucial importance in the postgenomics age.

In this chapter we review recent progress on the prediction of targeting signals as well as subcellular localization in general. As a practical guide for the readers of this series, an example of how to use our latest prediction server is also given. For older, well-established methods, readers may consult previous reviews *(8–14)*.

2. Prediction Methods of Targeting Signals
2.1. Signal Peptides
2.1.1. General Issues

Signal peptides are N-terminal extensions of polypeptides that target them from the cytosol to the cytoplasmic membrane of prokaryotes and to the endoplasmic reticulum (ER) membrane of eukaryotes *(15,16)*. After translocation across the membrane, they are cleaved off in most cases. In eukaryotes, the ER is not always the final localization site of the targeted proteins; they are usually transported through vesicles to various places including the extracellular space (i.e., the proteins may be secreted). In Gram-positive bacteria, having a cleavable signal peptide means that the targeted protein is most likely secreted, while in Gram-negative bacteria the protein may be further targeted into either the periplasm, the outer membrane, or the extracellular space. Basically, both prokaryotic and eukaryotic signal peptides share a common structure and so most prediction programs are applicable to both types, although they use different parameters for each. The most prominent feature of signal peptides is the presence of its central hydrophobic region (called the H-region). In addition, they often show a weak consensus sequence around the cleavage site (von Heijne's -3, -1 rule) *(17)*.

Several reviews on the prediction of signal peptides have been published *(15, 18)*. One of the oldest prediction methods is a so-called weight matrix reflecting the hydrophobic nature of the H-region and the -3, -1 rule *(19)*. Since then, a number of more sophisticated methods have been developed, but their biological basis seems to be unchanged. Probably, the most famous predictor is SignalP by Nielsen et al., which is based on two kinds of artificial neural network (or a combination of neural network and hidden Markov model) *(20)*. Recently, a new version (SignalP 3.0) was released *(21)*. Another new algorithm (PrediSi) employs a weight matrix corrected by considering the amino acid frequency bias present in proteins *(22)*. Both methods are available through the Internet. Because both signal peptides and transmembrane segments (TMSs) are characterized by their hydrophobicity, their distinction is important. In fact, there sometimes exist uncleavable signal peptides in integral membrane proteins and to predict the cleavage of the most N-terminal hydrophobic segment is essential for the prediction of subcellular localization and membrane topology.

It has been proposed that the main difference between signal peptides and TMSs is the degree (length) of their hydrophobicity *(23)*. The charge balance between the left and right side of each region also seems important in light of the positive-inside rule (the observation that the cytosolic side of transmembrane proteins have more positive charges than loops on the noncytosolic side). Recently, a program (Phobius) that combines the prediction of transmembrane topology and signal peptides was developed as an extension of TMHMM, one of the standard programs for topology prediction *(24)*. Phobius is based on a rather complex hidden Markov model (HMM), and the authors report that the false classification rate between signal peptides and TMSs was significantly reduced. In another work, Teasdale's group discriminated sequence features from their hydrophobic regions (amino acid frequency, hydrophobicity, and the start position) *(25)*. These methods need to be verified with more data in the future because there are not enough membrane proteins with experimentally characterized sequence features to be used for training and testing.

As described above, the presence of an N-terminal signal peptide itself does not specify the localization site of a protein. However, if a protein has a (cleavable) signal peptide and does not have any other TMSs, it is very likely to be secreted to the extracellular space in the case of Gram-positive bacterial proteins. Even in the case of eukaryotes, such proteins are good candidates for secreted proteins, because the number of proteins that remain in the lumen of the vesicular pathway (ER, Golgi, lysosome, etc.) seems to be relatively small. Thus, predictions of the secretome (all secreted proteins) from genome data have often been attempted. As a recent example, Chen et al. constructed a putative human set of 5235 proteins and Grimmond et al. identified 2033 candidates of the mouse secretome *(26,27)*. The comparison of such prediction results with proteomic data will be described later. One important issue to be considered is that all secreted proteins do not always have typical signal peptides. Therefore, for more accurate secretome predictions, it is necessary to predict such nonclassical/leaderless secretions. Bendtsen et al. constructed SecretomeP, which predicts such nonclassical secreted proteins as well as classical ones *(28)*. Their idea is to incorporate various general sequence features (amino acid composition, secondary structure, disordered regions, etc.) of the mature protein into an artificial neural network. In version 2.0, SecretomeP become applicable to bacterial sequences as well *(29)*. Although such methods may not be used for elucidating the mechanisms of nonclassical secretion, they can be practically useful for the analysis of novel genome sequence data, which continues to be generated at a rapid rate.

2.1.2. Specific Issues for Bacteria

In the past, the function of all signal peptides was thought to be the same, but recent studies have clarified variations of the theme in bacterial signal peptides, and some were reflected in the development of novel prediction methods *(30)*.

The existence of one variation, lipoprotein signal peptides, has been known for many years, but recently there were some advances in their prediction. It is recognized that the basic structure of signal peptides for bacterial lipoproteins is almost the same as that of ordinary signals except that it contains the lipobox, LA(G/A) | C, where "|" denotes the cleavage site of the signal *(31)*. The cleavage is performed by a specialized enzyme, signal peptidase II. After the cleavage, the cysteine at the new N-terminus is linked to a lipid molecule, enabling the protein to be anchored at the membrane. In 2003, Juncker et al. released a novel HMM-based prediction server, LipoP, for Gram-negative bacterial proteins *(32)*. They also constructed a neural network-based predictor and these two methods gave similar good results, which is not surprising because the prediction of lipoproteins has not been regarded as a tough problem. However, such a method should be effective for screening candidate lipoproteins from a number of genome sequences. Gonnet et al. also developed a prediction method and constructed a curated set of 81 proven and 44 predicted lipoproteins of *Escherichia coli* K-12 by combining various data *(33)*. The prediction of lipoproteins can also be important for medical research because they are often related to pathogenesis processes. Recently, Setubal et al. developed a prediction algorithm, SpLip, for spirochaetes, a distinct group in Gram-negative bacteria *(34)*. Because of their isolated position in phylogeny, the consensus sequence of the lipobox is rather different in these bacteria, and consequently currently available methods are not useful for their prediction. SpLip may be useful for analysis of emerging spirochaetal genome sequence data in the future.

Another variation of typical bacterial signal peptides is the twin-arginine signal peptides *(35)*. Proteins that have this signal are translocated through a specialized pathway, the Tat (twin arginine translocation) pathway. The Tat signal is similar to ordinary bacterial signal peptides but it has the twin-arginine motif immediately upstream of the h-region. The consensus sequence of the motif is (S/T)RRXFLK, where X represents an arbitrary residue. Recently, Bendtsen et al. published a paper on their new predictor, TatP *(36)*. It first filters input sequences with regular expressions, and then an artificial neural network is used to examine the hydrophobic segment. The authors claim that TatP shows better results (fewer false positives) than a rule-based method.

Some other types of bacterial signal peptides have been reported as well. For example, some signal peptides are known to be recognized not by the

SRP (signal recognition particle)-dependent pathway but the SecB-dependent pathway *(37)*. Although it is thought that SRP-dependent signal peptides are more hydrophobic in their h-region, no objective prediction methods have been released so far. Bacteria have developed various types of secretion/export systems to secrete proteins including toxins. Some use part of the ordinary protein translocation system, which is characterized with the use of SecYEG proteins and their signals look similar to ordinary signal peptides. For example, a variant of signal peptides cleaved by another peptidase PilD (XcpA) is known *(38)*. Signals that lead to other types of secretion systems are not well characterized, but at least some of them are predictable with SecretomeP.

2.2. Mitochondrial Targeting Signals

2.2.1. Recent Advances in Cell Biological Studies

Apart from signal peptides, the mitochondrial targeting signal is perhaps the best known sorting signal. In all textbooks on molecular cell biology, its amphiphilic structure is described. Nevertheless, the prediction (detection) of mitochondrial targeting signals from a given set of amino acid sequences is not so easy because their real structure varies extensively. Moreover, a significant number of mitochondrial proteins do not have the typical cleavable presequence on their N-terminus. Thus, it is clearly insufficient to screen the candidates of all nuclear-encoded mitochondrial proteins with the prediction of the typical mitochondrial targeting signal *(4)*. In fact, it turned out that there are multiple translocases in both the outer membrane and the inner membrane *(39,40)*. The famous signal is for translocating through the TOM-TIM complex into the matrix space, and all currently available prediction methods are designed for this type of signal. Thus, the elucidation of novel targeting signals for each specific pathway should be explored. Recently, there have been advances in the understanding of the sorting pathway/signal of outer-membrane proteins. In 2003 Rapoport reviewed the findings on the signals for amino-terminally anchored proteins as well as tail-anchored proteins *(41)*. According to him, the signal for both is contained within the single transmembrane domain and its flanking regions. Namely, the transmembrane domain must be short and moderately hydrophobic, while the flanking regions must be rich in positively-charged residues. More recently, Rapoport and colleagues wrote another review on the biogenesis of β-barrel membrane proteins in the outer membrane of mitochondria *(42)*. Although the sorting signal of β-barrel proteins has not been well characterized, it turns out that they first go into the intermembrane space through the ordinary TOM complex and that they are integrated into the outer membrane with the activity of the TOB complex. Interestingly, this machinery is conserved in Gram-negative bacteria and in chloroplasts. Thus, this knowledge should also be useful for developing a prediction algorithm for outer proteins of Gram-negative bacteria.

2.2.2. Recent Advances in Databases and Predictive Methods

To develop effective prediction algorithms, it is very important to know the complete list of proteins that are known to have a certain signal or the list of proteins known to be localized at a certain site. With recent advances in proteomic studies, various kinds of protein sets from particular organelles have been published. Thus, databases that contain the latest information of various sources for defining the complete members of, for example, mitochondrial proteins in various organisms will become quite useful resources for bioinformaticians. MitoP2 and AMPDB are recent examples of such projects: MitoP2 contains the latest reference list of mitochondrial proteins in yeast, human, and mouse, while AMPDB contains the list of mitochondrial protein in *Arabidopsis thaliana (43–45)*. A recently published list of 31 human mitochondrial proteins that have multiple localizations may also be a resource for developing a novel algorithm *(46)*.

As noted above, accurate prediction of TOM-dependent mitochondrial targeting signals is not easy. As a pioneering attempt, Nakai and Kanehisa constructed a simple predictor of the signal as a component of a more general prediction system, PSORT *(47)*. The component was a linear discriminant function where amino acid composition and some values of the hydrophobic moment were used as variables. But probably the most famous predictor is TargetP, which is based on the artificial neural network technique *(48)*. TargetP tries to detect N-terminal sorting signals, which include signal peptides, mitochondrial targeting signals, and, in the case of plants, chloroplast transit peptides. Recently, a similar program, Predotar, which is also based on the neural network method, was released *(49)*. Other programs that are specifically designed for the prediction of mitochondrial proteins include: MITOPRED, which is based on Pfam domain occurrence patterns, and MitPred, which is primarily based on a support vector machine (SVM) classifier applied to the local frequencies of dipeptides *(50–52)*. Recently, a predicted set of 361 human mitochondrial proteins were derived using an old prediction program Mitoprot, and the sequence homology between human and mouse *(53,54)*. Although the algorithm itself may not be theoretically interesting, their method is probably the most conservative and practical. Thus, future experimental verification of its prediction results will be intriguing to assess the limitation of homology-based methods.

2.3. Chloroplast Transit Peptides

For analyses of amino acid sequences from plants, prediction of chloroplast transit peptides is another big topic. Chloroplasts are the most well-known example of plant-specific organelles called plastids *(55)*. Chloroplasts are relatively complex, containing three types of membranes (the outer membrane, the inner membrane, and the thylakoid membrane), and the signals for their detailed sorting processes are not well characterized. However, a variety of translocators identified from homology with bacteria are now being studied intensively.

Except for outer envelope membrane proteins, many chloroplast proteins seem to have a chloroplast transit peptide as a cleavable N-terminal extension. The signal works as a stromal targeting signal; that is, proteins that carry the signal are not only targeted to chloroplasts, but also translocated through two envelope membranes by default. There are not many prediction programs covering this signal. In the first version of PSORT, a predictor based on the amino acid composition of N-terminal 20-residue segment is included (the second version of PSORT is not applicable to plant sequences) *(47)*. In 1999, Emanuelsson et al. released a novel predictor, ChloroP *(56)*. Like SignalP, ChloroP is based on artificial neural networks that are used to detect the signal and its cleavage site. Later, it was included in a more general predictor, TargetP, which became a standard tool in this field *(48)*. One of the criticisms of the neural network-based method is the difficulty in grasping the sequence determinants that are biologically important for signal recognition. To overcome this difficulty, Schein et al. developed another algorithm, PCLR, which processes the amino acid composition of the N-terminal 20 amino acids with a principal component logistic regression method *(57)*. With a similar motivation, Bannai et al. developed a rule-based prediction method, iPSORT *(58)*. It is questionable, however, whether these methods succeeded in extracting essential sequence features. Indeed, according to Richly and Leister's (relatively simple) assessment, the accuracy of four predictors (iPSORT, PCLR, Predotar, and TargetP) were found to be substantially lower than previously reported *(59)*. The same authors also report that a combination of these methods showed a better result than any one of them alone.

Signals necessary for more precise sorting in chloroplasts are not known except for the one for the thylakoid lumen (or membrane). Like the signal for mitochondrial intermembrane space, the signal for chloroplast thylakoid lumen shows a bipartite structure where a signal peptide-like hydrophobic segment follows the N-terminal transit peptide. The second part is evolutionarily related to bacterial signal peptides, and they share many homologous translocation systems. A primitive implementation of its prediction was attempted in PSORT, but there was very little training data at that time *(47)*. More recently, another new member of neural network-based predictors, LumenP, was reported *(60)*. When used with TargetP, LumenP reaches a significantly better performance than PSORT; mostly because of fewer false positives.

2.4. Nuclear Localization Signals and Nuclear Export Signals

2.4.1. Prediction of Nuclear Localization Signals

The targeting mechanism of nuclear proteins seems to be exceptional in several points. The most notable difference is that proteins are transported into the nucleus through the nuclear pore and thus proteins do not need to be translocated across any membranes. Second, probably because nuclear proteins need to be retargeted after cell division, the nuclear localization signal (NLS) exists in the middle of amino acid sequences and is not cleaved off after the localization. In any case, the prediction of NLSs is probably the most difficult among the various targeting signals. Nevertheless, because there are a large number of proteins in the nucleus, the prediction of nuclear proteins is very important. Several reviews highlight the complexity of nuclear targeting signals and pathways *(61–62)*.

Probably the oldest predictor for nuclear proteins is PSORT *(47)*. It uses both a few rules to extract NLSs and the amino acid composition as input information. In 2000, Cokol et al. developed a prediction method (PredictNLS) *(63)*. Their method uses invented NLSs produced by iterated *in silico* mutagenesis from a set of 91 experimentally verified NLSs. The derived set matched 43% of all nuclear proteins without any false positives. Later, the same group released a database of nuclear localization signals (NLSdb), which is essentially an extension of their previous study based on an enlarged original collection of 114 NLSs *(64)*. More recently, another study on the prediction of NLSs was published by Heddad et al. *(65)*. Their idea is to evolve regular expressions using genetic programming. The accuracy of their program (NucPred) is reported to be similar to other tools.

2.4.2. Prediction of Nuclear Export Signals

It is now evident that there is heavy traffic of proteins between the cytoplasm and the nucleus. Thus, some proteins need to be not only transported into the nucleus but also exported out of it. This export is mediated by a variety of exportins, the most typical being CRM1/Exp1. Among export signals, the leucine-rich nuclear export signal (NES) is well characterized *(66)*. Its consensus sequence is represented as $\Phi\text{-}x_{2-3}\text{-}\Phi\text{-}x_{2-3}\text{-}\Phi\text{-}x\text{-}\Phi$ ($\Phi = $ L, I, V, F,

M; x: arbitrary), although NESs that match the consensus too well seem to be avoided. La Cour et al. constructed a database of leucine-rich NESs (NESbase) and a prediction server (NetNES) based on it *(67,68)*. NetNES combines the results of HMM and artificial neural network matching. They report that not only the hydrophobic residues in the consensus pattern but also some of their intervening residues are important.

2.5. Peroxisomal Targeting Signals

Peroxisomes are ubiquitous organelles enclosed by a single membrane. Like mitochondria and plastids, proteins destined to be localized in peroxisomes are transported from the cytosol (without passing through the ER). In recent years, our understanding of the peroxisomal targeting signals has advanced significantly (although this has also raised more questions, as always), and a number of papers on their prediction were published *(69)*. So far, two kinds of targeting signals are well known: PTS1 and PTS2.

2.5.1. Peroxisomal Targeting Signal 1 (PTS1)

PTS1 is unique in that its consensus sequence (SKL or its relatives) is found in the C terminus of proteins. According to Neuberger et al.'s comparative analysis, at least the 12 C-terminal residues look more or less conserved *(70)*. A refined motif description (PTS1) was implemented in a prediction tool reflecting their taxon-specific difference (metazoa, fungi, et al.) *(71)*. The authors claim that their predictor could distinguish nonperoxisomal protein amino acid sequences, even with an additional SKL motif artificially added to their C-termini. Emanuelsson et al. also performed a systematic search of potential PTS1-bearing peroxisomal proteins using neural networks and (SVMs) vector machines after some preprocessing *(72)*. As a result, it was found that plants have the highest number of predicted peroxisomal proteins. More recently, another group reported a systematic survey of novel peroxisomal proteins with the PTS1 motif in human and rodents *(73)*.

As an interesting application of the prediction, Neuberger et al. confirmed that false-positive patterns found in nonperoxisomal proteins interact with the native receptor, peroxin 5, in a yeast two-hybrid test *(74)*. Some of these proteins were even targeted into peroxisomes in vivo if their original signals are impaired. In fact, such a competition between different signals seems to be explicitly used in the differential sorting between isoforms. Nakao et al. predicted a number of such examples from their analysis of full-length cDNA data *(3)*.

2.5.2. Peroxisomal Targeting Signal 2 (PTS2)

PTS2 is a nine-residue pattern: $(R/K)(L/V/I)X_5(H/Q)(L/A)$, which is found in the N-terminal 20–30 residues. At least some of them are cleaved after targeting. The number of proteins targeted with PTS2 is fewer than those with PTS1, and it has fewer prediction methods as well. Petriv et al. proposed two more elaborate consensus patterns, such as $R(L/V/I/Q)XX(L/V/I/H)(L/S/G/A)X(H/Q)(L/A)$, with a detailed analysis of known examples *(75)*. Reumann also specified both PTS1 and PTS2 of plant proteins *(76)*. The pattern for PTS2 was simple: $R(L/I)X_5HL$. Then he and his colleagues constructed a database of putative *Arabidopsis* peroxisomal proteins (AraPerox) based on motif search as well as additional analysis *(77)*.

2.6. Other Specific Predictions

There remain several other minor signals/predictions for some specific organelles. For a comprehensive review of these up to the year 2000, see Nakai's review *(8)*. Here, only recently published papers are introduced. Prediction of lipid-anchored proteins is also mentioned because this process is closely related to their subcellular localization.

2.6.1. Cell Wall of Gram-Positive Bacteria

Although Gram-negative bacteria are surrounded by only one membrane, they are equipped with a cell wall made mainly of a peptidoglycan matrix for protecting themselves *(78)*. One class of surface proteins covalently bound to the cell wall is characterized by a cell wall-sorting motif called LPXTG, located at the C terminus. This motif is known to be the recognition site of a membrane-bound enzyme called sortase. Sortase cleaves the position between T and G of the motif and covalently attaches the threonine to the peptido-glycan. Boekhorst et al. collected known substrates of sortase and put them in a database (sortase substrate database or LPxTG-DB) *(79)*. They further derived HMMs for various organisms and combined them with other information into a prediction system. Consequently, they identified 732 putative substrates in 49 genomes.

2.6.2. Cell Wall of Yeasts

For the development of antibiotics, the information of potential cell wall proteins is useful. For this purpose, systematic predictions of proteins linked to GPI have been performed. Recently, Terashima et al. tried another approach:

they regarded the proteins predicted to be secreted with PSORT II to be candidates for cell wall proteins *(80)*. Of the 51 candidates novel cell wall proteins identified in *Saccharomyces cerevisiae*, 11 were proved to be associated with the cell wall.

2.6.3. Golgi Apparatus

Proteins that have N-terminal signal peptides are transported to various sites through the vesicular pathway. But their subsequent localization signals are not well understood except for the lumenal and some membrane proteins in the ER. Because the signals for targeting proteins to the Golgi apparatus, which also belongs to the secretory pathway, are still uncharacterized, empirical approaches have been attempted for the prediction of Golgi proteins *(81)*. For example, to predict type II membrane proteins of the Golgi, Yuan and Teasdale developed a linear discriminant function taking the hydrophobicity value and frequencies of different residues within transmembrane domains as variables *(82)*. Their Golgi predictor server is accessible through the Internet.

2.6.4. Lipid Anchors

Systematic prediction of lipid anchors was first included as a component of PSORT *(47)*. Three types of modifications were adopted: N-myristoylation, glycosylphosphatidylinositol (GPI) lipid anchoring, and farnesylation. Recently, the latter two modifications have been extensively studied by Eisenhaber's group *(83)*. For both modifications, the authors first collected a dataset of amino acid sequences from the literature and then analyzed the conservation as well as differences between taxons *(84,85)*. Next, taxon-specific prediction servers were constructed *(86)*. The NMT server for the prediction of glycine N-myristoylation sites is applicable to higher eukaryotes (including their viruses) and fungi, while the big-Π server for the prediction of the C-terminal signal for GPI lipid anchor attachment is applicable to animals, plants, and fungi *(87–89)*.

3. Localization Prediction Methods

Here we switch from discussion of individual localization signals to general, i.e., relatively comprehensive, localization predictors. We roughly categorize the information used by most prediction methods followed by a brief description of some common classification methods employed and a discussion of causality issues for the various types of information. We then tabulate some of the localization prediction resources available and describe the user experience of one particular tool, WoLF PSORT.

An impressive number and variety of methods has been applied to localization prediction. Indeed, the literature on localization prediction supplies a nearly comprehensive review of the classification methods applied to sequence analysis problems in general. Many of these methods have been reviewed elsewhere *(13,14)*.

The information used for prediction can generally be summarized into five types: (generalized) amino acid content, localization sequence signal detectors (signal and transit peptide prediction), sequence similarity, domain signatures, and nonsequence information. The prediction of localization signals has been treated in detail in the previous section. Here we briefly discussion prediction based on the other types of information.

3.1. Amino Acid Content

The correlation with amino acid was noted as early as 1982 *(90)*. In a statistical analysis of globular proteins localizing to the cytosol, nucleus, or extracellular space, Andrade et al. *(91)* found that the amino acid bias is almost completely a result of surface residues. They concluded that the bias is most likely a result of an evolutionary advantage obtained by proteins whose surface composition is well suited to the chemical environment found in their respective localization sites. For example, ionic residues are scarce in proteins exported to the extracellular space, with its relatively low ionic concentration, whereas positively charged residues are more common in proteins found in the nucleus, whose DNA gives it a high anion concentration. In the last few years, a large number of predictors using machine learning techniques, especially SVMs, which rely on amino acid content or simple generalizations such as dipeptide content, have been developed *(92–104)*. Many other works also use amino acid content in combination with other features *(105,106)* or overall physicochemical properties (such as number of positively charged residues), which are determined by amino acid content *(28)*.

The dipeptide or higher-order *n*-peptide features used in some works have the potential to leverage secondary structure, surface accessibility, and some information from sorting signals. However, if applied without caution in datasets containing pairs of similar sequences, these higher-order features may also encode information that would be much more naturally represented with standard sequence similarity.

3.2. Signal and Transit Peptide Prediction

The prediction of these localization signals has been treated in detail in the previous section.

3.3. Sequence Similarity

Orthologous proteins typically have similar sequences and similar localization patterns. Thus, high sequence similarity correlates well with localization site. Significant but low sequence similarity, indicating distant homology, does not necessarily imply co-localization. This effect has been quantitatively studied by Nair and Rost *(107)*. In any case, sequence similarity is used as a feature in many classification schemes *(102–105,107)*. Horton et al. found that predicting the localization site of the most similar sequence by BLAST e-value is sufficient to achieve 94% accuracy on a Swiss-Prot annotation derived set of 12,771 animal proteins. The simple strategy of using WoLF PSORT instead of sequence similarity for hits with low e-value reduced the error rate from 6 to 4.4%. It must be stressed, however, that the dataset was highly redundant, including many well-conserved orthologous sequences. Thus, a high degree of sequence similarity to a protein with known localization is a reliable predictor. The main drawback of this method is that the proteins for which there is the greatest need for more annotation are precisely the proteins that lack well-characterized highly similar orthologs. Also, it is known that a single amino acid substitution in a localization signal can change the localization of a protein. Thus, sequence similarity is a largely noncausal feature in terms of localization and would have to be used with caution when applied to artificial or nonnative sequences (especially between very distant organisms such as bacterial and mammalian).

3.4. Domain Signatures

Different organelles house proteins with different functions and therefore tend to contain different functional and structural motifs. Several works have used protein motifs, such as PROSITE or Pfam patterns, as features for localization prediction *(105,108–110)*. The use of PSI-BLAST-derived PSSMs *(111)* captures local sequence similarity information that may correspond to domain signatures in some cases.

3.5. Non-Sequence-Derived Features

A variety of non-sequence-derived features have been used. One of them is structural information of the folded protein. Proteins are localized to the nuclear in a folded state, and NLSs *(112)* are required to be surface accessible to function. Also, the general correlation of amino acid content and localization site has been shown to be largely a result of the biased composition of surface residues *(91)*. LOC3D *(113)* is an example of a method that uses structural information.

Although the most careful studies, e.g., SignalP *(20)*, have included extensive hand curation by the developers, localization datasets have typically been constructed based on the SWISS-PROT *(114)* subcellular localization field annotation. However, SWISS-PROT contains much localization-related annotation in other fields. For example, the keyword "transit peptide" in an animal would indicate a mitochondrial protein. Several methods use text information for prediction *(99,115–117)*. Some other non-sequence-derived features that have been used in automated classifiers include fluorescence microscopy images *(118,119)* and mRNA expression data *(120)*.

3.6. Typical Classification Methods

3.6.1. Support Vector Machines

In previous sections, neural network and HMM classifiers have been briefly mentioned in the context of predicting specific localization signals. Neural networks have also been used by general localization predictors, but more often SVMs have been used *(96,98,100–105,108,121)*. Support vector machines *(122–124)* are linear binary classifiers whose learning computation involves the dot product of pairs of training set example feature vectors. When the dot product is replaced with a suitable function, known as a kernel function, SVMs can learn nonlinear decision boundaries by implicitly mapping the feature space onto a higher dimensional space. An example of nonlinear mapping that allows for linear separation of originally nonlinearly separable data is shown in **Fig. 1**. Another important characteristic of the SVM classifier is that it is a so-called large margin classifier, which maximize the minimum distance from a point to the decision boundary rather than, for example, the mean distance, as illustrated in **Fig. 2**. In some sense this learning technique focuses on separating the difficult-to-classify examples (which hopefully are not difficult simply because

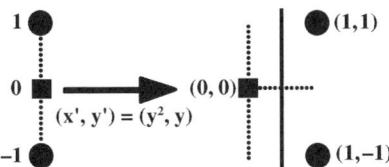

Fig. 1. An example of obtaining linear separability through a nonlinear mapping to higher dimensional space is shown. The left-hand side shows the original one-dimensional feature space, in which the square cannot be separated from the circles with one line. On the right-hand side, the feature values have been mapped to two dimensions in a nonlinear fashion and the two shapes are easily separated.

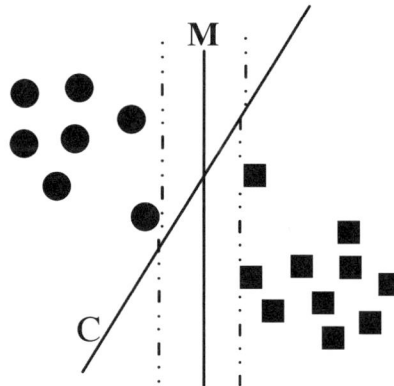

Fig. 2. Two potential linear decision boundaries are shown separating the circles and the squares. The boundary "C" is optimized to be far away from the average point in each class. The large-margin boundary marked "M" instead is optimized to maximize the distance between the boundary is its closest point. This distance, known as the margin, is the distance between "M" and the two parallel dashed lines.

they are incorrectly labeled). The work on localization prediction involving SVMs has mainly used one of two standard software packages: SVM-light *(125)* or LIBSVM *(126)*. Almost all studies have used the radial basis function (RBF) kernel function, which seems to work better than other common kernel functions for this problem. RBF has two adjustable parameters, C and γ, but it has generally been reported that the accuracy is not overly sensitive to the values used for those parameters. One more detail is the method for combining binary classifying SVMs into a multiclass localization site classifier. This is often *(see,* e.g., **refs.** *100,121*) done with the one-vs-rest method, in which one binary classifier is learned for each site, e.g., one nuclear vs nonnuclear classifier, one mitochondria vs nonmitochondrial classifier, etc., is learned. A protein is predicted to belong to the class the associated SVM of which gives the highest scoring positive prediction.

3.6.2. Nearest Neighbor Classifiers

The k nearest neighbors (kNN) *(127)* classifier has been used in several *(107,116,128,129)* general localization predictors. kNN requires defining a similarity measure between any two protein sequences and predicts a query protein based on the k most similar proteins in the dataset. This is more or less the same as what is done with a sequence similarity search, and thus homology-based predictions methods could also be considered to be nearest neighbor classifiers.

3.6.3. Hierarchical Classifiers Reflecting Sorting Pathways

A few prediction methods use structured classifiers designed to reflect sorting pathways, such as the secretory pathway. To our knowledge, the earliest of these were the PSORT expert system *(47,130)* and a subsequent study using machine learning *(131)*. Because of the amount of data or particular methodology applied, the accuracy obtained was inferior to a nonstructured classifier using the simple *k*NN method *(128)*. Roughly 10 years later, the developers of LOCtree *(132)* found that the use of a protein-sorting savvy decision tree structure improved accuracy. The amount of structure used in each of these attempts is modest: the main information encoded in the decision trees were the secretory pathway and the fact that nuclear, mitochondrial, and chloroplastic proteins are imported directly from the cytosol without transit through any other organelle. Still, we feel this approach has an appeal as a first step towards more realistic modeling of the localization process.

3.7. Correlation and Causality

An important issue in considering prediction schemes for localization is the distinction between sequence features that reflect causal influences on localization vs those that merely reflect correlation with localization due to evolutionary fitness constraints common to proteins that localize to particular sites. In this section we use a particular example to discuss causality in the context of localization features. **Figure 3** illustrates some actual and potential causal influences involving NLSs and DNA-binding motifs found in nuclear proteins.

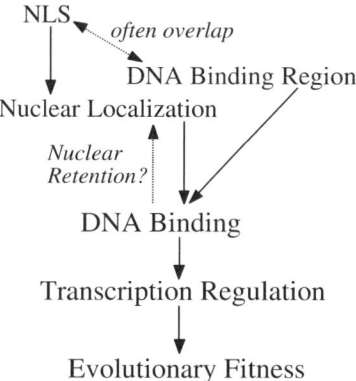

Fig. 3. A schematic diagram illustrating some causal influences is shown. NLS, nuclear localization signal. Many omitted variables also exert causal influence on function and fitness.

NLSs *(112)* in proteins cause them to be selectively imported into the nucleus by importins. Indeed, a classic study by Goldfarb and colleagues *(133)* showed that not only is nuclear localization impaired by mutations in the NLS, but also that nonnuclear proteins are imported into the nucleus when modified to include artificial NLSs. Thus, the presence of an NLS has a causal relationship with nuclear localization and naturally correlates with nuclear localization.

The vast majority of the DNA in a eukaryotic cell is found in the nucleus. Thus, proteins whose function is to interact with DNA are generally imported to the nucleus, and therefore DNA-binding motifs such as the zinc finger-binding motif *(134)* also correlate with nuclear localization. There is, however, a fundamental difference between these two correlations. The zinc finger-binding motif does not necessarily exert causal influence on nuclear localization; for example, Mingot et al. *(135)* created a mutant form of a nuclear protein in which DNA binding was abolished but nuclear localization was retained.

The above discussion and **Fig. 3** (without the dashed arrows) are taken from an earlier review of ours *(13)*. Upon further reflection, however, we noted that LaCasse and Lefebvre *(136)* observed (and Cokol et al. *[63]* further confirmed) that the DNA-binding region of nuclear proteins often overlap their NLSs. In this case a DNA-binding region may exert a causal influence on nuclear localization by acting as part of the NLS. LaCasse and Lefebvre *(136)* also discussed the possibility that DNA binding may be important for nuclear retention. Lim and Li *(137)* show that a single basic region (PKAAR) of the human DNA repair protein O^6-methylguanine–DNA methyltransferase can bind to DNA and is necessary for nuclear retention and stable nuclear localization.

Thus, the separation of causal and noncausal sequence features is not always simple. Still signal sequence-based methods generally seem more likely to capture causal features, while global sequence features such as amino acid content, or functional sequence motifs, are more likely to correlate to localization site only indirectly through evolutionary selection to adapt to the chemical environment of, or perform related functions at a specific localization site.

It should be kept in mind that noncausal correlations may not be robust when applied to mutant or nonnative proteins. It is known that a single amino acid substitution at a particularly important position can destroy the function of a signal peptide or NLS. Since a single amino acid substitution has almost no affect on either overall amino acid composition, sequence similarity scores, or the presence of functional domain signatures, the only methods that have any hope of giving the correct answer in this case are localization signal-based methods.

On the other hand, naturally occurring proteins have co-evolved to have consistency between their localization signals and functional or structural constraints reflected in amino acid composition. Indeed information based on such global or distributed sequence characteristics is clearly useful in this case, especially for genome annotation projects in which the first exons of predicted protein products are often incorrect.

4. Localization Prediction Tools

4.1. Tool Sorting

Dozens of programs covering various aspects of protein subcellular localization prediction are available. We do not attempt to cover them here. Instead, in this section we briefly introduce some of the better known tools with relatively long histories of sustained development.

Like the sequences of localization signals themselves, the names of localization prediction methods are often similar in form but not necessarily similar in function. The three most common tool name motifs are the "∗P" tools from the Danish Technical Institute, the "LOC∗" tools from the Rost group in Columbia University, and the "∗PSORT∗" tools originating from Prof. Nakai's work, but with development now dispersed amongst three institutes. These tools are briefly summarized in **Table 1**. Of course, many important tools have been developed from other research groups. Some of them are listed in **Table 2**.

Table 1
Public Localization Prediction Servers That Can Be Clustered Into Groups of Related Tools

Name *(ref.)*	Description
TargetP *(48)*	Neural network. Prediction based on N-terminal sequence region. Eukaryotes.
SignalP *(20,21)*	Neural network and HMM. Predicts the presence and location of signal peptide cleavage sites. Eukaryotes, Gram-positive and Gram-negative bacteria
ChloroP *(56)*	Neural network. Predicts the presence of chloroplast transit peptides (cTP) and the location of potential cTP cleavage sites.
SecretomeP *(28,29)*	Meta-classifier. Neural network. Predicts secreted proteins without reliance on N-terminal signal peptides. Eukaryote, Gram-positive and Gram-negative.

(Continued)

Table 1. (*Continued*)

Name *(ref.)*	Description
LipoP *(32)*	HMM. Prediction of N-terminal localization sites including SPaseII recognition sites. Bacteria.
LOCtree *(132)*	Meta-classifier. SVM decision tree. Uses PredictNLS, LOChom, LOCkey, and LOCnet. Eukaryotic soluble proteins.
LOCtarget *(164)*	Meta-classifier. Uses PredictNLS, LOChom, LOCkey, and LOCnet. Eukaryotic proteins.
LOCthree-dimensional *(113)*	Meta-classifier. Uses PredictNLS, LOChom, LOCkey, and LOCthree-dimensionalini. Eukaryotic proteins with three-dimensional structure.
LOCkey *(116)*	Uses SWISS-PROT keywords. Eukaryotes.
PredictNLS *(63)*	Template based classifier. Predicts nuclear proteins by extrapolating from known nuclear localization signal sequences.
LOChom *(107)*	Sequence similarity based classifier. Eukaryotes.
PSORT features *(47,130)*	Set of localization signal sequence feature detectors; many borrowed some original. Any method using these features is a meta-classifier.
PSORT *(47,130)*	Expert system combining PSORT features with a protein sorting savvy structure.
PSORT II *(128,165)*	k-nearest neighbors (kNN) classifier using PSORT features.
iPSORT *(58)*	Machine learned residue and residue physicochemical patterns of signal peptides.
PSORTb *(105,166)*	Naive Bayes meta-clasifier. Combining sequence similarity, localization correlating motifs, SVM predictions. Bacteria.
WoLF PSORT *(106)*	kNN with feature selection. Extension of PSORT II using the PSORT features, some iPSORT features, and amino acid content. Eukaryotes.

SVM, support vector machine; HMM, hidden Markov model. Meta-classifiers are methods that use the output of other prediction programs as input.

4.2. Which Sites Can Be Predicted?

The gene ontology (GO) *(138)* subcellular compartment ontology provides a fine-grained definition of organellar and sub-organellar (such as the thylokoid in chloroplasts or the nucleolus in the nucleus) sites. The signals involved in

Table 2
Public Localization Prediction Servers

Name *(ref.)*	Description
PLOC *(100)*	SVM using amino acid and fixed-length gapped pair content features. Eukaryotes.
SubLoc *(96)*	SVM classifier. Amino acid content features.
CELLO *(101)*	SVM classifier. Uses n-peptide composition. Eukaryotic and bacterial.
Predotar *(49)*	Neural network. Predicts N-terminal localization signals. Plants.
PSLpred *(103)*	SVM classifier. Uses generalized amino acid content and PSI-BLAST features. Gram-negative bacteria.
HSLpred *(104)*	SVM classifier. Uses generalized amino acid content and PSI-BLAST features. Human.
ESLpred *(102)*	SVM classifier. Uses generalized amino acid content and PSI-BLAST features. Eukaryotes.
LOCSVMPSI *(111)*	SVM classier using PSI-BLAST derived features. Eukaryotes.
Proteome Analyst *(117)*	Naive Bayes with feature selection. Uses Swiss-Prot keywords of similar sequences.

SVM, support vector machine.

localization to some of these sites have been described in previous sections. However, current tools do not provide accurate prediction at the level of granularity of the most specific GO cellular components. Instead, most "general" predictors for eukaryotic proteins divide the cell into three to six compartments. A typical example would be: secreted, nuclear, mitochondrial, chloroplastic (when applicable), and cytoplasmic. Many methods do not distinguish between the ER, the Golgi body, peroxisomes, or vacuoles. Some methods do, such as the *PSORT* programs for eukaryotic predictions and CELLO. However, as least for the *PSORT* programs, the sensitivity for these organelles is very low (well under 50% for novel proteins).

4.3. Using WoLF PSORT

We close this part of the chapter with a description of the information one may obtain from the WoLF PSORT server, using the protein treacle as a running example. We do not claim that WoLF PSORT is the best server available, but it is the one we are the most qualified to describe.

Mutant forms of treacle have been shown to cause a serious disease known as Treacher Collins syndrome (TCS) in humans *(139)*. On the basis of sequence similarity and putative nuclear and nucleolar localization signals, treacle was hypothesized to be a nucleolar phosphoprotein *(140)*. This was supported experimentally for exogenously *(141,142)* and endogenously expressed *(143)* wild-type treacle. In contrast, exogenously expressed C-terminally truncated treacle mutants, mimicking those found in TCS patients, localized to the cytoplasm (in depots near the nuclear envelope) *(142,143)*.

In the UniProt entry for treacle (id TCOF_HUMAN), the subcellular localization fields contains the word "potential," which caused it to be excluded from our dataset. This makes it a reasonable, albeit anecdotal, test for WoLF PSORT. In gene ontology it is now annotated as nucleolar, with a traceable author statement evidence code, and therefore it will be included in our next dataset update.

Figure 4 shows the largely self-explanatory submission page, which simply requires choosing an organism type and submitting one or more sequences in multifasta format. As shown, a single sequence may also be submitted as just a raw sequence. Nonalphabetic characters are ignored, which is useful when cutting and pasting sequences from UniProt (SWISS-PROT *(114)*) entries.

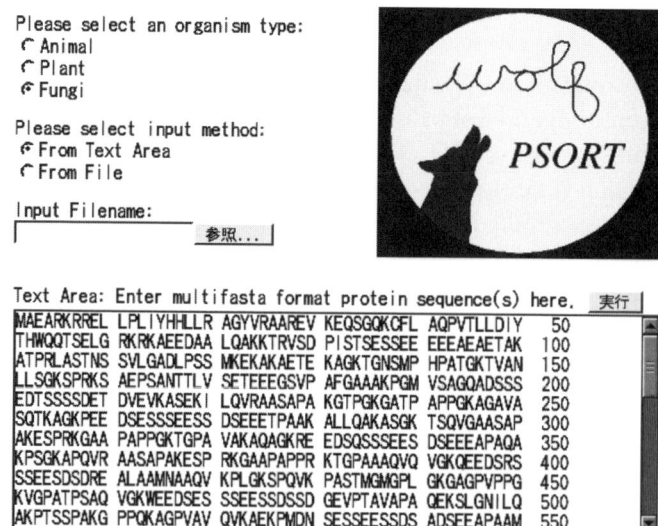

Fig. 4. The sequence submission screen of `wolfpsort.org` is shown. Sequences can also be submitted in multifasta format.

Upon submitting a query a prediction summary page is returned with one line for each query sequence submitted. The line obtained when the treacle sequence is used as a query looks like:

<u>details</u> nucl: 27, cyto_nucl: 16.333, cyto: 3.5, ...

where "nucl" and "cyto" are the abbreviations for nucleus and cytosol, respectively. Abbreviations joined with an underscore such as "cyto_nucl" indicate dual localization. Currently WoLF PSORT makes animal predictions based on the 32 most similar dataset entries. The scores for each localization site are calculated from the localization sites of the nearest neighbors, e.g., in this case the score of "nucl" is calculated as the number of "nucl" neighbors (26) plus one-half the number of "cyto_nucl" neighbors (2). We must mention that the dual localization predictions are still at the experimental stage and should be viewed with strict skepticism. Conversely, most proteins that in fact should be predicted as "cyto_nucl" are currently predicted as either "nucl" or "cyto" (*106*).

Figure 5 shows part of the query result screen, which gives a list of the "most similar" sequences from the, largely UniProt-derived, WoLF PSORT dataset to the query. At this point the user experience is similar to that when using a BLAST server except that the similarity measure is optimized to give proteins with the same localization patterns a high similarity (equivalently small distance). The main practical difference between this similarity score and sequence similarity is that sequence similarity is only a reliable indicator of co-localization for highly

id	site	distance	identity	
ABL1_HUMAN	cyto	1319.27	15.5918%	[Uniprot] SWISS-PROT45:Cytoplasmic.
ENL_HUMAN	nucl	2298.21	11.7647%	[Uniprot] SWISS-PROT45:Nuclear. GO:0005634;
YEMA_DROME	nucl	2632.19	15.7335%	[Uniprot] SWISS-PROT45:Nuclear.
GOA4_MOUSE	cyto_golg	2827.26	13.6729%	[Uniprot] SWISS-PROT45:Cytoplasmic; periphe
CPD1_DROME	nucl	3116.14	8.57548%	[Uniprot] SWISS-PROT45:Nuclear.
FYB_MOUSE	cyto_nucl	3173.5	14.6704%	[Uniprot] SWISS-PROT45:Nuclear and cytoplas
SFRC_RAT	nucl	3218.32	9.92204%	[Uniprot] SWISS-PROT45:Nuclear.
SFRC_HUMAN	nucl	3588.4	10.0638%	[Uniprot] SWISS-PROT45:Nuclear.
NLFE_HUMAN	nucl	3639.37	8.07938%	[Uniprot] SWISS-PROT45:Nuclear. GO:0005634;
TFH1_MOUSE	nucl	3725.68	11.5521%	[Uniprot] SWISS-PROT45:Nuclear.
FMR2_APLCA	extr	3783.06	10.7725%	[Uniprot] SWISS-PROT45:Secreted.
PAP2_XENLA	nucl	3825.43	10.0638%	[Uniprot] SWISS-PROT45:Nuclear.
NLFE_MOUSE	nucl	3910.68	7.79589%	[Uniprot] SWISS-PROT45:Nuclear.
K179_HUMAN	nucl	3941.44	12.8278%	[Uniprot] SWISS-PROT45:Nuclear; nucleolar.

Fig. 5. The top hit list portion of the `wolfpsort.org` detailed result page is shown. Because of space limitations, only part of the list is shown.

similar sequences *(107)*. In contrast, the WoLF PSORT similarity measure has some predictive power even for sequences without strong sequence similarity hits *(106)*. The third column, labeled "distance," is the distance based on our similarity measure. The closest protein, human c-Abl, has a UniProt subcellular annotation of "cytoplasmic" and GO nuclear annotation but with only the weak "nontraceable author statement" evidence code. It is in fact, believed to shuttle between the cytosol and the nucleus *(144)*. Despite the mislabeling as cytosolic in the dataset, the prediction of c-Abl by WoLF PSORT is: `nucl 26, cyto_nucl 16.333, cyto 4.5,` ... which would be: `nucl 26.5, cyto_nucl 16.833, cyto 4.0,` ... if the dataset label of c-Abl were corrected. This illustrates the robustness to annotation error gained by voting among several (in this case 32) good hits, instead of just looking at the single most similar protein. The righthand "comments" column is a catchall, which currently provides some links and a summary of the UniProt and GO localization annotation.

Figure 6 shows a table of localization feature values, found further down the screen on the same detailed result page. As in the previous figure, the rows are the nearest neighbors of the query. The columns represent sequence features used for the similarity measure. In our current version, animal predictions use 22 features, 15 from PSORT, 2 from iPSORT, the amino acid content of C, I, K, S, and the sequence length. The numbers are shown normalized as percentiles relative to the distribution in the dataset. The feature names are not self-explanatory, but a link to a minimal documentation page is provided. In some cases the individual features can provide corroborating evidence. For

id	site	iPSORT		PSORT Features									
		-1_25	MxHy1_30	act	alm	dna	gvh	leu	mNt	mip	mit	myr	nuc
query	nucl?	87	51	50	74	44	48	46	49	26	67	49	99
ABL1_HUMAN	cyto	87	60	50	65	44	37	46	49	26	54	49	97
ENL_HUMAN	nucl	87	15	50	89	44	36	46	49	26	42	49	95
YEMA_DROME	nucl	78	42	50	58	44	41	46	49	65	45	49	97
GOA4_MOUSE	cyto_golg	78	22	50	92	44	45	46	49	26	49	49	85
CPD1_DROME	nucl	93	43	50	85	44	23	46	49	26	25	49	97
FYB_MOUSE	cyto_nucl	64	17	50	92	44	34	46	49	26	16	49	96
SFRC_RAT	nucl	46	64	50	57	44	66	46	49	26	63	49	99
SFRC_HUMAN	nucl	46	59	50	58	44	76	46	49	26	50	49	99
NLFE_HUMAN	nucl	64	25	50	86	44	45	46	49	26	4	49	91

Fig. 6. Part of the details page localization feature table of `wolfpsort.org` is shown. There are actually 22 columns in this display, but some have been truncated for space.

example, the "nucl" feature of treacle has a very high 97 percentile value. The "nucl" feature is designed to detect nuclear localization signals, and therefore a high value is consistent with a nuclear localization. For users familiar with PSORT and PSORT II, another table with the raw (unnormalized) feature values is included along with a link, "PSORT features and traditional PSORT II prediction," which gives the output of PSORT II in verbose mode.

5. Proteomic Verification of Prediction Results

5.1. Eubacteria

Proteomic or systematic studies for determining subcellular localization are very useful for assessing the prediction accuracy as well as for finding novel sorting mechanisms. In bacteria, the main issue is the distinction between excreted proteins (i.e., proteins secreted into the outside medium) and the others remaining within the cell.

In a Gram-negative bacterium, *Bacillus subtilis*, 82 extracellular proteins were identified from about 200 visualized by two-dimensional gel electrophoresis *(145,146)*. Only half of them concurred with the prediction results of cleavable signal peptidase I. The rest include signal peptide-less proteins (even cytoplasmic proteins) and lipoproteins. In their later study, proteins that remained in the cell were also examined *(147)*. A significant number of proteins that are predicted to have a cleavable signal peptide were included. It was also found that some lipoproteins that were originally anchored to the membrane were released by proteolysis. The authors concluded that the accuracy of current signal peptide prediction algorithms is around 50–60% (except LipoP for lipoproteins) and that applying the majority vote method to the output of several prediction methods shows a significant improvement. Because the signal peptides of *B. subtilis* deviate somewhat from typical examples, mainly those obtained from *Escherichia coli*, a simple optimization of currently available algorithms may be quite effective *(148)*.

In another study using a Gram-negative pathogen, *Pseudomonas aeruginosa*, the localization site of 310 proteins predicted to have cleavable/uncleavable/SPase II-dependent signal peptides by PSORTb were tested with a PhoA fusion screen *(149)*. Only 9 of them seem to be cytoplasmic proteins (false-positives), which is an impressive result, but cannot be regarded as the accuracy of *ab initio* prediction because PSORTb (a novel bacterial sequence predictor by Brinkman's group, which uses some PSORT features) incorporates the result of a BLAST search *(105)*. Brinkman and colleagues also performed a systematic comparison of their PSORTb prediction results with 10 available proteomic studies determining the subcellular localization of various bacteria

(150). They estimated the rate of contamination in fractionation in these subproteomic studies and discussed the complementary roles that computational and laboratory approaches can play.

5.2. Eukaryotes

5.2.1. Localisome and Secretome

A number of efforts have been made to systematically determine eukaryotic protein localization. Because of the complexity of such localisomes, the experimental techniques are still under development and a certain degree of error seems to be inevitable. For a recent review, see Warnock et al. *(151)*. In yeast, rather comprehensive studies have been undertaken. For example, Snyder and colleagues determined the localization of 2744 yeast proteins with an epitope-tagging method in 2002 *(152)*. Huh et al. determined the localization of 75% of the yeast proteome with the green fluorescent protein fusion technique *(153)*. In contrast, the discovery of potentially secreted proteins in multicellular organisms still seems to be attempted by somewhat more primitive means: Clark et al. reported their Secreted Protein Discovery Initiative, in which novel human secreted and transmembrane proteins are explored with various methods, including the signal sequence trap method and the prediction of N-terminal signal peptides as well as transmembrane segments applied to expressed sequence tag data *(154)*.

5.2.2. Specific Organelles

It seems that the efforts to collect the subcellular localization of all proteins are particularly active in the research community of *Arabidopsis thaliana*, a model organism of plants. For example, as described above, Millar's group identified 416 mitochondrial proteins in *Arabidopsis* using lipquid chromatography–tandem mass spectrometry *(4,155)*. Only approximately half of them were predicted to be mitochondrial, indicating the limitations of *in silico* approaches in estimating subproteomes. The same group then released a subcellular location database of *Arabidopsis* proteins (SUBA) by integrating the results of predictions, comparative genomic studies, and proteomic experiments *(156)*. SUBA contains the data of more than 4400 proteins. As for proteins located at the mitochondrial outer membrane, a proteome analysis from *Neurospora* identified 30 proteins *(157)*. Van Wilk's group has contributed much to the determination of the chloroplast subproteome in *Arabidopsis*: In 2002 they identified 81 proteins located in the lumen (including peripheral membrane proteins) of the thylakoid and assessed the accuracy of prediction programs *(158)*. The performance on proteins translocated through the twin-arginine pathway was satisfactory. Later, they also identified 154 thylakoid membrane proteins *(159)*. Of these, 49% were α-helical integral membrane proteins. Based on these data, they constructed a plastid proteome database (PPDB). Using the PPDB, sequence features that characterize proteins at each localization site were sought *(160)*. For example, it was found that cysteine was rather less common in thylakoid proteins than those in the inner envelope, which may have implications regarding the role of the thylakoid membrane. Proteome analyses were also applied to the identification of *Arabidopsis* proteins tightly bound to the cell wall as well as its membrane proteins *(161,162)*. As a recent example of a nuclear proteomic study, Hwang et al. identified 1174 putative nuclear proteins from human T leukemia cells. They tried to characterize the change of their expression during apoptosis and used three prediction tools (PredictNLS, NucPred, and PSORT II) to further analyze the genes *(163)*.

6. Conclusion

In this chapter we have reviewed the various subcellular localization signals found in the amino acid sequences of proteins and the methods used to predict them computationally. We have also discussed general prediction methods, which are not closely associated with a particular localization signal. Although our coverage is not comprehensive, we have described the kind of information generally used and a few popular classifier paradigms. In that portion of the chapter we also included a discussion highlighting the issue of causal vs. noncausal correlations. As a practical guide for the reader, we have provided a list of available prediction methods and described the use of one server, WoLF PSORT, by walking through an example query. We closed with a discussion of the likely effects of proteomic scale localization measurements on the localization prediction field and concluded that although the impact of proteomics will be large, localization prediction is likely to remain an important topic for the foreseeable future.

Acknowledgments

This work was partly supported by the National Project on Protein and Functional Analysis and Grant-in-Aid for Scientific Research on Priority Areas from the Ministry of Education, Culture, Sports, Science and Technology of Japan.

References

1. Petrey, D. and Honig, B. (2005) Protein structure prediction: inroads to biology. *Mol.Cell* **20**, 811–819.
2. Chandonia, J. M. and Brenner, S. E. (2006) The impact of structural genomics: expectations and outcomes. *Science* **311**, 347–351.
3. Nakao, M., Barrero, R. A., Mukai, Y., Motono, C., Suwa, M., and Nakai, K. (2005) Large-scale analysis of human alternative protein isoforms: pattern classification and correlation with subcellular localization signals. *Nucleic Acids Res.* **33**, 2355–2363.
4. Heazlewood, J. L., Tonti-Filippini, J. S., Gout, A. M., Day, D. A., Whelan, J., and Millar, A. H. (2004) Experimental analysis of the Arabidopsis mitochondrial proteome highlights signaling and regulatory components, provides assessment of targeting prediction programs, and indicates plant-specific mitochondrial proteins. *Plant Cell* **16**, 241–256.
5. Wu, L. F., Chanal, A., and Rodrigue, A. (2000) Membrane targeting and translocation of bacterial hydrogenases. *Arch. Microbiol.* **173**, 319–324.
6. Margeot, A., Blugeon, C., Sylvestre, J., Vialette, S., Jacq, C., and Corral-Debrinski, M. (2002) In Saccharomyces cerevisiae, ATP2 mRNA sorting to the vicinity of mitochondria is essential for respiratory function. *EMBO J.* **21**, 6893–6904.
7. Muslin, A. J. and Xing, H. (2000) 14-3-3 proteins: regulation of subcellular localization by molecular interference. *Cell Signal* **12**, 703–709.
8. Nakai, K. (2000) Protein sorting signals and prediction of subcellular localization. *Adv. Protein Chem.* **54**, 277–344.
9. Nakai, K. (2001) Review: prediction of in vivo fates of proteins in the era of genomics and proteomics. *J. Struct. Biol.* **134,** 103–116.
10. Emanuelsson, O. and von Heijne, G. (2001) Prediction of organellar targeting signals. *Biochim. Biophys. Acta* **1541**, 114–119.
11. Emanuelsson, O. (2002) Predicting protein subcellular localisation from amino acid sequence information. *Brief Bioinform.* **3**, 361–376.
12. Donnes, P. and Hoglund, A. (2004) Predicting protein subcellular localization: past, present, and future. *Genom. Proteom. Bioinform.* **2**, 209–215.
13. Horton, P., Mukai, Y., and Nakai, K. (2004) Protein subcellular localization prediction, in *Practical Bioinformatician* (Wong, L., ed.), World Scientific Publishing Co., pp. 193–216.
14. Schneider, G. and Fechner, U. (2004) Advances in the prediction of protein targeting signals. *Proteomics* **4**, 1571–1580.
15. Nakai, K. (2002) Signal peptides, in *Cell-Penetrating Peptides: Processes and Applications* (Langel, U., ed.), CRC Press, Boca Raton, FL, pp. 295–324.
16. Halic, M. and Beckmann, R. (2005) The signal recognition particle and its interactions during protein targeting. *Curr. Opin. Struct. Biol.* **15**, 116–125.

17. von Heijne, G. (1983) Patterns of amino acids near signal-sequence cleavage sites. *Eur. J. Biochem.* **133**, 17–21.

18. Chou, K. C. (2002) Prediction of protein signal sequences. *Curr. Protein Pept. Sci.* **3**, 615–622.

19. von Heijne, G. (1986) A new method for predicting signal sequence cleavage sites. *Nucleic Acids Res.* **14**, 4683–4690.

20. Nielsen, H., Engelbrecht, J., Brunak, S., and von Heijne, G. (1997) Identification of prokaryotic and eukaryotic signal peptides and prediction of their cleavage sites. *Protein Eng.* **10**, 1–6.

21. Bendtsen, J. D., Nielsen, H., von Heijne, G., and Brunak, S. (2004) Improved prediction of signal peptides: SignalP 3.0. *J. Mol. Biol.* **340**, 783–795.

22. Hiller, K., Grote, A., Scheer, M., Munch, R., and Jahn, D. (2004) PrediSi: prediction of signal peptides and their cleavage positions. *Nucleic Acids Res.* **32**, W375–379.

23. von Heijne, G. (1998) Life and death of a signal peptide. *Nature 396*, 111, 113.

24. Kall, L., Krogh, A., and Sonnhammer, E. L. (2004) A combined transmembrane topology and signal peptide prediction method. *J. Mol. Biol.* **338**, 1027–1036.

25. Yuan, Z., Davis, M. J., Zhang, F., and Teasdale, R.D. (2003) Computational differentiation of N-terminal signal peptides and transmembrane helices. *Biochem. Biophys. Res.Commun.* **312**, 1278–1283.

26. Chen, Y., Yu, P., Luo, J., and Jiang, Y. (2003) Secreted protein prediction system combining CJ-SPHMM, TMHMM, and PSORT. *Mamm. Genome* **14**, 859–865.

27. Grimmond, S. M., Miranda, K. C., Yuan, Z., Davis, M. J., et al. (2003) The mouse secretome: functional classification of the proteins secreted into the extracellular environment. *Genome Res.*13, 1350–1359.

28. Bendtsen, J. D., Jensen, L. J., Blom, N., Von Heijne, G., and Brunak, S. (2004) Feature-based prediction of non-classical and leaderless protein secretion. *Protein Eng. Des Sel.* **17**, 349–356.

29. Bendtsen, J.D., Kiemer, L., Fausboll, A., and Brunak, S. (2005) Non-classical protein secretion in bacteria. *BMC Microbiol.* **5**, 58.

30. Martoglio, B. and Dobberstein, B. (1998) Signal sequences: more than just greasy peptides. *Trends Cell Biol.* **8**, 410–415.

31. von Heijne, G. (1989) The structure of signal peptides from bacterial lipoproteins. *Protein Eng.* **2**, 531–534.

32. Juncker, A. S., Willenbrock, H., Von Heijne, G., Brunak, S., Nielsen, H., and Krogh, A. (2003) Prediction of lipoprotein signal peptides in Gram-negative bacteria. *Protein Sci.* **12**, 1652–1662.

33. Gonnet, P., Rudd, K. E., and Lisacek, F. (2004) Fine-tuning the prediction of sequences cleaved by signal peptidase II: a curated set of proven and predicted lipoproteins of Escherichia coli K-12. *Proteomics* **4**, 1597–1613.

34. Setubal, J. C., Reis, M., Matsunaga, J., and Haake, D. A. (2006) Lipoprotein computational prediction in spirochaetal genomes. *Microbiol.ogy* **152**, 113–121.

35. Berks, B. C., Palmer, T., and Sargent, F. (2005) Protein targeting by the bacterial twin-arginine translocation (Tat) pathway. *Curr. Opin. Microbiol.* **8**, 174–181.

36. Bendtsen, J. D., Nielsen, H., Widdick, D., Palmer, T., and Brunak, S. (2005) Prediction of twin-arginine signal peptides. *BMC Bioinform.* **6**, 167.

37. de Gier, J. W., and Luirink, J. (2001) Biogenesis of inner membrane proteins in Escherichia coli. *Mol. Microbiol.* **40**, 314–322.

38. Peabody, C. R., Chung, Y. J., Yen, M. R., Vidal-Ingigliardi, D., Pugsley, A. P., and Saier, M. H., Jr. (2003) Type II protein secretion and its relationship to bacterial type IV pili and archaeal flagella. *Microbiology* **149**, 3051–3072.

39. Koehler, C. M. (2004) New developments in mitochondrial assembly. *Annu. Rev. Cell Dev. Biol.* **20**, 309–335.

40. Taylor, R. D. and Pfanner, N. (2004) The protein import and assembly machinery of the mitochondrial outer membrane. *Biochim. Biophys. Acta* **1658**, 37–43.

41. Rapaport, D. (2003) Finding the right organelle. Targeting signals in mitochondrial outer-membrane proteins. *EMBO Rep.* **4**, 948–952.

42. Paschen, S. A., Neupert, W., and Rapaport, D. (2005) Biogenesis of beta-barrel membrane proteins of mitochondria. *Trends Biochem. Sci.* **30**, 575–582.

43. Andreoli, C., Prokisch, H., Hortnagel, K., et al. (2004) MitoP2, an integrated database on mitochondrial proteins in yeast and man. *Nucleic Acids Res.* **32**, D459–462.

44. Prokisch, H., Andreoli, C., Ahting, U., et al. (2006) MitoP2: the mitochondrial proteome database–now including mouse data. *Nucleic Acids Res.* **34**, D705–711.

45. Heazlewood, J. L. and Millar, A. H. (2005) AMPDB: the Arabidopsis Mitochondrial Protein Database. *Nucleic Acids Res.* **33**, D605–610.

46. Mueller, J. C., Andreoli, C., Prokisch, H., and Meitinger, T. (2004) Mechanisms for multiple intracellular localization of human mitochondrial proteins. *Mitochondrion* **3**, 315–325.

47. Nakai, K. and Kanehisa, M. (1992) A knowledge base for predicting protein localization sites in eukaryotic cells. *Genomics* **14**, 897–911.

48. Emanuelsson, O., Nielsen, H., Brunak, S., and von Heijne, G. (2000) Predicting subcellular localization of proteins based on their N-terminal amino acid sequence. *J. Mol. Biol.* **300**, 1005–1016.

49. Small, I., Peeters, N., Legeai, F., and Lurin, C. (2004) Predotar: A tool for rapidly screening proteomes for N-terminal targeting sequences. *Proteomics* **4**, 1581–1590.

50. Guda, C., Fahy, E., and Subramaniam, S. (2004) MITOPRED: a genome-scale method for prediction of nucleus-encoded mitochondrial proteins. *Bioinformatics* **20**, 1785–1794.

51. Guda, C., Guda, P., Fahy, E., and Subramaniam, S. (2004) MITOPRED: a web server for the prediction of mitochondrial proteins. *Nucleic Acids Res.* **32**, W372–374.

52. Kumar, M., Verma, R., and Raghava, G.P. (2006) Prediction of mitochondrial proteins using support vector machine and hidden markov model. *J. Biol. Chem.* **281**, 5357–5363.

53. Claros, M. G. and Vincens, P. (1996) Computational method to predict mitochondrially imported proteins and their targeting sequences. *Eur. J. Biochem.* **241**, 779–786.

54. Cameron, J. M., Hurd, T., and Robinson, B. H. (2005) Computational identification of human mitochondrial proteins based on homology to yeast mitochondrially targeted proteins. *Bioinformatics* **21**, 1825–1830.

55. Reumann, S., Inoue, K., and Keegstra, K. (2005) Evolution of the general protein import pathway of plastids (review). *Mol. Membr. Biol.* **22**, 73–86.

56. Emanuelsson, O., Nielsen, H., and von Heijne, G. (1999) ChloroP, a neural network-based method for predicting chloroplast transit peptides and their cleavage sites. *Protein Sci.* **8**, 978–984.

57. Schein, A. I ., Kissinger, J. C., and Ungar, L. H. (2001) Chloroplast transit peptide prediction: a peek inside the black box. *Nucleic Acids Res.* **29**, E82.

58. Bannai, H., Tamada, Y., Maruyama, O., Nakai, K., and Miyano, S. (2002) Extensive feature detection of N-terminal protein sorting signals. *Bioinformatics* **18**, 298–305.

59. Richly, E. and Leister, D. (2004) An improved prediction of chloroplast proteins reveals diversities and commonalities in the chloroplast proteomes of Arabidopsis and rice. *Gene* **329**, 11–16.

60. Westerlund, I., Von Heijne, G., and Emanuelsson, O. (2003) LumenP–a neural network predictor for protein localization in the thylakoid lumen. *Protein Sci.* **12**, 2360–2366.

61. Christophe, D., Christophe-Hobertus, C., and Pichon, B. (2000) Nuclear targeting of proteins: how many different signals? *Cell Signal* **12**, 337–341.

62. Pemberton, L. F. and Paschal, B. M. (2005) Mechanisms of receptor-mediated nuclear import and nuclear export. *Traffic* **6**, 187–198.

63. Cokol, M., Nair, R., and Rost, B. (2000) Finding nuclear localization signals. *EMBO Rep.* **1**, 411–415.

64. Nair, R., Carter, P., and Rost, B. (2003) NLSdb: database of nuclear localization signals. *Nucleic Acids Res.* **31**, 397–399.

65. Heddad, A., Brameler, M., and MacCallum, R. M. (2004) Evolving regular expression-based sequence classifiers for protein nuclear localisation. *Lecture Notes Computer Sci.* **3005**, 31–40.

66. Kutay, U. and Guttinger, S. (2005) Leucine-rich nuclear-export signals: born to be weak. *Trends Cell Biol.* **15**, 121–124.

67. la Cour, T., Gupta, R., Rapacki, K., Skriver, K., Poulsen, F. M., and Brunak, S. (2003) NESbase version 1.0: a database of nuclear export signals. Nucleic Acids Res. **31**, 393–396.

68. la Cour, T., Kiemer, L., Molgaard, A., Gupta, R., Skriver, K., and Brunak, S. (2004) Analysis and prediction of leucine-rich nuclear export signals. *Protein Eng. Des. Sel.* **17**, 527–536.
69. Baker, A. and Sparkes, I.A. (2005) Peroxisome protein import: some answers, more questions. *Curr. Opin. Plant Biol.* **8**, 640–647.
70. Neuberger, G., Maurer-Stroh, S., Eisenhaber, B., Hartig, A., and Eisenhaber, F. (2003) Motif refinement of the peroxisomal targeting signal 1 and evaluation of taxon-specific differences. *J. Mol. Biol.* **328**, 567–579.
71. Neuberger, G., Maurer-Stroh, S., Eisenhaber, B., Hartig, A., and Eisenhaber, F. (2003) Prediction of peroxisomal targeting signal 1 containing proteins from amino acid sequence. *J. Mol. Biol.* **328**, 581–592.
72. Emanuelsson, O., Elofsson, A., von Heijne, G., and Cristobal, S. (2003) In silico prediction of the peroxisomal proteome in fungi, plants and animals. *J. Mol. Biol.* **330**, 443–456.
73. Kurochkin, I. V., Nagashima, T., Konagaya, A., and Schonbach, C. (2005) Sequence-based discovery of the human and rodent peroxisomal proteome. *Appl. Bioinform.* **4**, 93–104.
74. Neuberger, G., Kunze, M., Eisenhaber, F., Berger, J., Hartig, A., and Brocard, C. (2004) Hidden localization motifs: naturally occurring peroxisomal targeting signals in non-peroxisomal proteins. *Genome Biol.* **5**, R97.
75. Petriv, O. I., Tang, L., Titorenko, V. I., and Rachubinski, R. A. (2004) A new definition for the consensus sequence of the peroxisome targeting signal type 2. *J. Mol. Biol.* **341**, 119–134.
76. Reumann, S. (2004) Specification of the peroxisome targeting signals type 1 and type 2 of plant peroxisomes by bioinformatics analyses. *Plant Physiol.* **135**, 783–800.
77. Reumann, S., Ma, C., Lemke, S., and Babujee, L. (2004) AraPerox. A database of putative Arabidopsis proteins from plant peroxisomes. *Plant Physiol.* **136**, 2587–2608.
78. Ton-That, H., Marraffini, L. A., and Schneewind, O. (2004) Protein sorting to the cell wall envelope of Gram-positive bacteria. *Biochim. Biophys. Acta* **1694**, 269–278.
79. Boekhorst, J., de Been, M. W., Kleerebezem, M., and Siezen, R .J. (2005) Genome-wide detection and analysis of cell wall-bound proteins with LPxTG-like sorting motifs. *J. Bacteriol.* **187**, 4928–4934.
80. Terashima, H., Fukuchi, S., Nakai, K., et al. (2002) Sequence-based approach for identification of cell wall proteins in Saccharomyces cerevisiae. *Curr. Genet.* **40**, 311–316.
81. Rodriguez-Boulan, E., and Musch, A. (2005) Protein sorting in the Golgi complex: shifting paradigms. *Biochim. Biophys. Acta* **1744**, 455–464.
82. Yuan, Z. and Teasdale, R.D. (2002) Prediction of Golgi Type II membrane proteins based on their transmembrane domains. *Bioinformatics* **18**, 1109–1115.

83. Eisenhaber, B., Eisenhaber, F., Maurer-Stroh, S., and Neuberger, G. (2004) Prediction of sequence signals for lipid post-translational modifications: insights from case studies. *Proteomics* **4**, 1614–1625.

84. Maurer-Stroh, S., Eisenhaber, B., and Eisenhaber, F. (2002) N-terminal N-myristoylation of proteins: refinement of the sequence motif and its taxon-specific differences. *J. Mol. Biol.* **317**, 523–540.

85. Eisenhaber, B., Maurer-Stroh, S., Novatchkova, M., Schneider, G., and Eisenhaber, F. (2003) Enzymes and auxiliary factors for GPI lipid anchor biosynthesis and post-translational transfer to proteins. *Bioessays* **25**, 367–385.

86. Maurer-Stroh, S., Eisenhaber, B., and Eisenhaber, F. (2002) N-Terminal N-myristoylation of proteins: prediction of substrate proteins from amino acid sequence. *J. Mol. Biol.* **317**, 541–557.

87. Eisenhaber, F., Eisenhaber, B., Kubina, W., et al. (2003) Prediction of lipid posttranslational modifications and localization signals from protein sequences: big-Pi, NMT and PTS1. *Nucleic Acids Res.* **31**, 3631–3634.

88. Eisenhaber, B., Wildpaner, M., Schultz, C. J., Borner, G. H., Dupree, P., and Eisenhaber, F. (2003) Glycosylphosphatidylinositol lipid anchoring of plant proteins. Sensitive prediction from sequence- and genome-wide studies for Arabidopsis and rice. *Plant Physiol.* **133**, 1691–1701.

89. Eisenhaber, B., Schneider, G., Wildpaner, M., and Eisenhaber, F. (2004) A sensitive predictor for potential GPI lipid modification sites in fungal protein sequences and its application to genome-wide studies for Aspergillus nidulans, Candida albicans, Neurospora crassa, Saccharomyces cerevisiae and Schizosaccharomyces pombe. *J. Mol. Biol.* **337**, 243–253.

90. Nishikawa, K. and Ooi, T. (1982) Correlation of the amino acid composition of a protein to its structural and biological characters. *J. Biochem. (Tokyo)* **91**, 1821–1824.

91. Andrade, M. A., O'Donoghue, S. I., and Rost, B. (1998) Adaptation of protein surfaces to subcellular location. *J. Mol. Biol.* **276**, 517–525.

92. Nakashima, H. and Nishikawa, K. (1994) Discrimination of intracellular and extracellular proteins using amino acid composition and residue-pair frequencies. *J. Mol. Biol.* **238**, 54–61.

93. Cedano, J., Aloy, P., Perez-Pons, J. A., and Querol, E. (1997) Relation between amino acid composition and cellular location of proteins. *J. Mol. Biol.* **266**, 594–600.

94. Reinhardt, A. and Hubbard, T. (1998) Using neural networks for prediction of the subcellular location of proteins. *Nucleic Acids Res.* **26**, 2230–2236.

95. Yuan, Z. (1999) Prediction of protein subcellular locations using Markov chain models. *FEBS Lett.* **451**, 23–26.

96. Hua, S. and Sun, Z. (2001) Support vector machine approach for protein subcellular localization prediction. *Bioinformatics* **17**, 721–728.

97. Feng, Z. P. and Zhang, C. T. (2001) Prediction of the subcellular location of prokaryotic proteins based on the hydrophobicity index of amino acids. *Int. J. Biol. MacroMol.* **28**, 255–261.

98. Cai, Y. D., Liu, X. J., Xu, X. B., and Chou, K. C. (2000) Support vector machines for prediction of protein subcellular location. *Mol. Cell Biol. Res. Commun.* **4**, 230–233.

99. Stapley, B. J., Kelley, L. A., and Sternberg, M. J. (2002) Predicting the subcellular location of proteins from text using support vector machines. *Pac. Symp. Biocomput.* 374–385.

100. Park, K. J., and Kanehisa, M. (2003) Prediction of protein subcellular locations by support vector machines using compositions of amino acids and amino acid pairs. *Bioinformatics* **19**, 1656–1663.

101. Yu, C. S., Lin, C. J., and Hwang, J. K. (2004) Predicting subcellular localization of proteins for Gram-negative bacteria by support vector machines based on n-peptide compositions. *Protein Sci.* **13**, 1402–1406.

102. Bhasin, M. and Raghava, G.P. (2004) ESLpred: SVM-based method for subcellular localization of eukaryotic proteins using dipeptide composition and PSI-BLAST. *Nucleic Acids Res.* **32**, W414–419.

103. Bhasin, M., Garg, A., and Raghava, G. P. (2005) PSLpred: prediction of subcellular localization of bacterial proteins. *Bioinformatics* **21**, 2522–2524.

104. Garg, A., Bhasin, M., and Raghava, G. P. (2005) Support vector machine-based method for subcellular localization of human proteins using amino acid compositions, their order, and similarity search. *J. Biol. Chem.***280**, 14427–14432.

105. Gardy, J. L., Spencer, C., Wang, K., et al. (2003) PSORT-B: improving protein subcellular localization prediction for Gram-negative bacteria. *Nucleic Acids Res.* **31**, 3613–3617.

106. Horton, P., Park, K. J., Kobayashi, T., and Nakai, K. (2006) Protein subcellular localization prediction with WoLF PSORT, in *4th Asia-Pacific Bioinformatics Conference* (T. Jiang, et al., eds.), Imperial College Press, London, pp. 39–48,

107. Nair, R. and Rost, B. (2002) Sequence conserved for subcellular localization. *Protein Sci.* **11**, 2836–2847.

108. Chou, K. C. and Cai, Y. D. (2002) Using functional domain composition and support vector machines for prediction of protein subcellular location. *J. Biol. Chem.* **277**, 45765–45769.

109. Cai, Y. D. and Chou, K. C. (2003) Nearest neighbour algorithm for predicting protein subcellular location by combining functional domain composition and pseudo-amino acid composition. *Biochem. Biophys. Res.Commun.* 305, 407–411.

110. Guda, C. and Subramaniam, S. (2005) pTARGET (corrected) a new method for predicting protein subcellular localization in eukaryotes. *Bioinformatics* **21**, 3963–3969.

111. Xie, D., Li, A., Wang, M., Fan, Z., and Feng, H. (2005) LOCSVMPSI: a web server for subcellular localization of eukaryotic proteins using SVM and profile of PSI-BLAST. *Nucleic Acids Res.* **33**, W105–110.
112. Gorlich, D. (1997) Nuclear protein import. *Curr. Opin. Cell Biol.* **9**, 412–419.
113. Nair, R. and Rost, B. (2003) LOC3D: annotate sub-cellular localization for protein structures. *Nucleic Acids Res.* **31**, 3337–3340.
114. Boeckmann, B., Bairoch, A., Apweiler, R., et al. (2003) The SWISS-PROT protein knowledgebase and its supplement TrEMBL in 2003. *Nucleic Acids Res.* **31**, 365–370.
115. Eisenhaber, F. and Bork, P. (1999) Evaluation of human-readable annotation in biomolecular sequence databases with biological rule libraries. *Bioinformatics* **15**, 528–535.
116. Nair, R. and Rost, B. (2002) Inferring sub-cellular localization through automated lexical analysis. *Bioinformatics* **18 (Suppl. 1)**, S78–86.
117. Lu, Z., Szafron, D., Greiner, R., et al. (2004) Predicting subcellular localization of proteins using machine-learned classifiers. *Bioinformatics* **20**, 547–556.
118. Murphy, R. F., Boland, M. V., and Velliste, M. (2000) Towards a systematics for protein subcelluar location: quantitative description of protein localization patterns and automated analysis of fluorescence microscope images. *Proc. Int. Conf. Intell. Syst. Mol. Biol.* **8**, 251–259.
119. Boland, M. V. and Murphy, R. F. (2001) A neural network classifier capable of recognizing the patterns of all major subcellular structures in fluorescence microscope images of HeLa cells. *Bioinformatics* **17**, 1213–1223.
120. Drawid, A. and Gerstein, M. (2000) A Bayesian system integrating expression data with sequence patterns for localizing proteins: comprehensive application to the yeast genome. *J. Mol.Biol.* **301**, 1059–1075.
121. Matsuda, S., Vert, J. P., Saigo, H., Ueda, N., Toh, H., and Akutsu, T. (2005) A novel representation of protein sequences for prediction of subcellular location using support vector machines. *Protein Sci.* **14**, 2804–2813.
122. Vapnik, V. (1998) *Statistical Learning Theory*, Wiley-Interscience, New York.
123. Cristianini, N. and Shawe-Taylor, J. (2000) *An Introduction to Support Vector Machines*, Cambridge University Press, Cambridge, UK.
124. Scholkopf, B. and Smola, A. J. (2002) *Learning with Kernels*, MIT Press, Cambridge, MA.
125. Joachims, T. (1999) Making large-scale SVM learning practical, in *Advances in Kernel Methods—Support Vector Learning* (Scholkopf, B., Burges, C., and Smola, A., eds.), MIT Press, Cambridge, MA.
126. Chang, C.-C. and Lin, C.-J. (2001) LIBSVM: a library for support vector machines.
127. Duda, R. O., Hart, P. E., and Stork, D. G. (2000) *Pattern Classification*, 2nd ed., John Wiley & Sons, New York.

128. Horton, P. and Nakai, K. (1997) Better prediction of protein cellular localization sites with the k nearest neighbors classifier. *Proc. Int. Conf. Intell. Syst. Mol.Biol.* **5**, 147–152.

129. Huang, Y. and Li, Y. (2004) Prediction of protein subcellular locations using fuzzy k-NN method. *Bioinformatics* **20**, 21–28.

130. Nakai, K. and Kanehisa, M. (1991) Expert system for predicting protein localization sites in gram-negative bacteria. *Proteins* **11**, 95–110.

131. Horton, P. and Nakai, K. (1996) A probabilistic classification system for predicting the cellular localization sites of proteins. *Proc. Int. Conf. Intell. Syst. Mol. Biol.* **4**, 109–115.

132. Nair, R. and Rost, B. (2005) Mimicking cellular sorting improves prediction of subcellular localization. *J. Mol. Biol.* **348**, 85–100.

133. Goldfarb, D. S., Gariepy, J., Schoolnik, G., and Kornberg, R. D. (1986) Synthetic peptides as nuclear localization signals. *Nature* **322**, 641–644.

134. Klug, A. and Schwabe, J. W. (1995) Protein motifs 5. Zinc fingers. *FASEB J.* **9**, 597–604.

135. Mingot, J. M., Espeso, E. A., Diez, E., and Penalva, M. A. (2001) Ambient pH signaling regulates nuclear localization of the Aspergillus nidulans PacC transcription factor. *Mol. Cell Biol.* **21**, 1688–1699.

136. LaCasse, E. C. and Lefebvre, Y. A. (1995) Nuclear localization signals overlap DNA- or RNA-binding domains in nucleic acid-binding proteins. *Nucleic Acids Res.* **23**, 1647–1656.

137. Lim, A. and Li, B. F. (1996) The nuclear targeting and nuclear retention properties of a human DNA repair protein O6-methylguanine-DNA methyltransferase are both required for its nuclear localization: the possible implications. *EMBO J.* **15**, 4050–4060.

138. Ashburner, M., Ball, C. A., Blake, J. A., et al. (2000) Gene ontology: tool for the unification of biology. The Gene Ontology Consortium. *Nat. Genet.* **25**, 25–29.

139. The Treacher Collins Syndrome Collaborative Group. (1996) Positional cloning of a gene involved in the pathogenesis of Treacher Collins syndrome. *Nat. Genet.* **12**, 130–136.

140. Wise, C. A., Chiang, L. C., Paznekas, W. A., et al. (1997) TCOF1 gene encodes a putative nucleolar phosphoprotein that exhibits mutations in Treacher Collins Syndrome throughout its coding region. *Proc. Natl. Acad. Sci. USA* **94**, 3110–3115.

141. Winokur, S. T. and Shiang, R. (1998) The Treacher Collins syndrome (TCOF1) gene product, treacle, is targeted to the nucleolus by signals in its C-terminus. *Hum. Mol.Genet.* **7**, 1947–1952.

142. Marsh, K. L., Dixon, J., and Dixon, M. J. (1998) Mutations in the Treacher Collins syndrome gene lead to mislocalization of the nucleolar protein treacle. *Hum. Mol.Genet.* **7**, 1795–1800.

143. Isaac, C., Marsh, K. L., Paznekas, W. A., et al. (2000) Characterization of the nucleolar gene product, treacle, in Treacher Collins syndrome. *Mol. Biol. Cell* **11**, 3061–3071.

144. Taagepera, S., McDonald, D., Loeb, J. E., et al. (1998) Nuclear-cytoplasmic shuttling of C-ABL tyrosine kinase. *Proc. Natl. Acad. Sci. USA* **95**, 7457–7462.

145. Antelmann, H., Tjalsma, H., Voigt, B., et al. (2001) A proteomic view on genome-based signal peptide predictions. *Genome Res.***11**, 1484–1502.

146. Tjalsma, H., Antelmann, H., Jongbloed, J. D., et al. (2004) Proteomics of protein secretion by Bacillus subtilis: separating the "secrets" of the secretome. *Microbiol. Mol. Biol. Rev.* **68**, 207–233.

147. Tjalsma, H. and van Dijl, J.M. (2005) Proteomics-based consensus prediction of protein retention in a bacterial membrane. *Proteomics* **5**, 4472–4482.

148. Nakai, K. (1996) Refinement of the prediction methods of signal peptides for the genome analyses of Saccharomyces cerevisiae and Bacillus subtilis, in *Genome Informatics Workshop* (Akutsu, T., et al., eds.), Universal Academy Press, Tokyo, pp. 72–81.

149. Lewenza, S., Gardy, J. L., Brinkman, F. S., and Hancock, R. E. (2005) Genome-wide identification of Pseudomonas aeruginosa exported proteins using a consensus computational strategy combined with a laboratory-based PhoA fusion screen. *Genome Res.* **15**, 321–329.

150. Rey, S., Gardy, J. L., and Brinkman, F. S. (2005) Assessing the precision of high-throughput computational and laboratory approaches for the genome-wide identification of protein subcellular localization in bacteria. *BMC Genomics* **6**, 162.

151. Warnock, D. E., Fahy, E., and Taylor, S. W. (2004) Identification of protein associations in organelles, using mass spectrometry-based proteomics. *Mass Spectrom. Rev.* **23**, 259–280.

152. Kumar, A., Agarwal, S., Heyman, J. A., et al. (2002) Subcellular localization of the yeast proteome. *Genes Dev.* **16**, 707–719.

153. Huh, W. K., Falvo, J. V., Gerke, L. C., et al. (2003) Global analysis of protein localization in budding yeast. *Nature* **425**, 686–691.

154. Clark, H. F., Gurney, A. L., Abaya, E., et al. (2003) The secreted protein discovery initiative (SPDI), a large-scale effort to identify novel human secreted and transmembrane proteins: a bioinformatics assessment. *Genome Res.* **13**, 2265–2270.

155. Millar, A. H., Heazlewood, J. L., Kristensen, B. K., Braun, H. P., and Moller, I. M. (2005) The plant mitochondrial proteome. *Trends Plant Sci.* **10**, 36–43.

156. Heazlewood, J. L., Tonti-Filippini, J., Verboom, R. E., and Millar, A. H. (2005) Combining experimental and predicted datasets for determination of the subcellular location of proteins in Arabidopsis. *Plant Physiol.* **139**, 598–609.

157. Schmitt, S., Prokisch, H., Schlunck, T., et al. (2006) Proteome analysis of mitochondrial outer membrane from Neurospora crassa. *Proteomics* **6**, 72–80.

158. Peltier, J..B., Emanuelsson, O., Kalume, D. E., et al. (2002) Central functions of the lumenal and peripheral thylakoid proteome of Arabidopsis determined by experimentation and genome-wide prediction. *Plant Cell* **14**, 211–236.

159. Friso, G., Giacomelli, L., Ytterberg, A. J., et al. (2004) In-depth analysis of the thylakoid membrane proteome of Arabidopsis thaliana chloroplasts: new proteins, new functions, and a plastid proteome database. *Plant Cell* **16**, 478–499.

160. Sun, Q., Emanuelsson, O., and van Wijk, K. J. (2004) Analysis of curated and predicted plastid subproteomes of Arabidopsis. Subcellular compartmentalization leads to distinctive proteome properties. *Plant Physiol.* **135**, 723–734.

161. Bayer, E. M., Bottrill, A. R., Walshaw, J., et al. (2006) Arabidopsis cell wall proteome defined using multidimensional protein identification technology. *Proteomics* **6**, 301–311.

162. Schwacke, R., Flugge, U. I., and Kunze, R. (2004) Plant membrane proteome databases. *Plant Physiol. Biochem.* **42**, 1023–1034.

163. Hwang, S. I., Lundgren, D. H., Mayya, V., et al. (2006) Systematic characterization of nuclear proteome from human T leukemia cells: a quantitative proteomic study during apoptosis by differential extraction and stable isotope labeling. *Mol. Cell Proteomics.* **5**, 1131–1145.

164. Nair, R. and Rost, B. (2004) LOCnet and LOCtarget: sub-cellular localization for structural genomics targets. *Nucleic Acids Res.* **32**, W517–521.

165. Nakai, K. and Horton, P. (1999) PSORT: a program for detecting sorting signals in proteins and predicting their subcellular localization. *Trends Biochem. Sci.* **24**, 34–36.

166. Gardy, J. L., Laird, M. R., Chen, F., et al. (2005) PSORTb v.2.0: expanded prediction of bacterial protein subcellular localization and insights gained from comparative proteome analysis. *Bioinformatics* **21**, 617–623.

30

Phylogenetic Analysis to Uncover Organellar Origins of Nuclear-Encoded Genes

Bernardo J. Foth

Summary

Most proteins that are located in mitochondria or plastids are encoded by the nuclear genome, because the organellar genomes have undergone severe reduction during evolution. In many cases, although not all, the nuclear genes encoding organelle-targeted proteins actually originated from the respective organellar genome and thus carry the phylogenetic fingerprint that still bespeaks their evolutionary origin. Phylogenetic analysis is a powerful *in silico* method that can yield important insights into the evolutionary history or molecular kinship of any gene or protein and that can thus also be used more specifically in the context of organellar targeting as one means to recognize protein candidates (e.g., from genome data) that may be targeted to mitochondria or plastids. This chapter provides protocols for creating multiple sequence alignments and carrying out phylogenetic analysis with the robust and comprehensive software packages Clustal and PHYLIP, which are both available free of charge for multiple computer platforms. Besides presenting step-by-step instructions on how to run these computer programs, this chapter also covers topics such as data collection and presentation of phylogenetic trees.

Key Words: Phylogenetics; molecular evolution; endosymbiosis; gene transfer; PHYLIP, Clustal.

1. Introduction

Mitochondria and plastids are double membrane-bound organelles that originally derive from endosymbiotic events whereby a eukaryotic cell engulfed and retained a bacterium, and their organellar genomes are the relics of the original bacterial genome of the respective endosymbiont. Over time many organellar genes were transferred to the host cell's nucleus and lost from the organellar genome, whereas others were simply lost. A large number of extant

From: *Methods in Molecular Biology, Vol. 390: Protein Targeting Protocols: Second Edition*
Edited by: M. van der Giezen © Humana Press Inc., Totowa, NJ

mitochondrion- and plastid-resident proteins thus originally derive from the organelle, are encoded in the nuclear genome, translated in the cytoplasm, and targeted posttranslationally back into the organelle (*see*, e.g., **refs. *1,2***).

Researchers who have nuclear-encoded genes or whole nuclear genome sequences at their disposal often want to predict whether a gene product is targeted to the mitochondrion, the plastid, or another subcellular compartment. The most straightforward approach is to try to identify a corresponding targeting signal in the protein sequence, e.g., an N-terminal mitochondrial or plastidic transit peptide *(3)*. In the case of secondarily derived plastids, which are usually situated within the secretory pathway of the cell, organellar targeting is achieved by a combination of a signal and transit peptide *(4)*. Frequently, though, such targeting predictions may be inconclusive or simply not feasible, e.g., when the N-terminus of a protein sequence is uncertain or unknown, in the case of membrane proteins, or when a protein is targeted to the mitochondrial intermembrane space (and not the matrix). The risk of failing to recognize a protein's N-terminal targeting signal from genomic sequence data is aggravated by the fact that such signals usually show very little sequence conservation even between closely related species, that they are sometimes encoded by a separate exon, and that alternative splicing may create multiple gene products. Because of these difficulties—but in general also to corroborate the prediction of a putative targeting sequence—it is often useful to substantiate one's supposition with an independent prediction method.

Because many organelle-resident proteins are encoded by genes that originated from the organellar genome, they carry a corresponding phylogenetic signal. α-Proteobacteria and cyanobacteria represent the groups of extant prokaryotes that are most closely related to the organisms that gave rise to eukaryotic mitochondria and plastids, respectively. Mitochondrion- and plastid-derived protein sequences may therefore often be recognized in phylogenetic analyses because they appear closely related to homologous proteins from α-proteobacteria (e.g., *Rickettsia*) and cyanobacteria (e.g., *Synechocystis*), respectively. For a few select examples, *see* **Fig. 1** and **refs. *5–10***.

One particular example for the application of phylogenetic analysis in the context of organellar targeting is the plastid proteome of the malaria parasite *Plasmodium falciparum* and related apicomplexans. These protists contain a relic, secondary plastid (called "apicoplast") that was only recognized as such *(11–13)* a few years before the first genome of an apicomplexan was sequenced *(14)*. Because the apicoplast is seen as a rich source for potential new drug targets against these important pathogens, one immediate question was whether one could predict the apicoplast proteome from the—at the time—growing nuclear

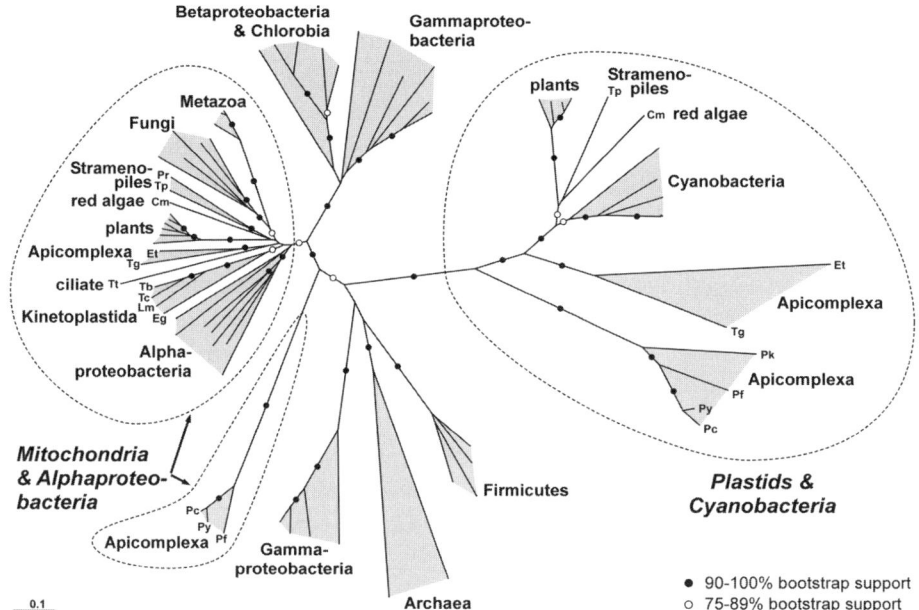

Fig. 1. An unrooted Neighbor-Joining phylogeny of dihydrolipoamide dehydro-genase, the E3 subunit of the pyruvate dehydrogenase, and similar enzyme complexes. The corresponding multiple sequence alignment was created with ClustalX, and both the tree shown as well as corresponding Neighbor-Joining bootstrap analysis (300 pseudosamples) were carried out with PHYLIP's "protdist" (JTT substitution model, no γ approximation) and "neighbor" programs. Plants and other plastid-bearing organisms contain two different E3 proteins, one derived from the plastid and the other from the mitochondrion, whereas organisms lacking a plastid only contain a mitochondrial E3. While the tree shows the expected monophyletic clustering of plastid- and most mitochondrion-derived E3 proteins with cyanobacteria and α-proteobacteria, respec-tively, it also provides an example for sequences that do not cluster as anticipated: the mitochondrial E3 subunits of the genus *Plasmodium* do not form a monophyletic group with the other mitochondrial and α-proteobacterial sequences, a result that is probably artifactual as a result of divergent evolution of the *Plasmodium* sequences. The names of select eukaryotes are indicated as follows: Cm, *Cyanidioschyzon merolae*; Eg, *Euglena gracilis*; Et, *Eimeria tenella*; Lm, *Leishmania major*; Pc, *Plasmodium chabaudi*; Pf, *Plasmodium falciparum*; Pk, *Plasmodium knowlesi*; Pr, *Phytophthora ramorum*; Py, *Plasmodium yoelii*; Tb, *Trypanosoma brucei*; Tc, *Trypanosoma cruzi*; Tg, *Toxoplasma gondii*; Tp, *Thalassiosira pseudonana*; Tt, *Tetrahymena thermophila*. All sequences, except those of Et, were taken from GenBank.

genome data. In these early days, phylogenetic analysis proved to be a crucial aid in assembling first data sets of putative apicoplast proteins, which then served as a springboard to investigate the mechanism of protein targeting to the apicoplast and to predict the proteome of this organelle *(15,16)*. In particular, phylogenetic analyses, the presence of N-terminal targeting signals, and a comparison with the situation in plants suggested an organellar affiliation for many metabolic enzymes and ribosomal proteins, some of which were confirmed experimentally to actually reside in the apicoplast (e.g., **refs. 8,17,18**). In addition to its primary application in evolutionary studies, phylogenetic analysis can thus be used as a versatile and very convenient method to help direct and to play a supporting role for research in other areas, including protein targeting studies.

Finally, a word of caution: despite the fact that the idea of gene transfer from organelle to nucleus and of gene product retargeting is generally accepted and has been confirmed in many individual cases, one must not forget that not all nuclear-encoded organelle-derived genes encode proteins that are also retargeted to that same organelle, and that nuclear-encoded organelle-targeted gene products may derive from any genetic source including the host cell nucleus, another organelle, or even another organism via lateral gene transfer *(2)*. A strong phylogenetic affiliation of a nuclear-encoded protein with α-proteobacteria or cyanobacteria, or the lack thereof, must therefore not be overinterpreted, and—where available—independent evidence that may corroborate or conflict with the phylogenetic results must always be considered.

2. Materials

1. Sequence data: usually amino acid sequences (*see* **Note 1**) that derive from one's own data, from public databases, or—in many cases—a combination of both. Data set size can have a crucial impact on the duration of downstream analyses (*see* **Note 2**).
2. A computer: for many analyses, a "regular" modern desktop or notebook computer is sufficient. Obviously, faster processor speed and larger memory may result in drastically reduced computation times, depending on the specific analysis. For most of the software presented in this chapter, any of the three common computer platforms or operating systems (OS) will work. Below, "Windows" refers to the Windows OS by Microsoft Corporation run on a PC, "Mac" to the Macintosh OS by Apple Computer, and "Unix" to Unix and/or Linux OS.
3. Software to generate multiple sequence alignments, e.g., Clustal *(19,20)* (*see* **Notes 3** and **4**). Clustal can be downloaded for free for use under Windows, Mac, and Unix, e.g., from EBI's software repository (ftp://ftp.ebi.ac.uk/pub/software/).
4. Software to manipulate multiple sequence alignments, e.g., BioEdit (Windows only; http://www.mbio.ncsu.edu/BioEdit/bioedit.html) or the Java application

Jalview *(21)* (Windows, Mac, Unix; http://www.jalview.org/), which can both be downloaded for free (*see* **Note 3**).
5. Software to generate phylogenetic trees, e.g., Clustal or PHYLIP *(22)* (*see* **Notes 3**, **5**, and **6**). Like Clustal, PHYLIP is available free of charge for Windows, Mac, and Unix (http://evolution.genetics.washington.edu/phylip.html).
6. Software to visualize phylogenetic trees, e.g., TreeView *(23)* (*see* **Notes 3** and **7**), which is available for Windows, Mac, and Unix and can be downloaded for free from the author's web page (http://taxonomy.zoology.gla.ac.uk/rod/treeview.html).

3. Methods

The step-by-step protocols provided below offer an introduction to phylogenetic analysis with two of the most popular software packages (*see* **Note 3** and **8**). An overview of the main procedures is given in **Fig. 2**. They do not assume prior practical experience in phylogenetics but at least some familiarity with biological sequences, sequence databases such as GenBank, and with phylogenetic analysis and trees in general. Unfortunately, this chapter cannot provide a comprehensive discussion of how to design one's analysis or theory regarding phylogenetic inference and its common pitfalls, e.g., long branch attraction or the influence of "rogue" sequences. For such issues the reader is referred to the pertinent literature, including **ref. 24**, a highly recommended short tutorial; **ref. 25**, an excellent troubleshooting guide; relevant books (e.g., **ref. 26**); and last but not least the often very informative descriptions, "Help" Subheadings, and manuals that accompany each program and that should always be consulted. The reader is also explicitly encouraged to creatively explore the software beyond the recommendations given here, which represent just one of many possible ways to achieve similar outcomes. In case of unexpected results and before drawing sweeping conclusions, the novice is strongly advised to also seek expert advice.

Below, **this font type** (Courier bold) denotes program items such as menu options or input the user has to enter, as it appears on the screen. For example, **File – Load Sequences** in relation to ClustalX instructs the user to click on menu option **File** followed by choosing the command **Load Sequences**. Names of files and PHYLIP programs are referred to in quotation marks and in normal font (e.g., "infile" or "seqboot").

Finally, a few frequently used technical terms shall briefly be defined: **alignment** = multiple sequence alignment, i.e., a data set in which all sequences have been aligned to one another such that, ideally, every column represents a homologous sequence position or gaps across all sequences; **bootstrapping** = a statistical way of testing how well the data support any given branch in the tree

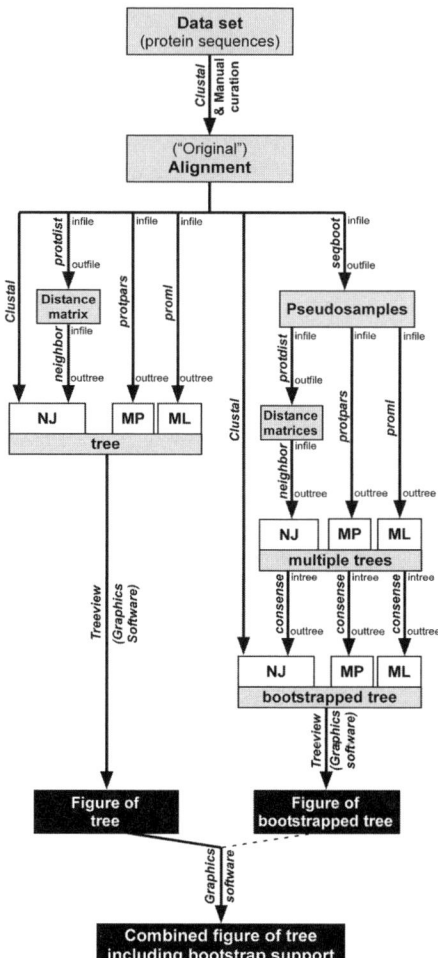

Fig. 2. A schematic overview of the main procedures for creating phylogenetic trees and respective figures as described in this chapter. Grey boxes represent text files, and names of computer programs are indicated in italics next to the arrows, with all lower case names referring to programs of the PHYLIP package. Relevant input and output files of PHYLIP programs are indicated to the right of the arrows. Importantly, conclusions should never be based on single trees without corresponding bootstrap analysis. A common way to present phylogenetic results is to show a combined figure of a tree that is based on one's original alignment to which corresponding bootstrap information has been added manually. ML, Maximum Likelihood; MP, Maximum Parsimony; NJ, Neighbor-Joining.

(*see,* e.g., **ref. 24**)—one should never base conclusions on phylogenetic trees without bootstrapping; **consensus tree** = a tree that summarizes the information of many phylogenetic trees; **data set** = amino acid sequence collection (*see* **Note 1**); **distance matrix** = a table of pairwise distances between sequences of an alignment calculated according to a particular amino acid substitution model (*see* **Note 9**); **pseudosample** or **replicate** = an "artificial" alignment that has been generated from an original, "real" alignment by resampling with replacement (used for bootstrapping; *see* **ref. 24**); **tree file** = a text file that contains the formalized, exact, and concise description of one or many phylogenetic trees.

3.1. Data Collection and Preparation

1. Data collection and preparation can be very time-consuming. Therefore, if at all possible, one should base one's data set on an already assembled sequence collection to which one's own sequences of interest can be added. It can be very worthwhile to check in the literature whether an appropriate data set or alignment can be downloaded—often from a journal's web page as supplementary material to an article, or from a laboratory's web site—or to directly contact other researchers.

2. In most cases one's data set will include sequences that have been downloaded from a database via the internet, e.g., from one of the three major sequence repositories NCBI (National Center for Biotechnology Information), EBI (European Bioinformatics Institute), and DDBJ (DNA Data Bank of Japan), and/or from a more specialised database like Swiss-Prot. It is advisable to sample broadly from diverse taxonomic lineages, and in particular to make sure to include homologs of the sequence of interest from α-proteobacteria and cyanobacteria (*see* **Note 10**).

3. Collect sequences in FASTA format (*see* **Note 11**). Most sequence databases allow one to download multiple sequences in FASTA format in one go. Make use of such options, whicht can save a lot of time and effort otherwise spent moving the mouse and clicking buttons.

4. To make the resulting phylogenetic trees easily interpretable, it is very helpful to label sequences with meaningful names. In general, keep sequence names short and avoid spaces and other nontext characters, because most tree-making programs either cut off sequence names after a given number of letters or after certain characters: e.g., PHYLIP considers only the first 10 characters of a name, whereas Clustal cuts off names after spaces.

5. One should always save one's data set as a plain TEXT file (file extension ".txt") and not in a word processor-typical file format such as Rich Text Format (".rtf") or as a Microsoft Word document (".doc"). Downstream software only accepts data sets that have been saved as plain text files.

6. Decide which part(s) of the sequence of interest should be used for the analysis. For example, many organelle-targeted proteins have N-terminal

extensions that cannot be aligned and compared in a meaningful way across diverse species and that may be absent from prokaryotic homologs. Multiple sequence alignments (*see* **Subheading 3.2.**) and pairwise BLASTing *(27)* (e.g., at http://www.ncbi.nlm.nih.gov/blast/bl2seq/wblast2.cgi) can be used to confidently identify N- or C-terminal extensions or internal loops that may be present in only some or individual sequences in the data set. Removing such extensions and loops before further analysis may significantly accelerate and improve the resulting multiple sequence alignment and phylogenetic analysis (*see* **Fig. 3A**).

3.2. Multiple Sequence Alignment With ClustalX

1. Start ClustalX and open your data set file via **File – Load Sequences** (*see* **Notes 12** and **13**).
2. Generate the alignment in ClustalX (**Alignment – Do Complete Alignment – Align**). Depending on data set size, Clustal settings, and computer speed, this step can take from a second to more than an hour. By default, the resulting output alignment file is saved in Clustal format (file extension ".aln").
3. Always carefully inspect the alignment to check whether it is satisfactory and that Clustal has not made obvious mistakes (which does happen at times, depending on the data set). If the alignment is not satisfactory, change one or some of the many alignment parameters (*see* **Note 14**) or look for (and delete) extensions or internal loops that may be present in certain sequences (*see* **Fig. 3A** and **Subheading 3.1.**) and redo the alignment. If it is known that one's sequences contain individual amino acids or whole stretches (protein domains) that are highly conserved, one should—as a control—check whether they have been aligned correctly across all sequences. A phylogenetic analysis can only ever be as good as the alignment it is based on, so it is critical to ensure that the alignment is good and does not contain misaligned sequences.
4. If the alignment includes blocks of positions that are badly aligned (*see* **Fig. 3B**), it may be advisable to delete such blocks prior to a definitive phylogenetic analysis, e.g., by using a multiple sequence alignment editor such as BioEdit or Jalview (*see* **Subheading 2.4.**). If badly aligned blocks are not removed, they may add noise to the phylogenetic analysis or contribute to artifactual clustering of sequences (e.g., because of shared amino acid bias in otherwise distantly related sequences).
5. In order to generate a phylogenetic tree with ClustalX, you can go straight on to do so, following instructions given in **Subheading 3.3.** (**Fig. 2**). Alternatively, save your alignment from within ClustalX (**File – Save Sequences as...**)—specifying an appropriate sequence format—and calculate the phylogenetic tree using other programs such as PHYLIP or PAUP, for which the alignment has to be in PHYLIP or NEXUS format, respectively.

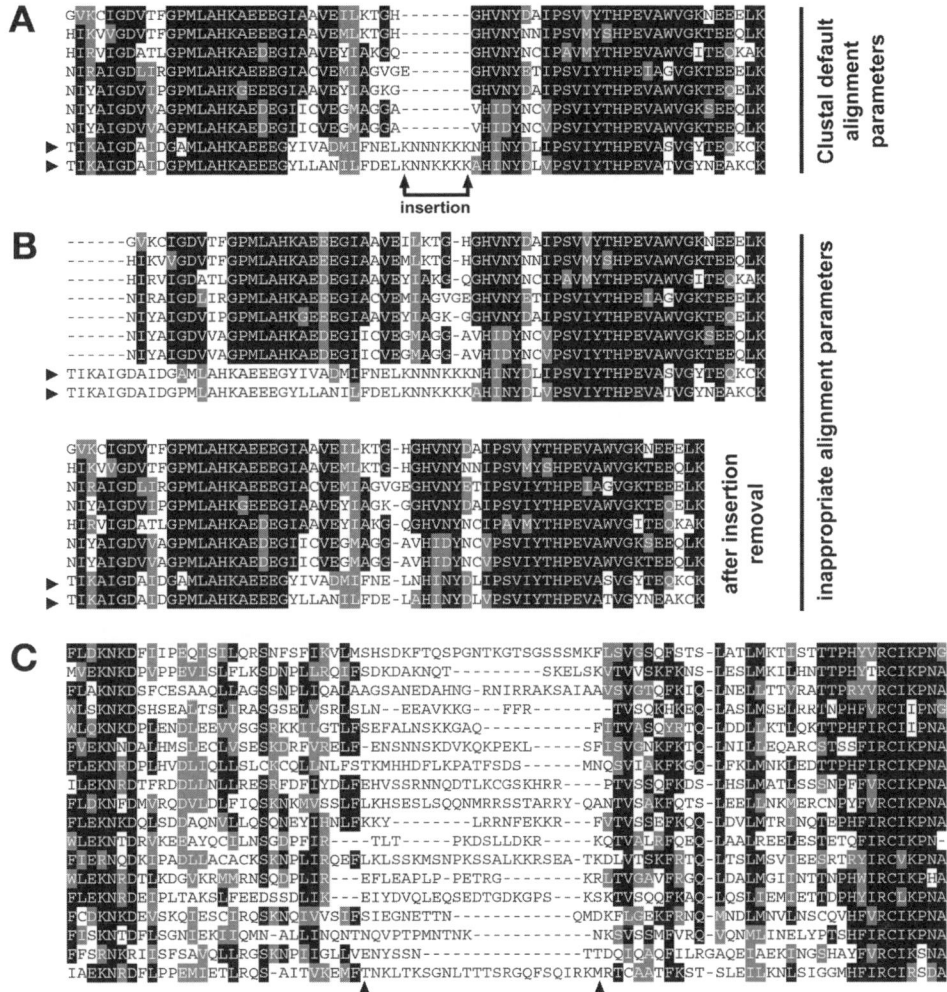

Fig. 3. Two examples of problems that may be encountered in making multiple protein sequence alignments. Alignments were created with Clustal, and highlighting of conserved residues was performed with BoxShade (at http://www.ch.embnet.org/software/BOX_form.html). (**A**) All sequences in this alignment are correctly aligned, i.e., Clustal has correctly recognized the short insertion present in the two more divergent sequences (of *Plasmodium* spp.) at the bottom of the alignment. (**B**) Choosing inappropriate alignment parameters (excessively high gap opening and gap extension parameters) causes misalignment of the two divergent sequences (upper panel), whereas manual removal of the insertions prior to generating the alignment yields again a correct alignment (lower panel). While for

3.3. Generating Phylogenetic Trees With ClustalX

1. If not already open, open your alignment file in ClustalX (**File – Load Sequences**) (*see* **Note 13**).
2. To generate a single, nonbootstrapped phylogenetic tree from the alignment, go to **Trees – Draw NJ Tree** and click **OK** (*see* **Note 15**). By default, the resulting output tree file (*see* **Note 16**) has the same name as the alignment input file plus the file extension ".ph".
3. To perform bootstrap analysis, one first has to adjust the output parameters such that the results can be later directly visualized with TreeView (*see also* **Subheading 3.7.**): go to **Trees – Output Format Options** and for the option "**Bootstrap labels on:**" choose **NODE**. Clustal will remember these settings until the program is quit.
4. To start the bootstrap analysis (*see* **Note 15**), go to **Trees – Bootstrap NJ Tree**, enter a random number (between 1 and 1000) in the field **Random number generator seed**, and choose an appropriate **Number of bootstrap trials** (*see* **Note 17**). Start the bootstrap analysis by clicking **OK**. By default, the resulting output tree file (*see* **Note 16**) has the same name as the alignment input file plus the file extension ".phb".

3.4. Generating a Phylogenetic Tree With PHYLIP: Distance Matrix Analysis With "protdist" and "neighbor"

This section is divided into two parts. It first describes how to generate a distance matrix from an alignment using the program "protdist" and second how to calculate a phylogenetic tree based on the distance matrix according to the Neighbor-Joining method *(28)* using the program "neighbor" (*see* **Fig. 2** and **Note 18**).

1. First, make sure your alignment file, which needs to have been saved in PHYLIP format (*see* **Note 19**), is located in the same folder as the PHYLIP programs (*see* **Note 20**).

Fig. 3. *(Continued)* this very simplistic example the problem of misaligned sequences was provoked artificially, similar problems (using normal alignment parameters) may persistently occur with real data sets where insertions in individual sequences may be very long (e.g., more than 100 amino acids). (**C**) Beginning and end of this alignment are satisfactory, whereas a block in the middle appears badly aligned. In this case, the problem is not a shortcoming of the alignment procedure, but the presence of a variable region in the protein that is located between conserved areas. If such blocks are not removed (e.g., by using a multiple sequence alignment editor), they may add noise to or exert an artifactual influence during the ensuing phylogenetic analysis.

2. Start the PHYLIP program "protdist." To get started, use the default settings for all options (*see* **Note 21**) including the use of the Jones-Taylor-Thornton (JTT) matrix as model for amino acid change *(29)* (*see* **Notes 9** and **22**).
3. Confirm that all settings are correct and start the analysis by entering **Y**. The output of the program "protdist" consists of a distance matrix that is saved as "outfile".
4. Second, the output of "protdist" is used as input for the program "neighbor": rename "outfile" into "infile," which first requires the previous "infile" to be moved, deleted, or renamed (*see* **Note 20**).
5. Start the PHYLIP program "neighbor." Choose to randomize the order in which the sequences are entered into the analysis by entering **J** (for the **Randomize input order of species** option) and provide a **Random number seed** (*see* **Note 23**)
6. Confirm that all settings are correct and start the analysis by entering **Y**. The output of the program "neighbor" consists of two files: "outfile" contains a semi-graphical version of the resulting phylogeny which can be readily viewed in a text editor, whereas "outtree" is a tree file (*see* **Note 16**).

3.5. Generating a Phylogenetic Tree With PHYLIP: Maximum Likelihood and Maximum Parsimony Analysis

This section describes how to calculate a phylogenetic tree from a multiple amino acid sequence alignment employing a Maximum Likelihood or a Maximum Parsimony method with the programs "proml" and "protpars," respectively. The procedure for both types of analysis is very similar and is in fact simpler than generating a distance matrix-based phylogeny with PHYLIP (**Fig. 2**). But beware that Maximum Likelihood calculations can take very long to complete.

1. Make sure your alignment file, which needs to have been saved in PHYLIP format (*see* **Note 19**), is located in the same folder as the PHYLIP program (*see* **Note 20**).
2. Start the PHYLIP program "protml" or "protpars." To get started, use the default settings for most options (*see* **Note 21**) including the use of the JTT matrix in "proml" as model for amino acid change *(29)* (*see* **Notes 9** and **22**).
3. Choose to randomize the order in which the sequences are entered into the analysis by entering **J** (for the **Randomize input order of species** option), provide a **Random number seed**, and specify the desired **Number of times to jumble** (*see* **Note 23**).
4. Confirm that all settings are correct and start the analysis by entering **Y**. The output of either program consists of two files: "outfile" contains a semi-graphical version of the resulting phylogenetic tree(s), which can be readily viewed in a text editor, whereas "outtree" is a tree file (*see* **Note 16**). Note that "protpars" (but not "proml") may frequently report more than one "equally best" tree at the end of the analysis, in which case a consensus tree can be calculated by following instructions given in **Subheading 3.6.**

3.6. Bootstrapping With PHYLIP

In PHYLIP, bootstrapping has to be done "manually" in three steps (**Fig.** 2). First, bootstrapped pseudosamples of one's original alignment are created by using the program "seqboot." Second, multiple phylogenetic trees are generated using one's method of choice (this is essentially the same procedure as detailed in **Subheadings 3.4.** and **3.5.**). Third, the resulting tree file, which contains multiple phylogenetic trees, is converted by the program "consense" into a consensus tree that indicates the bootstrap support value for each branch in the tree.

1. Make sure your alignment file, which needs to have been saved in PHYLIP format (*see* **Note 19**), is located in the same folder as the PHYLIP programs (*see* **Note 20**).
2. First, start the PHYLIP program "seqboot." To get started, use the default settings for all options (*see* **Note 21**). One frequently adjusted option is the number of replicates (pseudosamples) that are to be created (option **R**) (*see* **Note 17**).
3. Confirm that all settings are correct and start the resampling process by entering **Y**. The resulting output file "outfile" contains the specified number of pseudosamples.
4. Second, generate phylogenetic trees for all pseudosamples created by "seqboot," i.e., use the "outfile" produced by "seqboot" as input for the tree-making program(s) of your choice by renaming it to "infile." Then run either "protdist" followed by "neighbor," or "proml," or "protpars" as described in **Subheadings 3.4.** and **3.5.** The only difference is that you now have to toggle option **M** (**Analyze multiple data sets**) and specify the appropriate number of data sets (alignments) to prepare the respective program for the fact that "infile" contains more than one alignment. The programs "protdist," "proml," and "protpars" at this point also ask whether "infile" contains **Multiple data sets or multiple weights**, which is answered by entering **D**. For the other options that may automatically be offered at this step, *see* **Note 23**.
5. Confirm that all settings are correct and start the respective analysis by entering **Y**. (After having run "protdist," do not forget to also run "neighbor").
6. As usual (*see* **Subheadings 3.4.** and **3.5.**), the output of the tree-generating programs "neighbor," "proml," and "protpars" consists of two files: "outfile" contains a semi-graphical version of the resulting phylogenetic trees, which can be readily viewed in a text editor, whereas "outtree" is a tree file (*see* **Note 16**). Importantly, both files contain the multiple output for all pseudosamples that went into the analysis. To make the information of these multiple trees easily accessible, the file "outtree" has to be run through yet another PHYLIP program, whereas the "outfile" is usually of little interest.
7. Third, create one consensus tree from the multiple trees present in the file "outtree" from the previous step by using the program "consense." To do so, rename the file "outtree" to "intree" (not "infile") and make sure "consense" is located in the same directory.

8. Run "consense" and start the analysis by entering **Y**. This program generates two output files: "outfile" contains a semi-graphical version of the consensus tree that can be readily viewed in a text editor, and the tree file "outtree" (*see* **Note 16**).

3.7. Presenting Phylogenetic Trees

There are different ways of presenting phylogenetic trees and countless programs that could be used to prepare a respective figure, and the reader is explicitly encouraged to adapt the following protocol according to his or her preferences. This section briefly describes one way of preparing a publishable figure that shows a tree derived from an analysis of an original alignment to which indicators of bootstrap support are added manually (*see* **Fig. 1** and **refs. 5–10** for a few select examples).

1. Start by preparing printouts of the bootstrap analysis/analyses for easy reference later on. Open the tree file(s) containing the trees with the bootstrap information (e.g., the output of Clustal or of PHYLIP's "consense" after bootstrapped Neighbor-Joining, Maximum Likelihood, or Maximum Parsimony analysis; *see* **Subheadings 3.3. and 3.6.**) in Treeview (*see* **Note 3**). Select the options **Show internal edge labels** to display the bootstrap values, choose **Rectangular cladogram** or **Phylogram** as tree type, and adjust font size as desired. Print the tree including the bootstrap information by using the **File – Print...** command (*see* **Note 16**).
2. Open the tree file containing the tree (e.g., the output of Clustal or of PHYLIP's "protdist", "proml", or "protpars") from an analysis of your original alignment in Treeview and adjust the program options as desired (*see* **Notes 3** and **16**).
3. Save the tree as a graphics file, either by using the **File – Save as graphic...** command, which will save a picture in an OS-specific file format, or by using the **File – Print...** command and by choosing an appropriate printer driver that—depending on the computer—allows you to save the picture in PostScript or PDF format (*see* **Note 16**).
4. Open the graphics file in the vector-based illustrations software of your choice (e.g., Adobe's Illustrator, or ACD's Canvas).
5. Referring to the printouts of the bootstrap analysis/analyses prepared earlier, manually add (indicators of) the bootstrap values to the tree. Take great care that bootstrap values are placed only at branches for which the topology (the distribution of all sequences relating to whether they are found on one side or the other of a particular branch) is exactly the same in both the bootstrap consensus tree and in the original single tree. Bootstrap values are usually only shown for branches that are supported by bootstrap values greater than 50%. It is often useful to visually highlight meaningful clusters in the tree (e.g., protein (sub)families or taxonomic lineages; *see*, e.g., **Fig. 1**).

Notes

1. When generating phylogenetic trees from sequences that derive from distantly related organisms, it is much more appropriate to use amino acid sequences (and not DNA sequences). In particular, when trying to establish whether a eukaryotic, nuclear-encoded gene may be of organellar origin, one should always include homologous sequences from prokaryotes, including those of α-proteobacteria and cyanobacteria.

2. As a rule of thumb, a data set of up to 50 sequences with sequences of 100–1000 amino acids in length should present no serious problems in regard to file sizes and duration of analyses (except for Maximum Likelihood, which is by far the most time-consuming type of analysis and may require long computation times even with small alignments). Bigger data sets may—depending on computer hardware, software, and type of analysis—lead to excessively long computing times or huge file sizes (*see also* **ref. *30***).

3. The core descriptions in this chapter have been based on ClustalX version 1.83 and PHYLIP version 3.65 (primarily the Windows versions). In addition to these and other programs listed in this chapter, a huge range of other software is available for making multiple sequence alignments and for inferring and visualizing phylogenetic trees, either as standalone programs to be run on one's own computer or as internet-based services. As a starting point in the search for other relevant software, check, e.g., Joe Felsenstein's comprehensive web page (http://evolution.gs.washington.edu/phylip/software.html), the EMBL-EBI site (http://www.ebi.ac.uk/FTP/), the Indiana University Biology Department's software listing (http://iubio.bio.indiana.edu/), or Bioinformatics.org (http://bioinformatics.org/). Clustal and PHYLIP have been chosen to be presented in this chapter for many reasons: they offer a huge range of functions and options, have been in development and "matured" over a long time (PHYLIP since 1980, Clustal since 1988) and keep being updated, are extremely widely used, are available for several OS (Windows, Mac, Unix), and are downloadable free of charge. It is nevertheless not implied that they are necessarily "the best" or the most user-friendly programs.

4. Today, Clustal is available in two versions: ClustalW is run from the command-line or is accessed as a web-based service (e.g., at http://www.ebi.ac.uk/clustalw/), whereas ClustalX has a mouse-driven user interface. ClustalX is therefore more user-friendly to use on one's own computer, but the results should be the same with both programs. For the latest version of Clustal, check ftp://ftp-igbmc.u-strasbg.fr/pub/ and http://www.embl-heidelberg.de/~chenna/.

5. The tree-making functionality of Clustal is very convenient, fast, and sufficient for simple and/or preliminary analyses, but it does not offer many options. In contrast, PHYLIP is a package of many individual programs that offer a huge range of functionality in regard to phylogenetic inference. Unfortunately, PHYLIP is not very user-friendly in the sense that the different programs are not smoothly

integrated with one another, and the settings for each program are controlled via simple menu screens with command-line interface, i.e., not by mouse-click navigation. But one should definitely not shy away from using PHYLIP, since its *modus operandi* may be awkward but is not difficult, and one easily gets used to it.

6. PAUP is another very popular program. It is very user-friendly under Mac OS, allows for many options, utilizes the powerful NEXUS file format *(31)*, and provides a good manual. But it is not considered further in this chapter because it does not (yet) offer as much functionality for amino acid data as PHYLIP, is user-friendly only under Mac OS, and is not available for free (*see* http://paup.csit.fsu.edu/).

7. NJplot including "unrooted" is another software option for visualizing trees that is available for free for multiple computer platforms. *See* http://pbil.univ-lyon1.fr/software/njplot.html.

8. The ease and speed with which software can produce alignments and phylogenetic trees may lure one into thinking that phylogenetic analysis is always a simple, trouble-free endeavor. In general, however, the bigger problems with phylogenetic inference lie not in simply generating trees, but either in designing the analysis (in regard to size and scope of the data set, producing a good sequence alignment, and choosing a suitable type of phylogenetic analysis) such that the resulting trees are reliable and relevant in regard to the question being asked, or in an unfavorable phylogenetic signal-to-noise ratio intrinsic to the data that leads to inconclusive or even misleading phylogenies. Phylogenetic inference and the interpretation of results can thus easily turn into an uncertain and very complex undertaking. Yet for the purpose of investigating potential organellar origins of proteins by using phylogenetic analysis, I personally tend to think that for the latter reason alone—possible bad signal-to-noise ratio in the data—the following may (arguably!) be seen as a healthy attitude to adopt: if simple analyses (e.g., Neighbor-Joining analyses as described in this chapter) do not yield a reasonably clear-cut result—either confirming or refuting an organellar origin of a protein—or at least a fairly well-resolved tree with decent bootstrap values, then further work to improve the analysis or the choice of more sophisticated methods (e.g., Maximum Likelihood) is not very likely to produce an unequivocal result and may not be worth the extra effort.

9. A nonsilent genetic mutation leads to the substitution of one amino acid by another in a protein, but not all possible substitutions occur at the same frequency because of the nature of the genetic code (amino acid changes that may arise from a single nucleotide change will occur more frequently than those requiring two or three nucleotide changes) or functional selection (amino acids under functional selection will more often be substituted by a similar amino acid—e.g., a hydrophobic residue by another hydrophobic residue—than by a completely different one). An amino acid substitution matrix or model is a table that lists a score or probability for every possible amino acid change, i.e., for every possible substitution of a given amino acid *x* by another amino acid *y*. Such matrices or

models, e.g., the Jones-Taylor-Thornton, PAM, and BLOSUM matrices, usually derive from empirical data. For more information, *see*, e.g., the EBI Help web pages (http://www.ebi.ac.uk/help/matrix_frame.html).

10. In choosing sequences from databases to be included in the data set, one should not just rely on their annotation, which in many cases may be incomplete, different than expected, or simply wrong. BLASTing *(32)* one's sequence of interest against a general database (e.g., at NCBI, EBI, DDBJ) will yield a more unbiased and complete list of sequences that may potentially be included in the data set.

11. In FASTA format every sequence consists of a one-line header followed by one or several lines of sequence data. The header is preceded by the symbol "{>}", and blank lines and spaces in the sequence data are ignored. For more information, *see*, e.g., the EBI Help web pages (http://www.ebi.ac.uk/help/formats_frame.html).

12. Randomization of sequence order prior to generating the alignment eliminates any potentially artifactual bias due to the order in which sequences are listed in one's data set. It may be a good idea to perform sequence order randomization at least once on one's data. A simple way of randomizing is to use a spreadsheet program such as, Microsoft's Excel. First, import the sequence names and sequences into two adjacent columns of a spreadsheet. Second, add a third column with random numbers (e.g., by using the **=RAND()** function of Excel). Third, sort all three columns by the column containing the random numbers. Finally, convert the randomized data set back to FASTA format. To turn a FASTA data set into a tab-delimited set of sequence names and sequences for easy pasting into a spreadsheet, start by converting it into ACEDB format by using readseq (*see* **Note 19**) followed by deleting all paragraph marks/end-of-line characters; replacing **Peptide :** " (the word "Peptide" followed by a space, a colon, a space, and a quotation mark) by paragraph marks/end-of-line characters; and by replacing all remaining " (quotation marks) by tabs. To convert a tab-delimited list of sequence names and sequences back into FASTA, simply convert all tabs into paragraph marks/end-of-line characters.

13. Text files generated on a computer running one of the three major OS (Windows, Mac, and Unix) can readily be used on another computer running a different OS, with one exception: the three different platforms use different paragraph marks/end-of-line characters, i.e., characters that denote the end of a line ("line feed" and/or "carriage return"). For example, a data set containing sequences in FASTA format created on an Apple computer may not open properly in ClustalX running under Windows, or ClustalX v1.83.1 for Mac OSX may correctly open only data sets that have been saved using Unix-specific (!) end-of-line characters. This type of problem can easily be solved by saving one's data set with the end-of-line characters appropriate for the OS and version of Clustal being used, which can conveniently be done with a capable text editor. There are numerous good text editors available, and freely downloadable examples include Crimson Editor (for Windows; http://www.crimsoneditor.com/), Vim/Cream (for Windows

and Unix; http://cream.sourceforge.net/index.html), and TextWrangler (for Mac OSX; http://www.barebones.com/products/textwrangler/).

14. Clustal generates alignments in two steps: it first compares pairs of sequences ("Pairwise Alignment," this often takes up most of the time of the whole alignment procedure) followed by creating the actual multiple sequence alignment ("Multiple Alignment"), and parameters for both steps can be changed independently of one another (**Alignment — Alignment Parameters**). In my experience (but it all depends on the data set!), the default alignment parameters that are active upon starting the program can be a bit too "keen" to introduce small gaps in too many sequences. One solution to this problem is to increase the **Gap Opening** or **Gap Extension** parameters, especially in the **Multiple Alignment Parameters**. Alternatively, I sometimes got good results by changing the **Pairwise Alignment Parameters** from **Slow-Accurate** to **Fast-Approximate** and choosing the following parameters: **Gap Penalty** = 4, **K-Tuple Size** = 1, **Top Diagonals** = 50, **Window Size** = 50. But again, everything depends on the particular data set, and the reader is explicitly encouraged to experiment with the many parameters available in Clustal to arrive at a satisfactory alignment. For more information on Clustal's options and parameters, please refer to the "Help" and the documentation files that accompany the program (e.g., "clustalx.html" and "clustalw.doc" in the Windows version, and "clustalw_help" and "clustalx_help" in the Mac version). For further tips, *see also* **ref. *33***, while **refs. *34*** and ***35*** provide some more technical information. It cannot be stressed enough that an accurate alignment is an essential prerequisite for any meaningful phylogenetic analysis.

15. Trees calculated by Clustal are based on pairwise distances between sequences (distance matrix) that are converted into a tree by the Neighbor-Joining method *(28)*. There are two options available: **Exclude Positions with Gaps** will do exactly what it says, i.e., all alignment positions at which one or more sequences contain a gap (–) will be omitted from the analysis. This option is not useful when gaps are present in many alignment positions, even if the gaps occur only in individual sequences. The second option is **Correct for Multiple Substitutions**: this option tries to compensate for the fact that many positions in a biological sequence accumulate more than one mutation over evolutionary timescales. Except for data sets that contain very divergent sequences, it is generally recommended to enable this option, which employs the algorithm of Kimura *(36)*.

16. A tree file is a text file that contains the formalized, exact description of one or several phylogenetic trees, e.g., the output of Clustal or of PHYLIP's "neighbor," "proml," "protpars," and "consense." Tree files are not meant to be viewed directly (e.g., in a text editor) but should be visualized using a program such as TreeView, NJplot, PAUP, or PHYLIP's "drawgram" and "drawtree." Treeview can handle tree files containing one or multiple trees and offers many options, e.g., to depict the tree rooted or unrooted, to draw it as a cladogram (all branches in the tree

are equally long) or a phylogram (branch lengths are representative of the amount of inferred evolutionary change between sequences), and to adjust font type, font size, and branch line width. The display of bootstrap values can be toggled on and off using the **Tree — Show internal edge labels** command. Note that bootstrap values from a Clustal-generated file will only be displayed if Clustal had been instructed to place the labels on the nodes and not on the branches (the latter is the default; *see* **Subheading 3.3.**). With tree files generated by PHYLIP's "consense," Treeview (at least version 1.6.6 for Windows) has the minor problem that the display of bootstrap values can only be toggled by using the **Tree — Show internal edge labels** command via the menu bar, whereas the corresponding button below the menu bar, which usually has the same functionality, is greyed out and cannot be used. In this case, Treeview also omits the bootstrap values when the tree is printed via the **Tree — Print trees...** command, but it correctly includes the values when the **File — Print...** command is used. Saving a tree by using the **File — Save as graphic...** command will save a picture in a Metafile format under Windows and as a Macintosh picture file (PICT) on a Mac. *See also* **Subheading 3.7.**

17. For a preliminary bootstrap analysis 100 bootstrap replicates (pseudosamples) are usually enough. For a thorough analysis one may want to run between 300 and 1000 replicates, depending on the size of the alignment.

18. Instead of using "neighbor" it is also possible to use the programs "fitch" or "kitch," which employ the Fitch-Margoliash or Minimum Evolution methods (instead of Neighbor-Joining) to calculate a phylogenetic tree from the distance matrix. *See* the PHYLIP documentation for further details.

19. To interconvert different sequence formats, use, e.g., Clustal or ReadSeq (downloadable for free from http://iubio.bio.indiana.edu/soft/molbio/readseq/ and available as an internet-based service provided by many bioinformatic centers, e.g., at http://www.ebi.ac.uk/cgi-bin/readseq.cgi).

20. The input and output of individual PHYLIP programs is handled via files that are located in the same folder as the PHYLIP programs themselves. By default, most PHYLIP programs (e.g., "protdist," "neigbor," "proml," "protpars," "seqboot") expect the input file (e.g., a file containing your sequence alignment or distance matrix) to be called "**infile**," whereas the standard output of a program is saved to a file called "**outfile**." Programs that require a tree file as input (e.g., "consense") expect this file to be called "**intree**," whereas programs that (also) generate tree files (e.g., "neighbor," "proml," "protpars," "consense") save such files under the name "**outtree**" (*see also* **Fig. 2**). If one does not supply an input file to a PHYLIP program whose name conforms to this convention, the program will prompt the user for an alternative name. Also, if the output file of one program (e.g., of "protdist") serves as input file for another program (e.g., for "neighbor" or "fitch"), one simply has to rename the appropriate "outfile" to "infile." It is advisable to always keep, e.g., Windows Explorer (Windows) or Finder (Mac) open displaying

the working directory/folder to facilitate this frequently required renaming of files. While one can quickly get used to these (seemingly archaic) procedures, one has to beware that they may quickly lead to confusion as to the exact contents of a given file or to the unwanted deletion or overwriting of important files. It is therefore highly recommended to be disciplined with one's files and file names and in particular (1) to carry out PHYLIP work in a dedicated working directory (folder) into which one copies both the required input files and the PHYLIP programs one wants to use, (2) to always keep (copies of) important files—especially the original data set and alignment files—separate from this working directory, and (3) to instantly rename output files one intends to keep with a meaningful label (specifying the settings under which the file was generated) and to move them away from the working directory.

21. If you want to change any of the options listed by a PHYLIP program, type the letter (or number) corresponding to the option you want to change and confirm by hitting Enter/Return. This will toggle the setting for the respective option and in some cases prompt you for further input. To come back to a previous setting for an option, just toggle the option through all available settings (the number depends on the option) until the desired setting is displayed again.

22. It is plausible to assume that different amino acid positions in a protein change at unequal rates over evolutionary time because of different selective pressures at different positions. One advanced option for inferring phylogenies is to approximate the distribution of rates of change among amino acid positions to a distribution, available in PHYLIP's "protdist" and "proml" (by default, the programs assume the rate of evolutionary change to be equal for all positions). Very briefly, to approximate a γ distribution including a class of invariant sites, one first has to determine the specific "coefficient of variation" and "fraction of invariant sites" for the alignment. One possibility is to use Tree-Puzzle *(37)* (http://www.tree-puzzle.de/), which expects an "infile" containing an alignment in PHYLIP format and also has a similar user interface as PHYLIP. In Tree-Puzzle, toggle option **k** to **Pairwise distances only**, option **m** to the desired amino acid substitution model, option **w** to **Mixed**, and set option **c** to the desired number of rate categories (e.g., four). After the Tree-Puzzle analysis, note the relevant "Gamma distribution parameter alpha" and "Fraction of invariable sites" near the end of the "outfile." Returning to PHYLIP, toggle option **G** (**Gamma distribution of rates among positions**) in "protdist" or option **R** (**Rate variation among sites**) in "proml" until it is set to **Gamma+Invariant**. Once the analysis is started by entering **Y**, the program prompts for the **Coefficient of variation**, which is 1/(square root of α), with α being the γ distribution parameter determined by Tree-Puzzle. "Proml" (but not "protdist") then asks for the number of rate categories (**Rates in HMM**, **Number of categories (1–9)?**—enter, e.g., five, which includes one rate for invariant sites). Finally, enter the **Fraction of invariant sites** (as determined by Tree-Puzzle).

The number of rate categories relating to the approximation of a discrete γ distribution in "proml" (**Rates in HMM**) should not be confused with option **C** (**One category of sites?**, available in both "proml" and "protdist"), which allows user-defined rate categories. For this and other program options see the PHYLIP documentation.

23. To randomize sequence input order, the programs "neighbor," "proml," and "protpars" ask for a **Random number seed**, for which one should enter a random, integer number (no fraction) between 1 and 32767 of the form $4n+1$, i.e., the number must give a remainder of 1 when divided by 4 (if the last two digits of a number are of the form $4n + 1$, then the whole number conforms to this requirement). This random number seed is used to provide a truly random addition order of sequences during the analysis (option **J, Randomize input order of sequences/species**), a sensible option that should be invoked where available. In addition, the "jumble" option (**Number of times to jumble**) is automatically offered by "proml" and "protpars" at this point and determines how often the program will run the entire phylogenetic analysis, each time randomizing the order in which one's sequences are entered into the analysis. In the end, the program only reports the best tree (or the "equally best" trees) found during the repeated analyses.

Acknowledgments

I wish to thank D. Bhattacharya and G. I. McFadden for having introduced me to molecular evolution, and P. J. Keeling and J. Felsenstein for helpful comments. Thanks are also due to the *Eimeria* Genome Sequencing Consortium, Institute for Animal Health, Compton, UK, for unpublished sequence data of *Eimeria tenella*, which were accessed via the Sanger Institute's web site ("These sequence data were produced by the Protozoan Pathogen Sequencing Group at the Sanger Institute and can be obtained from ftp://ftp.sanger.ac.uk/pub/pathogens/Eimeria/tenella/.") The author gratefully acknowledges the financial support of a postdoctoral long-term fellowship from the European Molecular Biology Organization (EMBO) and, more recently, of the Département de l'Instruction Publique (DIP) of Geneva, Switzerland.

References

1. Dyall, S. D., Brown, M. T., and Johnson, P. J. (2004) Ancient invasions: from endosymbionts to organelles. *Science* **304,** 253–257.
2. Timmis, J. N., Ayliffe, M. A., Huang, C. Y., and Martin, W. (2004) Endosymbiotic gene transfer: organelle genomes forge eukaryotic chromosomes. *Nat. Rev. Genet.* **5,** 123–135.
3. Emanuelsson, O. and von Heijne, G. (2001) Prediction of organellar targeting signals. *Biochim. Biophys. Acta* **1541,** 114–119.

4. Nassoury, N. and Morse, D. (2005) Protein targeting to the chloroplasts of photosynthetic eukaryotes: getting there is half the fun. *Biochim. Biophys. Acta* **1743**, 5–19.

5. Schnarrenberger, C. and Martin, W. (2002) Evolution of the enzymes of the citric acid cycle and the glyoxylate cycle of higher plants. A case study of endosymbiotic gene transfer. *Eur. J. Biochem.* **269**, 868–883.

6. Harper, J. T. and Keeling, P. J. (2003) Nucleus-encoded, plastid-targeted glyceraldehyde-3-phosphate dehydrogenase (GAPDH) indicates a single origin for chromalveolate plastids. *Mol. Biol. Evol.* **20**, 1730–1735.

7. Huang, J., Mullapudi, N., Sicheritz-Ponten, T., and Kissinger, J.C. (2004) A first glimpse into the pattern and scale of gene transfer in Apicomplexa. *Int. J. Parasitol.* **34**, 265–274.

8. Foth, B. J., Stimmler, L. M., Handman, E., Crabb, B. S., Hodder, A. N., and McFadden, G. I. (2005) The malaria parasite *Plasmodium falciparum* has only one pyruvate dehydrogenase complex, which is located in the apicoplast. *Mol. Microbiol.* **55**, 39–53.

9. Obornik, M. and Green, B. R. (2005) Mosaic origin of the heme biosynthesis pathway in photosynthetic eukaryotes. *Mol. Biol. Evol.* **22**, 2343–2353.

10. Li, S., Nosenko, T., Hackett, J. D., and Bhattacharya, D. (2006) Phylogenomic analysis identifies red algal genes of endosymbiotic origin in the chromalveolates. *Mol. Biol. Evol.* **23**, 663–674.

11. Wilson, R. J., Williamson, D. H., and Preiser, P. (1994) Malaria and other Apicomplexans: the "plant" connection. *Infect. Agents Dis.* **3**, 29–37.

12. McFadden, G. I., Reith, M. E., Munholland, J., and Lang-Unnasch, N. (1996) Plastid in human parasites. *Nature* **381**, 482.

13. Kohler, S., Delwiche, C. F., Denny, P. W., et al. (1997) A plastid of probable green algal origin in Apicomplexan parasites. *Science* **275**, 1485–1489.

14. Gardner, M. J., Hall, N., Fung, E., et al. (2002) Genome sequence of the human malaria parasite *Plasmodium falciparum*. *Nature* **419**, 498–511.

15. Foth, B. J., Ralph, S. A., Tonkin, C. J., et al. (2003) Dissecting apicoplast targeting in the malaria parasite *Plasmodium falciparum*. *Science* **299**, 705–708.

16. Ralph, S. A., van Dooren, G. G., Waller, R. F., et al. (2004) Tropical infectious diseases: metabolic maps and functions of the *Plasmodium falciparum* apicoplast. *Nat. Rev. Microbiol.* **2**, 203–216.

17. Waller, R. F., Keeling, P. J., Donald, R. G. K., et al. (1998) Nuclear-encoded proteins target to the plastid in *Toxoplasma gondii* and *Plasmodium falciparum*. *Proc. Natl. Acad. Sci. USA* **95**, 12352–12357.

18. Waller, R. F., Reed, M. B., Cowman, A. F., and McFadden, G. I. (2000) Protein trafficking to the plastid of *Plasmodium falciparum* is via the secretory pathway. *EMBO J.* **19**, 1794–1802.

19. Higgins, D. G. and Sharp, P. M. (1988) CLUSTAL: a package for performing multiple sequence alignment on a microcomputer. *Gene* **73**, 237–244.

20. Chenna, R., Sugawara, H., Koike, T., et al. (2003) Multiple sequence alignment with the Clustal series of programs. *Nucl.. Acids Res.* **31,** 3497–3500.

21. Clamp, M., Cuff, J., Searle, S. M., and Barton, G. J. (2004) The Jalview Java alignment editor. *Bioinformatics* **20,** 426–427.

22. Felsenstein, J. (1989) PHYLIP—Phylogeny Inference Package (Version 3.2). *Cladistics* **5,** 164–166.

23. Page, R. D. (1996) TreeView: an application to display phylogenetic trees on personal computers. *Comput. Appl. Biosci.* **12,** 357–358.

24. Baldauf, S. L. (2003) Phylogeny for the faint of heart: a tutorial. *Trends Genet.* **19,** 345–351.

25. Sanderson, M. J. and Shaffer, H. B. (2002) Troubleshooting molecular phylogenetic analyses. *Annu. Rev. Ecol. Syst.* **33,** 49–72.

26. Higgs, P. G. and Attwood, T. K. (2004) *Bioinformatics and Molecular Evolution.* Blackwell Publishing, Oxford, UK.

27. Tatusova, T. A. and Madden, T. L. (1999) BLAST 2 Sequences, a new tool for comparing protein and nucleotide sequences. *FEMS Microbiol. Lett.* **174,** 247–250.

28. Saitou, N. and Nei, M. (1987) The neighbor-joining method: a new method for reconstructing phylogenetic trees. *Mol. Biol. Evol.* **4,** 406–425.

29. Jones, D. T., Taylor, W. R., and Thornton, J. M. (1992) The rapid generation of mutation data matrices from protein sequences. *Comput. Appl. Biosci.* **8,** 275–282.

30. Poe, S. and Swofford, D. L. (1999) Taxon sampling revisited. *Nature* **398,** 299–300.

31. Maddison, D. R., Swofford, D. L., and Maddison, W. P. (1997) NEXUS: an extensible file format for systematic information. *Syst. Biol.* **46,** 590–621.

32. Altschul, S. F., Madden, T. L., Schaffer, A. A., et al. (1997) Gapped BLAST and PSI-BLAST: a new generation of protein database search programs. *Nucleic Acids Res.* **25,** 3389–3402.

33. Jeanmougin, F., Thompson, J. D., Gouy, M., Higgins, D. G., and Gibson, T. J. (1998) Multiple sequence alignment with Clustal X. *Trends Biochem. Sci.* **23,** 403–405.

34. Thompson, J. D., Higgins, D. G., and Gibson, T. J. (1994) CLUSTAL W: improving the sensitivity of progressive multiple sequence alignment through sequence weighting, position-specific gap penalties and weight matrix choice. *Nucleic Acids Res.* **22,** 4673–4680.

35. Thompson, J. D., Gibson, T. J., Plewniak, F., Jeanmougin, F., and Higgins, D. G. (1997) The CLUSTAL_X windows interface: flexible strategies for multiple sequence alignment aided by quality analysis tools. *Nucleic Acids Res.* **25,** 4876–4882.

36. Kimura, M. (1983) *The Neutral Theory of Molecular Evolution.* Cambridge University Press, Cambridge, UK.

37. Schmidt, H. A., Strimmer, K., Vingron, M., and von Haeseler, A. (2002) TREE-PUZZLE: maximum likelihood phylogenetic analysis using quartets and parallel computing. *Bioinformatics* **18,** 502–504.

Index